Biofuels and Biorefineries

Volume 8

Aims and Scope of the Series

The Biofuels and Biorefineries Series aims at being a comprehensive and integrated reference for biomass, bioenergy, biofuels, and bioproducts. The series provides leading global research advances and critical evaluations of methods for converting biomass to biofuels and chemicals. Scientific and engineering challenges in biomass production and conversion are covered that show technological advances and approaches for creating new bio-economies in a format that is suitable for both industrialists and environmental policy decision-makers.

The Biofuels and Biorefineries Series provides readers with clear and concisely written chapters that are peer-reviewed on significant topics in biomass production, biofuels, bioproducts, chemicals, catalysts, energy policy, economics, and processing technologies. The text covers major fields of plant science, green chemistry, economics and economy, biotechnology, microbiology, chemical engineering, mechanical engineering, and energy.

Series Description

Annual global biomass production is about 220 billion dry tons or 4,500 EJ, equivalent to 8.3 times the world's energy consumption in 2014 (543 EJ). On the other hand, world-proven oil reserves at the end of 2011 reached 1652.6 billion barrels, which can only meet just over 50 years of global production. Therefore, alternative resources are needed to both supplement and replace fossil oils as the raw material for transportation fuels, chemicals, and materials in petroleum-based industries. Renewable biomass is a likely candidate, because it is prevalent over the Earth and is readily converted to other products. Compared with coal, some of the advantages of biomass are (i) its carbon-neutral and sustainable nature when properly managed, (ii) its reactivity in biological conversion processes, (iii) its potential to produce bio-oil (ca. yields of 75%) by fast pyrolysis because of its high oxygen content, (iv) its low sulfur and lack of undesirable contaminants (e.g., metals, nitrogen content), (v) its wide geographical distribution, and (vi) its potential for creating jobs and industries in energy crop productions and conversion plants. Many researchers, governments, research institutions, and industries are developing projects for converting biomass including forest woody and herbaceous biomass into chemicals, biofuels, and materials, and the race is on for creating new "biorefinery" processes needed for future economies. The development of biorefineries will create remarkable opportunities for the forestry sector and biotechnology, materials, and chemical processing industry and stimulate advances in agriculture. It will help to create a sustainable society and industries that use renewable and carbon-neutral resources.

More information about this series at http://www.springer.com/series/11687

Zhen Fang • Richard L. Smith Jr.
Hu Li
Editors

Production of Biofuels and Chemicals with Bifunctional Catalysts

Springer

Editors
Zhen Fang
College of Engineering
Nanjing Agricultural University
Nanjing, Jiangsu, China

Hu Li
College of Engineering
Nanjing Agricultural University
Nanjing, Jiangsu, China

Richard L. Smith Jr.
Graduate School of Environmental Studies,
Research Center of Supercritical Fluid
Technology
Tohoku University
Sendai, Japan

ISSN 2214-1537 ISSN 2214-1545 (electronic)
Biofuels and Biorefineries
ISBN 978-981-10-5136-4 ISBN 978-981-10-5137-1 (eBook)
https://doi.org/10.1007/978-981-10-5137-1

Library of Congress Control Number: 2017961301

Printed on acid-free paper

This Springer imprint is published by Springer Nature
The registered company is Springer Nature Singapore Pte Ltd.
The registered company address is: 152 Beach Road, #21-01/04 Gateway East, Singapore 189721, Singapore

Preface

Catalytic processes in multiple steps are typically required in the selective conversion of biomass derivatives into value-added chemicals and biofuels. Much effort has been made in the past decades toward the integration of different types of catalytic transformations with a bifunctional catalyst to improve reaction efficiency, enhance product selectivity, and promote specific reaction pathways. This text provides state-of-the-art reviews, current research, prospects, and challenges of production of platform chemicals (e.g., hexose, 5-hydroxymethylfurfural, 2,5-furandicarboxylic acid, furfural, and levulinic acid) and liquid biofuels (e.g., biodiesel, 2,5-dimethylfuran, 2-methylfuran, and γ-valerolactone) from sustainable biomass resources with cooperative catalytic processes that include heterogeneous bifunctional chemocatalysis (acid-base, Brønsted-Lewis acid, and metal-acid) and combined bio-/chemo-catalytic routes. Fundamentals of bifunctional catalysis and catalysts, characterization reaction mechanism/pathways, methods for heterogeneous catalysts, the impact of catalyst design on the reactivity, and catalytic process integration are introduced. The application of biomass-derived compounds for the synthesis of commodity chemicals and liquid transportation fuels via various catalytic routes is also covered.

This book is the eighth book of the series entitled "Biofuels and Biorefineries," and it contains 12 chapters contributed by leading experts in the field. The text is arranged into four key areas:

Part I: Fundamentals (Chaps. 1 and 2)
Part II: Bifunctional Chemocatalysis (Chaps. 3, 4, 5, 6, 7, 8 and 9)
Part III: Production of Biodiesel (Chaps. 10 and 11)
Part IV: Chemoenzymatic Catalysis (Chap. 12)

Chapter 1 presents a brief introduction on the fundamentals of bifunctional catalysis to understand some of the specific roles in designing catalytic materials and a perspective outlook on bifunctional catalytic pathways for biomass transformations. **Chapter 2** provides an introduction to characterization methods for heterogeneous catalysts for the general reader and an overview on the fundamentals of the solid acid, metal, and metal-acid bifunctional catalysts for cellulose conversion.

Chapter 3 focuses on the application of Brønsted-Lewis bifunctional catalysts for efficient biomass conversion and emerging sustainable biorefining processes. **Chapter 4** describes the pathways of biofuel production from lignocellulosic biomass and the processes involved in transforming lignocellulosic biomass to biosyngas, the research activities on converting syngas into biofuels over bifunctional Fischer-Tropsch synthesis catalysts, and some of the selectivity control strategies that can be adopted with bifunctional catalysts. **Chapter 5** illustrates the effect of catalyst functions on product selectivity in the reductive conversion of furfurals in aqueous solutions using heterogeneous catalysts along with some studies on homogeneous catalysts. **Chapter 6** reviews key strategies used to convert biomass into polyols, with special emphasis on the emerging processes aimed to valorize biomass through a cascade approach. **Chapter 7** introduces the impact of catalyst design on the activity and the selectivity to lactic acid transformation with particular attention to the production of specific and important products such as γ-valerolactone and valeric fuels. **Chapter 8** provides state-of-the-art overview on furanic biofuel production from biomass with a brief description of the 2,5-dimethylfuran and 2-methylfuran production process. **Chapter 9** provides an overview on the recent achievements made with bifunctional catalysts for the upgrading of biomass-derived furans via selective hydrogenation, selective oxidation, Diels-Alder reaction, and others, as well as accompanying reaction pathways and plausible reaction mechanisms. **Chapter 10** describes the synthesis of biodiesel via simultaneous transesterification and esterification catalyzed by mixed metal oxides, acidic ionic liquids, carbon-based solid acids, magnetic solid acids, and hybrid solid acids, the advantages of heterogeneous catalytic reaction systems, and the guidance on improving the performance of heterogeneous acid catalyst for biodiesel production. **Chapter 11** introduces the continuous production of biodiesel fuel from sunflower oil, nonedible oil, and waste oil in a fixed-bed multiphase reactor packed with bifunctional heterogeneous catalyst. **Chapter 12** highlights a series of combined bio-/chemo-catalytic routes that have the potential to facilitate the transition away from petroleum feedstock and toward sustainable and environmentally benign feeds.

The text should be of interest to students, researchers, academicians, and industrialists who are working in the areas of catalysis, catalyst design, renewable energy, environmental and chemical sciences, engineering, resource development, biomass processing, sustainability, materials, biofuels, and chemical industries.

Nanjing, China — Zhen Fang
Sendai, Japan — Richard L. Smith Jr.
Nanjing, China — Hu Li

Acknowledgments

First and foremost, we would like to cordially thank all the contributing authors for their great efforts in writing and revising the chapters and ensuring the reliability of the information given in their chapters. Their contributions have really made this project realizable.

Apart from the efforts of authors, we would also like to acknowledge the individuals listed below for carefully reading the book chapters and giving many constructive comments that significantly improved the quality of the chapters:

Prof. Stefania Albonetti, Bologna University, Italy
Dr. Christophe Bengoa, Universitat Rovira i Virgili, Spain
Dr. Luis Francisco Bobadilla, Universidad de Sevilla, CSIC, Spain
Dr. Giuseppe Bonura, Consiglio Nazionaledelle Ricerche, Istituto di Tecnologie Avanzate per l'Energia "Nicola Giordano," CNR-ITAE, Italy
Prof. Nicolas Brosse, Université de Lorraine, France
Dr. Shih-Yuan Chen, National Institute of Advanced Industrial Science and Technology (AIST, Tsukuba), Japan
Prof. Dr. Johannes G. de Vries, Leibniz Institute for Catalysis (LIKAT Rostock), Germany
Dr. Nikolaos Dimitratos, Cardiff University, UK
Dr. Paul-Henri Ducrot, INRA/AgroParisTech, France
Dr. Lucía García, Universidad de Zaragoza, Spain
Dr. Chao He, Nanyang Technological University, Singapore
Prof. Juan Antonio Melero Hernández, Universidad Rey Juan Carlos, Madrid, Spain
Dr. Masato Kouzu, Tokyo City University, Japan
Prof. Pedro Jesús Maireles-Torres, University of Málaga, Spain
Dr. Paivi Maki-Arvela, Åbo Akademi University, Finland
Dr. Yoshinao Nakagawa, Tohoku University, Japan
Dr. Manuel Fernando R. Pereira, University of Porto (FEUP), Portugal
Dr. Martin Rebros, Slovak University of Technology, Slovakia
Dr. S. Saravanamurugan, Center of Innovative and Applied Bioprocessing (CIAB), Punjab, India

Dr. Takafumi Sato, Utsunomiya University, Japan
Dr. Jesús Canales Vázquez, Universidad de Castilla-La Mancha, Spain
Dr. Bin Wang, University of Oklahoma, USA
Dr. Ziqiang Xu, Hubei University, China
Dr. Aritomo Yamaguchi, National Institute of Advanced Industrial Science and Technology (AIST, Sendai), Japan
Professor Hiromi Yamashita, Osaka University, Japan
Dr. Ning Yan, National University of Singapore, Singapore
Prof. Ji-Jun Zou, Tianjin University, China

We are also grateful to Ms. Becky Zhao (senior editor) and Ms. Abbey Huang (editorial assistant) for their encouragement, assistance, and guidance during the preparation of the book.

Finally, we would like to express our deepest gratitude toward our families for their love, understanding, and encouragement, which help us in the completion of this project.

Zhen Fang, August 25, 2017, in Nanjing (Zhen Fang)

Richard L. Smith, Jr., August 25, 2017, in Sendai (Richard L. Smith, Jr.)

Hu Li, August 25, 2017, in Nanjing (Hu Li)

Contents

Contributors

Ion Agirre Engineering Faculty of Bilbao, University of Basque Country (UPV/EHU), Bilbao, Spain

M.E. Borges Chemical Engineering Department, University of La Laguna, Tenerife, Spain

Pedro Luis de Hoyos Chemical and Environmental Engineering Department, University of the Basque Country UPV/EHU, Donostia-San Sebastián, Spain

Xabier Erdocia Chemical and Environmental Engineering Department, University of the Basque Country UPV/EHU, Donostia-San Sebastián, Spain

P. Esparza Inorganic Chemistry Department, University of La Laguna, Tenerife, Spain

Zhen Fang College of Engineering, Nanjing Agricultural University, Nanjing, Jiangsu, China

Javier Fernández-Rodríguez Chemical and Environmental Engineering Department, University of the Basque Country UPV/EHU, Donostia-San Sebastián, Spain

Naijia Guan School of Materials Science and Engineering & National Institute for Advanced Materials, Nankai University, Tianjin, People's Republic of China

Jun Huang Laboratory for Catalysis Engineering, University of Sydney, Sydney, NSW, Australia

Aitziber Iriondo Engineering Faculty of Bilbao, University of Basque Country (UPV/EHU), Bilbao, Spain

Andrea Jouve Dipartimento di Chimica, Università degli Studi di Milano, Milan, Italy

Xiao Kong Biomass Group, College of Engineering, Nanjing Agricultural University, Nanjing, Jiangsu, China

Jalel Labidi Chemical and Environmental Engineering Department, University of the Basque Country UPV/EHU, Donostia-San Sebastián, Spain

Hu Li College of Engineering, Nanjing Agricultural University, Nanjing, Jiangsu, China

Landong Li School of Materials Science and Engineering & National Institute for Advanced Materials, Nankai University, Tianjin, People's Republic of China

Fabio Lorenzini School of Chemistry and Chemical Engineering, Queen's University Belfast, Belfast, UK

Andrew C. Marr School of Chemistry and Chemical Engineering, Queen's University Belfast, Belfast, UK

P. Martín-Zarza Inorganic Chemistry Department, University of La Laguna, Tenerife, Spain

Patrick J. Morgan School of Chemistry and Chemical Engineering, Queen's University Belfast, Belfast, UK

Junya Ohyama Materials Chemistry, Graduate School of Engineering, Nagoya University, Nagoya, Japan

Hu Pan State-Local Joint Engineering Lab for Comprehensive Utilization of Biomass, State Key Laboratory Breeding Base of Green Pesticide and Agricultural Bioengineering (Ministry of Education), Center for R&D of Fine Chemicals, Guizhou University, Guiyang, China

Yan Pei Collaborative Innovation Center of Chemistry for Energy Materials, Department of Chemistry and Shanghai Key Laboratory of Molecular Catalysis and Innovative Materials, Fudan University, Shanghai, People's Republic of China

Laura Prati Dipartimento di Chimica, Università degli Studi di Milano, Milan, Italy

Minghua Qiao Collaborative Innovation Center of Chemistry for Energy Materials, Department of Chemistry and Shanghai Key Laboratory of Molecular Catalysis and Innovative Materials, Fudan University, Shanghai, People's Republic of China

Jesús Requies Engineering Faculty of Bilbao, University of Basque Country (UPV/EHU), Bilbao, Spain

J.C. Ruiz-Morales Inorganic Chemistry Department, University of La Laguna, Tenerife, Spain

Atsushi Satsuma Materials Chemistry, Graduate School of Engineering, Nagoya University, Nagoya, Japan

Ane Sequeiros Chemical and Environmental Engineering Department, University of the Basque Country UPV/EHU, Donostia-San Sebastián, Spain

Richard L. Smith Jr. Graduate School of Environmental Studies, Research Center of Supercritical Fluid Technology, Tohoku University, Sendai, Japan

Song Song School of Materials Science and Engineering & National Institute for Advanced Materials, Nankai University, Tianjin, People's Republic of China

Alberto Villa Dipartimento di Chimica, Università degli Studi di Milano, Milan, Italy

Hao Wang Collaborative Innovation Center of Chemistry for Energy Materials, Department of Chemistry and Shanghai Key Laboratory of Molecular Catalysis and Innovative Materials, Fudan University, Shanghai, People's Republic of China

Zichun Wang Laboratory for Catalysis Engineering, University of Sydney, Sydney, NSW, Australia

Guangjun Wu School of Materials Science and Engineering & National Institute for Advanced Materials, Nankai University, Tianjin, People's Republic of China

Song Yang State-Local Joint Engineering Lab for Comprehensive Utilization of Biomass, State Key Laboratory Breeding Base of Green Pesticide and Agricultural Bioengineering (Ministry of Education), Center for R&D of Fine Chemicals, Guizhou University, Guiyang, China

Heng Zhang State-Local Joint Engineering Lab for Comprehensive Utilization of Biomass, State Key Laboratory Breeding Base of Green Pesticide and Agricultural Bioengineering (Ministry of Education), Center for R&D of Fine Chemicals, Guizhou University, Guiyang, China

Yifeng Zhu State Key Laboratory of Catalysis, Dalian Institute of Chemical Physics, Chinese Academy of Sciences, Dalian, China

Baoning Zong State Key Laboratory of Catalytic Materials and Chemical Engineering, Research Institute of Petroleum Processing, SINOPEC, Beijing, People's Republic of China

About the Editors

Zhen Fang is professor and leader of the biomass group at Nanjing Agricultural University. He is the inventor of the "fast hydrolysis" process. He is listed in the "Most Cited Chinese Researchers" in energy for 2014, 2015, and 2016 (Elsevier-Scopus). Professor Fang specializes in thermal/biochemical conversion of biomass, nanocatalyst synthesis and its applications, and pretreatment of biomass for biorefineries. He obtained his PhDs from China Agricultural University (biological and agricultural engineering, Beijing) and McGill University (materials engineering, Montreal). Professor Fang is associate editor of *Biotechnology for Biofuels* and is serving on editorial boards of major international journals in energy.

Richard L. Smith Jr. is professor of chemical engineering at the Graduate School of Environmental Studies, Research Center of Supercritical Fluid Technology, Tohoku University, Japan. Professor Smith has a strong background in physical properties and separations and obtained his PhD in chemical engineering from the Georgia Institute of Technology (USA). His research focuses on developing green chemical processes especially those that use water and carbon dioxide as the solvents in their supercritical state. He has expertise in physical property measurements and in separation techniques with ionic liquids and has published more than 200 scientific papers, patents, and reports in the field of chemical engineering. Professor Smith is the Asia regional editor for the *Journal of Supercritical Fluids* and has served on editorial boards of major international journals associated with properties and energy.

Hu Li is a researcher of agricultural engineering in the group of Professor Zhen Fang at Nanjing Agricultural University, China. Dr. Li obtained his PhD from Guizhou University (China). His research focuses on the catalytic conversion of biomass into chemicals and biofuels with functional catalytic materials. Dr. Li has coauthored more than 50 peer-reviewed papers on biomass valorization and is a guest editor of *Current Organic Chemistry*.

Part I
Fundamentals

Chapter 1
Fundamentals of Bifunctional Catalysis for Transforming Biomass-Related Compounds into Chemicals and Biofuels

Hu Li, Xiao Kong, Zhen Fang, and Richard L. Smith Jr.

Abstract Multi-step catalytic processes are generally required in chemoselective conversion of biomass derivatives into chemical products. Integration of different types of transformations into a single catalytic system with a bifunctional catalyst can be used to improve reaction efficiency, enhance product selectivity and to promote specific reaction pathways. In this chapter, the fundamentals of bifunctional catalysis are introduced to understand some of the specific roles in designing catalytic materials for biomass transformations. Acid-mediated hydrolysis of polysaccharides to mono- and oligosaccharides in combination with successive isomerization is one of the effective strategies for valorization of sugar-containing solutions, wherein the type of acid catalyst (i.e., Lewis and Brønsted acids) can be used to control product distribution. Catalytic production and upgrading of furanic compounds (e.g., 5-hydroxymethylfurfural, 2,5-furandicarboxylic acid, and 2,5-dimethylfuran) with bifunctional materials via corresponding tandem reaction pathways is another effective strategy. Emphasis can be placed on the synthesis of organic acids (e.g., levulinic acid and lactic acid) from carbohydrates through either anaerobic or aerobic oxidation, integrated with preprocessing reactions. This chapter provides an overview of bifunctional catalytic systems and a perspective outlook on bifunctional catalytic pathways for biomass transformations.

H. Li • Z. Fang (✉)
College of Engineering, Nanjing Agricultural University,
Nanjing, Jiangsu, China
e-mail: zhenfang@njau.edu.cn

X. Kong
Biomass Group, College of Engineering, Nanjing Agricultural University,
Nanjing, Jiangsu, China

R.L. Smith Jr.
Graduate School of Environmental Studies, Research Center of Supercritical Fluid Technology, Tohoku University, Sendai, Japan

Z. Fang et al. (eds.), *Production of Biofuels and Chemicals with Bifunctional Catalysts*, Biofuels and Biorefineries 8, https://doi.org/10.1007/978-981-10-5137-1_1

1.1 Bifunctional Catalysis

In comparison with monofunctional catalytic systems, bifunctional catalysts with Brønsted/Lewis acid, acid-base, or metal-containing acid/base sites have been demonstrated to exhibit remarkably high performance in both gas-phase (e.g., syn-gas to hydrocarbons or alcohols) and liquid-phase (e.g., aqueous-organic) conversions [1–7]. Bifunctional catalysis may be in use, more commonly than is generally recognized, because a certain number of monofunctional catalysts are most likely bifunctional in nature [8]. For instance, the surfaces of transition metal oxides (e.g., MoO_2 and Cr_2O_3) have been found to possess both types of hydrogenation-dehydrogenation and acidic sites, implying that the apparently single-site catalysts possibly have bifunctional activities, in spite of their relatively lower activity than their tailor-designed two-site counterparts [8].

To mimic enzymatic catalysts, materials can be designed with functionalities (e.g., metal centers, basic and acidic sites, nucleophiles, hydrogen bond donors/accepter groups) that are spartially organized around active sites to promote efficient transformations [9]. Some reported heterogeneous catalysts have unprecedented performance unattainable with homogeneous catalysts [9]. Typically, the cooperation between the active sites or groups (T1/T2 as in Fig. 1.1) of a bifunctional material surface in catalyzing a reaction with two substrates (S1/S2 as in Fig. 1.1) can be divided into four different modes (Fig. 1.1) that possibly include dual activation (A), sequential activation (B), self activation (C), and multiple point transition state stabilization (D) [9, 10]. All of these strategies have been adopted as references for the design of synthetic catalysts that are able to take full advantage of multifunctional cooperation in catalytic processes [9].

From the point of view of organic chemistry, bifunctional catalysis is defined as both nucleo- and electrophilic substrates being separately activated by discrete active sites present on a single catalytic material (Fig. 1.2a) [11]. In line with bifunctional catalysis, both cascade and double activation catalysis may be present,

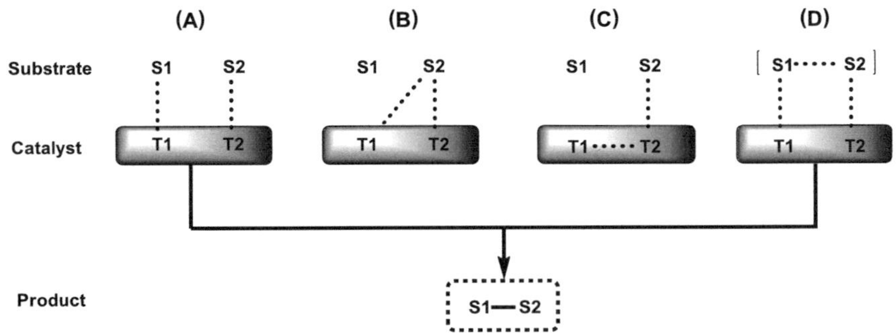

Fig. 1.1 Cooperative modes between active sites or groups (T1/T2) of a bifunctional catalyst in the activation of two substrates (S1/S2) (Adapted with permission from Ref. [9], Copyright © 2008 Royal Society of Chemistry) (**A**) Dual activation, (**B**) Sequential activation, (**C**) Self activation, (**D**) Multiple point transition state stabilization

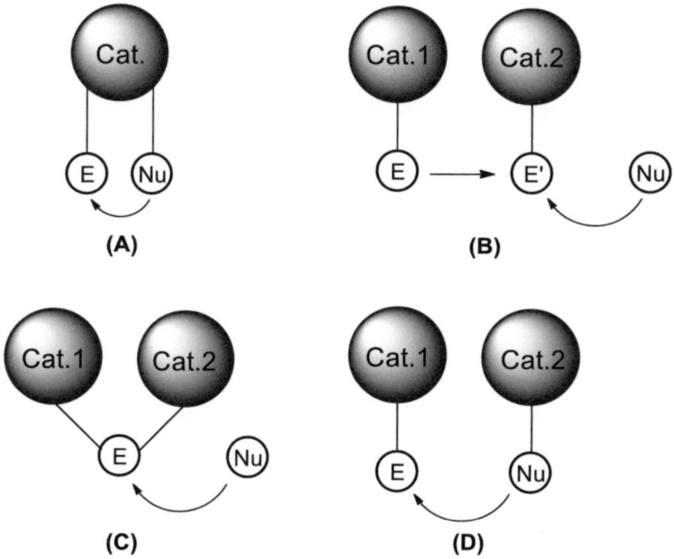

Fig. 1.2 Schematic representation of bifunctional catalysis (Adapted with permission from Ref. [11], Copyright © 2014 Royal Society of Chemistry) (**A**) Bifunctional catalysis, (**B**) Cascade catalysis, (**C**) Double activation catalysis, (**D**) Synergistic catalysis

wherein two or more catalysts activate the identical substrate sequentially or in a concerted manner, respectively (Fig. 1.2 b–c). On the contrary, the nucleo- and electrophilic substrates may be concurrently activated by two or multiple separate catalysts such that synergistic catalysis occurs (Fig. 1.2d). In this chapter, acid-participated bifunctional catalytic hydrolysis of polysaccharides to mono- and oligosaccharides will be introduced. The key step of glucose isomerization involved in the valorization of sugar-containing solutions mediated by dual functional groups is depicted. Further, discussion is made on application of the techniques to the production of biofuranic platform molecules (e.g., 5-hydroxymethylfurfural and furfural) and the upgrading of these compounds to oxygenated and hydrogenated furans (e.g., 2,5-furandicarboxylic acid and 2,5-dimethylfuran) tandem reaction pathways. Finally, the synthesis of organic acids (e.g., levulinic acid and lactic acid) from carbohydrates is introduced through either anaerobic or aerobic oxidation via bifunctional catalysis.

1.2 Hydrolysis of Oligo- and Polysaccharides

Simple sugars hydrolyzed from biomass over chemical and enzymatic catalysts have been used to study the production of liquid/gas fuels (e.g., ethanol, butanol, hydrogen, and methane) or formation of chemical products (e.g., organic acids, furanic/aromatic compounds, and polyols), as illustrated in Fig. 1.3 [12, 13]. In the

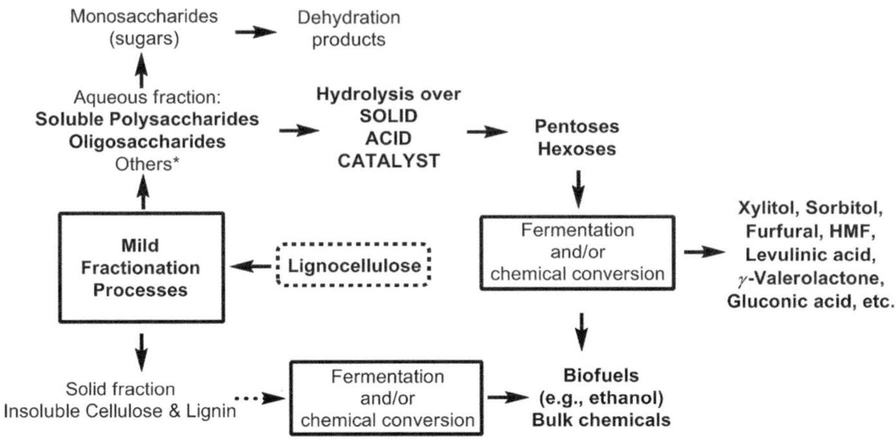

Fig. 1.3 Schematic representation of a catalytic strategy for biomass upgrading (Adapted with permission from Ref. [12], Copyright © 2012, Elsevier)

conversion of oligo- and polysaccharides (e.g., cellobiose and cellulose) into monosaccharides such as glucose via hydrolysis, homogeneous chemical catalysts generally show good activity, but they require harsh reaction conditions (> 180 °C), and have recycle and waste treatment issues especially for when mineral acids (e.g., H_2SO_4, H_3PO_4 and HNO_3) are used. Enzymes, on the other hand, have good selectivity but are generally inefficient, expensive and are susceptible to impurities [14]. In this regard, the use of solid acids or supported catalysts that are stable and recoverable could address the issues inherent in both homogeneous catalysts and enzyme catalysts to promote hydrolysis reactions that are selective and efficient that can be performed at mild conditions without environmental wastes.

1.2.1 Brønsted/Lewis Acid Bifunctional Solid Catalysts

In the organosolv process, Schwiderski et al. [15] demonstrated that a Lewis acid $AlCl_3$ was able to catalyze both ketol-endiol-tautomerism and dehydration to produce furfural from beech wood chips, showing comparable and even superior activity to that of a Brønsted acid such as HCl [15]. For both Lewis and Brønsted acid sites, the increase in acidity was found to be favorable for the cleavage of β-1,4-glycosidic bonds in cellobiose and in cellulose, thus enhancing their hydrolysis [16]. Typically, the distribution of Brønsted and Lewis acid sites on mixed metal oxides can be adjusted by simply changing the molar ratio of metal species or calcination temperature of the catalyst. In an in-depth study on the role of WO_x clusters in WO_3/ZrO_2 catalysts, it was observed that a high WO_x loading resulted in an increase in the number of Brønsted acid sites, and prevailed over the number of Lewis acid sites of ZrO_2 even at calcination temperatures greater than 700 °C [17].

One type of commercially available Brønsted/Lewis acid bifunctionalized catalysts is the H-form zeolites that show moderate activity and stability for hydrolysis of hemicellulose and cellulose. H-form zeolites forms that have relatively higher acidic (e.g., ferrierite) and hydrophobic sites with high Si/Al ratios have superior performance to others for hydrolysis of carbohydrates [18, 19]. With these types of catalysts, the polymerization of monosaccharides (e.g., glucose) to humins with zeolites is significantly inhibited, which possibly can be attributed to shape-selectivity, in comparison with other solid and homogeneous acids (e.g., sulphated zirconia and heteropoly acids) [20]. Over H-USY-15 (Si/Al = 15) in water, a moderate yield of xylose and arabinose (56%) with >90% carbon balance is obtained from hemicellulose at 170 °C in 1 h [21], while the sulfonated H-USY zeolite gives improved yields (ca. 78%) at 140 °C after 9 h [22]. These results highlight the effect of Brønsted acid sites and the importance of Lewis acid sites. Meso- and macropores created on zeolites by dealumination (e.g., treated with oxalic acid) promote the accessibility of the acid sites to the reactants, thus improving the mass transfer, with the yield of total reduced sugars increasing by tenfold [23]. Under microwave heating conditions, the hydrolysis of cellulose to glucose gives moderate yields (ca. 37%) over H-form zeolites, but only requires reaction times on the order of minutes (e.g., 8 min) [24]. The integration of bifunctional catalytic materials with auxiliaries such as microwave and ultrasound offers yet another approach to improve reaction efficiency.

As a biomass-derived polymer that has aromatic monomeric units rather than glucose monomeric units, the hydrolytic degradation of lignin into aromatic monomers can also be achieved with well-designed solid acid catalysts based on zeolites [25]. For example, H-USY affords 60% yields of monomeric aromatics from lignin at 250 °C, further demonstrating the potential of Brønsted/Lewis acid bifunctional solids (particularly zeolites) for conversion of biomass-related compounds into small components or fractions.

1.2.2 Base-Acid Bifunctional Solid Catalysts

The pretreatment of biomass with base (e.g., $Ca(OH)_2$, NaOH and KOH) can be used to remove lignin at ambient to 100 °C temperatures, while the remaining polysaccharides can be hydrolyzed with acidic or enzymatic catalysts [26]. In an improved process, Chandel et al. found that crude lignocelluloses (e.g., sugarcane bagasse) consecutively pretreated with dilute H_2SO_4 and NaOH could surmount the structural recalcitrance of the biomass, thus enhancing subsequent cellulase accessibility and hydrolysis efficiency for ethanol production [27]. Likewise, an acid-base mixture can be employed to simplify the catalytic steps for one-pot production of cellulosic ethanol, without solid/liquid separation, neutralization and detoxification (Fig. 1.4) [28]. This process integrates acid-base catalyzed hydrolysis of cellulose and subsequent fermentation of ethanol with *in situ* produced cellulase into a single step. Gratifyingly, the yield of ethanol reached 70.7% from the fermentation

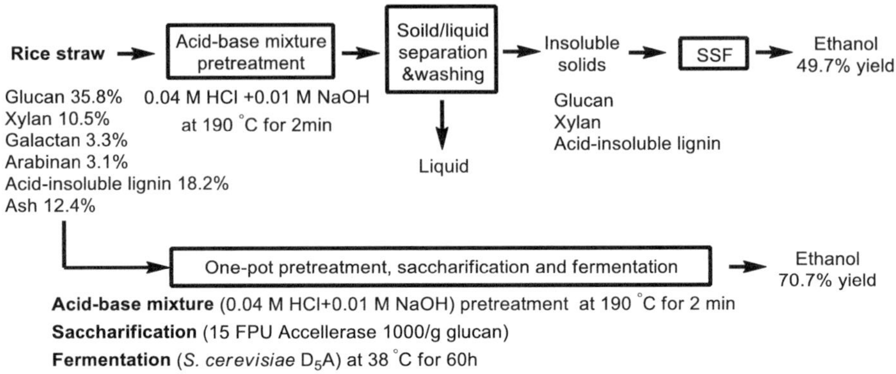

Fig. 1.4 Schematic pathway for integrated processing of rice straw with acid-base pretreatment/enzyme catalysis to produce ethanol (Adapted with permission from Ref. [28], Copyright © 2014 Royal Society of Chemistry)

of the whole rice straw slurry after 60 h, which was relatively higher than that (49.7% yield) obtained by simultaneous saccharification and fermentation (SSF) of washed solids of pretreated rice straw (Fig. 1.4) [28]. These results demonstrate both the advantages of an integrated process in the production of ethanol from rice straw.

The activation of hydrotalcite (HT) with $Ca(OH)_2$ simultaneously increases its basicity and acidity from 0.42 and 0.21 mmol/g to 1.76 and 1.17 mmol/g, as measured by CO_2- and NH_3-TPD, respectively [29]. The resulting HT-OH_{Ca} catalyst shows a greatly enhanced glucose yield (39.8%) and selectivity (85.3%) in the hydrolysis of ball-milled cellulose, as compared with pristine HT (11.0% yield and 40.6% selectivity), $Ca(OH)_2$ (10.6% yield and 39.8% selectivity) and reference experiments without a catalyst (0.4% yield and 3.5% selectivity) at 150 °C in 24 h (Table 1.1). The HT-OH_{Ca} catalyst was able to be recycled for at least four times (sequentially reused catalysts denoted as HT-OH_{Ca2}, HT-OH_{Ca3} and HT-OH_{Ca4}), with glucose yield and selectivity slightly decreasing to 35.0% and 80.5%, respectively (Table 1.1, entries 5–7) [29]. In addition, $Ba(OH)_2$, KOH and NaOH-activated HTs were found to show moderate glucose selectivities and yields (Table 1.1, entries 8–10), further confirming that the catalyst performance was positively correlated with the content of both acid and base sites.

1.3 Sugar Isomerization

Enzymes (e.g., glucose isomerase) are commonly active for the isomeric conversion of simple monosaccharides. The stability of an enzyme can be increased by immobilization onto solid supportssuch as silica or chitosan [30, 31]. Solid basic and acidic chemocatalysts that have comparable reactivity as their homogeneous counterparts and which have favorable reaction times and recycle characteristics have

Table 1.1 Hydrolysis of cellulose in water with hydrotalcite (HT) and $Ca(OH)_2$activated HT ($HT\text{-}OH_{Ca}$) or other activated HT forms

		Glucose (%)		Acid-base content (mmol/g)	
Entry	Catalyst	Selectivity	Yield	Basicity	Acidity
1	None	3.5	0.4	–	–
2	HT	40.6	11.0	0.42	0.21
3	$Ca(OH)_2$	39.8	10.6	Solution pH = 12.84	
4	$HT\text{-}OH_{Ca}$	85.3	39.8	1.76	1.17
5	$HT\text{-}OH_{Ca2}$	83.4	36.6	1.54	1.09
6	$HT\text{-}OH_{Ca3}$	82.3	36.0	1.48	1.06
7	$HT\text{-}OH_{Ca4}$	80.5	35.0	1.47	1.05
8	$HT\text{-}OH_{Ba}$	84.6	38.2	1.57	1.12
9	$HT\text{-}OH_{K}$	72.9	29.6	0.87	0.64
10	$HT\text{-}OH_{Na}$	69.0	27.2	0.75	0.52

Reaction conditions: 0.45 g ball-milled cellulose, 0.5 g catalyst, 150 mL distilled water, and 150 °C for 24 h reaction time. Values after Ca (2, 3, 4) refer to recycle number

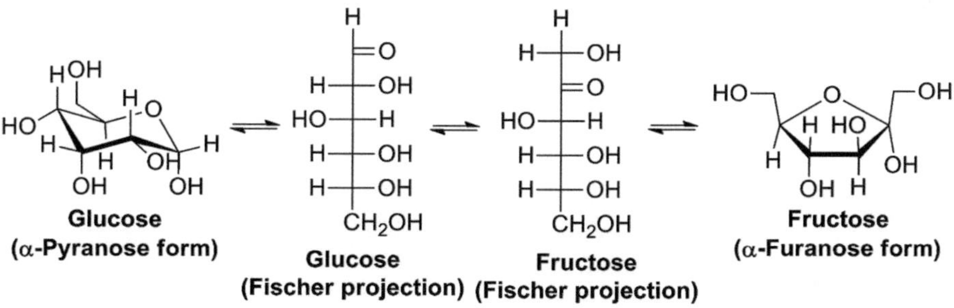

Fig. 1.5 Schematic illustration of the isomerization of glucose to fructose

been proposed for the selective isomerization of sugars [32, 33]. Solid bases such as TiO_2, ZrO_2 and relevant hybrids normally have low catalytic activity, giving no more than 20% yields of fructose in the case of glucose isomerization [34, 35]. However, Sn-beta, which typically acts as a Lewis acid, is able to isomerize glucose to fructose with high yields of 31% in water at 110 or 140 °C in 30 min or 12 min reaction time, respectively [36]. Glucose in the pyranose form partitions into the zeolite, followed by ring-opening to the acyclic form over the Lewis acid center and isomerizes into fructose in the acyclic form, which finally proceeds to ring closure to give the furanose form of fructose (Fig. 1.5) [37].

Besides the Lewis acid center of Sn species being located in the framework of BEA zeolite, the adjacent silanol group synergistically promotes the hydride transfer in the glucose-to-fructose isomerization, and allows the single-step reaction to proceed with a low activation energy barrier while the epimerization of glucose to mannose via a Bilik-type rearrangement is impeded [38]. The proximate hydroxyl group is possibly derived from either co-adsorbed water or intrinsic silanol, and the

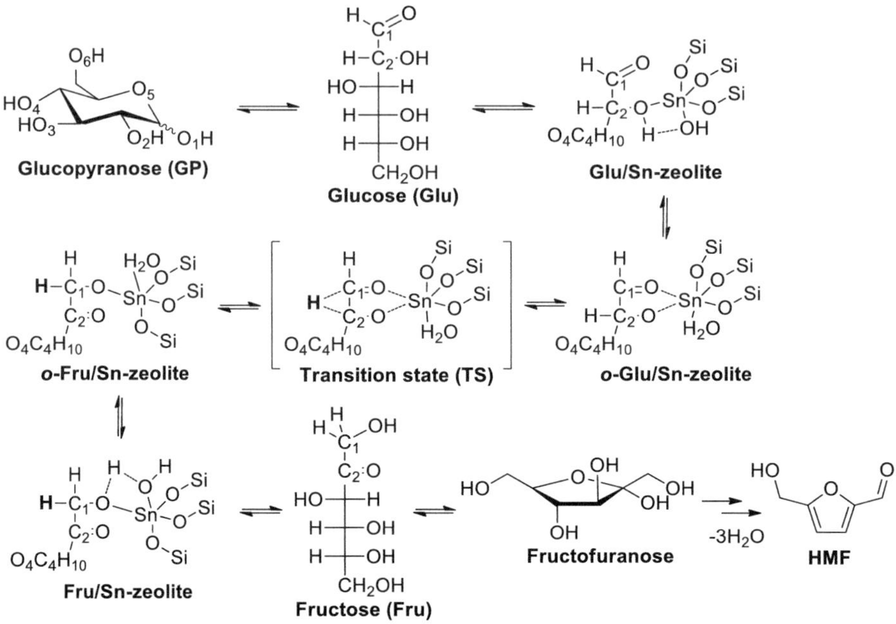

Fig. 1.6 Possible reaction mechanism for the catalytic isomerization of glucose (Glu) to fructose (Fru) with partially hydrolyzed $(SiO)_3Sn$-OH site in a BEA-type zeolite framework [39, 40]

cooperative effect of the Sn center and –OH group compensates the negative charge more effectively on the O1 atom of glucose during the rate-controlling step of a hydride shift from the C2 carbon to the C1 carbon (transition state (TS) in Fig. 1.6) [39, 40]. In the isomerization of glucose to fructose over tungstite ($WO_3 \cdot H_2O$), cooperative action of tungsten species (Lewis acid site) with a neighboring proton donor occurs [41]. Thc conversion of glucose increases when methanol is used as solvent instead of water, probably due to the differences in the solvation of the hydrophobic pores [42]. However, the epimerization of glucose to mannose as the dominant side reaction may take place by either two subsequent 1,2-intramolecular hydride transfer steps over Sn-Beta or one 1,2 intramolecular carbon shift step on Na-exchanged Sn-Beta in methanol [42]. Gao et al. conducted a series of glucose isomerization reactions in subcritical aqueous straight- and branched-chain alcohols, and found that primary and secondary alcohols could promote glucose conversion to fructose in high selectivity, while the increase of *t*-butyl alcohol concentration resulted in a decrease of both glucose conversion and fructose yield [43, 44]. In addition, the ordered mesoporosity of the Sn-containing zeolite was favorable for the isomerization of C5 and C6 sugars (e.g., xylose and glucose) because it improved mass transfer and increased reaction rate [45]. Further, even the aldose component in the disaccharide could be isomerized to ketose (e.g., lactose-lactulose and cellobiose-cellobiulose) over Sn-beta zeolites in water or subcritical aqueous ethanol [46, 47]. Unexpectedly, the beta zeolite containing Ti^{4+}centers (Ti-beta) were

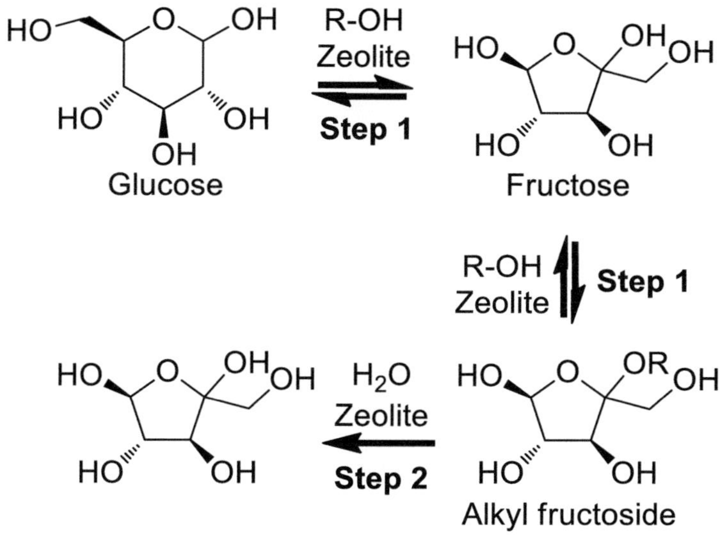

Fig. 1.7 Reaction pathway for converting glucose to fructose in a two-step process (Step 1, in R-OH; Step 2, water addition) (Adapted with permission from Ref. [49], Copyright © 2013, American Chemical Society)

found to preferentially catalyze the isomerization of glucose to sorbose through an intramolecular C5-C1 rather than C2-C1 hydride-shift [48].

By using commercially available zeolites as Brønsted-Lewis acidic catalysts, efficient isomerization of glucose to fructose is possible with remarkable yields (up to 55%) via a two-step methanol-water process that uses methanol in the initial reaction for the isomerization and etherification of glucose to a mixture of fructose and methyl fructoside at 120 °C for 1 h (step 1), and that is followed by water addition in the second step proceeding hydrolysis to regenerate fructose from methyl fructoside (step 2) as shown in Fig. 1.7 [49]. In this catalytic process, the isomerization of glucose to fructose and trapping of fructose as fructoside in the alcohol are catalyzed by Lewis and Brønsted acid sites, respectively [50]. It is found that the H-form zeolite (e.g., H-USY) promotes the glucose-to-fructose isomerization via an intramolecular hydride-shift rather than through solvent exchange [51]. Likewise, catalytic isomerization of xylose to xylulose (up to 47% yield) via the two-step methanol-water process, and erythrose to erythrulose (45% yield) in water can be realized over the commercial large-pore zeolites such as H-USY-6 (Si/Al = 6) and H-beta (Si/Al = 12.5) [52, 53]. In the presence of a single-unit-cell Sn-MFI with both Brønsted and Lewis acid sites, a maximum fructose yield of 65% is obtained from glucose via an identical two-step process in ethanol and water [54]. The coupling of Lewis and Brønsted acidity with newly developed catalytic routes allows an increase in reaction efficiency.

1.4 Synthesis of Biofuranic Compounds

In the past decade, a large number of catalytic systems and functional materials have been reported that are selective for the transformation of biomass-derived hexoses and pentoses into furans such as 5-hydroxymethylfurfural (HMF), furfural, 2,5-dimethylfuran, γ-valerolactone, 5-ethoxymethylfurfural, 2,5-diformylfuran, 2,5-furandicarboxylic acid and maleic anhydride via cascade reactions such as hydrolysis, dehydration, etherification, hydrogenation and oxidation (Fig. 1.8) [55]. These biofuranic compounds have high potential for use in industrial applications. In this section, catalytic strategies used for production of furanic products from bio-based compounds are briefly discussed.

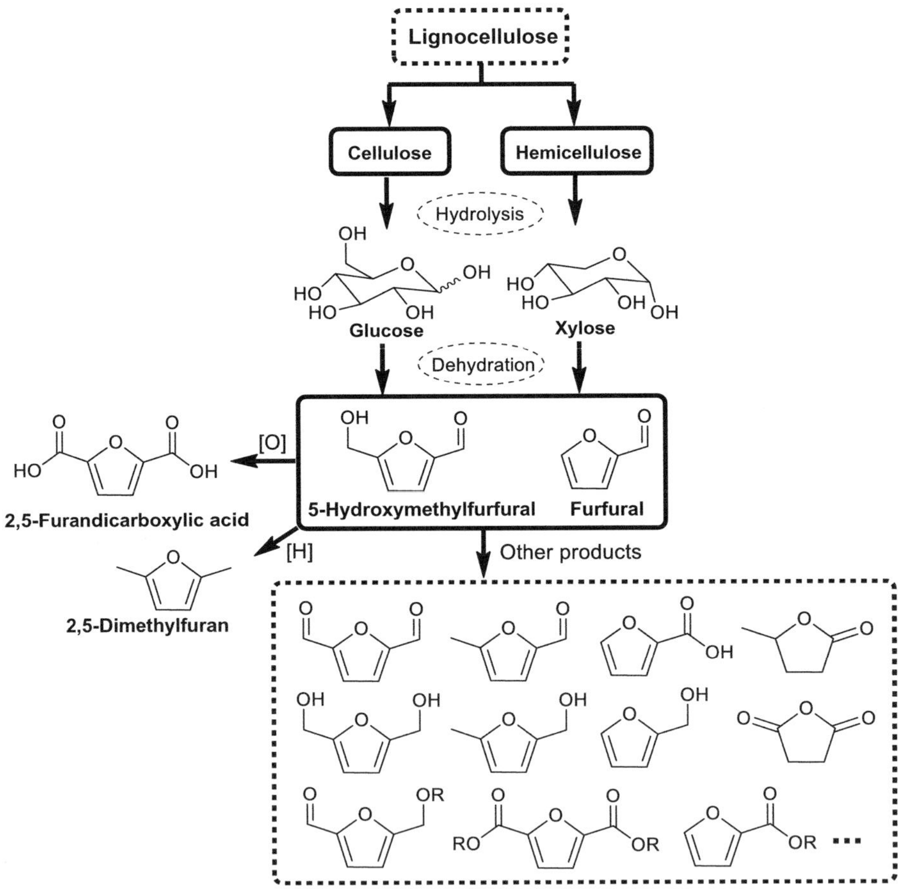

Fig. 1.8 Schematic routes for producing biofuranic compounds from lignocellulose

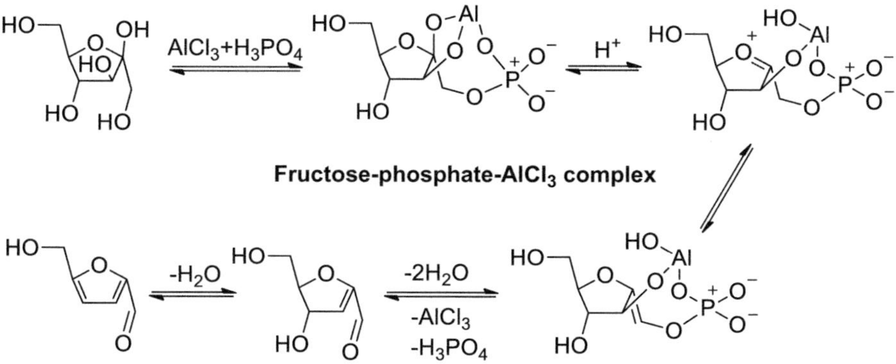

Fig. 1.9 Reaction pathway for producing HMF from fructose stabilized by a fructose-phosphate-$AlCl_3$complex (Adapted with permission from Ref. [56], Copyright © 2014 Royal Society of Chemistry)

1.4.1 Brønsted-Lewis Acid Bifunctional Catalysis

In the dehydration of fructose, either Lewis acid (e.g., $AlCl_3$) or Brønsted acid (e.g., HCl, H_2SO_4 and H_3PO_4) alone promote formation of HMF with high efficiency (20 min), but in only fair yields (< 60%) at moderate reaction conditions (120 °C). However, a mixture of Lewis and Brønsted acids (e.g., $AlCl_3$ and H_3PO_4) in a single catalytic system is able to achieve HMF from fructose in yields as high as 92.6% under identical reaction condtions [56]. The *in situ* formed fructose-phosphate-$AlCl_3$ complex is proposed as the reason for the enhanced stability of the fructofuranose during the reaction through its shifting of the balance from fructopyranose to fructose-phosphate-$AlCl_3$ complex (Fig. 1.9), followed by dehydration over the protonic acid to give HMF. Binder and Raines showed that glucose, cellulose and even untreated corn stover could also be converted into HMF (48–81% yields) at temperatures of 100–140 °C in a single step with $CrCl_3$ and HCl catalysts [57].

For producing HMF from glucose in cascade reactions using $CrCl_3$together with HCl in aqueous media, Lewis acidic catalysts are active for the isomerization of glucose to fructose, while Brønsted acid catalysts are favorable for subsequently promoting the dehydration reaction to give HMF (Fig. 1.10) [58]. The glucose-to-fructose isomerization is inhibited by HCl which decreases the equilibrium $[Cr(H_2O)_5OH]^{2+}$concentration, while the overall consumption rate of fructose and HMF over Lewis acid species most likely promotes side reactions. Several other studies show that both the fructose dehydration rate and HMF selectivity (up to 62% yield) are enhanced in a mixed solvent of water and tetrahydrofuran or 2-*sec*-butylphenol at 160–170 °C by introducing a certain amount of HCl into the Lewis acidic Al^{3+}aqua/hydroxo complexes [59, 60]. In a similar manner, the combined use of Sn-beta with HCl [61], or metal salts (e.g., $CrCl_3$, $InCl_3$, $CoSO_4$, $FeCl_2$ and $MnCl_2$) with SO_3 H-functionalized ionic liquids [62–66] is able to catalyze glucose and glucose-containing saccharides (e.g., cellulose) to obtain HMF in good yields

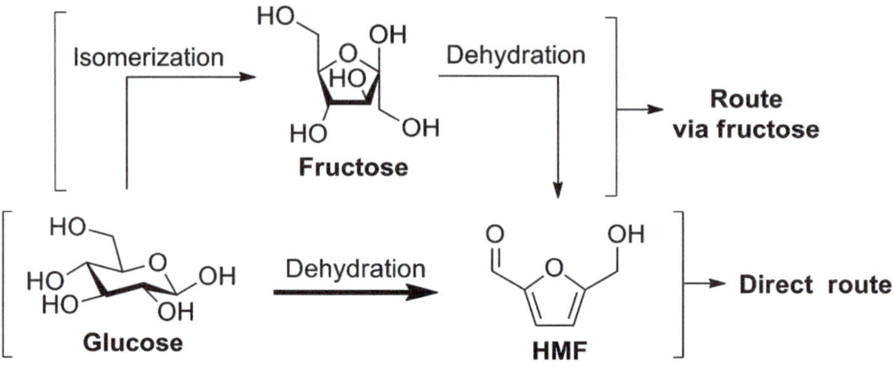

Fig. 1.10 Reaction pathway for the synthesis of HMF from glucose via fructose

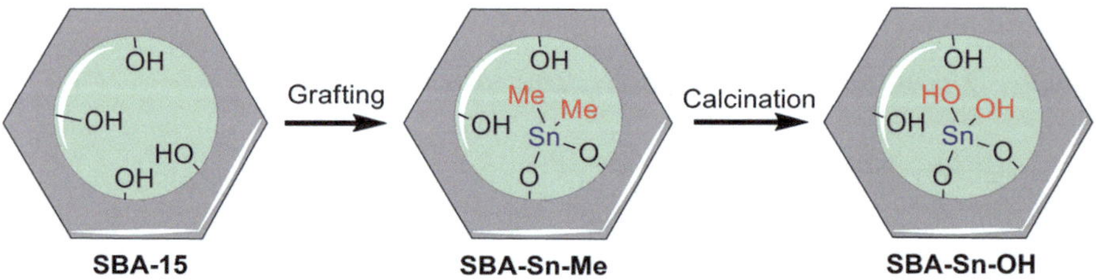

Fig. 1.11 Schematic route for synthesis of Lewis-Brønsted bifunctional SBA-Sn-OH (Adapted with permission from Ref. [77], Copyright © 2014 Royal Society of Chemistry)

(up to >80%). The recyclability of the acidic ionic liquids coupled with metal salts can be improved by immobilizing both active species into a solid polymer that retains catalytic activity [67].

Metal oxides are typically Lewis acid catalysts, and they rarely exhibit high selectivity and reactivity in the conversion of glucose to HMF [68]. However, the introduction of Brønsted acid sites into metal oxides by treatment with mineral acids (e.g., phosphoric acid and sulfuric acid) or acidic metal species (e.g., tungsten), and the resulting Brønsted-Lewis acid bifunctional hybrids (e.g., zirconium, aluminum, titanium, tantalum and niobium phosphates, sulfated zirconia, and W-Zr or W-Sn oxide) increase the efficiency of the direct conversion of glucose to HMF as compared with pristine metal oxides [69–76]. Some other neoteric strategies have been explored for the introduction of Brønsted acid sites into metal oxides. For example, the SBA-Sn-OH Lewis and Brønsted bifunctional catalyst can be prepared by grafting of dimethyldichlorostannane onto the SBA-15 surface (SBA-Sn-Me) that is followed by calcination to transform the $-CH_3$ into –OH groups (Fig. 1.11) [77].

The synthesis of heteropolyacid salts via exchange with metal ions (Cs^+, $Ag^+$and Cr^{3+}) is an efficient way to solidify homogeneous heteropolyacids such as 12-tungstophosphoric acid and silicotungstic acid for obtaining Lewis-Brønsted dual sites and porous structure [78–80]. Notably, the replacement of W in 12-tungstophosphoric acid with Cr makes the resulting acid $H_3PW_{11}CrO_{39}$ simulta-

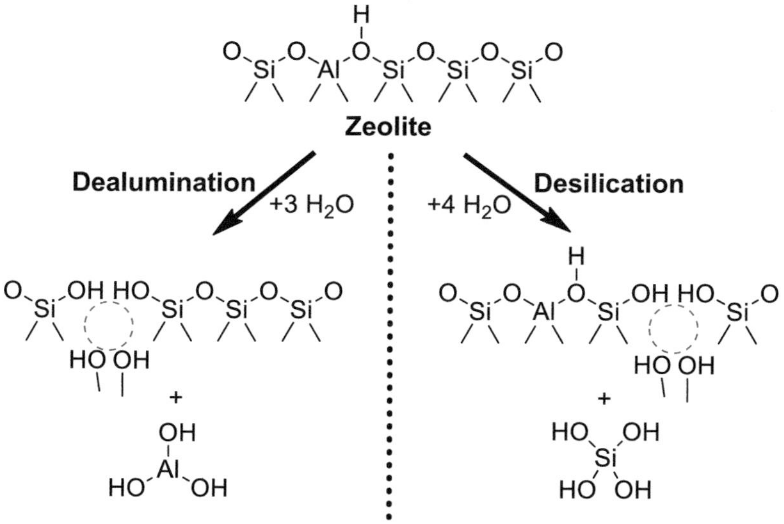

Fig. 1.12 Schematic representation for the framework dealumination (left) under steaming or acidic aqueous conditions, and desilication (right) under neutral or basic aqueous conditions

neously bear strong Brønsted and Lewis acid centers, while the further introduction of hexadecyltrimethylammonium bromide (CTAB) forms a highly water-tolerant salt [81]. Due to the dual acidity and hydrophobicity of the catalyst, more than 90% hexose conversion has been reported for the production of HMF [81]. The framework dealumination and desilication of H-form zeolites are efficient for the extension of pore diameters and for the adjustment of Lewis-Brønsted acid sites distribution [82]. The resulting desilicated H-ZSM-5 and dealuminated H-beta zeolites display pronounced performance in the transformation of either cellulose or glucose (67–81% conversion) into HMF (up to ca. 50% yield) in water or ionic liquids [83–85] (Fig. 1.12).

Lewis-Brønsted bifunctional catalysts like cesium salts of 12-tungstophosphoric acid, silicoaluminophosphate, partially hydroxylated magnesium fluoride, Zr-(W,Al) mixed oxides, sulfonated carbonaceous TiO_2, and H-form zeolites promote formation of furfural in moderate to good yields from pentose sugars (e.g., hemicellulose, xylan, xylose, arabinose and ribose) [21, 86–93]. The catalysts with high molar ratios of Brønsted to Lewis acid sites have selectivity towards furfural as much as 30 times higher than that for relatively higher concentrations of Lewis acid sites [93]. Notably, the transformation of hexoses (e.g., glucose) into furfural (up to 37% yield) is realized by involving a retro-aldol reaction to form pentoses over H-form zeolites (e.g., H-ZSM-5 and H-beta) in γ-valerolactone with 10 wt% water, while the strong Brønsted acid Amberlyst-70 is exclusively selective for the dehydration of glucose and fructose to HMF rather than furfural [94]. These results show that the close interplay of Lewis acid and Brønsted acid centers in the solid catalyst seems to be crucial for achieving good activity in the synthesis of either HMF or furfural from biomass-derived sugars.

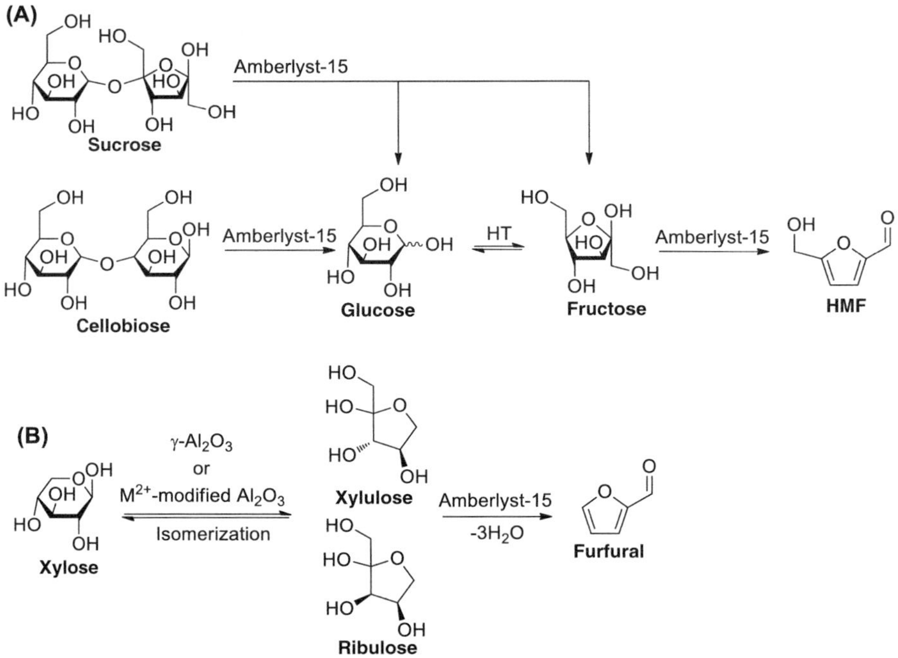

Fig. 1.13 One-pot transformation of glucose- and xylose-containing sugars into HMF and furfural catalyzed by solid base and acid: (**a**) via fructose and (**b**) xylulose as key intermediate, respectively

1.4.2 Base-Acid Bifunctional Catalysis

Besides the interplay of Lewis-Brønsted dual acidic sites, the reactivity of acids in the catalytic transformation of glucose into HMF is also affected by the presence of a solid base catalyst [95]. Combination of a solid base catalyst with an acidic catalyst allows glucose to be efficiently converted into HMF through glucose-to-fructose isomerization over HT, and the subsequent dehydration is catalyzed by Amberlyst-15 under mild conditions (Fig. 1.13a) [96]. After reaction in *N,N*-dimethylformamide at 80 °C for 9 h, 58% selectivity towards HMF for a glucose conversion of 73% is obtained, while no HMF forms in the presence of either Amberlyst-15 or HT. These results clearly show the synergistic role of solid acid and solid base in the synthesis of HMF from glucose. The mixed catalytic systems can be further extended to disaccharides including sucrose and cellobiose for producing HMF to obtain yields of 54% and 35% at 120 °C in 3 h reaction time, respectively [96]. The formation of anhydroglucose from glucose over an acid (i.e., Amberlyst-15) is evidently limited by HT (base)-catalyzed isomerization of glucose to fructose at relatively low reaction temperatures [97]. By using this solid acid-base mixed catalyst system, xylose can be directly transformed into furfural (up to 41% yield) at 100 °C for 3 h, with the isomer xylulose as a key intermediate (Fig. 1.13b) [98, 99]. Likewise, a mixture of arabinose, rhamnose and lactose over Amberlyst-15 and HT is selectively converted to yield furfural (30.5%), HMF (29.1%) and 5-methyl-2-furaldehyde (32.2%) at 110 °C after 6 h, respectively [100]. Catalytically upgrading sugar components of crude biomass with solid base and acid mixtures is most likely to be a promising approach for one-pot synthesis of biofuranic compounds especially HMF and furfural.

TiO_2, ZrO_2 and their mixed oxides possess both basic and acidic sites for the conversion of glucose components to HMF via the key intermediate fructose [101]. Moreover, the use of these single catalysts greatly simplify the separation and recovery process, although hot compressed water (ca. 200–300 °C) is required to transform glucose into HMF (<20% yields) [102, 103]. The relatively low HMF yields can be ascribed to the lack of strong Brønsted acid sites. In connection to this, Qi et al. found that the acidity of ZrO_2 could be significantly enhanced by impregnation with H_2SO_4, and the resulting sulfated ZrO_2 exhibited good activity in the dehydration of fructose (93.6% conversion) to HMF (72.8% yields) under microwave heating in acetone/dimethylsulfoxide (DMSO) at 180 °C for 20 min reaction time [104]. Under mild conditions of 100 °C and 6 h, the sulfated ZrO_2 catalyst could also afford HMF (<10% selectivity) but with a large portion of fructose (ca. 80% selectivity) from glucose [105]. It can be speculated that tetragonal ZrO_2 (Lewis base) should be active for the isomerization of glucose to fructose via an enol intermediate, while polydentate surface SO_4^{2-} species (Brønsted acid) acts on the subsequent dehydration of fructose to HMF, as illustrated in Fig. 1.14. Moreover, Lewis acidic Zr^{4+}might assist the stabilization of the enolate intermediate formed during the isomerization reaction. With the aim of combining both basic and acidic centers into a single catalyst, some other protocols have been examined. For example, the post-grafting of SO_3 H and NH_2 groups onto mesoporous silica nanoparticles via covalent bonds [106], the assembly of lysine with 12-tungstophophoric acid [107], and the incorporation of heteroatoms (e.g., Sn, Zr, Cr and Ti) into the framework of molecular sieves make the resulting hybrids highly disperse and to have compatible dual active sites [108], which are active for producing HMF in yields of reaching >80% from hexose sugars.

1.4.3 *Metal-Base Bifunctional Catalysis*

Selective oxidation of biomass derivatives to carboxylic acids is an important route to petroleum-based chemicals. In particular, 2,5-furandicarboxylic acid (FDCA) has been regarded as a potential replacement for terephthalic acid to produce polyethyleneterephthalate, which can be attained by the aerobic oxidation of HMF (Fig. 1.15) [109]. In most cases, the addition of a basic additive (e.g., NaOH) is necessary to facilitate the oxidation reaction by desorption of the acid product from the metal (e.g., Pt, Pb and Au) surface, wherein the reactivity of hydroxide is generally superior to that of carbonate [110, 111]. The alcohol oxidation of 5-hydroxym ethyl-2-furancarboxylic acid (HMFCA) into FDCA is found to be the

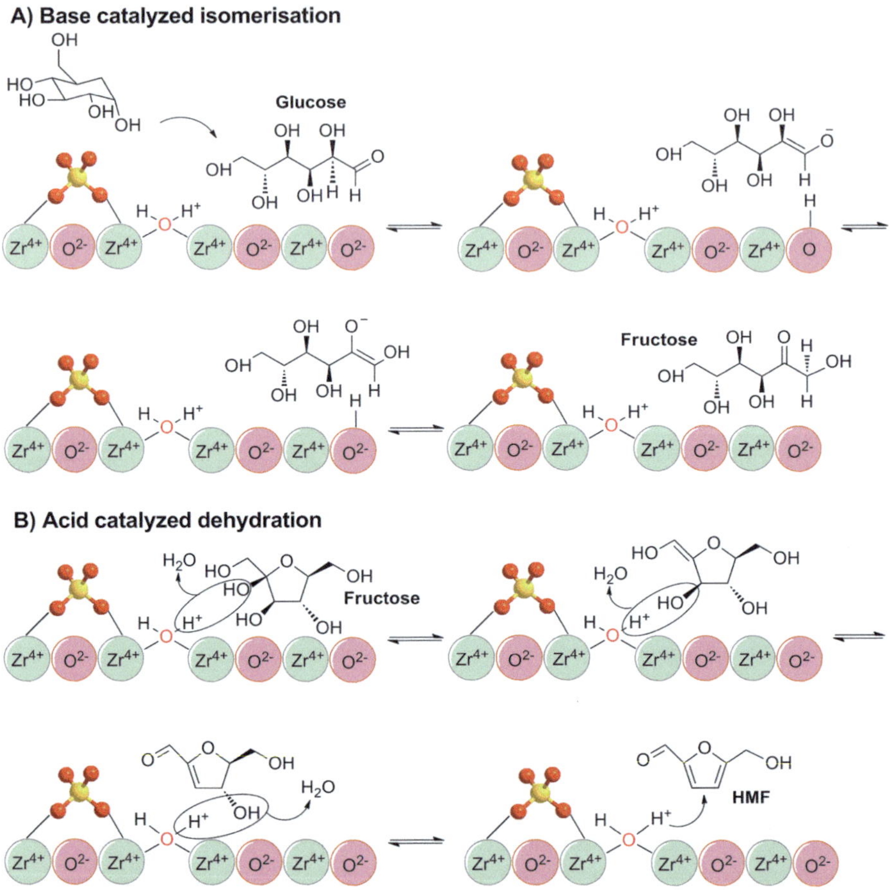

Fig. 1.14 Sulfated ZrO_2mediated reaction mechanism for (**a**) glucose-to-fructose isomerisation over basic sites (O^{2-}), and (**b**) fructose-to-HMF dehydration with Brønsted acid sites (–OH) (Adapted with permission from Ref. [105], Copyright © 2014 Royal Society of Chemistry)

rate-determining step [110, 111]. In the presence of 5 mol% NaOH, 1 wt% Au supported on TiO_2 catalyzes HMF being fully oxidized to FDCA (71% yield) at 30 °C and 2.0 MPa O_2 after 18 h [112]. The formation of partially oxidized intermediate HMFCA is proportional to the base dosage and O_2 pressure, implying that the reaction occurs mainly through initial oxidation of the –CHO moiety to 5-formyl-2-furancarboxylic acid (FFCA) followed by oxidation of the –OH group in HMF to yield the final product FDCA (Fig. 1.15). The possible oxygenated intermediate 2,5-diformylfuran (DFF) is hardly observed during the reaction [113–115], which means that the reaction pathway is most likely that shown in Fig. 1.15.

To improve the recyclability of the catalyst system, porous solid base (e.g., HT) can be used as either a co-catalyst or a support [116, 117]. The combination of HT with Au/TiO_2, and Au supported on HT both show excellent performance in the

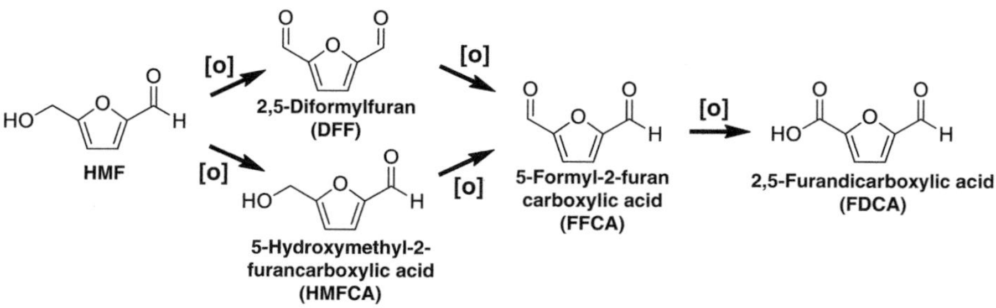

Fig. 1.15 Probable reaction pathway for the oxidation of HMF to FDCA

aerobic oxidation of HMF [116, 117], giving FDCA in maximum yield of 99% at 95 °C in water under flowing O_2 (50 mL/min). By simple deposition of a relatively low-cost ruthenium salt (e.g., $Ru(NO_3)_3$ or $RuCl_3$) onto a basic support (e.g., HT, MgO, La_2O_3 or spinel), supported ruthenium catalysts are active in the oxidation reaction, and $Ru(OH)_x$ with basic supports as the active center [118–120]. On the other hand, metal ions (e.g., $Fe^{3+}$and Co^{2+}) supported on porphyrin-based porous organic polymers exhibit good activity in the oxidation of HMF (>95% conversion) to yield FDCA (up to >90%) at 100 °C after 3–24 h using either oxygen/air or peroxide (e.g., *tert*-butyl hydroperoxide), wherein basic porphyrin subunits and metal ion centers are dual active species [121, 122]. Unlike catalytic systems mediated by zero-valent metal particles (e.g., Pt, Pd and Au), DFF instead of HMFCA is observed to be the key intermediate for producing FDCA from HMF over supported metal-porphyrin catalysts [122]. The reactivity of metal-base bifunctional catalytic materials is greatly enhanced by their concerted interplay, but the reaction pathways might be affected by the type of active species.

1.4.4 Metal-Acid Bifunctional Catalysis

Hydrogenation and hydrodeoxygenation (HDO) over metal particles accompanied by acidic species are the frequently used routes to decrease the oxygen contents of biomass components. Much attention has been placed on the selective hydrogenation of HMF to 2,5-dimethylfuran (DMF), which is deemed to be a promising alternative liquid fuel for transportation [123, 124]. A Brønsted acid (e.g., HCl and H_2SO_4) can be used to cleave the furanic C-O bond, and then PdAu particles stabilized on carbon support (PdAu/C) show good performance for the complete hydrogenation of HMF to DMF (up to 96% yield) at 60 °C after 12 h under an H_2 atmosphere [125]. The 2-methylfurfural (5-MF) and 5-methylfurfuryl alcohol (MFA) are both observed while ring-hydrogenated products such as 2,5-dimethyltetrahydrofuran (DMTHF) and 2,5-bis(hydromethyl)tetrahydrofuran (DHTHF) do not seem to form (Fig. 1.16), indicating that the reaction proceeds through intermediates containing alcohol groups. Apart from the promotional role of negatively

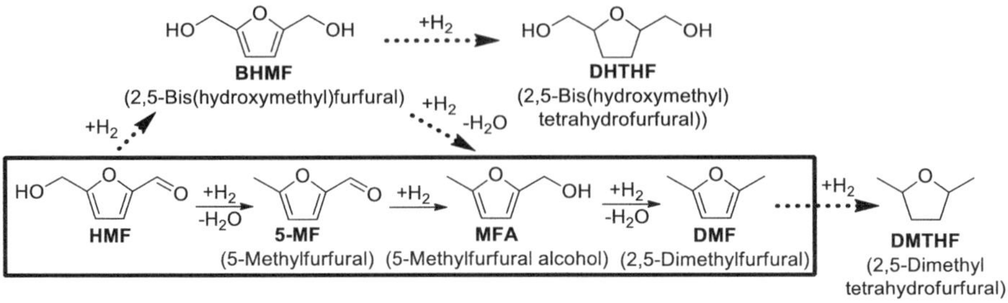

Fig. 1.16 Reaction route toward DMF from HMF hydrogenation

charged Au formed by the co-existence of Pd, HCl with strong acidity ($pK_a < -5$) and chloride ion are necessary as co-catalyst for effective and selective production of DMF by preventing undesirable ring-hydrogenation and facilitating hydrogenolysis through nucleophilic substitution on the alcohol groups of the reactants to form an active chlorinated intermediate [125].

Bimetallic catalysts containing a Lewis acidic center (e.g., $Zn^{2+}$and Co^{2+}) and a hydriding metal component (e.g., Pd and Ru) are more recyclable and have good reactivity for forming DMF from hydrogenation of HMF. For instance, the combination of Zn^{2+}in the form of ZnO, with Pd, Ru or Cu particles catalyzes HMF conversion to DMF with yields of >85% under 0.8–1.5 MPa H_2 at 150–220 °C after 5–8 h, which were more active than $ZnCl_2$ (no DMF observed) and a single metallic catalyst such as Pd/C and Ru/C (<30% DMF yield) [126, 127]. Likewise, the introduction of acidic CoO_xspecies into Pt, Ru or Ni nanoparticles promotes the hydrogenation of HMF under mild conditions with reaction temperatures as low as 130 °C and no more than 1 MPa H_2 pressure, producing DMF in yields of up to 98% [128–130]. The co-added Lewis acidic species seems to be favorable for the substrate absorption and the C-O bond cleavage. Notably, the reusability of these bimetallic hybrids is greatly enhanced by the presence of secondary metal oxides (e.g., Co_3O_4 and ZnO) during hydrogenolysis. On the other hand, the use of acidic solid supports (e.g., zeolite and tungsten carbide) to immobilize and disperse metal particles provides another efficient way to efficiently hydrogenate HMF to produce DMF [131, 132]. Further integration of a strong Brønsted acid (e.g., Amberlyst-15) with a solid catalyst containing metal-acid dual sites allows effective conversion of fructose to yield DMF via the intermediate of HMF in a two-step process involving fructose dehydration and subsequent hydrogenation [133]. The design of metal-acid bifunctional catalytic materials seems to be one of the most promising ways to directly valorize upstream biomass derivatives to desired products having low oxygen content.

1.5 Synthesis of Bio-Based Organic Acids

Instead of producing oxygenates (e.g., gluconic acid, glycolic acid, and acetic acid) from biomass-derived sugars under aerobic conditions, the anaerobic oxidation process leads to the formation of either levulinic acid (LA) or lactic acid (LAC), as shown in Fig. 1.17, which is closely dependent on the type of catalysts and reaction conditions. For the synthesis of LA or its ester from hexose sugars, HMF is the key intermediate, while the subsequent rehydration removes one carbon atom as formic acid [134]. However, glucose over a Brønsted acid is not converted to LA or its ester, but yields anhydroglucose in dipolar aprotic solvents (e.g., *N,N*-dimethyl formamide or dimethylsulfoxide) or alkyl glucopyranoside in alcohols (e.g., methanol or ethanol) as the dominant product [135]. This obstacle is overcome by the introduction of Lewis acidic species into Brønsted acid-catalyzed systems [136]. In this regard, the solid zeolitic materials seem to be one of the best options that can be used as bifunctional catalysts. Saravanamurugan and Riisager illustrate that the commerical zeolites especially large-pore H-USY(Si/Al = 6) with both Brønsted and Lewis acid sites produce methyl and ethyl levulinates (ML and EL) with yields of >40% from glucose at 160 °C after 20 h, wherein the cascade reactions including isomerization, dehydration, and rehydration take place in sequence [137]. Further doping of Lewis acidic species (e.g., $Cr^{3+}$and Fe^{3+}) onto the surface framework of Y-type faujasite zeolites results in the formation of hybrids (e.g., Cr/HY and Fe/HY) with enhanced Lewis acidity, which exhibit pronounced LA yields (ca. 62%) at 180 °C within 3 h [138, 139]. Similarly, a number of solid Brønsted-Lewis acid bifunctional catalysts such as sulfated Dowex 50 modified by cation exchange with $FeCl_3$ [140] and sulfated or phosphorylated metal oxides (e.g., SO_4/ZrO_2, SO_4/TiO_2, and PO_4/ZrO_2) [141–143] are active for the conversion of glucose to LA, and the appropriate molar ratio of Brønsted to Lewis acid sites (ca. 1.2–2.0) play a crucial role in the enhancement of glucose conversion and LA yields.

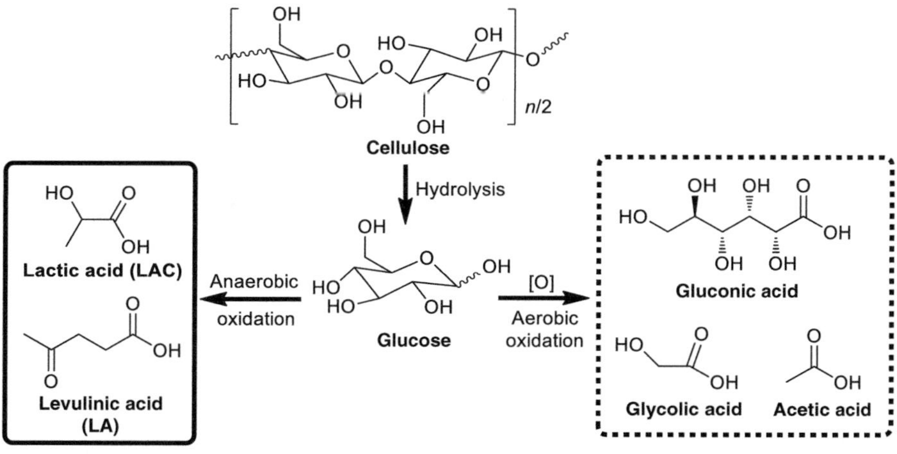

Fig. 1.17 Synthesis of organic acids from sugars via anaerobic and aerobic oxidation

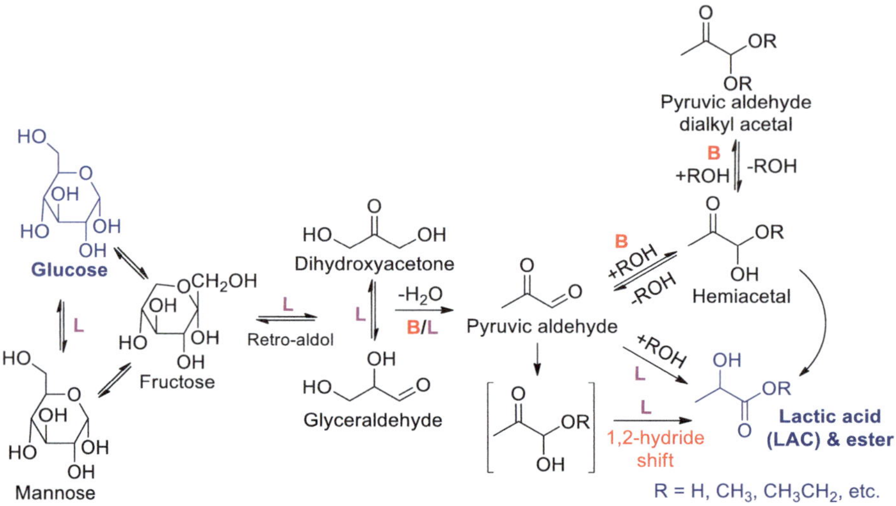

Fig. 1.18 Reaction pathway for transforming glucose into lactic acid (LAC) and ester; L = Lewis acid, B = Brønsted acid [144]

Besides the occurrence of aldose-ketose isomerization, the presence of excess Lewis acidic sites results in the retro-aldol reaction of hexose sugars (e.g., glucose and fructose) to trioses including glyceraldehyde and dihydroxyacetone, followed by dehydration to pyruvaldehyde over Lewis acid sites, finally giving LAC or its ester via the addition of one molecule water or alcohol and succeeding the 1,2-hydride shift (Fig. 1.18) [144]. The Sn grafted carbon-silica (MCM-41) has both Lewis acidity (Sn^{4+}) and Brønsted acidity (e.g., −OH and –COOH derived from carbon part) that can be changed by adjusting the carbon loading amount [145]. The carbon-containing Sn-Si-MCM-41 catalyst with balanced Lewis/Brønsted acidic sites greatly improves the conversion of sucrose in methanol at 155 °C after 20 h to yield methyl lactate (45%), and is superior to $SnCl_4 \cdot 5H_2O$ (26%) and carbon-free Sn-Si-MCM-41 (18%) [145]. Zeolitic materials with both Lewis acidic Sn and Brønsted acidic Al centers exhibit a faster formation of ethyl lactate (>100 h^{-1}) from dihydroxyacetone in comparison with Al-free Sn-beta (72 h^{-1}), which is possibly due to the dehydration reaction being promoted by Brønsted acid sites [146]. When cellobiose or even cellulose is employed as substrate, the hydrolysis step integrated with retro-aldol, dehydration and hydration reactions is able to be accelerated by the coexistence of strong Brønsted acidity and weak or medium Lewis acidity [16], thus directly giving the desired lactic acid product in moderate yields (e.g., 18–27% over AlW and ZrW) in water under hydrothermal conditions (e.g., at 190 °C for 24 h) via a single-pot process [147]. These results emphasize the significant role that bifunctional catalysts will play in the efficient valorization of low-cost upstream biomass components to value-added products with high selectivity.

1.6 Conclusions and Future Outlook

Bifunctional catalytic pathways have been demonstrated to be efficient for multi-step transformations of biomass derivatives into value-added chemcials and biofuels. The combination of Lewis acid or base centers with Brønsted acid sites can selectively promote sequential hydrolysis, isomerization and dehydration of sugars to produce HMF, furfural or organic acids (e.g., LA and LAC) with moderate to good yields, however, the results depend on the sugar type (hexose or pentose) as well as the distribution and charateristics of active sites on solid catalysts. With respect to the oxidation of furanic compounds (e.g., HMF) mediated by metal particles (e.g., Pt, Pb and Au), the catalytic process is greatly accelerated by the presence of basic species that promotes desorption of the acidic product from the metal surface. For the hydrogenation or HDO, the presence of acidic speices is able to promote the cleavage of the C-O bond in biomass derivatives, thus facilitating the removal of oxygen content over metal particles.

Some issues regarding the establishment of well-defined reaction systems and solid catalytic materials may be explored for further research on bifunctional catalytic pathways: (1) increasing the incompatibility of different active sites in a single solid catalyst; (2) suppressing the formation of undesirable products or humins; (3) controlling the sequence of cascade reactions as specified; (4) improving the uniformity, stability and recyclability of bifunctional solid catalysts; and (5) employing computational methods to comprehend preferred configurations or arrangements of active species, thus guiding the exploration of suitable approaches for the inclusion of catalytic sites.

A great number of opportunities exist for the development of bifunctional catalytic pathways mediated by well-defined solid materials in both fundamental and applied fields. The examples illustrated in this chapter provide an introduction into some of the ways for practical realization of multi-catalytic processes for biomass valorization.

Acknowledgements This work is financially supported by Nanjing Agricultural University (68Q-0603), International Postdoctoral Exchange Fellowship Program of China (20170026), Postdoctoral Science Foundation of China (2016M600422), and Jiangsu Postdoctoral Research Funding Plan (1601029A).

References

1. Udaya V, Rao S, Gormley RJ. Bifunctional catalysis in syngas conversions. Catal Today. 1990;6:207–34.
2. Li H, Fang Z, Smith RL, Yang S. Efficient valorization of biomass to biofuels with bifunctional solid catalytic materials. Prog Energy Combust Sci. 2016;55:98–194.
3. Xiao K, Bao Z, Qi X, Wang X, Zhong L, Fang K, Lin M, Sun Y. Advances in bifunctional catalysis for higher alcohol synthesis from syngas. Chin J Catal. 2013;34:116–29.

4. Tanabe K, Yamaguchi T. Acid-base bifunctional catalysis by ZrO_2 and its mixed oxides. Catal Today. 1994;20:185–98.
5. Huang Y, Long B, Tang M, Rui Z, Balogun MS, Tong Y, Ji H. Bifunctional catalytic material: an ultrastable and high-performance surface defect CeO_2 nanosheets for formaldehyde thermal oxidation and photocatalytic oxidation. Appl Catal B Environ. 2016;181:779–87.
6. Bauer G, Kirchner KA. Well-defined bifunctional iron catalysts for the hydrogenation of ketones: iron, the new ruthenium. Angew Chem Int Ed. 2011;50:5798–800.
7. Roessner F, Roland U. Hydrogen spillover in bifunctional catalysis. J Mol Catal A Chem. 1996;112:401–12.
8. Sinfelt JH. Bifunctional catalysis. Adv Chem Eng. 1964;5:37–74.
9. Margelefsky EL, Zeidan RK, Davis ME. Cooperative catalysis by silica-supported organic functional groups. Chem Soc Rev. 2008;37:1118–26.
10. Li H, Wu H, Zhang Q, Liu J, Liu X, Liu Y, Yang S. Solid acid-base bifunctional catalysts in organic transformations. Curr Catal. 2013;2:173–212.
11. Li H, Bhadury PS, Riisager A, Yang S. One-pot transformation of polysaccharides via multi-catalytic processes. Cat Sci Technol. 2014;4:4138–68.
12. Guo F, Fang Z, CC X, Smith RL. Solid acid mediated hydrolysis of biomass for producing biofuels. Prog Energy Combust Sci. 2012;38:672–90.
13. Vilcocq L, Castilho PC, Carvalheiro F, Duarte LC. Hydrolysis of oligosaccharides over solid acid catalysts: a review. ChemSusChem. 2014;7:1010–9.
14. Hu L, Lin L, Wu Z, Zhou S, Liu S. Chemocatalytic hydrolysis of cellulose into glucose over solid acid catalysts. Appl Catal B Environ. 2015;174:225–43.
15. Schwiderski M, Kruse A, Grandl R, Dockendorf D. Comparison of the influence of a Lewis acid $AlCl_3$ and a Brønsted acid HCl on the organosolv pulping of beech wood. Green Chem. 2014;16:1569–78.
16. Shimizu KI, Furukawa H, Kobayashi N, Itaya Y, Satsuma A. Effects of Brønsted and Lewis acidities on activity and selectivity of heteropolyacid-based catalysts for hydrolysis of cellobiose and cellulose. Green Chem. 2009;11:1627–32.
17. Kourieh R, Bennici S, Marzo M, Gervasini A, Auroux A. Investigation of the WO_3/ZrO_2 surface acidic properties for the aqueous hydrolysis of cellobiose. Catal Commun. 2012;19:119–26.
18. Carà PD, Pagliaro M, Elmekawy A, Brown DR, Verschuren P, Shiju NR, Rothenberg G. Hemicellulose hydrolysis catalysed by solid acids. Cat Sci Technol. 2013;3:2057–61.
19. Onda A, Ochi T, Yanagisawa K. Selective hydrolysis of cellulose into glucose over solid acid catalysts. Green Chem. 2008;10:1033–7.
20. Lanzafame P, Temi DM, Perathoner S, Spadaro AN, Centi G. Direct conversion of cellulose to glucose and valuable intermediates in mild reaction conditions over solid acid catalysts. Catal Today. 2012;179:178–84.
21. Sahu R, Dhepe PL. A one-pot method for the selective conversion of hemicellulose from crop waste into C_5 sugars and furfural by using solid acid catalysts. ChemSusChem. 2012;5:751–61.
22. Zhou L, Liu Z, Shi M, Du S, Su Y, Yang X, Xu J. Sulfonated hierarchical H-USY zeolite for efficient hydrolysis of hemicellulose/cellulose. Carbohydr Polym. 2013;98:146–51.
23. Zhou L, Shi M, Cai Q, Wu L, Hu X, Yang X, Chem C, Xu J. Hydrolysis of hemicellulose catalyzed by hierarchical H-USY zeolites–the role of acidity and pore structure. Microporous Mesoporous Mater. 2013;169:54–9.
24. Zhang Z, Zhao ZK. Solid acid and microwave-assisted hydrolysis of cellulose in ionic liquid. Carbohydr Res. 2009;344:2069–72.
25. Bhaumik P, Deepa AK, Kane T, Dhepe PL. Value addition to lignocellulosics and biomass-derived sugars: an insight into solid acid-based catalytic methods. J Chem Sci. 2014;126:373–85.
26. Perego C, Bianchi D. Biomass upgrading through acid-base catalysis. Chem Eng J. 2010;161:314–22.

27. Chandel AK, Antunes FA, Anjos V, Bell MJ, Rodrigues LN, Polikarpov I, de Azevedo ER, Bernardinelli OD, Rosa CA, Pagnocca FC, da Silva SS. Multi-scale structural and chemical analysis of sugarcane bagasse in the process of sequential acid-base pretreatment and ethanol production by Scheffersomyces shehatae and Saccharomyces cerevisiae. Biotechnol Biofuels. 2014;7:63.
28. Jung YH, Park HM, Kim IJ, Park YC, Seo JH, Kim KH. One-pot pretreatment, saccharification and ethanol fermentation of lignocellulose based on acid-base mixture pretreatment. RSC Adv. 2014;4:55318–27.
29. Fang Z, Zhang F, Zeng HY, Guo F. Production of glucose by hydrolysis of cellulose at 423K in the presence of activated hydrotalcite nanoparticles. Bioresour Technol. 2011;102:8017–21.
30. Lee DG, Choi DJ, Park JK. Ketoisomeric conversion of glucose derived from microalgal biomasses. Process Biochem. 2015;50:941–7.
31. Zhao H, Cui Q, Shah V, Xu J, Wang T. Enhancement of glucose isomerase activity by immobilizing on silica/chitosan hybrid microspheres. J Mol Catal B Enzym. 2016;126:18–23.
32. Delidovich I, Palkovits R. catalytic isomerization of biomass-derived aldoses: a review. ChemSusChem. 2016;9:547–61.
33. Marianou AA, Michailof CM, Pineda A, Iliopoulou EF, Triantafyllidis KS, Lappas AA. Glucose to fructose isomerization in aqueous media over homogeneous and heterogeneous catalysts. ChemCatChem. 2016;8:1100–10.
34. Kitajima H, Higashino Y, Matsuda S, Zhong H, Watanabe M, Aida TM, Smith RL. Isomerization of glucose at hydrothermal condition with TiO_2, ZrO_2, CaO-doped ZrO_2 or TiO_2-doped ZrO_2. Catal Today. 2016;274:67–72.
35. Souza RO, Fabiano DP, Feche C, Rataboul F, Cardoso D, Essayem N. Glucose-fructose isomerisation promoted by basic hybrid catalysts. Catal Today. 2012;195:114–9.
36. Moliner M, Román-Leshkov Y, Davis ME. Tin-containing zeolites are highly active catalysts for the isomerization of glucose in water. Proc Natl Acad Sci U S A. 2010;107:6164–8.
37. Bermejo-Deval R, Assary RS, Nikolla E, Moliner M, Román-Leshkov Y, Hwang SJ, Palsdottir A, Silverman D, Lobo RF, Curtiss LA, Davis ME. Metalloenzyme-like catalyzed isomerizations of sugars by Lewis acid zeolites. Proc Natl Acad Sci U S A. 2012;109:9727–32.
38. Rai N, Caratzoulas S, Vlachos DG. Role of silanol group in Sn-beta zeolite for glucose isomerization and epimerization reactions. ACS Catal. 2013;3:2294–8.
39. Li G, Pidko EA, Hensen EJ. Synergy between Lewis acid sites and hydroxyl groups for the isomerization of glucose to fructose over Sn-containing zeolites: a theoretical perspective. Cat Sci Technol. 2014;4:2241–50.
40. Li YP, Head-Gordon M, Bell AT. Analysis of the reaction mechanism and catalytic activity of metal-substituted beta zeolite for the isomerization of glucose to fructose. ACS Catal. 2014;4:1537–45.
41. Li G, Pidko EA, Hensen EJ. A periodic DFT study of glucose to fructose isomerization on tungstite ($WO_3 \cdot H_2O$): influence of group IV–VI dopants and cooperativity with hydroxyl groups. ACS Catal. 2016;6:4162–9.
42. Christianson JR, Caratzoulas S, Vlachos DG. Computational insight into the effect of Sn-beta Na exchange and solvent on glucose isomerization and epimerization. ACS Catal. 2015;5:5256–63.
43. Gao DM, Kobayashi T, Adachi S. Kinetic effect of alcohols on hexose isomerization under subcritical aqueous conditions. Chem Eng Res Des. 2015;104:723–9.
44. Gao DM, Kobayashi T, Adachi S. Promotion or suppression of glucose isomerization in subcritical aqueous straight-and branched-chain alcohols. Biosci Biotechnol Biochem. 2015;79:470–4.
45. Cho HJ, Dornath P, Fan W. Synthesis of hierarchical Sn-MFI as Lewis acid catalysts for isomerization of cellulosic sugars. ACS Catal. 2014;4:2029–37.
46. Gounder R, Davis ME. Monosaccharide and disaccharide isomerization over Lewis acid sites in hydrophobic and hydrophilic molecular sieves. J Catal. 2013;308:176–88.
47. Soisangwan N, Gao DM, Kobayashi T, Khuwijitjaru P, Adachi S. Kinetic analysis for the isomerization of cellobiose to cellobiulose in subcritical aqueous ethanol. Carbohydr Res. 2016;433:67–72.

48. Gounder R, Davis ME. Titanium-beta zeolites catalyze the stereospecific isomerization of D-glucose to L-sorbose via intramolecular C_5–C_1 hydride shift. ACS Catal. 2013;3:1469–76.
49. Saravanamurugan S, Paniagua M, Melero JA, Riisager A. Efficient isomerization of glucose to fructose over zeolites in consecutive reactions in alcohol and aqueous media. J Am Chem Soc. 2013;135:5246–9.
50. Saravanamurugan S, Riisager A, Taarning E, Meier S. Combined function of Brønsted and Lewis acidity in the zeolite-catalyzed isomerization of glucose to fructose in alcohols. ChemCatChem. 2016;8:3107–11.
51. Saravanamurugan S, Riisager A, Taarning E, Meier S. Mechanism and stereoselectivity of zeolite-catalysed sugar isomerisation in alcohols. Chem Commun. 2016;52:12773–6.
52. Paniagua M, Saravanamurugan S, Melian-Rodriguez M, Melero JA, Riisager A. Xylose isomerization with zeolites in a two-step alcohol-water process. ChemSusChem. 2015;8:1088–94.
53. Saravanamurugan S, Riisager A. Zeolite-catalyzed isomerization of tetroses in aqueous medium. Cat Sci Technol. 2014;4:3186–90.
54. Ren L, Guo Q, Kumar P, Orazov M, Xu D, Alhassan SM, Mkhoyan KA, Davis ME, Tsapatsis M. Self-pillared, single-unit-cell Sn-MFI zeolite nanosheets and their use for glucose and lactose isomerization. Angew Chem Int Ed. 2015;54:10848–51.
55. Li H, Yang S, Riisager A, Pandey A, Sangwan RS, Saravanamurugan S, Luque R. Zeolite and zeotype-catalysed transformations of biofuranic compounds. Green Chem. 2016;18:5701–35.
56. Liu Y, Li Z, Yang Y, Hou Y, Wei Z. A novel route towards high yield 5-hydroxymethylfurfural from fructose catalyzed by a mixture of Lewis and Brønsted acids. RSC Adv. 2014;4:42035–8.
57. Binder JB, Raines RT. Simple chemical transformation of lignocellulosic biomass into furans for fuels and chemicals. J Am Chem Soc. 2009;131:1979–85.
58. Choudhary V, Mushrif SH, Ho C, Anderko A, Nikolakis V, Marinkovic NS, Frenkel AI, Sandler SI, Vlachos DG. Insights into the interplay of Lewis and Brønsted acid catalysts in glucose and fructose conversion to 5-(hydroxymethyl) furfural and levulinic acid in aqueous media. J Am Chem Soc. 2013;135:3997–4006.
59. Yang Y, Hu C, Abu-Omar MM. The effect of hydrochloric acid on the conversion of glucose to 5-hydroxymethylfurfural in $AlCl_3$-H_2O/THF biphasic medium. J Mol Catal A Chem. 2013;376:98–102.
60. Pagan-Torres YJ, Wang T, Gallo JMR, Shanks BH, Dumesic JA. Production of 5-hydroxymethylfurfural from glucose using a combination of Lewis and Brønsted acid catalysts in water in a biphasic reactor with an alkylphenol solvent. ACS Catal. 2012;2:930–4.
61. Nikolla E, Román-Leshkov Y, Moliner M, Davis ME. "One-pot" synthesis of 5-(hydroxymethyl) furfural from carbohydrates using tin-beta zeolite. ACS Catal. 2011;1:408–10.
62. Wu L, Song J, Zhang B, Zhou B, Zhou H, Fan H, Yang Y, Han B. Very efficient conversion of glucose to 5-hydroxymethylfurfural in DBU-based ionic liquids with benzenesulfonate anion. Green Chem. 2014;16:3935–41.
63. Li H, Zhang Q, Liu X, Chang F, Hu D, Zhang Y, Xue W, Yang S. $InCl_3$-ionic liquid catalytic system for efficient and selective conversion of cellulose into 5-hydroxymethylfurfural. RSC Adv. 2013;3:3648–54.
64. Tao F, Song H, Chou L. Catalytic conversion of cellulose to chemicals in ionic liquid. Carbohydr Res. 2011;346:58–63.
65. Tao F, Song H, Chou L. Hydrolysis of cellulose by using catalytic amounts of $FeCl_2$ in ionic liquids. ChemSusChem. 2010;3:1298–303.
66. Tao F, Song H, Yang J, Chou L. Catalytic hydrolysis of cellulose into furans in $MnCl_2$-ionic liquid system. Carbohydr Polym. 2011;85:363–8.
67. Li H, Zhang Q, Liu X, Chang F, Zhang Y, Xue W, Yang S. Immobilizing $Cr^{3+}$with SO_3H-functionalized solid polymeric ionic liquids as efficient and reusable catalysts for selective transformation of carbohydrates into 5-hydroxymethylfurfural. Bioresour Technol. 2013;144:21–7.
68. Nakajima K, Baba Y, Noma R, Kitano M, Kondo JN, Hayashi S, Hara M. $Nb_2O_5{\cdot}nH_2O$ as a heterogeneous catalyst with water-tolerant Lewis acid sites. J Am Chem Soc. 2011;133:4224–7.

69. Ordomsky VV, van der Schaaf J, Schouten JC, Nijhuis TA. Glucose dehydration to 5-hydroxymethylfurfural in a biphasic system over solid acid foams. ChemSusChem. 2013;6:1697–707.
70. Ordomsky VV, Sushkevich VL, Schouten JC, Van der Schaaf J, Nijhuis TA. Glucose dehydration to 5-hydroxymethylfurfural over phosphate catalysts. J Catal. 2013;300:37–46.
71. Jiménez-Morales I, Teckchandani-Ortiz A, Santamaría-González J, Maireles-Torres P, Jiménez-López A. Selective dehydration of glucose to 5-hydroxymethylfurfural on acidic mesoporous tantalum phosphate. Appl Catal B Environ. 2014;144:22–8.
72. Dutta A, Patra AK, Dutta S, Saha B, Bhaumik A. Hierarchically porous titanium phosphate nanoparticles: an efficient solid acid catalyst for microwave assisted conversion of biomass and carbohydrates into 5-hydroxymethylfurfural. J Mater Chem. 2012;22:14094–100.
73. Zhang Y, Wang J, Ren J, Liu X, Li X, Xia Y, Lu G, Wang Y. Mesoporous niobium phosphate: an excellent solid acid for the dehydration of fructose to 5-hydroxymethylfurfural in water. Cat Sci Technol. 2012;2:2485–91.
74. Joo JB, Vu A, Zhang Q, Dahl M, Gu M, Zaera F, Yin Y. A sulfated ZrO_2 hollow nanostructure as an acid catalyst in the dehydration of fructose to 5-hydroxymethylfurfural. ChemSusChem. 2013;6:2001–8.
75. Kourieh R, Rakic V, Bennici S, Auroux A. Relation between surface acidity and reactivity in fructose conversion into 5-HMF using tungstated zirconia catalysts. Catal Commun. 2013;30:5–13.
76. Yamaguchi K, Sakurada T, Ogasawara Y, Mizuno N. Tin-tungsten mixed oxide as efficient heterogeneous catalyst for conversion of saccharides to furan derivatives. Chem Lett. 2011;40:542–3.
77. Wang L, Zhang J, Wang X, Zhang B, Ji W, Meng X, Li J, DS S, Bao X, Xiao FS. Creation of Brønsted acid sites on Sn-based solid catalysts for the conversion of biomass. J Mater Chem A. 2014;2:3725–9.
78. Zhao Q, Wang L, Zhao S, Wang X, Wang S. High selective production of 5-hydroymethylfurfural from fructose by a solid heteropolyacid catalyst. Fuel. 2011;90:2289–93.
79. Jadhav AH, Kim H, Hwang IT. An efficient and heterogeneous recyclable silicotungstic acid with modified acid sites as a catalyst for conversion of fructose and sucrose into 5-hydroxymethylfurfural in superheated water. Bioresour Technol. 2013;132:342–50.
80. Zhao S, Cheng M, Li J, Tian J, Wang X. One pot production of 5-hydroxymethylfurfural with high yield from cellulose by a Brønsted-Lewis-surfactant-combined heteropolyacid catalyst. Chem Commun. 2011;47:2176–8.
81. Zheng H, Sun Z, Yi X, Wang S, Li J, Wang X, Jiang Z. A water-tolerant $C_{16}H_3PW_{11}CrO_{39}$ catalyst for the efficient conversion of monosaccharides into 5-hydroxymethylfurfural in a micellar system. RSC Adv. 2013;3:23051–6.
82. Gounder R. Hydrophobic microporous and mesoporous oxides as Brønsted and Lewis acid catalysts for biomass conversion in liquid water. Cat Sci Technol. 2014;4:2877–86.
83. Nandiwale KY, Galande ND, Thakur P, Sawant SD, Zambre VP, Bokade VV. One pot synthesis of 5-hydroxymethylfurfural by cellulose hydrolysis over highly active bimodal micro/mesoporous H-ZSM-5 catalyst. ACS Sustain Chem Eng. 2014;2:1928–32.
84. Hu L, Wu Z, Xu J, Sun Y, Lin L, Liu S. Zeolite-promoted transformation of glucose into 5-hydroxymethylfurfural in ionic liquid. Chem Eng J. 2014;244:137–44.
85. Otomo R, Yokoi T, Kondo JN, Tatsumi T. Dealuminated Beta zeolite as effective bifunctional catalyst for direct transformation of glucose to 5-hydroxymethylfurfural. Appl Catal A Gen. 2014;470:318–26.
86. Dias AS, Lima S, Pillinger M, Valente AA. Acidic cesium salts of 12-tungstophosphoric acid as catalysts for the dehydration of xylose into furfural. Carbohydr Res. 2006;341:2946–53.
87. Bhaumik P, Dhepe PL. Efficient, stable, and reusable silicoaluminophosphate for the one-pot production of furfural from hemicellulose. ACS Catal. 2013;3:2299–303.
88. Agirrezabal-Telleria I, Hemmann F, Jäger C, Arias PL, Kemnitz E. Functionalized partially hydroxylated MgF_2 as catalysts for the dehydration of D-xylose to furfural. J Catal. 2013;305:81–91.

89. Antunes MM, Lima S, Fernandes A, Candeias J, Pillinger M, Rocha SM, Ribeiro MA, Valente AA. Catalytic dehydration of D-xylose to 2-furfuraldehyde in the presence of Zr-(W,Al) mixed oxides; Tracing by-products using two-dimensional gas chromatography-time-of-flight mass spectrometry. Catal Today. 2012;195:127–35.
90. Mazzotta MG, Gupta D, Saha B, Patra AK, Bhaumik A, Abu-Omar MM. Efficient solid acid catalyst containing Lewis and Brønsted acid sites for the production of furfurals. ChemSusChem. 2014;7:2342–50.
91. Gallo JMR, Alonso DM, Mellmer MA, Yeap JH, Wong HC, Dumesic JA. Production of furfural from lignocellulosic biomass using beta zeolite and biomass-derived solvent. Top Catal. 2013;56:1775–81.
92. Dhepe PL, Sahu R. A solid-acid-based process for the conversion of hemicellulose. Green Chem. 2010;12:2153–6.
93. Weingarten R, Tompsett GA, Conner WC, Huber GW. Design of solid acid catalysts for aqueous-phase dehydration of carbohydrates: the role of Lewis and Brønsted acid sites. J Catal. 2011;279:174–82.
94. Gürbüz EI, Gallo JMR, Alonso DM, Wettstein SG, Lim WY, Dumesic JA. Conversion of hemicellulose into furfural using solid acid catalysts in γ-valerolactone. Angew Chem Int Ed. 2013;52:1270–4.
95. Setoyama T. Acid-base bifunctional catalysis: an industrial viewpoint. Catal Today. 2006;116:250–62.
96. Takagaki A, Ohara M, Nishimura S, Ebitani K. A one-pot reaction for biorefinery: combination of solid acid and base catalysts for direct production of 5-hydroxymethylfurfural from saccharides. Chem Commun. 2009;2009:6276–8.
97. Ohara M, Takagaki A, Nishimura S, Ebitani K. Syntheses of 5-hydroxymethylfurfural and levoglucosan by selective dehydration of glucose using solid acid and base catalysts. Appl Catal A Gen. 2010;383:149–55.
98. Takagaki A, Ohara M, Nishimura S, Ebitani K. One-pot formation of furfural from xylose via isomerization and successive dehydration reactions over heterogeneous acid and base catalysts. Chem Lett. 2010;39:838–40.
99. Shirotori M, Nishimura S, Ebitani K. One-pot synthesis of furfural from xylose using Al_2O_3–Ni-Al layered double hydroxide acid-base bi-functional catalyst and sulfonated resin. Chem Lett. 2015;45:194–6.
100. Tuteja J, Nishimura S, Ebitani K. One-pot synthesis of furans from various saccharides using a combination of solid acid and base catalysts. Bull Chem Soc Jpn. 2012;85:275–81.
101. Qi X, Watanabe M, Aida TM, Smith RL. Catalytical conversion of fructose and glucose into 5-hydroxymethylfurfural in hot compressed water by microwave heating. Catal Commun. 2008;9:2244–9.
102. Chareonlimkun A, Champreda V, Shotipruk A, Laosiripojana N. Catalytic conversion of sugarcane bagasse, rice husk and corncob in the presence of TiO_2, ZrO_2 and mixed-oxide TiO_2-ZrO_2 under hot compressed water (HCW) condition. Bioresour Technol. 2010;101:4179–86.
103. Watanabe M, Aizawa Y, Iida T, Nishimura R, Inomata H. Catalytic glucose and fructose conversions with TiO_2 and ZrO_2 in water at 473 K: relationship between reactivity and acid–base property determined by TPD measurement. Appl Catal A Gen. 2005;295:150–6.
104. Qi X, Watanabe M, Aida TM, Smith RL. Sulfated zirconia as a solid acid catalyst for the dehydration of fructose to 5-hydroxymethylfurfural. Catal Commun. 2009;10:1771–5.
105. Osatiashtiani A, Lee AF, Brown DR, Melero JA, Morales G, Wilson K. Bifunctional SO_4/ZrO_2 catalysts for 5-hydroxymethylfufural (5-HMF) production from glucose. Cat Sci Technol. 2014;4:333–42.
106. Peng WH, Lee YY, Wu C, KCW W. Acid-base bi-functionalized, large-pored mesoporous silica nanoparticles for cooperative catalysis of one-pot cellulose-to-HMF conversion. J Mater Chem. 2012;22:23181–5.
107. Zhao Q, Sun Z, Wang S, Huang G, Wang X, Jiang Z. Conversion of highly concentrated fructose into 5-hydroxymethylfurfural by acid-base bifunctional HPA nanocatalysts induced by choline chloride. RSC Adv. 2014;4:63055–61.

108. Xu Q, Zhu Z, Tian Y, Deng J, Shi J, Fu Y. Sn-MCM-41 as efficient catalyst for the conversion of glucose into 5-hydroxymethylfurfural in ionic liquids. Bioresources. 2013;9:303–15.
109. Zhang Z, Zhen J, Liu B, Lv K, Deng K. Selective aerobic oxidation of the biomass-derived precursor 5-hydroxymethylfurfural to 2, 5-furandicarboxylic acid under mild conditions over a magnetic palladium nanocatalys. Green Chem. 2015;17:1308–17.
110. Davis SE, Houk LR, Tamargo EC, Datye AK, Davis RJ. Oxidation of 5-hydroxymethylfurfural over supported Pt, Pd and Au catalysts. Catal Today. 2011;160:55–60.
111. Casanova O, Iborra S, Corma A. Biomass into chemicals: aerobic oxidation of 5-hydroxymethyl-2-furfural into 2,5-furandicarboxylic acid with gold nanoparticle catalysts. ChemSusChem. 2009;2:1138–44.
112. Gorbanev YY, Klitgaard SK, Woodley JM, Christensen CH, Riisager A. Gold-catalyzed aerobic oxidation of 5-hydroxymethylfurfural in water at ambient temperature. ChemSusChem. 2009;2:672–5.
113. Davis SE, Zope BN, Davis RJ. On the mechanism of selective oxidation of 5-hydroxymethylfurfural to 2, 5-furandicarboxylic acid over supported Pt and Au catalysts. Green Chem. 2012;14:143–7.
114. Pasini T, Piccinini M, Blosi M, Bonelli R, Albonetti S, Dimitratos N, Lopez-Sanchez JA, Sankar M, He Q, Kiely CJ, Hutchings GJ, Cavani F. Selective oxidation of 5-hydroxymethyl-2-furfural using supported gold-copper nanoparticles. Green Chem. 2011;13:2091–9.
115. Siankevich S, Savoglidis G, Fei Z, Laurenczy G, Alexander DT, Yan N, Dyson PJ. A novel platinum nanocatalyst for the oxidation of 5-hydroxymethylfurfural into 2, 5-furandicarboxylic acid under mild conditions. J Catal. 2014;315:67–74.
116. Zope BN, Davis SE, Davis RJ. Influence of reaction conditions on diacid formation during Au-catalyzed oxidation of glycerol and hydroxymethylfurfural. Top Catal. 2012;55:24–32.
117. Gupta NK, Nishimura S, Takagaki A, Ebitani K. Hydrotalcite-supported gold-nanoparticle-catalyzed highly efficient base-free aqueous oxidation of 5-hydroxymethylfurfural into 2, 5-furandicarboxylic acid under atmospheric oxygen pressure. Green Chem. 2011;13:824–7.
118. Gorbanev YY, Kegnæs S, Riisager A. Effect of support in heterogeneous ruthenium catalysts used for the selective aerobic oxidation of HMF in water. Top Catal. 2011;54:1318–24.
119. Gorbanev YY, Kegnæs S, Riisager A. Selective aerobic oxidation of 5-hydroxymethylfurfural in water over solid ruthenium hydroxide catalysts with magnesium-based supports. Catal Lett. 2011;141:1752–60.
120. Ståhlberg T, Eyjólfsdóttir E, Gorbanev YY, Sádaba I, Riisager A. Aerobic oxidation of 5-(hydroxymethyl)furfural in ionic liquids with solid ruthenium hydroxide catalysts. Catal Lett. 2012;142:1089–97.
121. Saha B, Gupta D, Abu-Omar MM, Modak A, Bhaumik A. Porphyrin-based porous organic polymer-supported iron(III) catalyst for efficient aerobic oxidation of 5-hydroxymethylfurfural into 2, 5-furandicarboxylic acid. J Catal. 2013;299:316–20.
122. Gao L, Deng K, Zheng J, Liu B, Zhang Z. Efficient oxidation of biomass derived 5-hydroxymethylfurfural into 2, 5-furandicarboxylic acid catalyzed by Merrifield resin supported cobalt porphyrin. Chem Eng J. 2015;270:444–9.
123. Hu L, Lin L, Liu S. Chemoselective hydrogenation of biomass-derived 5-hydroxymethylfurfural into the liquid biofuel 2, 5-dimethylfuran. Ind Eng Chem Res. 2014;53:9969–78.
124. Qian Y, Zhu L, Wang Y, Lu X. Recent progress in the development of biofuel 2, 5-dimethylfuran. Renew Sust Energ Rev. 2015;41:633–46.
125. Nishimura S, Ikeda N, Ebitani K. Selective hydrogenation of biomass-derived 5-hydroxymethylfurfural (HMF) to 2, 5-dimethylfuran (DMF) under atmospheric hydrogen pressure over carbon supported PdAu bimetallic catalyst. Catal Today. 2014;232:89–98.
126. Zhu Y, Kong X, Zheng H, Ding G, Zhu Y, Li YW. Efficient synthesis of 2, 5-dihydroxymethylfuran and 2, 5-dimethylfuran from 5-hydroxymethylfurfural using mineral-derived Cu catalysts as versatile catalysts. Cat Sci Technol. 2015;5:4208–17.
127. Saha B, Bohn CM, Abu-Omar MM. Zinc-assisted hydrodeoxygenation of biomass-derived 5-hydroxymethylfurfural to 2,5-dimethylfuran. ChemSusChem. 2014;7:3095–101.

128. Yang P, Cui Q, Zu Y, Liu X, Lu G, Wang Y. Catalytic production of 2,5-dimethylfuran from 5-hydroxymethylfurfural over Ni/Co_3O_4 catalyst. Catal Commun. 2015;66:55–9.
129. Zu Y, Yang P, Wang J, Liu X, Ren J, Lu G, Wang Y. Efficient production of the liquid fuel 2,5-dimethylfuran from 5-hydroxymethylfurfural over Ru/Co_3O_4 catalyst. Appl Catal B Environ. 2014;146:244–8.
130. Wang GH, Hilgert J, Richter FH, Wang F, Bongard HJ, Spliethoff B, Weidenthaler C, Schüth F. Platinum–cobalt bimetallic nanoparticles in hollow carbon nanospheres for hydrogenolysis of 5-hydroxymethylfurfural. Nat Mater. 2014;13:293–300.
131. Nagpure AS, Lucas N, Chilukuri SV. Efficient preparation of liquid fuel 2, 5-dimethylfuran from biomass-derived 5-hydroxymethylfurfural over Ru–NaY catalyst. ACS Sustain Chem Eng. 2015;3:2909–16.
132. Huang YB, Chen MY, Yan L, Guo QX, Fu Y. Nickel-tungsten carbide catalysts for the production of 2,5-dimethylfuran from biomass-derived molecules. ChemSusChem. 2014;7:1068–72.
133. Upare PP, Hwang DW, Hwang YK, Lee UH, Hong DY, Chang JS. An integrated process for the production of 2,5-dimethylfuran from fructose. Green Chem. 2015;17:3310–3.
134. Kruger JS, Choudhary V, Nikolakis V, Vlachos DG. Elucidating the roles of zeolite H-BEA in aqueous-phase fructose dehydration and HMF rehydration. ACS Catal. 2013;3:1279–91.
135. Saravanamurugan S, Riisager A. Solid acid catalysed formation of ethyl levulinate and ethyl glucopyranoside from mono-and disaccharides. Catal Commun. 2012;17:71–5.
136. Yang F, Fu J, Mo J, Lu X. Synergy of Lewis and Brønsted acids on catalytic hydrothermal decomposition of hexose to levulinic acid. Energy Fuel. 2013;27:6973–8.
137. Saravanamurugan S, Riisager A. Zeolite catalyzed transformation of carbohydrates to alkyl levulinates. ChemCatChem. 2013;5:1754–7.
138. Ya'aini N, Amin NAS, Endud S. Characterization and performance of hybrid catalysts for levulinic acid production from glucose. Microporous Mesoporous Mater. 2013;171:14–23.
139. Ramli NAS, Amin NAS. Fe/HY zeolite as an effective catalyst for levulinic acid production from glucose: characterization and catalytic performance. Appl Catal B Environ. 2015;163:487–98.
140. Yang H, Wang L, Jia L, Qiu C, Pang Q, Pan X. Selective decomposition of cellulose into glucose and levulinic acid over Fe-resin catalyst in NaCl solution under hydrothermal conditions. Ind Eng Chem Res. 2014;53:6562–8.
141. Peng L, Lin L, Zhang J, Shi J, Liu S. Solid acid catalyzed glucose conversion to ethyl levulinate. Appl Catal A Gen. 2011;397:259–65.
142. Peng L, Lin L, Li H, Yang Q. Conversion of carbohydrates biomass into levulinate esters using heterogeneous catalysts. Appl Energy. 2011;88:4590–6.
143. Weingarten R, Kim YT, Tompsett GA, Fernández A, Han KS, Hagaman EW, Conner WC, Dumesic JA, Huber GW. Conversion of glucose into levulinic acid with solid metal(IV) phosphate catalysts. J Catal. 2013;304:123–34.
144. Holm MS, Saravanamurugan S, Taarning E. Conversion of sugars to lactic acid derivatives using heterogeneous zeotype catalysts. Science. 2010;328:602–5.
145. de Clippel F, Dusselier M, Van Rompaey R, Vanelderen P, Dijkmans J, Makshina E, Giebeler L, Oswald S, Baron GV, Denayer JFM, Pescarmona PP, Jacobs PA, Sels BF. Fast and selective sugar conversion to alkyl lactate and lactic acid with bifunctional carbon-silica catalysts. J Am Chem Soc. 2012;134:10089–101.
146. Dijkmans J, Dusselier M, Gabriëls D, Houthoofd K, Magusin PC, Huang S, Pontikes Y, Trekels M, Vantomme A, Giebeler L, Oswald S, Sels BF. Cooperative catalysis for multistep biomass conversion with Sn/Al Beta zeolite. ACS Catal. 2015;5:928–40.
147. Chambon F, Rataboul F, Pinel C, Cabiac A, Guillon E, Essayem N. Cellulose hydrothermal conversion promoted by heterogeneous Brønsted and Lewis acids: remarkable efficiency of solid Lewis acids to produce lactic acid. Appl Catal B Environ. 2011;105:171–81.

Chapter 2
Introduction to Characterization Methods for Heterogeneous Catalysts and Their Application to Cellulose Conversion Mechanisms

Xiao Kong, Yifeng Zhu, Hu Li, Zhen Fang, and Richard L. Smith Jr.

Abstract To realize large-scale commercialization of cellulose conversion, design of catalysts based on understanding of their structure and detailed mechanisms are needed. Characterization methods for heterogeneous catalysts can be used to provide insight into the mechanisms of catalytic conversion of cellulose and other biomass-related components, regarding catalytic activity, selectivity and stability. This chapter provides an introduction to characterization methods for heterogeneous catalysts for the general reader and then provides an overview on the fundamentals of the solid acid, metal and metal-acid bifunctional catalysts for cellulose conversion. Solid acid catalysts for the hydrolysis of cellulose to glucose and 5-hydroxymethylfurfural are discussed. Metal catalysts that can be used to convert cellulose to syngas and hydrogen are introduced. Metal-acid bifunctional catalysts that can be used for converting cellulose to sugar alcohols, alkanes and ethylene glycol are covered. Techniques used to explore the mechanistic features of the catalysts are emphasized.

X. Kong
Biomass Group, College of Engineering, Nanjing Agricultural University, Nanjing, Jiangsu, China

Y. Zhu
State Key Laboratory of Catalysis, Dalian Institute of Chemical Physics, Chinese Academy of Sciences, Dalian, China

H. Li • Z. Fang (✉)
College of Engineering, Nanjing Agricultural University, Nanjing, Jiangsu, China
e-mail: zhenfang@njau.edu.cn

R.L. Smith Jr.
Graduate School of Environmental Studies, Research Center of Supercritical Fluid Technology, Tohoku University, Sendai, Japan

Z. Fang et al. (eds.), *Production of Biofuels and Chemicals with Bifunctional Catalysts*, Biofuels and Biorefineries 8, https://doi.org/10.1007/978-981-10-5137-1_2

2.1 Structure and Properties of Cellulose

As demand for energy and materials grows, there is greater need for protecting the environment. Academia is proposing to develop sustainable methods for energy and materials using renewable energies to reduce the dependence on fossil resources. Biomass, especially non-food or waste lignocellulose, is widely abundant in the form of plants or agriculture.

Lignocellulosic biomass contains cellulose, hemicellulose and lignin constituents and is estimated to contain from 35 to 50% by mass of cellulose [1, 2], which is a homo-polymer of glucose. Catalytic conversion of cellulose is popular in recent years as a way to develop sustainable feedstocks for future biorefineries.

Traditional catalytic processes that use petroleum, coal and gas resources as raw materials, mainly focus on the transformation of hydrocarbon-like compounds into products. However, the catalytic conversion of cellulose can hardly be based on traditional catalysis processes, because cellulose has a high oxygen content that requires new methods that will produce a new set of base chemical feedstocks, which are commonly referred to as platform chemicals [3–5]. New catalytic strategies are required to convert oxygen-containing raw materials into biofuels and chemical products through techniques such as hydrolysis, hydrogenation and hydrogenolysis [6]. Water-insoluble cellulose can be converted to saccharides through hydrolysis and subsequent hydrogenation or hydrogenolysis can afford products having low oxygen content. Substantial effort is being devoted to the development of appropriate catalysts for these processes [7–9] with possibly new additives such as ionic liquids [10].

Cellulose is a fibrous, tough and water-insoluble substance. It is a highly-abundant unbranched homo-polysaccharide and an important skeletal component in plants. Cellulose has six polymorphs (I, II, III_I, III_{II}, IV_I and IV_{II}) and it can be interconverted between these forms [11]. Cellulose is composed of a linear chain of several hundreds to thousands of β-1,4-D-glucose units with the formula of $(C_6H_{10}O_5)_n$ (Fig. 2.1). The chain terminated with a D-glucose unit with an normal

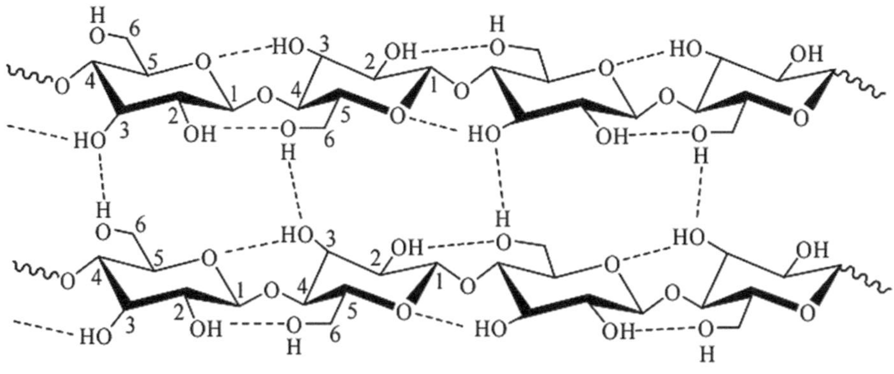

Fig. 2.1 Representative structure of cellulose (Reproduced from Ref. [10] with permission of The Royal Society of Chemistry)

C4-OH group (non-reducing end) at one end and an C1-OH group is in equilibrium with the aldehyde structure (reducing end) [12]. The degree of polymerization of cellulose can be as high as 10,000 [6]. In the structure of cellulose, abundant intra- and inter-molecular hydrogen bonds exist that protect the β-1,4-glycosidic bonds from attack by foreign molecules. The structure of cellulose makes it a substrate that is intrinsically resistant to modification or transformation. For example, cellulose is insoluble in most solvents and only can be depolymerized to a substantial degree with concentrated acids or at high-temperatures and high-pressures [13, 14]. Pyrolysis, which is the thermal decomposition of material in the absence of oxygen, is one of the most common conversion processes for renewable energy that can be used to obtain liquid products [15]. High temperatures (~500 °C) are needed for the process to obtain sufficient yields of liquid products. Thus, the conversion of cellulose with pyrolysis typically requires severe conditions.

For the conversion of cellulose to monomeric glucose units or to other chemicals with safe reaction solvents and under mild conditions, its chemical structure (C–C, C–O–C and C–O bonds) must be considered. Metal sites and acid sites in a catalytic material can target various functional groups and promote transformation of cellulose to the desired products. Thus, the design and characterization of such catalytic materials is key to understanding transformation mechanisms of cellulose.

Characterization methods discussed in this chapter include X-ray diffraction XRD, N_2 adsorption, X-ray photoelectron spectroscopy (XPS), Auger electron spectroscopy (AES), hydrogen temperature-programmed reduction (H_2-TPR), H_2 adsorption, CO adsorption, N_2O adsorption, temperature-programmed desorption (TPD), temperature-programmed oxidation (TPO), temperature-programmed surface reaction (TPSR), thermogravimetric analysis (TGA), infrared (IR) spectra, Raman spectra, transmission electron microscopy (TEM), scanning transmission electron microscopy (STEM), scanning electron microscopy (SEM), atomic force microscopy (AFM) and X-ray absorption spectroscopy (XAS). This chapter aims to highlight progress in the conversion of cellulose with heterogeneous catalysts with special emphasis on metal, acid and metal-acid bifunctional catalysts. Catalysts properties and reaction pathways are introduced in detail.

2.2 Characterizations of Acid and Metal Sites

Catalytic performance depends on many factors including active phase, electronic properties, and surface area of active sites. Metal and acid sites are common active sites of many solid materials. Thus, many kinds of characterization methods are carried out to determine the nature of the active sites and to understand their catalytic performance. Examples shown in this section are not limited to the catalysts used for cellulose conversion. The catalysts used in other areas are also included. Characterization provides insight into material design of solid acid catalysts, supported metal catalysts and metal-acid bifunctional catalysts. Based on the characterization of catalyst during different stages (e.g., preparation, before and after reaction, in-situ status), a more rational design of catalyst can be achieved.

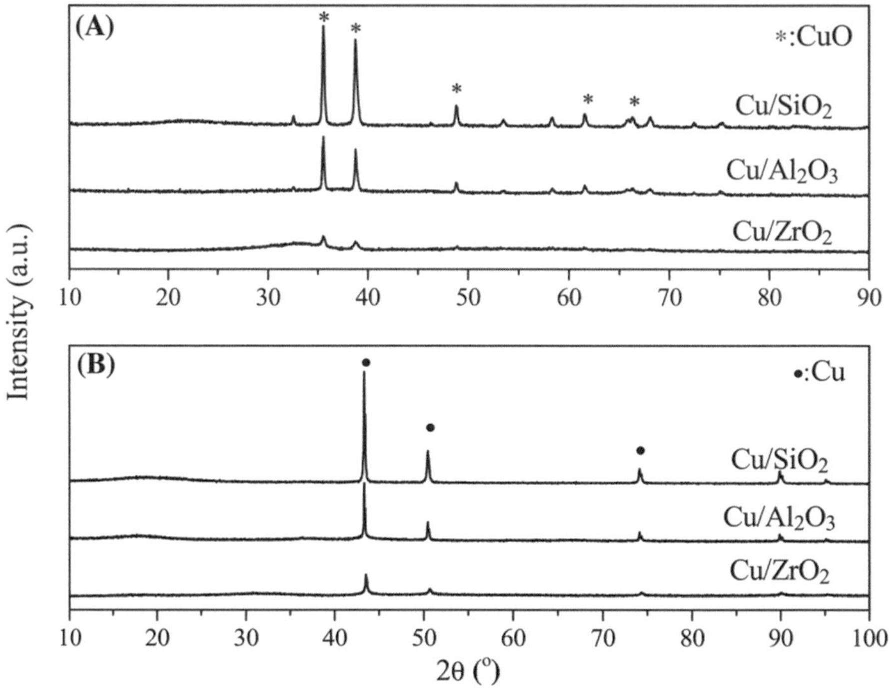

Fig. 2.2 XRD patterns of calcined copper catalysts (**A**) and reduced catalysts (**B**) (Reprinted with permission from Ref. [17], Copyright © 2013 Elsevier)

2.2.1 *X-Ray Diffraction*

X-ray diffraction (XRD) is one of the most frequently used bulk sensitive techniques for catalyst characterization. Normally, Cu Kα and Co Kα radiation from standard X-ray tubes with a power are used. Figure 2.2 shows XRD results collected on a D/Max-RA X-ray diffractometer (Rigaku, Japan) with Cu Kα radiation ($\lambda = 0.154$ nm) operated at 40 kV and 100 mA. By comparing the measured X-ray diffraction peaks with standard Power Diffraction File (PDF) in software (e.g., Jade), the crystalline phases in a catalyst can be identified. For calcined catalysts, it can be seen that the material is composed of CuO (PDF#48-1548), while the reduced catalyst is composed of metallic Cu (PDF#04-0836). The above results are obtained based on the lattice structural parameters and lattice spacing is derived from the Bragg equation [16]:

$$n\lambda = 2d\sin\theta; n = 1, 2, \ldots \tag{2.1}$$

where λ is the wavelength of the incoming X-ray, d is the distance between two lattice planes, θ is the angle between the incoming X-ray and the normal to the reflecting lattice plane, n is an integer called the order of the reflection.

Using X-rays of known λ, the θ value can be measured, and the d values can be calculated. The powder diffraction file (PDF) contained in XRD software has available d values. Thus, XRD can be used to determine the structure and the solid phase of a catalyst.

The crystalline size (not particle size) is calculated from line broadening of the characteristic peak at half its maximum intensity referred to as full width at half maximum (FWHM). The crystalline size measured by broadening analysis is often smaller than that measured by microscopy. The crystalline size is calculated based on the Scherrer equation [16]:

$$\mathrm{D} = \mathrm{K}\lambda / \beta \cos\theta \tag{2.2}$$

where D is the dimension of the crystalline domain, λ is the X-ray wavelength, β is the line broadening at half the maximum intensity, θ is the angle between the beam and the normal to the reflecting plane, and K is a dimensionless shape factor (for spherical crystallites, K = 0.94).

The larger peak width indicates a smaller crystalline size. However, XRD is unable to detect crystalline domains that are too small. The lower limit of sizes that can be measured by Scherrer equation is about 2 nm. Further, crystalline sizes that are too large that cause small peak broading cannot be determined with the Scherrer equation.

The XRD patterns in Fig. 2.2B show that the Cu crystalline size are 35.6, 29.6 and 21.0 nm for Cu/SiO_2, Cu/Al_2O_3 and Cu/ZrO_2 respectively, calculated by Scherrer eq. [17]. The angle of $2\theta = 43.3°$ for reduced catalysts is used to determine the crystalline size, with $\theta = 21.65$, $\lambda = 0.154$ nm, $K = 0.94$. β value of peak at 43.3° can be obtained from origin fitting or read from Jade software. The crystalline size (D value) is thus obtained.

XRD can be used to study catalysts *in situ* under actual conditions so that XRD can monitor changes in the phase of a catalytic material during reduction and reaction processes. *In situ* XRD is frequently used in the study of phase transformations of alloy catalysts. For *in situ* XRD measurements, a cell was used and the catalyst was reduced with diluted H_2 in a heated sample holder. Sharafutdinov et al. [18] (Fig. 2.3) conducted X-ray diffraction experiments with a PANalytical X'pert Pro diffractometer using Ni filtered Cu Kα. The evolution of phases during reduction is presented. With an increase in reduction temperature, the dynamics of Ni-Ga intermetallic formation is observed. At T > 200 °C, formation of NiO crystallites begins. At 300~350 °C, the intensities of the NiO reflections are become enhanced, indicating sintering during this step. Further increase of temperature leads to alloying of Ni with Ga, i.e. formation of α'-Ni_3Ga, δ-Ni_5Ga_3 and eventually formation of a β-NiGa phase, which corresponds well to the Ni/Ga ratio (1:1) of the precursor.

Rietveld refinement can be used for microstructure analysis based on XRD diffractograms, including the quantification of the relative amounts of phases. The "Handbook of Heterogeneous Catalysis" edited by Ertl et al. [16] is recommended for readers to obtain more details. To quantify the relative amounts of phases during reduction in Fig. 2.3a, Rietveld refinement performed using X'Pert HighScore Plus software and is shown in Fig. 2.3b. Refinement was done by fitting various combi-

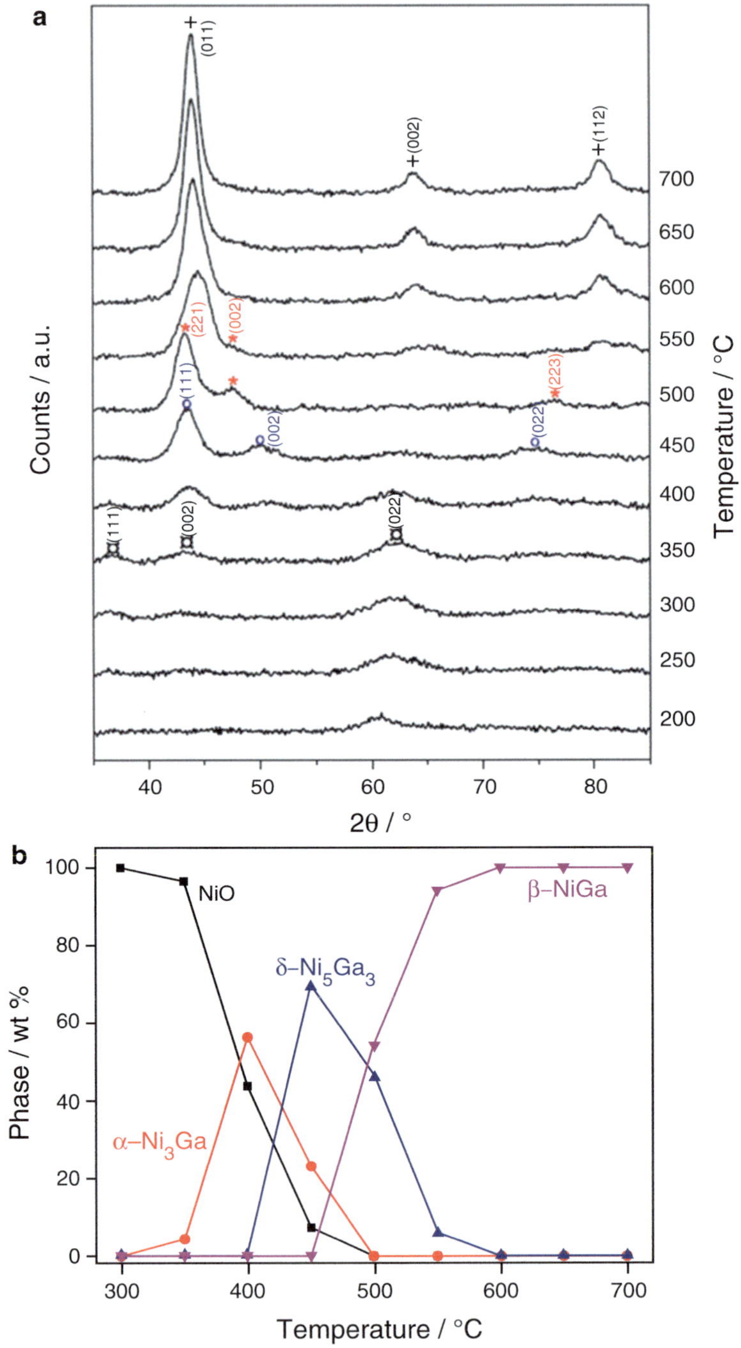

Fig. 2.3 Evolution of crystallographic phases during temperature-programmed reduction (TPR) to form NiGa/SiO_2 catalyst (Ni:Ga ratio 1:1 in the precursor) **(a)**. β-NiGa formation is analyzed by Rietveld refinement. Reflections are marked: + (β-NiGa),* (δ-Ni_5Ga_3), ¤ (NiO) and o (α'-Ni_3Ga). Gas composition: 90% H_2/He. Total flow rate: 40 ml/min, and (**b**) corresponding Rietveld refinement analysis of crystallographic phases during TPR (Reprinted with permission from Ref. [18], Copyright © 2014 Elsevier)

nations of Ni-Ga phases. Two or more phases coexist at intermediate temperatures and the β-NiGa phase is obtained at temperatures higher than 600 °C.

2.2.2 *N_2 Adsorption*

For solid catalysts, reactions take place on the surface or within the pores of the catalysts. High surface area of the solid generally contributes to sufficient dispersion of the active sites. Thus, methods to measure the surface area and porosity of a solid are important. N_2-adsorption is the most widely used method to measure the surface area (S_g) that is determined from Eq. (2.3):

$$S_g = \frac{NA_m V_m}{VW} \tag{2.3}$$

where N is Avogadro's number (6.023 × 10^{23}), A_m is the cross-sectional area of the adsorbate gas (e.g., N_2 is 0.162 $(nm)^2$) at 77 K. V_m is the volume of adsorbate at monolayer absorption, W is the mass of the sample, and V is the volume (22.4 × 10^3 cm^3). Although N_2 is commonly used to measure surface area and porosity of a solid, N_2-adsorption does not allow reliable estimate of the micropore volume or its size distribution so that other adsorptive molecules are used. For example, Ar-adsorption at 87 K can be advantageously used to characterize solids that having pore size diameters no wider than ~ 2 nm [19, 20]. The molecular area of Ar at 87 K is 0.142 nm^2 [16].

The Brunauer–Emmett–Teller (BET) model uses multilayer adsorption theory to estimate the surface area of a solid material (Eq. 2.4):

$$\frac{V}{V_m} = \frac{C(P/P_0)}{\left(1-\frac{P}{P_0}\right)\left[1-(1-C)(P/P_0)\right]} \tag{2.4}$$

where P is the pressure of the adsorbate (e.g., N_2) at equilibrium, P_0 is the saturated vapor pressure of adsorbate at that temperature and C is the BET constant value.

For most systems, the BET model can be used to determine the surface area, when P/P_0 is in the range of 0.05~0.35. A linearized form (Eq. 2.5) can be obtained by transformation of Eq. 2.4. By correlating the $P/V(P_0\text{-}P)$ and P/P_0, the slope and intercept value can be obtained so that the V_m value can be calculated. The surface area (S_{BET}) is thus obtained when the V_m value is introduced in Eq. 2.3:

$$\frac{P}{V(p_0 - P)} = \frac{1}{CV_m}\left[1+(C-1)\frac{p}{p_0}\right] \tag{2.5}$$

$$\text{Intercept} = \frac{1}{CV_m}$$

$$\text{Slope} = \frac{C-1}{CV_m}$$

N_2 adsorption also reflects the textural properties (e.g. pore volume and pore size distribution) of the material [16]. For conversion of large molecules (e.g., cellulose) with heterogeneous catalysts, the catalyst pore size plays key role. For example, many works have suggested that the ideal solid catalyst for depolymerization of cellulose should be a macroporous acidic material with a large surface area [14, 21]. Pore condensation can be observed in mesopores due to the interactions between the fluid molecules at high P/P_0 value (e.g., $P/P_0 = 0.4$). Pore volume is calculated from the condensation value of N_2 at which P/P_0 is 0.99. The Kelvin equation gives the curvature of the liquid meniscus in a pore to the relative pressure when condensation occurs (Eq. 2.6), which serves as the basis for mesopore analysis:

$$r_k = \frac{2\gamma v_1}{RT} \ln\left(P / P_0\right) \tag{2.6}$$

in which r_k is the Kelvin radius, ν_1 is the molar volume and γ is the surface tension of the liquid condensate.

For cylindrical pores, the Kelvin equation is modified since the pore walls are covered by a multilayer adsorbed film (thickness, t) at the onset of pore condensation ($r_p = r_k + t$). The Barrett–Joyner–Halenda (BJH) method is based on this model and the procedure is based on the notion of emptying of the pores by a stepwise reduction of P/P_0 and taking into account the thinning of the multilayer in those pores already emptied of the condensate.

2.2.3 X-Ray Photoelectron Spectroscopy (XPS) and Auger Electron Spectroscopy (AES)

X-ray photoelectron spectroscopy (XPS) is a surface sensitive method to determine the surface composition and electronic valence of species in a catalyst. XPS is based on the photoelectric effect: an atom absorbs an energy $h\nu$ where h is Planck's constant and ν is the frequency of the photon and thus a core or valence electron with binding energy E_b is ejected with kinetic energy of $E_k = h\nu - E_b - \varphi$, where φ represents the work function of the spectrometer [16]. Mg and Al anodes are common X-ray sources that give MgK_α (1486.6 eV) and AlK_α (1253.6 eV) photons respectively. Scanning the kinetic energy E_k of all photoelectrons gives characteristic values of E_b. Thus, the distinctive core electron binding energies of all elements with the exception of H and He allow characterization of the origin of the photoelectron.

For different atoms, the excited electron intensity is generally different. For example, Ag $3d_{3/2}$ and Ag $3d_{5/2}$ are the strongest peaks for the Ag atom, which can be used as the characteristic peak. However, Cu $2p_{3/2}$ and Cu $2p_{1/2}$ are used as the

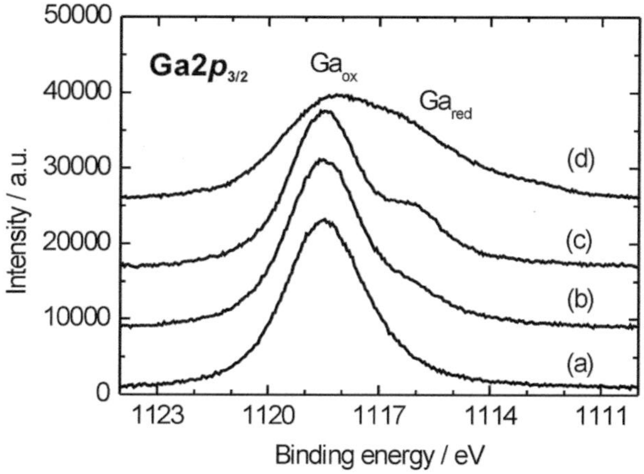

Fig. 2.4 XPS data of the $Ga2p_{3/2}$ peak of milled PdGa (**a**) untreated, (**b**) after H_2 treatment at 300 °C, (**c**) after H_2 treatment at 400 °C, and (**d**) after ISS (ion scattering spectroscopy) (Reprinted with permission from Ref. [22], Copyright © 2008 Elsevier)

characteristic peaks for the Cu atom. The XPS is a surface sensitive technique that can be used to provide surface elemental content and to reveal species valence and binding energy. The peak height or peak area, which is related to the content of each element, is used for quantification. The chemical shift of the binding energy is used to understand the chemical nature and environment of the species. High binding energy corresponds to high valence state of metals.

For example, as shown in Fig. 2.4, the XPS data of milled PdGa exhibits a broad $Ga2p_{3/2}$ peak at 1118.5 eV, which can be assigned to gallium oxide. With an increase in reduction temperature, a peak with lower binding energy emerges indicating the reduction of gallium species. From the XPS results, the gallium valence states during the reaction process or analysis procedure can be determined.

The charging effect, which is build-up of positive charge near the surface of a material due to continuous photoemission of insulating samples, must be considered since it can result in an apparent shift of the binding energies. A simple method to remove the charging effect is to use the C 1 s signal due to carbon contamination, because this signal appears for carbon on almost any dispersed sample [23, 24].

Auger electron spectroscopy (AES) is used to study the surface chemical states of solids. Detailed explanations of the principles and applications of AES are given in the literature [25, 26]. An excited atom with a hole in inner shell S (created by an electron or photon impact) tends to relax to an energetically lower state. Then the hole can be filled by an electron from an outer shell X, which is called the radiationless Auger process ($E_k = E_S - E_X - E'_Y$). The final state after relaxation is a free electron with kinetic energy E_k and an ionized atom E'_Y. The kinetic energy is associated with the binding energy of electrons in shells of the excited atom. KLL and LMM transitions are frequently investigated. For example, Cu-based catalysts usually consist of Cu^0 and/or Cu^+ sites after reduction. X-ray excited Auger spectroscopy of reduced xCuyAl catalysts is shown in Fig. 2.5. Surface Cu^0-Cu^+ species are distinguished with peaks at around 914 eV (Cu^+) and 918 eV (Cu^0) by deconvolution of the Cu LMM peaks [27]. Variations in the $Cu^+/(Cu^0+Cu^+)$ ratios in the surface of a solid material can be determined with AES spectroscopy.

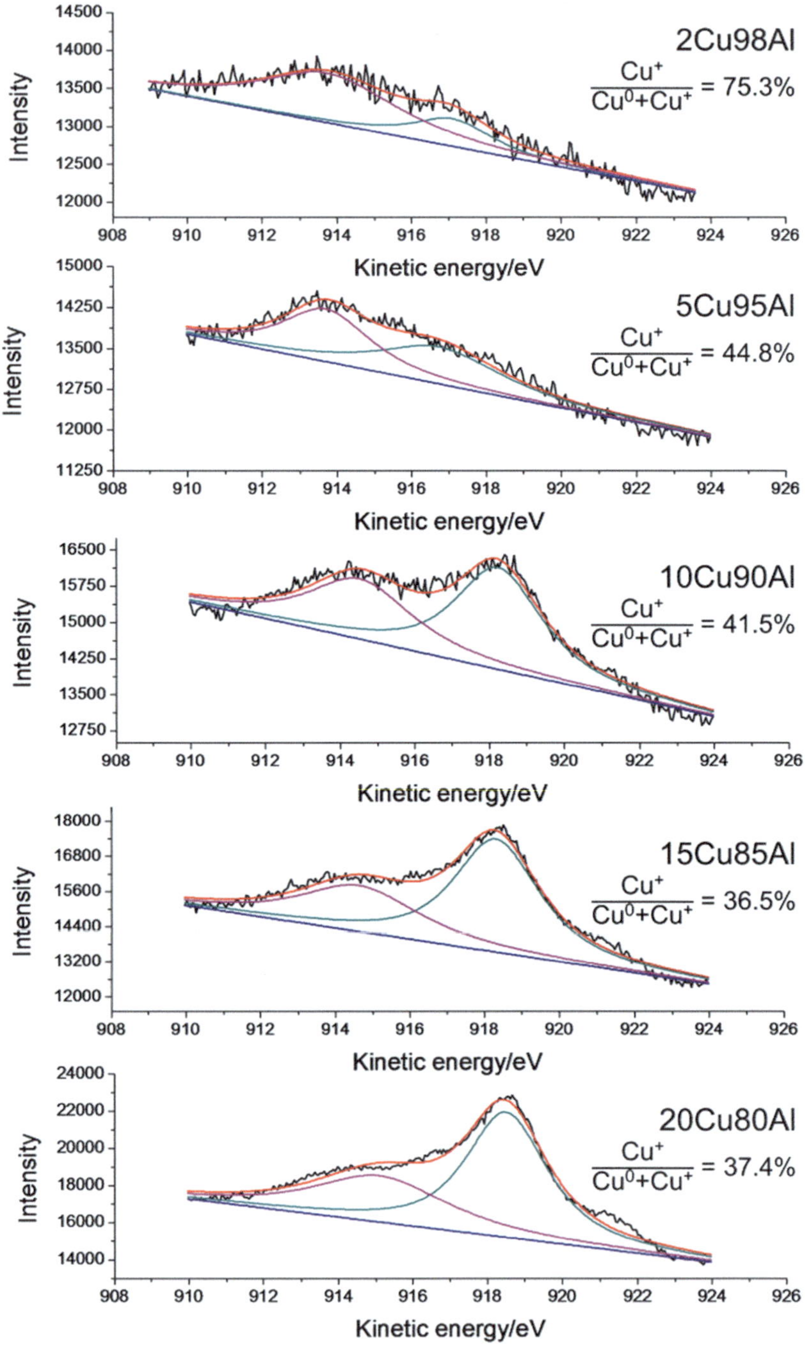

Fig. 2.5 Deconvolution of Cu LMM X-ray excited Auger spectroscopy of reduced *x*Cu*y*Al catalysts (Reprinted with permission from Ref. [27], Copyright (2015) American Chemical Society)

2.2.4 *Hydrogen Temperature-Programmed Reduction (H_2-TPR)*

Hydrogen temperature-programmed reduction (H_2-TPR) is the process of treating a catalyst under an atmosphere of H_2 or under a dilute H_2-containing atmosphere. The catalyst is reduced from a high valence state and to a low valence state and then to a zero-valence (metallic) state. For a typical H_2-TPR experiment, the catalyst is loaded into a quartz tube and heated in a dilute H_2 atmosphere with a programmed temperature. A thermal conductivity detector (TCD) is used to monitor changes in the H_2 atmosphere to obtain the variation between the TCD signal and the reduction temperature (Fig. 2.6) [28]. From typical H_2-TPR measurements, the reduction temperature needed for a catalyst can be determined. The method can also be used to study the interaction between metals and supports. As shown in Fig. 2.6, supported Pd catalyst is more easily reduced than Cu, while the presence of Pd increases the reducibility of Cu, leading to a shift towards lower temperatures. The results demonstrate the interactions between Pd and Cu, which can be further explored by combining other characterization methods.

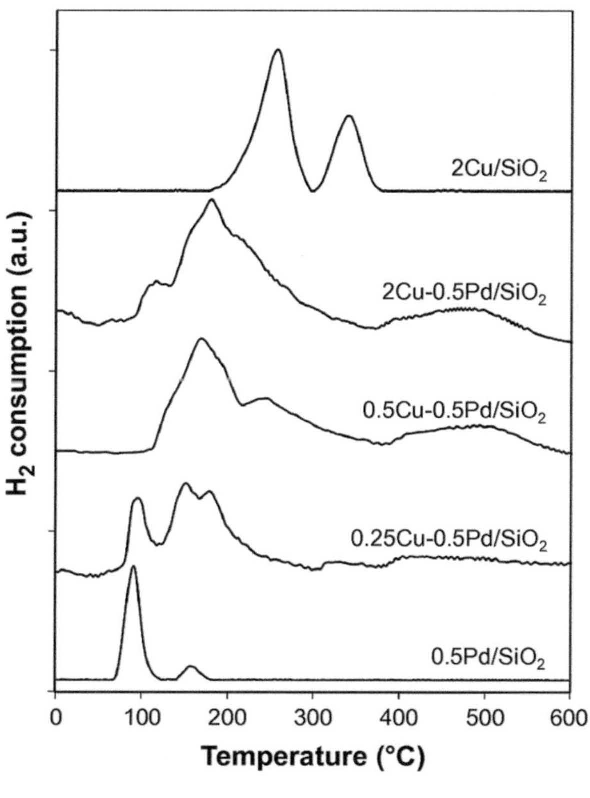

Fig. 2.6 H_2-TPR profiles of *x*Cu*y*Pd catalysts (Reprinted with permission from Ref. [28], Copyright © 2011 Elsevier)

2.2.5 H_2 Adsorption, CO Adsorption and N_2O Adsorption

H_2, CO and N_2O adsorption can be used to determine metal dispersion and particle size in a material [29, 30]. For different metals, different adsorption methods and adsorbates have been used with calculations being based on the assumption that the metal is present as spherical particles of uniform size. The active surface metal of the catalyst can be measured by chemisorption with either a pulse technique or by a conventional volumetric method [30, 31].

H_2 adsorption can be used to calculate the dispersion and particle size of non-noble metals (e.g. Ni, Co) [32]. The volumetric method is conventionally used with a Pyrex glass volumetric adsorption apparatus [33]. Reuel et al. [34] reported this method to calculate the dispersion and particle size of cobalt catalysts. The catalyst was first reduced and then the system was evacuated to less than 6.67×10^{-3} Pa (5×10^{-5} Torr) at a temperature 10–20 K below the reduction temperature and adsorption isotherms for H_2 were measured at 373 K. Adsorption isotherms were measured by the desorption (decreasing pressure) method after a 45 min equilibration at 300–350 Torr of adsorbate. The total gas uptake was determined by extrapolating the straight-line portion of the adsorption isotherm to zero pressure. Percentage dispersion (%D) was calculated according to the equation:

$$\%\mathrm{D} = 1.179 \cdot X / \left(W \cdot f\right) \tag{2.7}$$

where X is the total H_2 uptake in micromoles per gram of catalyst measured at the temperature of maximum adsorption, W is the weight percentage of cobalt, and f is fraction of cobalt reduced to the metal. Average crystalline diameter (in nanometers) is calculated from %D assuming spherical metal crystallites of uniform diameter d with a site density (s) of 14.6 atoms/nm^2 for supported cobalt crystallites. Thus

$$d = 6.59 \cdot s / \%\mathrm{D} = 96 / \%\mathrm{D} \tag{2.8}$$

CO adsorption is also widely used for characterizing Pt, Pd and Co based catalysts [35, 36]. In typical experiments, the catalyst is reduced in hydrogen and the atmosphere is purged with He at the same temperature. Then, the catalyst is cooled to room temperature and CO volumes are injected into the system from a calibrated on-line sampling valve. CO adsorption is assumed to be complete after three successive peaks showed the same peak areas. An equimolar CO/Pt stoichiometry is used in calculations. CO-IR results have shown that the linear CO-metal species, of which the CO-metal ratio is unity, is predominant on metals such as Pt [31]. Thus, CO adsorption can be used for CO/metal = 1 in the calculation of Pt based catalysts. However, the CO adsorption method is controversial due to the complicated CO-metal stoichiometric ratio for many metals.

N_2O adsorption at 90 °C is widely used for determining the dispersion and metal surface area of copper [37, 38]. Specifically, Cu catalysts are reduced at a defined temperature (e.g., 300 °C) for about 2 h. Then, N_2O/Ar flow is applied to oxidize the surface Cu^0 to Cu_2O. Finally, the catalyst is flushed with Ar and cooled to 40 °C to start a H_2-TPR program. Copper dispersion is calculated by dividing the amount of chemisorption sites into the total supported copper atoms; the copper surface area is calculated by assuming spherical shape of the copper metal particles and 1.46×10^{19} copper atoms/m^2 [39].

2.2.6 *Temperature-Programmed Desorption (TPD)*

Surface properties can be studied with temperature-programmed desorption (TPD), which is based on the removal of adsorbed molecules, including species such as NH_3 and CO_2 from a solid material.

NH_3-temperature programmed desorption (NH_3-TPD) is widely used for determining the acid strength and amounts of acidic sites in a material that is contacted with ammonia. NH_3-TPD is one of the basic characterizations for solid acid catalysts. The basic NH_3 molecule adsorbs onto the acidic sites of the catalyst and after a given time period, an increase in temperature with the flow of carrier gas causes desorption of NH_3, which is detected by TCD. The weakly adsorbed NH_3 is desorbed first, and then the strongly adsorbed NH_3 is desorbed. The higher desorption temperature of NH_3 indicates the stronger acid strength.

Mass spectrometry can also be applied to collect desorbed NH_3 signals (m/z = 16) more precisely. As shown in Fig. 2.7, H_2O has molecular fragments in the range of

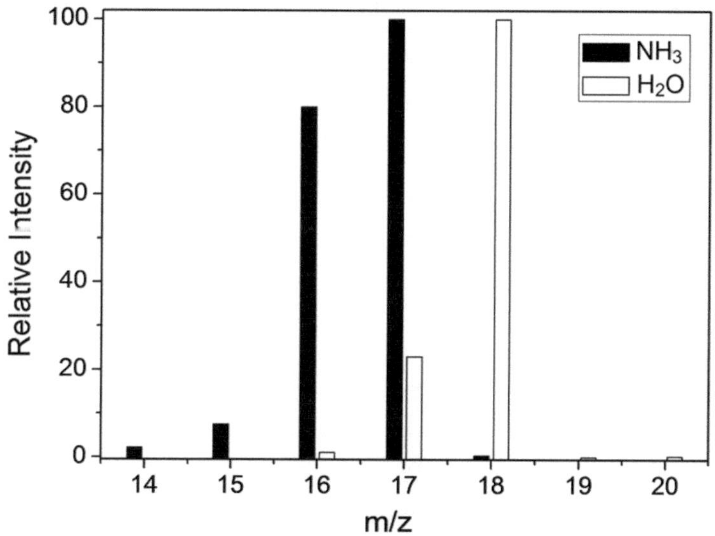

Fig. 2.7 Relative intensity of NH_3 and H_2O in the mass spectra of a Cu based catalyst (Reprinted with permission from Ref. [17], Copyright © 2013 Elsevier)

16–20 m/z whereas NH_3 has molecular fragments in the range of 14–18 m/z. For m/z = 17, although NH_3 has the highest intensity, the H_2O also contributes to this signal. Thus, to exclude the signal from water, m/z = 16 is frequently used in NH_3-TPD analyses with mass spectroscopy.

Figure 2.8 shows NH_3-TPD results of different catalysts, including H-ZSM-22, desilicated sample (DeZEO) and recrystallized materials (samples RZEO-1, RZEO-2, RZEO-3, and RZEO-4) [40]. The desorption peak at around 200 °C is attributed to the adsorbed NH_3 at the weak acid sites, while the peak around 400 °C is assigned to adsorbed NH_3 at the strong acid sites. By analyzing NH_3-TPD results, the acid strength and amounts can be obtained. The NH_3-TPD results show that recrystallization at different conditions (RZEO-1 to RZEO-4) leads to an enhanced trend in the decrease in the amount and strength of acid sites with the high temperature desorption peak shifting to lower temperatures. However, for NH_3 adsorbed onto super acid catalysts, the required high temperatures cause decomposition of NH_3 before desorption. Thus, the NH_3-TPD method is limited to catalysts that have

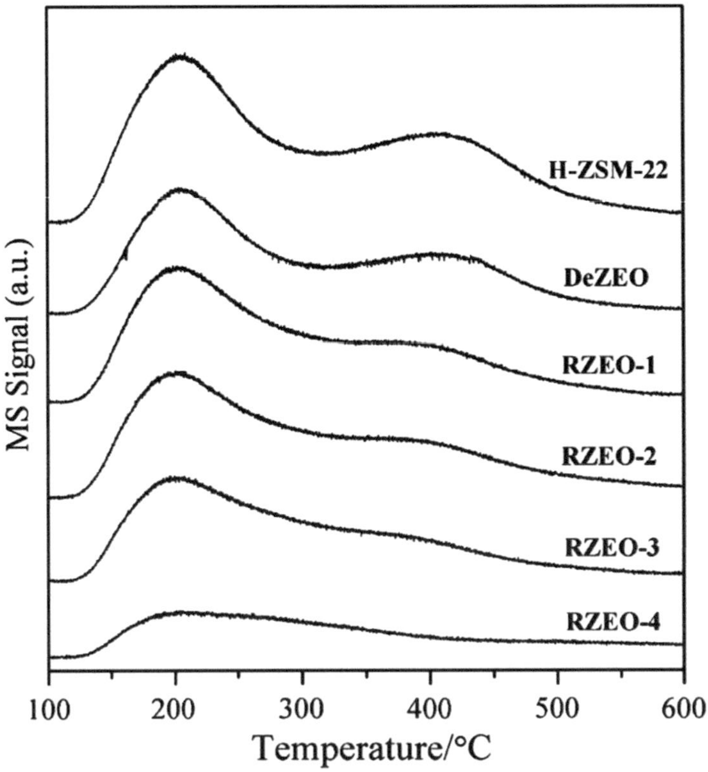

Fig. 2.8 NH_3-TPD profiles of various zeolite catalysts (Reprinted with permission from Ref. [40], Copyright © 2016 Elsevier)

weakly or medium strength acidic sites. When CO_2 is used (CO_2-TPD), the amount of surface basicity due to adsorption can be determined. The procedure for CO_2-TPD experiments is similar to that of NH_3-TPD. For reactions that are promoted by base catalyst or acid-base catalysts (e.g., transesterification reactions, glucose-fructose isomerization), CO_2-TPD characterization is essential in the determination of surface basic sites [41, 42].

2.2.7 Temperature-Programmed Oxidation (TPO)

Temperature-programmed oxidation (TPO) examines how a sample changes under conditions of oxidation. Surface carbonaceous materials are oxidized to CO_2, which are then detected by mass spectroscopy (m/e = 44). TPO can be used to investigate the types of carbon deposited on a material and to explore deactivation mechanisms. For example, supported Ni catalyst is frequently used in methane reforming with CO_2, but it readily deactivates due to carbon deposition. Hou et al. [43] investigated the properties of deposited carbon on Ni supported catalysts by TPO (Fig. 2.9).

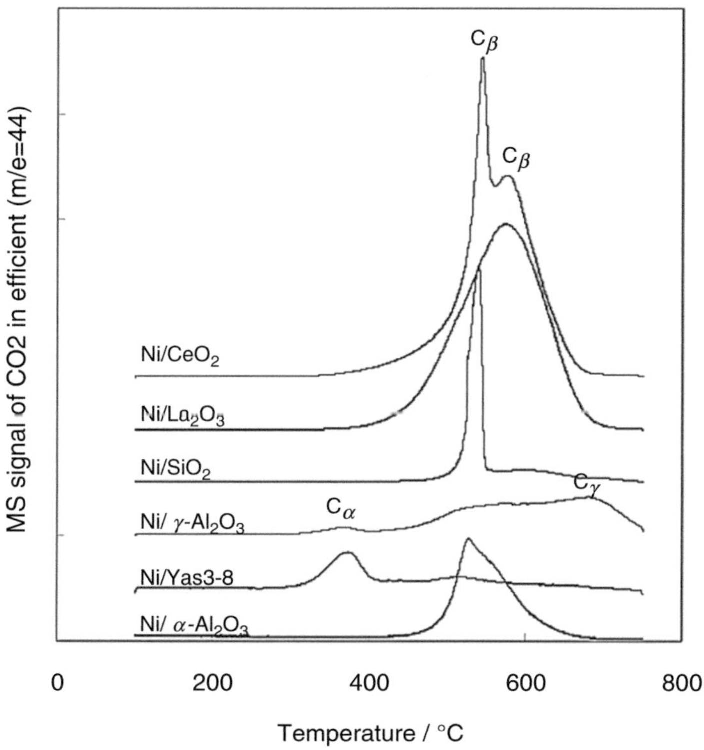

Fig. 2.9 TPO profiles of used catalysts (Reprinted from Ref. [43], Copyright © 2003 Springer)

Three types of carbons (C_α, C_β and C_γ) were detected on the catalysts. A highly reactive carbon (C_α) was detected at 380 °C and was found to be the active intermediate in the reforming reaction, while the C_β and C_γ deposited on the catalyst hindered access of the reactant to the Ni site on the surface. Thus, the Ni/Yas3-8 catalyst, which had a C_α peak, was determined to be the most active and stable catalyst.

2.2.8 Temperature-Programmed Surface Reaction (TPSR)

Temperature-programmed surface reaction (TPSR) is a technique in which reaction progress is monitored by mass spectroscopy in a temperature-programmed process. The catalyst sample is pretreated (degassed or reduced) and then reactants are introduced over the catalysts while recording mass spectrometry signals. This method can be used to study the properties of active sites and to infer reaction mechanisms [44–46]. Dong et al. reported a furfural TPSR experiment in a self-made reactor that was connected to an online mass spectrometer [47]. The Cu catalysts were reduced by pure H_2 and then purged with He at the same temperature and then cooled to 40 °C overnight. Thereafter, furfural/He flow was introduced at 40 °C for 2 h. After the absorption, samples were purged with He to remove the physical adsorbed furfural at 40 °C. Then a temperature-programmed desorption was performed with a heating rate of 10 °C/min in a flow of He stream from 40 to 800 °C. The desorbing products were detected with an online mass spectrometry.

Figure 2.10 shows the products during temperature-programmed process over different catalysts. Cu/SiO_2 catalyst has the highest copper dispersion, BET surface area and acid amounts. As shown in Fig. 2.10A, Cu/SiO_2 catalyst shows the highest desorption amount of furfural and has the highest desorption temperature. Furfuryl alcohol desorption showed similar trend with furfural (Fig. 2.10B). Moreover, Cu/SiO_2 catalyst exhibits a much higher amount of 2-methylfuran than the other catalysts, while the desorption temperature is relatively low (Fig. 2.10C). Figure 2.10D shows the desorption of water, which comes mainly from the hydrogenolysis of the intermediate furfuryl alcohol. During the conversion of furfuryl alcohol to 2-methylfuran, equimolar amounts of water are obtained. It is found that the desorption temperature of H_2O on Cu/SiO_2 catalysts is the lowest among the catalysts, which is consistent with the desorption of 2-methylfuran. The above results are evidence that Cu/SiO_2 catalysts has a sufficient adsorption amount of furfural and the fast desorption rate of the desired product, 2-methylfuran. Figure 2.10 shows that TPSR allows characterization of the adsorption-desorption surface species that influence catalytic performance of furfural hydrogenation. Thus, TPSR can provide mechanistic insight into the reaction processes of catalytic systems.

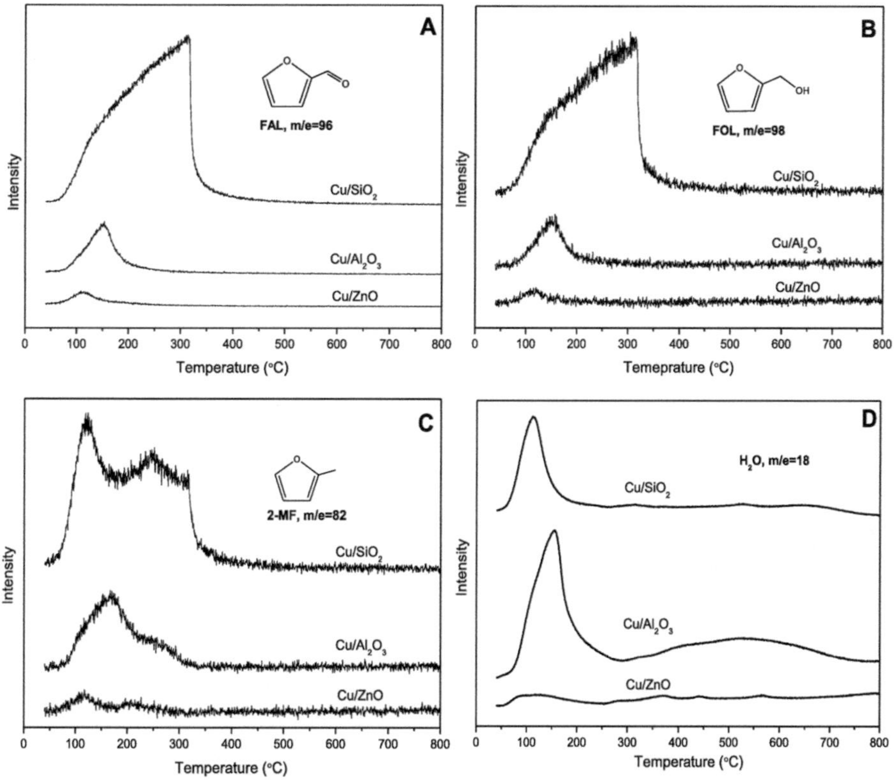

Fig. 2.10 (**A**) furfural desorption, m/e = 96; (**B**) furfuryl alcohol desorption from the reaction of furfural, m/e = 98; (**C**) 2-methylfuran desorption from the reaction of furfural, m/e = 82; and (**D**) H_2O desorption from the reaction of furfural, m/e = 18 (Reprinted with permission from Ref. [47], Copyright © 2015 Elsevier)

2.2.9 *Thermogravimetric Analysis (TGA)*

Thermogravimetric analysis (TGA) detects the amount and rate of change in the weight of the sample as a function of time or temperature in a specific atmosphere. TGA can be used to characterize the composition of materials and to predict their thermal stability due to decomposition, especially for catalyst precursors.

Based on TGA results, the decomposition process during calcination of catalyst precursors can be obtained under an air atmosphere [49, 50]. For example, hydrotalicte-like material is a widely used precursor for catalyst with general formula $[M^{2+}{}_{1-x}M^{3+}{}_x(OH)_2]^{x+}A^{z-}{}_{x/z}\bullet mH_2O$ (e.g., $M^{2+} = Ni^{2+}$, $M^{3+} = Al^{3+}$, $A^{z-} = CO_3{}^{2-}$). The calcination contributes to the formation of highly dispersed metal oxides. As shown in Fig. 2.11, the weight loss between 30 and 200 °C corresponds to the

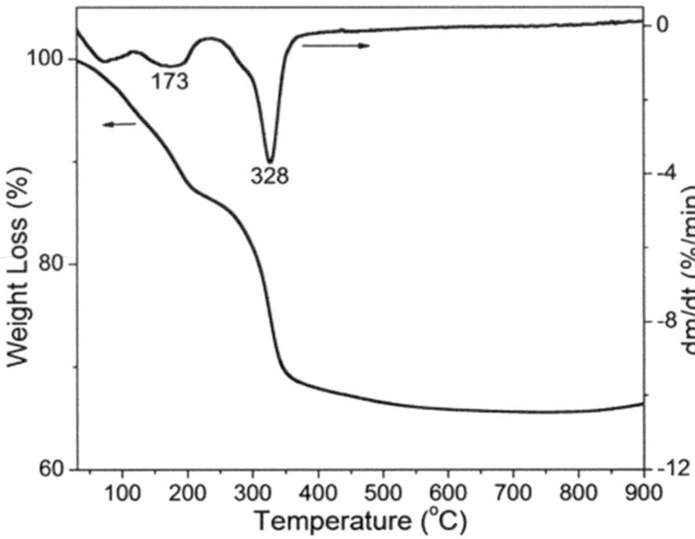

Fig. 2.11 TG curves of NiAl precursor (NiAl-HT) (Reproduced from Ref. [48] with permission of The Royal Society of Chemistry)

removal of physically adsorbed water and interlayer water. The rapid weight loss around 328 °C is probably due to the concomitant dehydroxylation of the laminae and expulsion of the carbonate anions located in the interlayer regions. At temperatures above 400 °C, only a slight weight loss is detected, indicating that the precursor has been totally decomposed. Based on the TGA results, the appropriate conditions for calcination of a catalytic material can be determined.

TGA can also be used to confirm the coke formation by comparing the fresh and used catalysts [48, 51]. As shown in Fig. 2.12, an increase in weight of the fresh catalyst is observed, which should be ascribed to Ni oxidation. However, the weight loss of used catalyst occurs around 345 °C, indicating the oxidization of organic compounds (about ~7% content in the used catalyst). Evidently, some carbonaceous species are strongly adsorbed onto the catalyst, which may cause the catalyst deactivation.

2.2.10 *Infrared (IR) Spectra*

Infrared (IR) transmission spectroscopy is a well-applied and powerful technique for detailed structural information of oxides (e.g., Al_2O_3, TiO_2, SiO_2, ZrO_2) and mineral materials, which are frequently used as catalytic supports and precursor. In the conventional technique, KBr and the sample are mixed and ground and then a thin disc is shaped into a matrix and placed in the IR cell. Phyllosilicates, which have highly-ordered structures (e.g., $Ni_3Si_2O_5(OH)_4$, $Ni_3(Si_2O_5)_2(OH)_4$, $Cu_2Si_2O_5(OH)_2$),

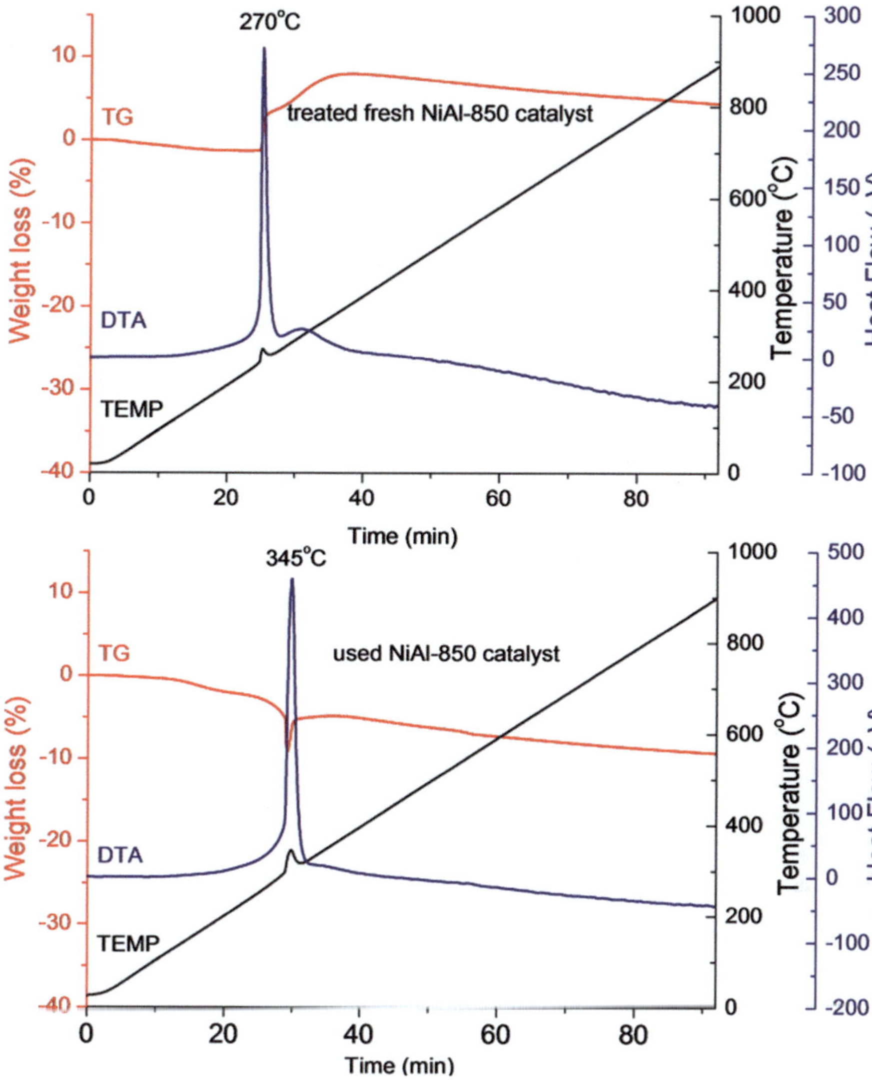

Fig. 2.12 TG results of treated fresh and used NiAl-850 catalyst (Reproduced from Ref. [48] with permission of The Royal Society of Chemistry)

are important catalyst precursor and can be studied with IR as a reference material due to the presence of structural OH groups [39, 52–55]. Phyllosilicate can be prepared by deposition precipitation methods [56, 57]. In Chen et al. [52], IR technique is used to confirm the formation of phyllosilicate in Cu/SiO_2 catalyst prepared by ammonia evaporation method. IR spectra of Cu/SiO_2 catalyst are shown in Fig. 2.13. The δ_{OH} band at 663 cm^{-1} and the ν_{SiO} band at 1040 cm^{-1} are characteristic of Cu phyllosilicate. The band at 1110 cm^{-1} is assigned to SiO asymmetric stretching. The

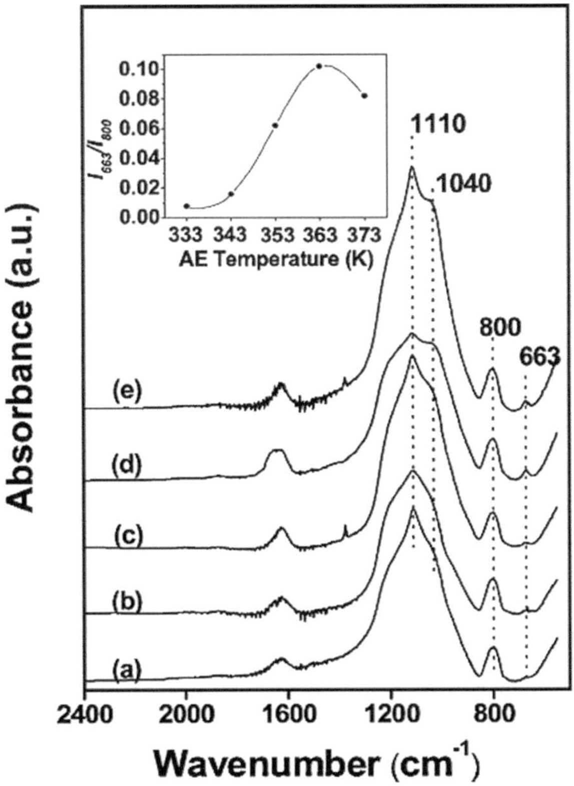

Fig. 2.13 IR spectra of the calcined Cu/SiO_2 catalysts. (**a**) CuSiO-333, (**b**) CuSiO-343, (**c**) CuSiO-353, (**d**) CuSiO-363, and (**e**) CuSiO-373. Inset shows the I_{663}/I_{800} intensity. The number indicates the ammonia evaporation temperature (K) (Reprinted with permission from Ref. [52], Copyright © 2008 Elsevier)

relative amount of Cu phyllosilicate in Cu/SiO_2 samples is calculated based on the ratio of integrated intensity of δ_{OH} band at 663 cm^{-1} and the ν_{SiO} symmetric stretching band of SiO_2 at 800 cm^{-1} (I_{663}/I_{800}). With IR spectroscopy, it can be determined that Cu phyllosilicate can be formed by ammonia evaporation.

Because the characteristic vibrational modes are closely related to the bond vibrations and interactions, molecule-adsorbed IR spectroscopy permits the determination of various types and properties of surface sites based on the interactions between surface sites and suitably-chosen probe molecules, for example pyridine-adsorbed infrared spectroscopy (Py-IR) or CO-IR [58].

Py-IR can be used to determine the acid type of sites on catalyst surface. Before experiments, samples are treated under vacuum at high temperatures (e.g., 300 °C) and the spectra of the pure sample are collected after cooling to low temperature (e.g., 50 °C). Then, a pyridine stream is introduced to allow adsorption to occur. Removal of physically-adsorbed pyridine is performed at around 150 °C and then the Py-IR spectra are collected under high vacuum. Py-IR will also be discussed in more detail in Sect. 2.3.

Carbon monoxide is sensitive to the state of a given metal in different coordination and oxidation states and its C–O stretching frequency can be well characterized [59–61]. For CO adsorbed IR (CO-IR), an IR cell that allows thermal treatment in vacuum or in a controlled atmosphere is used. Powder samples are compressed into thin discs and then the discs are placed in the cell. Layman et al. [62] investigated the surface chemistry of CO on reduced and sulfided Ni_2P/SiO_2 using IR spectroscopy. Four different ν_{CO} absorbances were observed for CO-IR results over reduced and sulfided Ni_2P/SiO_2 catalysts with the following assignments: (1) CO terminally bonded to *cus* $Ni^{\delta+}$ ($0 < \delta < 1$) sites (2083–2093 cm^{-1}); (2) CO adsorbed on $Ni^{\delta+}$ ($0 < \delta < 1$) bridge sites (1914 cm^{-1}); (3) CO terminally bonded to P (~2200 cm^{-1}); and (4) formation of $Ni(CO)_4$ (2056 cm^{-1}). As shown in Fig. 2.14, reduced Ni_2P/SiO_2 exhibits terminally bonded CO in the presence of CO (670 Pa CO partial pressure). The ν_{CO} absorbance at 2193 cm^{-1} due to the formation of P=C=O is observed.

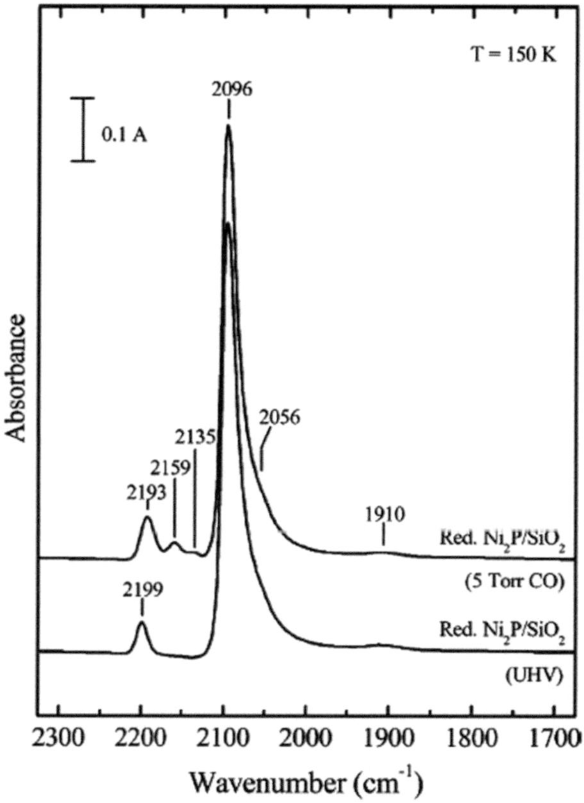

Fig. 2.14 IR spectra of adsorbed CO on reduced Ni_2P/SiO_2 catalysts in the presence CO (670 Pa or 5 Torr CO partial pressure) of CO and after evacuating to UHV pressures (at 150 K) (Reprinted with permission from Ref. [62], Copyright (2004) American Chemical Society)

The formation of $Ni(CO)_4$ and CO adsorption on Ni bridge sites can be confirmed. No clear change in the Ni_2P/SiO_2 catalyst after changing to ultrahigh-vacuum conditions occurred, indicating that CO is strongly adsorbed to the Ni sites. Based on the IR results, the mechanism and surface active sites of Ni_2P/SiO_2 catalyst for hydrodesulfurization reactions can be correlated. CO-IR is one of the most effective detection methods for surface metal active sites and change of electronic states and the technique is used in many works [59, 60, 63, 64]. With CO-IR, the adsorption of CO in different conformations, i.e., bridged (one CO molecule on two atom), linear (one CO molecule per metal surface atom) and subcarbonyl forms (two or more CO molecules per atom) can be determined. CO-IR measurements provide the variation in CO adsorption stoichiometry. Thus, CO adsorption is controversial for determining metal dispersion for many metals which showed complicated CO-metal stoichiometric ratios, as discussed in the subsection on CO adsorption.

2.2.11 Raman Spectroscopy

Raman spectroscopy is widely used to provide a structural fingerprint of catalytic materials [65] and relies on the scattering of incident photons (e.g., laser beam). The photon scattering is made up of elastic components and inelastic components, which are knowns as "Rayleigh scattering" and "Raman scattering" respectively. Due to the interaction between light and the molecular groups, an energy decrease is observed for inelastic scattering, which gives information about the vibrational modes of the molecule. Raman can be considered complementary to IR, in that the latter couples with anti-symmetric (ungerade) vibrational modes, while Raman couples with symmetric (gerade) vibrational modes.

Raman serves as the most important characterization tool for the analysis of graphene-based materials, giving information about the number of layers, quality of layers, doping levels and confinement [66]. The main features in Raman spectra of carbons are the G and D peaks. As shown in Fig. 2.15, the Raman spectrum of single-layer graphene (SLG) has two characteristic peaks of G and 2D. The G peak is around 1585 cm^{-1}, which is associated with the E_{2g} vibration mode of sp^2 bonds. The 2D peak is located around 2685 cm^{-1}, which is the second order of the D peak [67]. The D peak appears at 1350 cm^{-1} and is due to the A_{1g} mode breathing vibrations of six-membered sp^2 carbon rings. The D peak is absent in defect-free graphene. Thus, the D peak is an indicator of the defects in graphene. The weak D peak in pristine graphene indicates the high quality of the graphene.

The layer number (single- to five-layer) of few-layer graphene (FLG) can be distinguished by the full width at half-maximum (FWHM) of the 2D band based on Raman measurements [68]. Figure 2.15 shows the Raman spectra of FLG in which the FWHM of the SLG is 26.3 cm^{-1}. In bilayer graphene, the interaction of the two graphene planes causes the π and π^* electron bands to divide into four parabolic band structures denoted as π_1, π_2, π_1^*, and π_2^*. The 2D band is thus dispersive and fitted into four bands with slightly different frequencies having a wider FWHM of 52.1 cm^{-1}. For trilayer and four-layer graphene, the electronic bands split into more

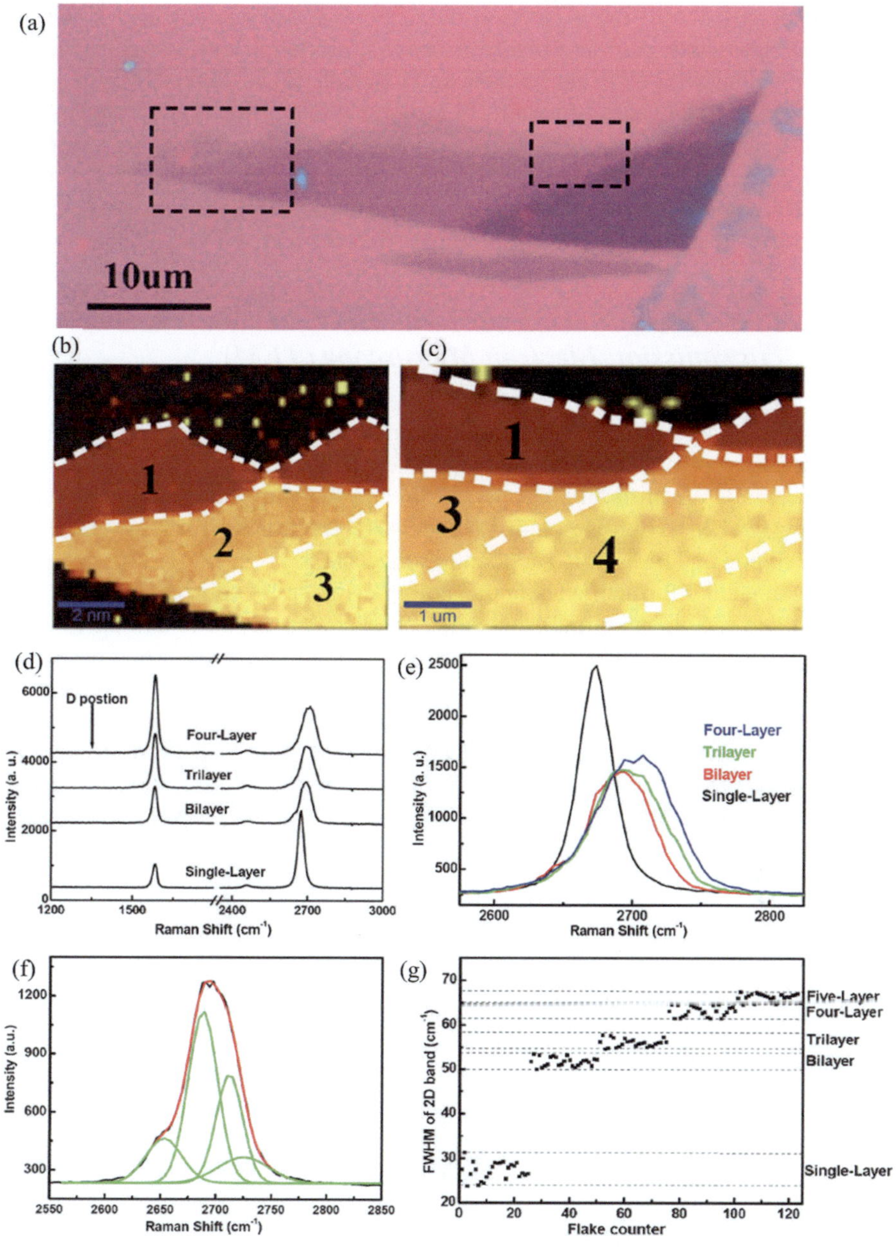

Fig. 2.15 Raman characterizations of AB-stacked FLG. (**a**) Optical image of graphene flakes with continuous layer number from 1 to 4. (**b**, **c**) Raman images from left box and right box in (**a**), respectively, according to FWHM of 2D band. Brighter color represents larger FWHM of Raman bands. Layer numbers are indicated in each part. (**d**) Raman spectra of each part of the FLG in (**a**). (**e**) The 2D bands from (**b**)are magnified to show different FWHM of each FLG with different layer number. (**f**)The fitted four components of the 2D band in bilayer graphene. The statistical data of FWHM with respect to different layer number is plotted in (**g**) (Reprinted with permission from Ref. [68], Copyright © 2010 Wiley)

complex signals due to the π electron interactions under stacking and therefore excited electron–hole pairs are involved in more scattering cycles, so that more resonant phonons with different frequencies contribute to the wider 2D bands. The typical FWHM of 2D peaks of various FLGs are plotted in Fig. 2.15g. Single-, bi-, tri-, four-, and five-layer graphenes exhibit distinguishable ranges for at $27.5 \pm 3.8\,cm^{-1}$, $51.7 \pm 1.7\,cm^{-1}$, $56.2 \pm 1.6\,cm^{-1}$, $63.1 \pm 1.6\,cm^{-1}$, and $66.1 \pm 1.4\,cm^{-1}$, respectively. The FLG with more than five layers has a similar band to that of the bulk graphite due to the continuous splitting of valence and conduction bands.

2.2.12 *Transmission Electron Microscopy (TEM)*

Electron microscopy offers the direct observation of catalyst morphology. Three types of microscopy are illustrated in this chapter: transmission electron microscopy (TEM), scanning electron microscopy (SEM) and scanning transmission electron microscopy (STEM). High-Resolution transmission electron microscopy (HRTEM) which is a mode of TEM, is also applied in some cases.

TEM can yield information about the atomic structures of solid catalysts. In this process, a thin specimen is irradiated with a parallel electron beam and the magnified images of the specimen are observed by combining the transmitted electrons using an electromagnetic objective lens. TEM instruments are widely operated in the 100–400 kV range. A higher energy will bring a better resolution at the expense of increasing beam damage to the sample. A setting of 200 kV in TEM analyses is popular since this provides a suitable resolution (about 0.18 nm) with low beam damage for many studies [16]. For sample preparation, the power is generally dispersed in a volatile liquid (e.g., ethanol) and then a drop of the suspension is placed on a support (e.g., Cu grid). TEM can be divided into conventional TEM and high-resolution TEM (HRTEM). The mean sizes of particles can be estimated from TEM images by directly counting hundreds of particles. Two mean particle sizes are usually considered as the length-number mean diameter (2.9) and the volume-area mean diameter (2.10) [16] with the volume-area mean diameter being the most useful parameter because it is related to the specific surface and therefore can be derived indirectly from surface area measurements by chemisorption.

$$d = \frac{\sum n_i d_i}{\sum n_i} \tag{2.9}$$

$$d = \frac{\sum n_i d_i^3}{\sum n_i d_i^2} \tag{2.10}$$

where d_i is the metal particle size, n_i is the number of metal particles in a certain range counted from the TEM images.

$$D = 6 \times (V_m / n_s) / d \tag{2.11}$$

$$v_m = \frac{M}{\rho N_A} \tag{2.12}$$

Metal dispersion (D) can be determined by Eqs. 2.11 and 2.12 [16, 27, 69]. In Eqs. 2.11 and 2.12, v_m is the volume occupied by an atom in the bulk of the metal, d is the mean particle diameter of metal nanoparticles, n_s is the amounts of atoms per unit area (e.g., 1.47×10^{19} m^{-2} for Cu), M and ρ indicate the molar mass and density of the metal, respectively and N_A represents the Avogadro's constant of 6.022×10^{23} molecules·mol^{-1}.

Figure 2.16 shows TEM micrographs of carbon nanotube-supported Ru catalysts (Ru/CNT). The plots (Figs. 2.16A–E) show that the mean sizes of Ru particles in these catalysts were almost the same (8.6–8.9 nm) despite of the concentration of HNO_3 used for CNT pretreatment from 5 to 68 wt. %, which were obtained by counting 150–200 particles [70].

HRTEM can provide direct information about the exposed surface and the metal-supported epitaxial relationship. Figure 2.17 shows HRTEM images of Au/ZnO-nanorod catalysts pretreated at different conditions [71]. The hemispherical shape of the particles indicates strong interactions between Au and ZnO that are widely discussed for Au/TiO_2 catalysts (Fig. 2.17b) [72]. Migration of ZnO onto the top surface of Au nanoparticles can be observed (Fig. 2.17c–d). Based on HRTEM (Fig. 2.17d), surface species can be determined to be the ZnO layer with a lattice spacing of 0.26 nm, corresponding to the (002) plane of ZnO. Further treatment of Au/ZnO under H_2 atmosphere does not change the structure significantly. HRTEM is widely used in the investigation of strong metal support interactions, which is a hot topic in heterogeneous catalysis [73–75].

2.2.13 Scanning Transmission Electron Microscopy (STEM)

STEM uses a focused beam to form a small spot that is raster-scanned over the sample while detecting scattered or transmitted electrons. High-angle annular dark field (HAADF) imaging is of great interest in heterogeneous catalysis. HAADF-STEM images are also known as Z-contrast images and are particularly effective for the study of catalysts containing small metal particles. Similar to preparation requirements for TEM, samples subjected to STEM analyses are typically be dispersed in a volatile solvent (e.g. ethanol), and then deposited onto a copper grid. HAADF-STEM images collected with a 24.5 mrad probe semiangle and ~25 pA probe current [76] are shown in Fig. 2.18, in which coverage of a thin layer of cobalt metal particles are detected. The thickness of the layer can be measured based on the STEM images. Prolonging the calcination and reduction process cause an increase of the size of layer.

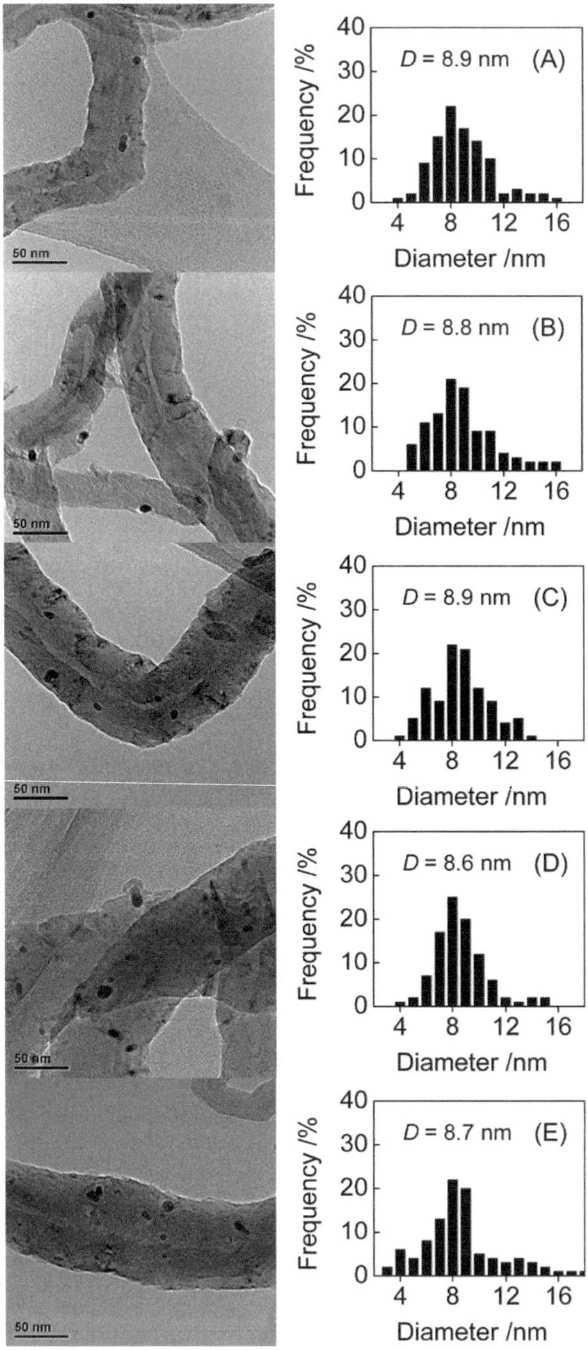

Fig. 2.16 TEM micrographs and Ru particle size distributions of the Ru/CNT catalysts with CNTs pretreated by HNO_3 with different concentrations. Concentration of HNO_3 used for CNT pretreatment: (**A**) 5 wt. %, (**B**) 19 wt. %, (**C**) 37 wt. %, (**D**) 52 wt. %, and (**E**) 68 wt. % (Reprinted with permission from Ref. [70], Copyright © 2010 Elsevier)

(a) (b) (c) (d) (e) (f)

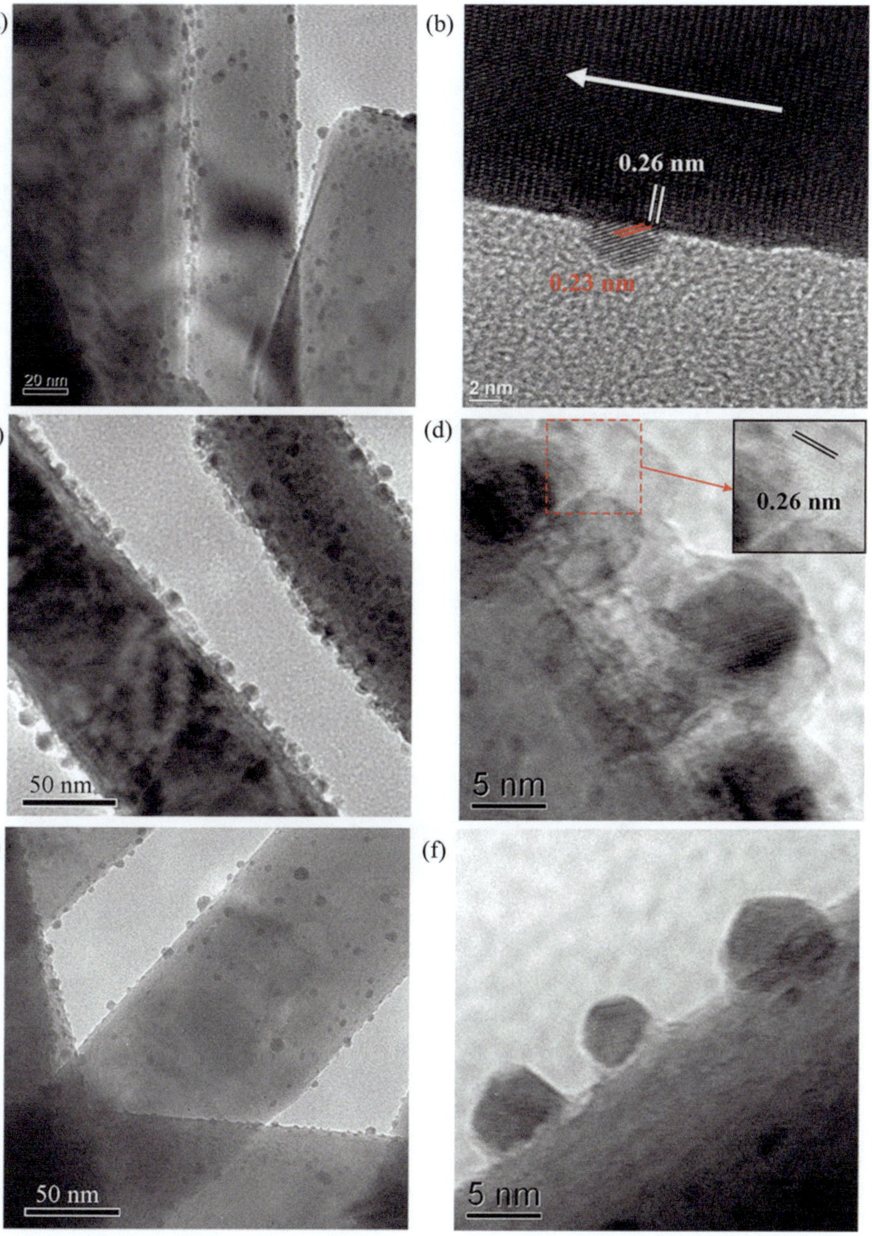

Fig. 2.17 HRTEM images of the Au/ZnO-nanorod catalysts pretreated at 200 °C (**a-b**) and 300 °C (**c-d**) in 10% O_2/He (flow rate: 30 mL/min) for 1 h and 300 °C in H_2 (flow rate: 30 mL/min) for 1 h (**e-f**) (Reprinted with permission from Ref. [71], Copyright (2012) American Chemical Society)

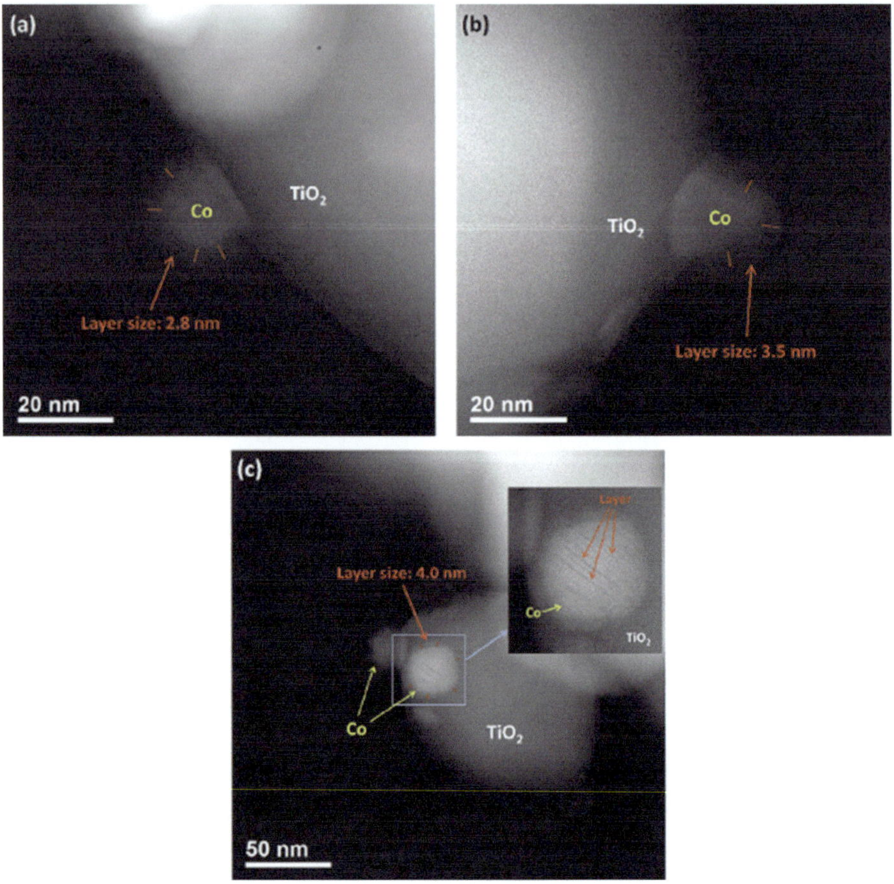

Fig. 2.18 STEM images of (**a**) Co/TiO_2 873R, (**b**) Co/TiO_2 673C-873R, and (**c**) Co/TiO_2 873C-873R. Labels: numbers indicate temperature in Kelvin, R indicates reduction, C indicates calcination (Reprinted with permission from Ref. [76], Copyright © 2015 Elsevier)

2.2.14 Scanning Electron Microscopy (SEM)

SEM images are obtained by scanning a finely focused probe in a raster pattern on the catalyst surface in step while writing the detector signal into a matrix. Making use of secondary electrons (SEs) that are emitted when the sample is excited by the electron beam is the most commonly used imaging mode. Due to the low energy of electrons (0–100 eV), SEM provides surface information of a specimen. It can be understood that a higher fraction of secondary electrons can be collected from the vertical surface than from the horizontal surface with the surface normal orthogonal to the beam direction. Thus, a difference in brightness is observed for different angles of the specimen surface, which reveals the sample morphology. Unlike TEM, in which the magnification is determined by excitation of the post-specimen lenses, the ratio of the area scanned to the display area determines the magnification of SEM.

SEM is mainly used to determine sample morphology and to examine aspects of a material that has been synthesized. Figure 2.19 shows a series of SEM images along with photographs of the materials [77]. Figure 2.19a shows the Ni foam before and after growth of graphene. Figure 2.19b shows graphene obtained in a single CVD progress after removal of the Ni foam. SEM images (Fig. 2.19c, d) provide information on the surface networks of graphene before and after Ni foam removal. With SEM, one can confirm that the graphene successfully copies the 3D network and porous structure of Ni foam after the removal of Ni foam. SEM is an effective method to characterize the surface morphology of materials and is used routinely in materials and catalysis studies.

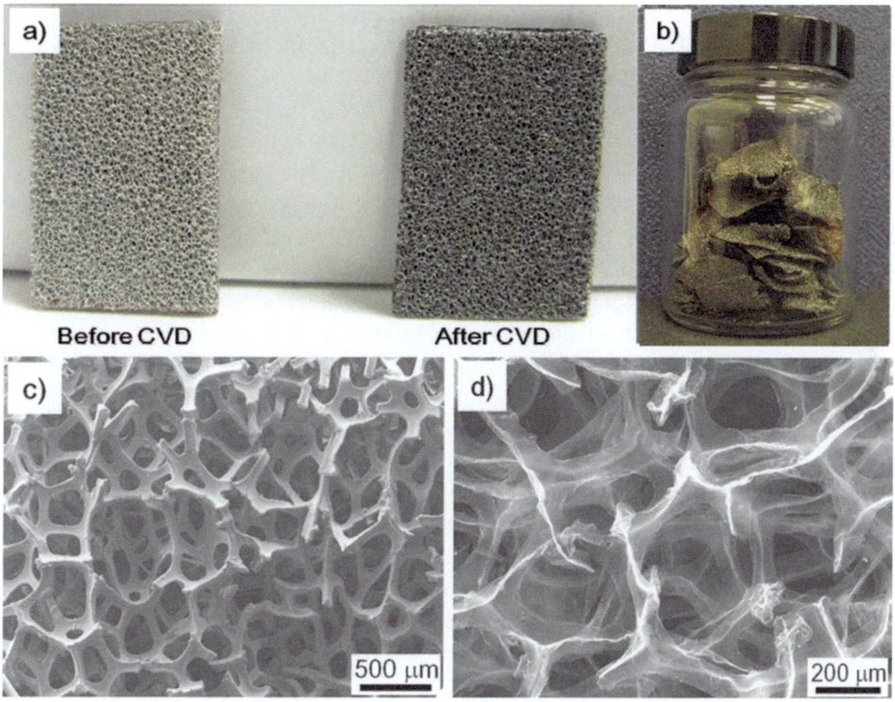

Fig. 2.19 Photographs of (**a**) Ni foam before and after growth of graphene, and (**b**) ~0.1 g 3D graphene networks obtained in a single CVD process after removal of the Ni foam. SEM images of (**c**) 3D graphene networks grown on Ni foam after CVD, and (**d**) 3D graphene networks after removal of Ni foam (Reprinted with permission from Ref. [77], Copyright © 2011 Wiley)

2.2.15 Atomic Force Microscopy (AFM)

Atomic force microscopy (AFM) allows study of the local properties of a sample surface and is based on probe tip-sample force interactions instead of electrons. Thus, AFM can supplement study of materials that are difficult to investigate by electron spectroscopy and other microscopies related to charge problems (e.g., insulators or poorly conducting oxide materials). Ideally, AFM is suitable for investigating model catalyst systems that contain insulating metal oxides and the technique can provide information on the sintering performance of nanoparticles and metal–support interactions. The most widely used AFM force sensor is the cantilever probe, which is made from a single-crystal Si wafer. The cantilever is driven to oscillate at its resonance frequency. During the close scanning of the surface, small changes in the tip-surface interactions are detected as resonance frequency deviations (Δf), and a digital image of the surface is assembled.

Figure 2.20 shows a schematic illustration of the AFM technique and imaging examples [78]. Figure 2.20b shows precise measurements of short-range chemical forces. Force spectroscopy is obtained by recording the Δf signal as a function of the tip-surface relative vertical displacement (Z). The interaction detected between the tip atom and the surface atom includes not only the short-range force associated with this chemical interaction, but also the long-range force contributions (i.e., van der Waals and electrostatic interactions) (Fig. 2.20c). The effect of the electrostatic interactions during imaging is usually minimized by choosing appropriate experimental set-up. Figures 2.20d-e show dynamic force microscopy topographic images of a single-atomic layer of Sn (Fig. 2.20d) and Pd (Fig. 2.20e) single-atomic layer grown over a Si(111) substrate, respectively.

AFM is a multifunctional technique and is considered to be in the molecular toolbox of many researchers and is applied to areas including catalysis [79], nanobiotechnology [80, 81] and nanotribology [82]. For catalysis, AFM is capable of resolving the chemical structure of a single molecule on a surface [83]. AFM can also be used to study surface reactions by analyzing reactants and products. For example, AFM images can directly reveal that diphenanthro[9,10-b:9′,10′-d]thiophene (DPAT) molecules undergo an on-surface planarization reaction upon annealing [84]. AFM can also be used to study the interaction between molecules. AFM has been used to compare the adhesion forces of cellulase-lignin and that of cellulase-cellulose by immobilizing a commercial trichoderma reesei (ATCC 26921) onto silicon wafers [85]. For example, Qin et al. [85] showed that average 45% higher forces are detected over Kraft lignin-cellulase than that of hydroxypropyl cellulose-cellulase.

2.2.16 X-Ray Absorption Spectroscopy (XAS)

X-ray absorption spectroscopy (XAS) is the only spectroscopic method that gives information about the electronic and structural properties of catalysts under reaction conditions [16]. XAS is mainly divided into conventional extended X-ray absorption

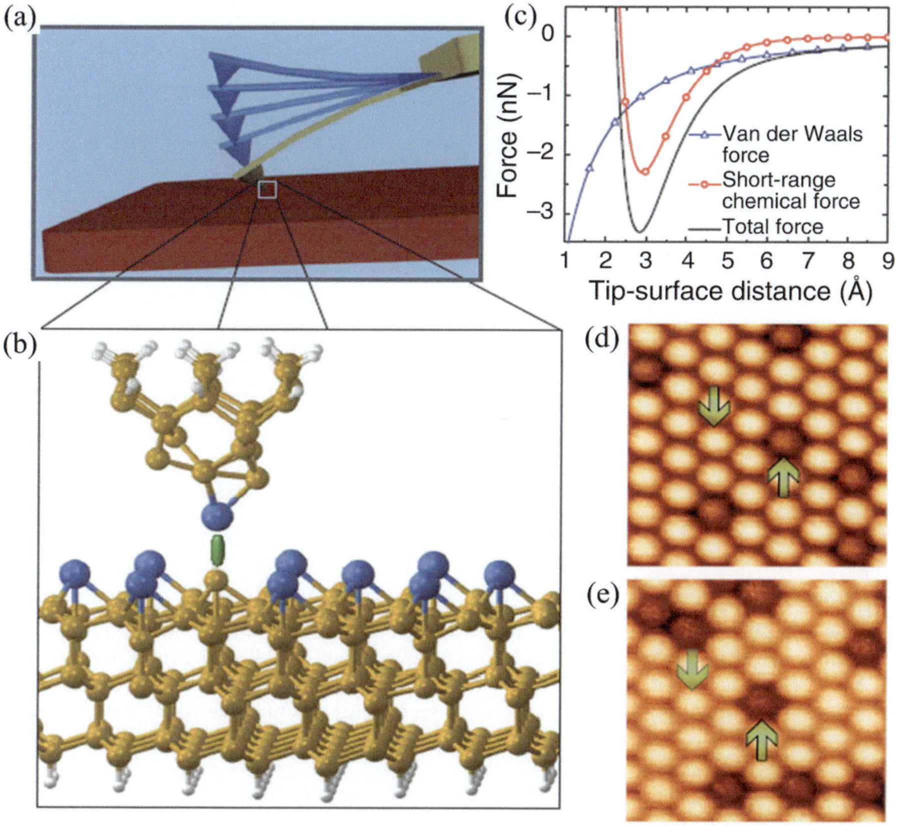

Fig. 2.20 (**a**) Schematic illustration of AFM experiment in dynamic mode. (**b**) Onset of chemical bonding between the outermost tip atom and a surface atom (highlighted by the green stick) that gives rise to the atomic contrast. (**c**) Curves are obtained with analytical expressions. (**d–e**), Dynamic force microscopy topographic images of Sn (**d**) and Pd (**e**) single-atomic layer grown over a Si(111) substrate respectively (Reprinted with permission from Ref. [78], Copyright © 2007 Nature)

fine structure (EXAFS) and X-ray absorption near-edge structure (XANES) [86]. XANES is referred to the region at energies below and around the ionization threshold. The excited electron is usually described using electron scattering that occurs around 20–30 eV above the absorption edge, and this region is referred to EXAFS.

EXAFS is an essential characterization technique for heterogeneous catalysts. The simplest theory for EXAFS is based on single scattering plane wave approximation. This theory assumes that the atomic radii are much smaller than the interatomic distances. EXAFS can be applied to effectively obtain structural information and changes [87–89]. As shown in Fig. 2.21, the reduced and aqueous phase reforming (APR) Pt exhibit different spectra with monometallic Pt being used as reference. The spectrum under H_2 shows higher peak intensity and a lower distance shift at ~2.1 A compared with that under APR, while the spectrum under APR exhibits an additional peak at ~2.9 A.

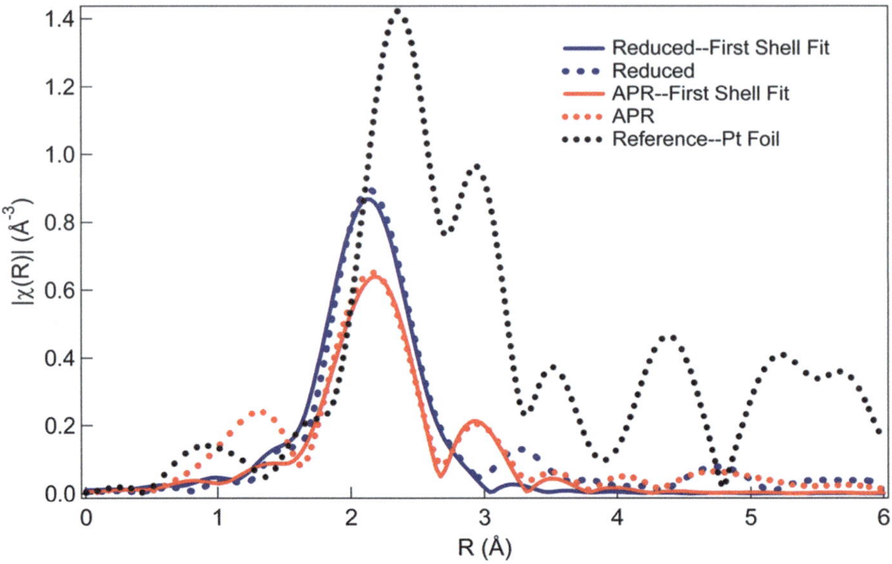

Fig. 2.21 Magnitude of k^2-weighted Fourier transform of Pt L_3-edge EXAFS data, dashed lines, and fits, solid lines, for NiPt/C catalyst under 10% H_2 at 225 °C (blue) and under APR reaction conditions (red, 10 v/v % ethylene glycol in H_2O at 225 °C and 3.2 MPa). For reference, Pt foil is in black. The catalyst was reduced at 350 °C in 10% H_2 for 2 h prior to EXAFS data scans (Reprinted with permission from Ref. [88], Copyright (2012) American Chemical Society)

Table 2.1 EXAFS fitting results of a NiPt/C catalyst for the Pt L_3-Edge

Conditions	10% H_2 at 50 °C	10% H_2 at 225 °C	APR at 225 °C	After APR 10% H_2 at 50 °C
CN(Pt-Pt)	1.9 ± 0.8	2.0 ± 0.7	6.0 ± 1.4	2.8 ± 0.9
CN(Ni-Pt)	3.5 ± 0.4	3.8 ± 0.3	1.9 ± 0.8	4.6 ± 0.4
R(Pt-Pt), Å	2.65 ± 0.03	2.64 ± 0.03	2.67 ± 0.02	2.69 ± 0.03
R(Ni-Pt), Å	2.548 ± 0.007	2.530 ± 0.008	2.64 ± 0.03	2.566 ± 0.008
σ(Pt-Pt), $Å^2$	0.008 ± 0.003	0.011 ± 0.002	0.010 ± 0.002	0.007 ± 0.002
σ(Ni-Pt), $Å^2$	0.007 ± 0.001	0.010 ± 0.004	0.011 ± 0.003	0.006 ± 0.001
ΔE_0 (eV)	2.5 ± 1.0	2.5 ± 0.6	4.0 ± 0.8	8.0 ± 0.9

The σ and ΔE_0 represent the Debye–Waller factor and the inner potential correction, respectively
Reprinted with permission from Ref. [88], Copyright (2012) American Chemical Society

Figure 2.21 reveals that an increase in the Pt-Pt coordination number and decease in Ni-Pt coordination number occur under APR. By fitting EXAFS results, the coordination number and Ni-Pt/Pt-Pt bond length can be determined (Table 2.1). After reduction, NiPt alloy is formed based on the detective coordination number of Ni-Pt. Moreover, a relatively larger Ni-Pt coordination number and a short Ni-Pt bond are obtained for reduced catalyst, indicating the core-shell structure in which the core is rich in Ni and the shell is rich in Pt. Under APR, the structures undergo

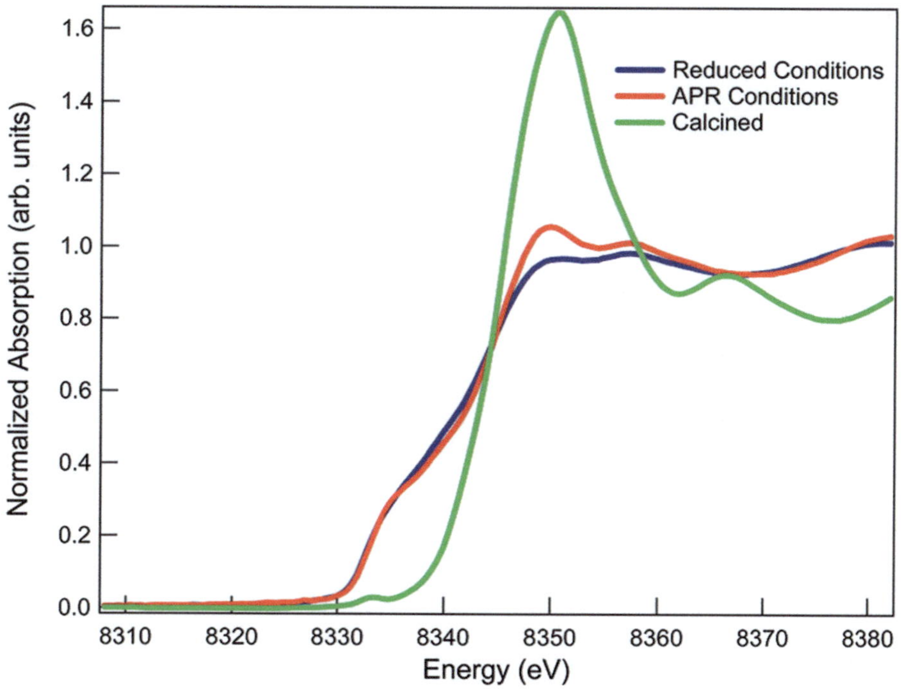

Fig. 2.22 Normalized absorption spectra at the Ni K-edge for NiPt/C catalyst before reduction (calcined), under 10% H_2 at 225 °C, and under APR reaction conditions (10 v/v % ethylene glycol in H_2O at 225 °C and 3.2 MPa) (Reprinted with permission from Ref. [88], Copyright (2012) American Chemical Society)

a reconstruction, with increased Pt-Pt coordination number (2.0–6.0) and decreased Ni-Pt coordination number (3.8–1.9). Further reduction of this catalyst causes a return to the Pt-rich shell/Ni-rich core structure, as indicated by the coordination number and length of Ni-Pt bond. The dynamic change of the catalyst is thus successfully obtained by the EXAFS technique.

XANES can be used to provide information on the change of species during reduction or reaction by using comparisons with standard samples. Figure 2.22 illustrates the Ni K-edge XANES spectra for calcined, reduced NiPt/C under H_2 and under APR. Under APR reaction conditions, oxidization of Ni is confirmed. Normalized Ni K-edge XANES spectra are scanned to investigate the formation of NiZn alloy species during a temperature-programmed reduction of 2 wt. % Ni/ZnO, demonstrating a gradual transformation of NiO to α-NiZn and β_1-NiZn [90]. The geometry and electronic structure of metal centers can also be revealed. Presence of linear Cu^+ species in O_{fw}-Cu-NH_3 or H_3N-Cu-NH_3 configurations can be determined using XANES scans [91]. The reader can consult several reviews that report on the application of XANES and EXAFS to heterogeneous catalysis to obtain further details [92–94].

2.3 Heterogeneous Acid Catalysts for Cellulose Conversion

In the previous sections, characterization techniques used in studying catalysts were introduced. In this section, the techniques are used to provide insight into catalytic mechanisms reported for the conversion of cellulose and related compounds such as glucose into target products. First, the properties of the catalysts are considered.

2.3.1 Properties of Acid Catalysts

The acid type (Brønsted or Lewis), strength, distribution, morphology and amount of acid sites in a solid material define its main catalytic properties. The characterization of the properties of the solid materials is necessary to understand mechanisms that are promoted by acid catalysis and to develop or design new catalysts.

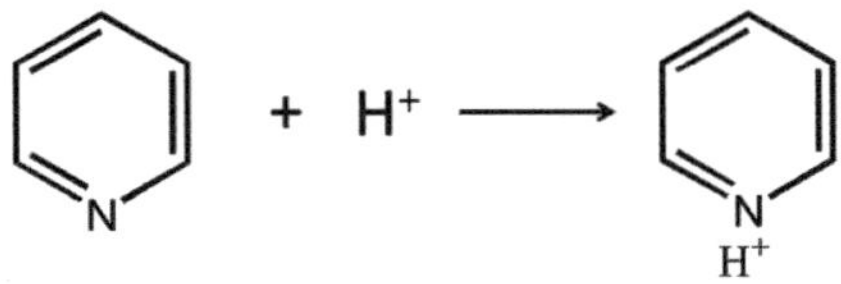

(2.13)

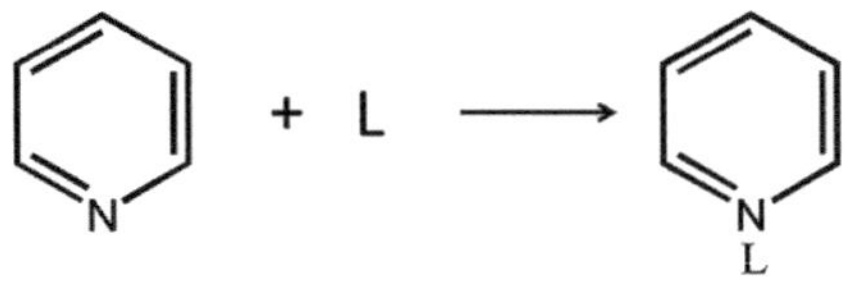

(2.14)

Acid types are broadly classified as Brønsted acid or Lewis acids. A Brønsted acid is a species that can donate a proton. A Lewis acid is a species that can accept a pair of electrons from another species. The Brønsted acid donates a proton, whereas the Lewis acid accepts a pair of electrons and forms a complex. Some reactions are catalyzed by Brønsted acid sites while other reactions are promoted by Lewis acid sites. Integrated catalytic processes that have proper design of Brønsted and Lewis acid sites have achieved a domino reaction effect to produce γ-valerolactone from furfural [95]. Py-IR is the most effective method to characterize the acid type due to the strong affinity of the nitrogen atom in the pyridine molecule with Brønsted and Lewis acids. The proton in the catalyst reacts with pyridine and forms a pyridine cation, generating an absorption band at 1540 cm^{-1} (Eq. 2.13). Lewis acid sites react with pyridine and form a Py-L ligand complex, producing an absorption band at 1450 cm^{-1} (Eq. 2.14). Thus, the bands at 1540 and 1450 cm^{-1} are generally accepted as characteristic bands for Brønsted and Lewis acid sites, respectively. Based on Py-IR results, the content of the two types of acid sites can be estimated by the corresponding calibrated peak area [96].

$$H_0 = pK_a + \log\left\{\frac{[B]_a}{[BH^+]_a}\right\} \tag{2.15}$$

The acid strength refers to the ability of the functional group to lose a proton or accept an electron pair. Indicator methods and temperature programmed desorption (e.g., NH_3-TPD) are commonly used methods to characterize the acid strength. The pH is a well-known indicator for acid strength of an aqueous solution. For solid acids, the Hammett acidity function (H_0) is used to determine acid strength. The Hammett acidity is defined using Eq. (2.15), where $[B]_a$ and $[BH^+]_a$ are the concentration of undissociated bases and conjugate acids, respectively ($[HA]_s$+ $[B]_a$→$[A^-]$ + $[BH^+]_a$). By choosing a series of indicators with different pK_a values, the H_0 can be measured by the interaction of indicators and solid acids. The indicators are substances with basic sites (e.g., methyl red, anthraquinone and *p*-nitrotoluene). The basic sites interact with the acidic sites in the solid, forming new species that undergo a change in color. If the indicator has an acid type of color, the H_0 value is equal or below the pK_a value of the indicator ($H_0 \leq pK_a$). Chai et al. [97] measured the catalyst acidity (strength and amount) using various Hammett indicators. In terms of the measured acid or base strength, the catalysts are classified into four groups and the locations of some different solid acid catalysts in the classified groups are obtained (Table 2.2).

An acidic surface is generally composed of a multitude of acid sites having different strengths rather than a single site that has the sole strength. The acid amount is a function of the acid strength. N-butylamine indicator is frequently used to determine the acid distribution. The acid distribution is widely used in analysis of zeolite frameworks. For example, there are three types of pores in H-MCM-22 (zeolite type) which are different in their catalytic performance, including sinusoidal channels, cylindrical supercages and d pockets on the external surface. Brønsted acid sites in the sinusoidal channels rather than in the surface pockets and supercages are beneficial for methanol to hydrocarbons over H-MCM-22 [98]. Thus, acid distribution is an important property of the zeolite framework.

Acid catalysts can be used for many reactions that occur through carbocation mechanisms, such as in the conversion of methanol into olefins, Friedel-Crafts

Table 2.2 Classification of widely used solid acid catalysts based on Hammett acidity

Group	Range	Classification	Example
1	$H_0 \geq +7.2$	base catalyst	CeO_2, La_2O_3, MgO
2	$-3.0 \leq H_0 \leq +6.8$	medium, strong and weak acid catalyst	SiO_2, ZrO_2, Nb_2O_5-700
3	$-8.2 \leq H_0 \leq -3.0$	strong acid catalyst	Al_2O_3, HZSM-5, Nb_2O_5-400, WO_3/ZrO_2, H_3PO_4/α-Al_2O_3, SAPO-34, $NiSO_4/\alpha$-Al_2O_3.
4	$H_0 \leq -8.2$	very strong acid catalyst	SO_4^{2-}/ZrO_2, SiO_2-Al_2O_3, Hβ, Nb_2O_5-350.

Reproduced from Ref. [97] with permission of The Royal Society of Chemistry

reactions and isomerization reactions [99–101]. Acidic materials can promote cellulose hydrolysis reactions by cleavage of the β-1,4-glycosidic bonds. By tuning the acidic properties (type, strength, distribution and amount), the catalytic performance of an acid catalyst can be adjusted.

Catalysts used for cellulose conversion can be divided into liquid acid catalysts and solid acid catalysts. Each type of catalyst will be discussed in detail in the following sections.

2.3.2 *Origin of Acid Sites*

Hydrolysis of cellulose via chemical processes can be divided into several methods: (i) concentrated acid hydrolysis (e.g., 72% H_2SO_4), (ii) dilute acid (e.g., 0.3~0.5% H_2SO_4) and (iii) solid acid [2, 102]. Concentrated acids cause swelling of cellulose and form an acid complex and are considered to act as a homogeneous process. Dilute acid hydrolysis of cellulose is a heterogeneous reaction and is limited by mass-transfer related to the particle size, degree of polymerization (DP) and crystallinity of cellulose. Pretreatment is necessary for dilute acid methods to be applied to cellulose hydrolysis [103]. However, both concentrated and dilute acid cellulose hydrolysis processes suffer from issues associated with reactor corrosion, product separation and acid recovery and recycle along with aqueous effluents that require waste treatment.

Solid acid catalysts suitable for promoting cellulose hydrolysis are metal oxides, zeolites, heteropolyacids, polymer based acids and sulfonated carbonaceous-based acids. In this section, the microstructures of typical solid acid catalysts are introduced so that the role of acid sites in a material can be understood and some conclusions can be made that may provide a guide for the design and development of acid catalysts.

Al_2O_3 is an oxide that has weak acid sites after heat treatment. The surface OH groups of Al_2O_3 become removed and form coordinative unsaturated Al^{3+}, which acts as a Lewis acid site. The surface OH groups are the origin of the acidity of Al_2O_3. SiO_2 is considered as an inert support with weak acidity. However, SiO_2-Al_2O_3 composite is a widely used as a strong acid catalyst for petroleum cracking, dehydration, and olefin polymerization and isomerization reactions. The binary SiO_2-Al_2O_3 oxide possesses strong acidity with Brønsted and Lewis acid sites. It could be thought of as the isomorphous substitution of trivalent Al onto a tetravalent SiO_2 lattice, which would cause the existence of redundant negative charges. Commonly, Al is hexacoordinated and constrained to adopt a tetracoordinated structure. Thus, Al tends to acquire a pair of electrons and behaves as a Lewis acid. In the presence of water, the H_2O molecule is adsorbed and forms a Brønsted acid site [104]. Many other oxides form acid sites by similar mechanisms, as the substitution of another atom with a different coordination number causes the formation of redundant charge and acid sites.

Zeolites are aluminosilicates with a highly-ordered structure. When water molecules and mobile alkali metal ions are introduced into zeolites, they can be exchanged with other cations that compensate for the excess negative charge in the anionic framework resulting from the aluminum content, as discussed for SiO_2-Al_2O_3 oxides. Zeolite synthesis starts from alkaline aqueous mixtures of aluminum and silicon compounds. The Si/Al atomic ratio is an important factor that affects the structure and acidity of the zeolite. The generation of acidic sites of zeolites can be expressed as shown in Fig. 2.23. Ion exchanged zeolites with ammonium is first obtained and then changed into the acid form with Brønsted acid sites after removal of ammonia by calcination. Hydroxyl groups that are present on the surface of zeolites can be regarded as another type of Brønsted acid site. With additional thermal treatment, the removal of water occurs, causing two Brønsted acid sites to form one Lewis acid site.

Heteropolyacids (HPAs) are composed of a well-defined heteropoly anion, cation and coordinated water molecules. The tunable acidity of HPAs allows them to be widely applicable to acid catalyzed reactions. The Keggin-type HPAs with the 12-structure are widely applied due to their high acidity, good thermal stability and facile preparation, such as $H_3PW_{12}O_{40}$•nH_2O, $H_3PMo_{12}O_{40}$•nH_2O, $H_4SiW_{12}O_{40}$•nH_2O and $H_4SiMo_{12}O_{40}$•nH_2O. Their chemical composition is typically described as $H_{8-x}XY_{12}O_{40}$ (with X = P, Si, B; Y = Mo, W, V; n = oxidation state of X). The metal Y makes them active in redox catalysis due to the different oxidation states. The primary Keggin type and secondary structure of typical heteropoly anions are shown in Fig. 2.24. The protonated $H_5O_2^+$ species forming the bridges between Keggin ions is responsible for the acidity of hydrated bulk heteropoly acid. In water, all protons are dissociated stepwise. Thus, HPAs have strong acid strength, which is significantly stronger than typical inorganic acids, including H_2SO_4, HBr, HCl, HNO_3 and $HClO_4$ [105]. The most acidic HPA corresponds to the Keggin structure $H_3PW_{12}0_{40}$. The differences in the acid strength between HPAs and typical inorganic acids can be explained within the framework of electrostatic theory based on size and charge of the anion.

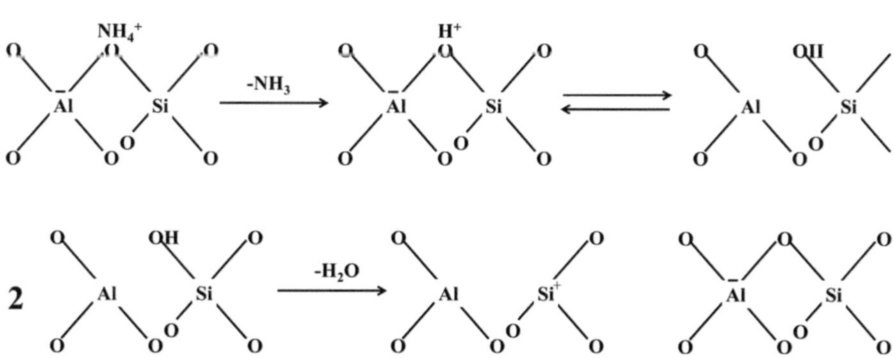

Fig. 2.23 Generation of acidic sites of zeolites (Reprinted with permission from Ref. [7], Copyright © 2002 American Chemical Society)

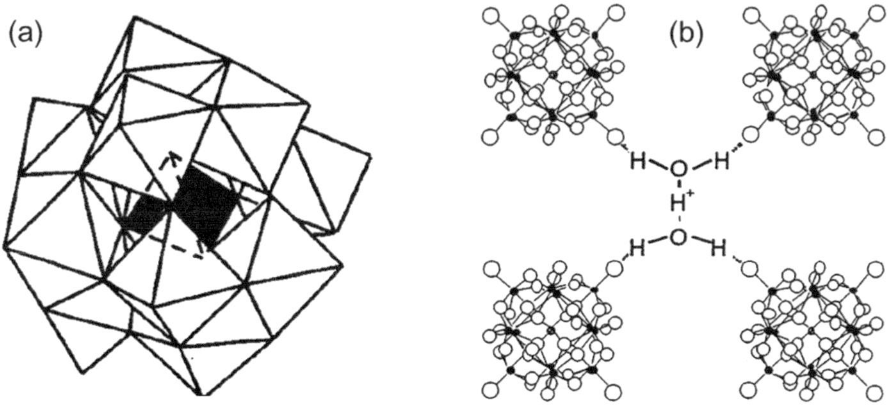

Fig. 2.24 Schematic structure of (**a**) heteropoly anion $(XY_{12}O_{40})^{m-}$ with the Keggin strucure; (**b**) secondary structure of the heteropoly acid (**b**) (Reprinted with permission from Ref. [7], Copyright © 2002 American Chemical Society)

However, HPAs are soluble in water and so their recovery hinders general application. Low surface areas (<10 $m^2 \cdot g^{-1}$) and low accessible surface acid sites also affect application of HPAs. However, HPAs loaded onto high surface area supports (e.g., zeolite, active carbon and silica) allow good dispersion and give HPAs high surface area and enhances their activity. For example, silica supported $H_3PW_{12}O_{40}$ in which the HPA is impregnated in a siliceous support forms crystalline heteropoly acid with an intact Keggin structure and an interacting form ($H_3PW_{12}O_{40}$ + ≡ Si-OH → (≡ Si-$OH_2)^+(H_2PW_{12}O_{40})^-$) [106]. The pure Keggin-type species on silica can be prepared at high loadings and the interacting structure forms at low loadings. Although HPA-supported catalysts are versatile, the leaching of some amounts of the HPA into the liquid reaction phase tends to be unavoidable.

HPA salts are good candidates for heterogeneous catalyst. Substitution of the proton by an alkaline cation (e.g., NH_4^+, K^+, and Cs^+) has large effects on the pore structure and surface area of the resulting material, as well as its aqueous solubility. Among the HPA salts, Cs salts substituted HPAs (e.g., $Cs_xH_{3-x}PW_{12}O_{40}$) are insoluble in water, exhibit high surface area and are attracting much attention. For example, materials with Cs content in the range of x = 2.0–2.7 are composed of dispersed crystallites (~100 m^2/g) and have high Brønsted acid strength, similar to the parent heteropoly acid [107]. Moreover, these catalysts are recoverable with no leaching of the soluble heteropoly acid into the aqueous reaction phase.

Polymer-based acids are mainly based on styrene-based sulfonic acids (e.g., Amberlyst) and the perfluor sulfonic acid-based (e.g., Nafion resin/silica nanocomposites). Harmer et al. [108] reviewed the microstructural features of two types of polymeric ion-exchange resins. As shown in Fig. 2.25, Amberlyst-15 is a macro reticular sulfonated polystyrene-based ion-exchange resin with 20% divinylbenzene. Nafion is a perfluorinated resin sulfonic acid copolymer. Both acids have different strengths. The perfluorinated resin sulfonic acid is a very strong acid with

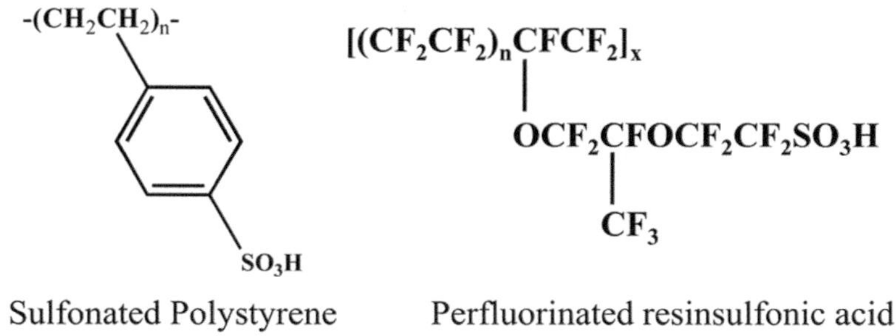

Fig. 2.25 Structure of two types of polymer-based acids (Reprinted with permission from Ref. [108], Copyright © 2001 Elsevier)

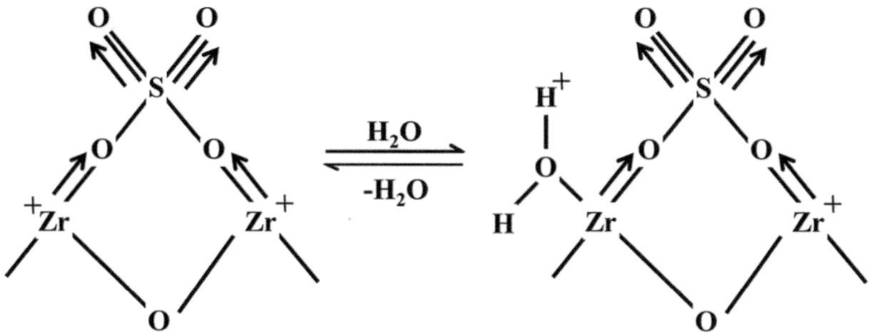

Fig. 2.26 Generation of acid on the sulfated ZrO_2

H_0 value from −11 to −13, while Amberlyst-15 has a H_0 value of about −2.2. Both acids have low surface area, especially for Nafion. Thus, high surface area Nafion resin/silica nanocomposites have been introduced to expand the application of perfluorinated resin sulfonic acids. For a given reaction, the acidity and catalytic performance can be tuned by the silica source and structure.

Sulfation of oxides (e.g., ZrO_2, TiO_2, and Fe_2O_3) enhances the surface acidity and allows formation of solid superacids. To understand the nature of acid sites, studies have been made to characterize the structure of sulfur-containing species [109] for which one kind of structure is shown in Fig. 2.26. Sulfate bridges across two Zr atoms and Py-IR analysis shows the existence of Lewis acid sites. The acidity of the sulfated oxide material is generated by electron induction of the S=O double bonds, which increases the electron deficient nature of Zr. In the presence of water, Lewis acid sites are converted to Brønsted acid sites via proton transfer. There are many structures proposed in the literature [109], but in many cases, there is no direct experimental evidence of the forms.

Sulfonated carbonaceous based acids are widely used strong solid acids that are more robust in hot water compared with resin-type or oxide-type solid acids. Carbonaceous material with SO_3H groups can be prepared by boiling carbon-containing materials in fuming sulfuric acid. The carbon materials can be natural

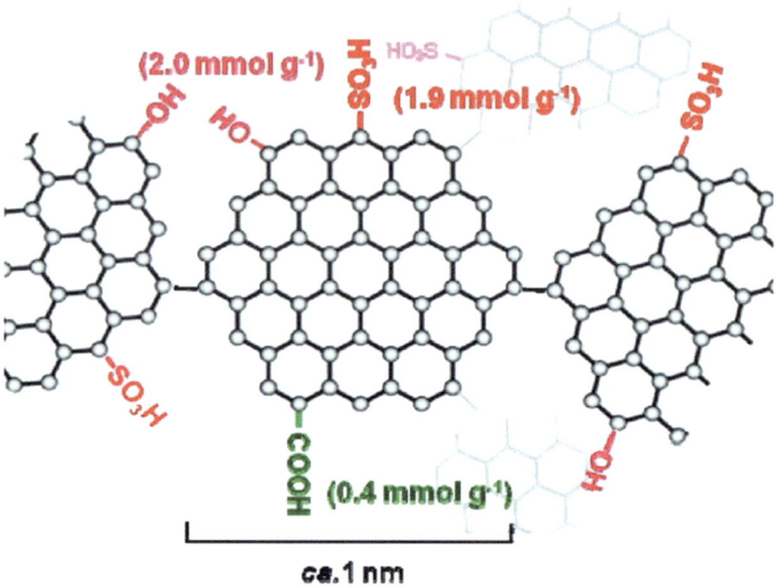

Fig. 2.27 Proposed schematic structure of a functionalized carbon material (Reprinted with permission from Ref. [110], Copyright © 2008 American Chemical Society)

organic compounds, such as sugar, cellulose and starch followed by sulfonation of the resulting amorphous carbon. As shown in Fig. 2.27, the amorphous carbon after treatment consists of flexible polycyclic carbon sheets with SO_3H, COOH, and phenolic hydroxyl (OH) groups in a three-dimensional network that is a typical solid Brønsted acid catalyst. The composition of the sample shown in Fig. 2.27 was determined to be $CH_{0.622}O_{0.540}S_{0.048}$, with the amounts of phenolic OH groups, SO_3H and COOH bonded to the graphene being 2.0, 1.9 and 0.4 mmol g^{-1} respectively as determined by elemental analysis and cation-exchange [110]. The surface acidity and hydrophilicity/hydrophobicity properties can be varied by both the sulfonation conditions and the carbon source [111].

2.3.3 Brønsted Acid Sites for Glucose Production

Cellulose is a biopolymer with glucose monomeric units that are linked by β-1,4-glycoside bonds. The selective hydrolysis of cellulose into glucose is a key process for the utilization of cellulose. The cellulose-to-glucose reaction is a tandem hydrolysis reaction from cellulose to oligosaccharides and then to glucose, which is usually catalyzed by acid catalysts with Brønsted acid sites (Fig. 2.28) [112].

Sulfated and sulfonated catalysts have been reported as efficient catalysts for production of glucose from cellulose. Onda et al. [14] reported that solid acid

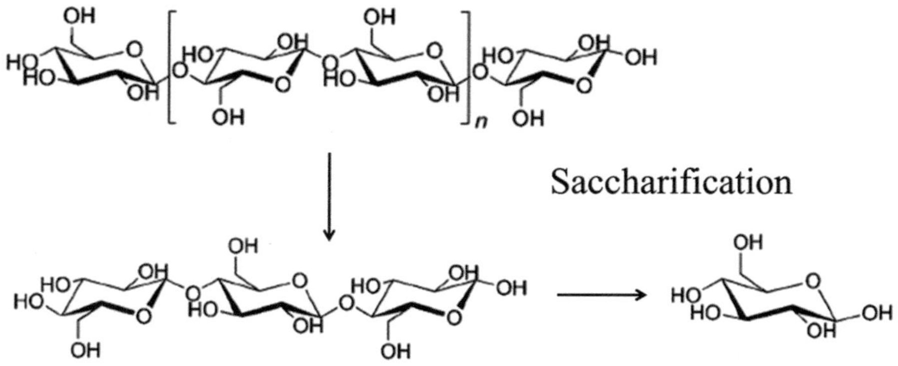

Fig. 2.28 Conversion of cellulose into glucose via soluble oligosaccharides oligomers

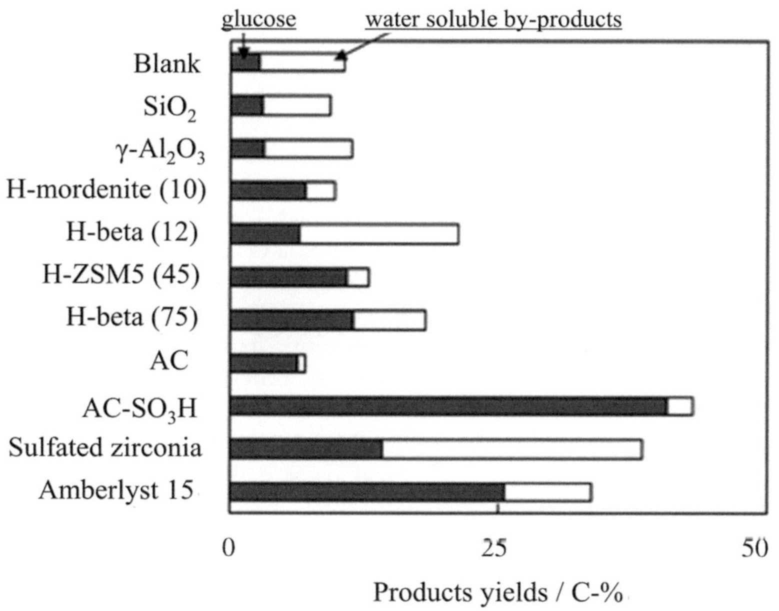

Fig. 2.29 Cellulose hydrolysis over solid acid catalysts at 150 °C. Reaction conditions: milled cellulose 45 mg, catalyst 50 mg, distilled water 5.0 mL, and 24 h (Reprinted with permission from Ref. [14], Copyright © 2008 The Royal Society of Chemistry)

catalysts hydrolyze cellulose (treated by ball-milling for 48 h) selectively into glucose and those authors screened many types of solid acids, including silica, alumina, H-zeolites, activated carbon, sulfated activated carbon, sulfated zirconia and Amberlyst-15 (Fig. 2.29). The silica and alumina showed almost no activity, while zeolite catalysts, especially the hydrophobic zeolites with high Si/Al ratios (H-ZSM5 (45), H-beta (75)), were found to catalyze the hydrolysis reaction efficiently. The hydrophobic character facilitates the catalysts with higher preference for adsorption of organic compounds than water, which may cause the relatively high glucose yields. Sulfated zirconia and Amberlyst-15 also gave a relatively high yield of glucose. However, sulfated zirconia gives a high fraction of water soluble byproducts and the SO_4^{2-} ions over the catalyst are leached. Among the screened solid catalysts, the sulfated activated carbon gave the highest glucose yield (40.5%) with minor water-soluble byproducts. Moreover, no loss of active SO_4^{2-} ions was observed. In comparison, activated carbon with acidic functional groups (e.g., carboxylic group) shows low reactivity for cellulose hydrolysis. The hydrolysis of cellulose to glucose proceeds over the Brønsted acid sites and probably can be promoted at high temperatures. Those authors investigated catalytic hydrothermal hydrolysis of the ball-milled cellulose using sulfonated activated-carbon (AC-SO_3H) catalyst, which is stable under hydrothermal conditions [113].

Carbon materials functionalized with groups of –COOH, –OH and –SO_3H were studied by Suganuma et al. [110], who reported that the pure microcrystalline cellulose could hardly be hydrolyzed into glucose or water-soluble β-1,4 glucan using conventional solid acid catalysts such as niobic acid, H-mordenite, Nafion and Amberlyst-15. In contrast, amorphous carbon containing -SO_3H, –COOH and phenolic -OH bearing groups on nanographene sheets exhibits remarkable hydrolysis performance (glucose yield of ~4% and β-1,4 glucan yield of ~64%) and is as effective as sulfuric acid. Interestingly, this carbon material has a small surface area and only one-tenth of the acid density of sulfuric acid. The high activity for cellulose hydrolysis can be attributed to the ability of the carbon material to adsorb β-1,4 glucan, and the large effective surface area in water and the SO_3H groups that are tolerable to hydration. Pang et al. [111] reported that the efficiency of sulfonated carbons can be greatly promoted by elevating the sulfonation temperature. The cellulose was pretreated by ball-milling for 48 h to lower its crystallinity. The cellulose is selectively converted into glucose with high yields (74.5%) over 250 °C-sulfonated catalyst. A sulfonated silica/carbon nanocomposite has been reported to have high cellulose (pretreated by ball-milling for 24 h) conversion (60.7%) and to give a glucose yield of 50.4% [114]. From the results, two merits including the strong and accessible Brønsted acid and the adsorption ability of β-1,4 glucan of the catalysts are crucial factors in determining the efficiency of cellulose hydrolysis. To facilitate the separation of catalysts and the reusability, Lai et al. [115] reported on a sulfonic group functionalized paramagnetic SBA-15 catalyst (Fe_3O_4-SBA-SO_3H) that promoted the hydrolysis of cellulose. The catalyst gave a glucose yield of 50% when amorphous cellulose was used as feed, and had high efficiency for other substrates such as starch or corn cob [115].

The catalytic performance of the heteropolyacid $H_3PW_{12}O_{40}$ was examined for the hydrolysis of microcrystalline cellulose with a crystallinity of 0.6 [116], where it was found that the catalyst promoted high conversion (92.3%) of cellulose and gave a glucose yield as high as 50.5%. This catalyst ($H_3PW_{12}O_{40}$) can be recycled by simple extraction with diethyl ether and it can be reused for at least six times. Shimizu et al. [105] reported HPA and metal salt catalysts for the hydrolysis of cellobiose and ball-milled cellulose into glucose or sugars, including $H_3PW_{12}O_{40}$, $H_4SiW_{12}O_{40}$ and $M_3(PW_{12}O_{40})_n$. Their work demonstrated that heteropoly acids have higher activity than mineral acids for the hydrolysis of cellulose to total reducing sugars (TRS) with values of the yields being in the following order: $H_3PW_{12}O_{40}$ > $H_4SiW_{12}O_{40}$ > $HClO_4$ > H_2SO_4 > H_3PO_4. The conversion of cellulose and the obtained TRS yields correspond well to the deprotonation enthalpy of these Brønsted acid catalysts, such that the stronger Brønsted acid is more favorable for the hydrolysis of β-1,4-glycosidicbonds of cellulose (Fig. 2.30).

Hegner et al. [117] reported the hydrolysis of cellobiose over amorphous silica supported Nafion and $FeCl_3$ catalysts at 130–190 °C. Both supported Nafion and $FeCl_3$ catalysts were active for the reaction, while the supported Nafion had better stability and did not deactivate for 3 cycles. Kobayashi et al. [112] reported a mesoporous carbon supported Ru catalysts to catalyze the hydrolysis of ball-milled cellulose. The support has large surface area and highly oxygen-functionalized surface and the catalyst was found to be efficient, water-tolerant and stable for the reaction. Those authors [112] proposed that a synergistic effect exists between carbon materials and Ru to catalyze the reaction efficiently, in which the carbon

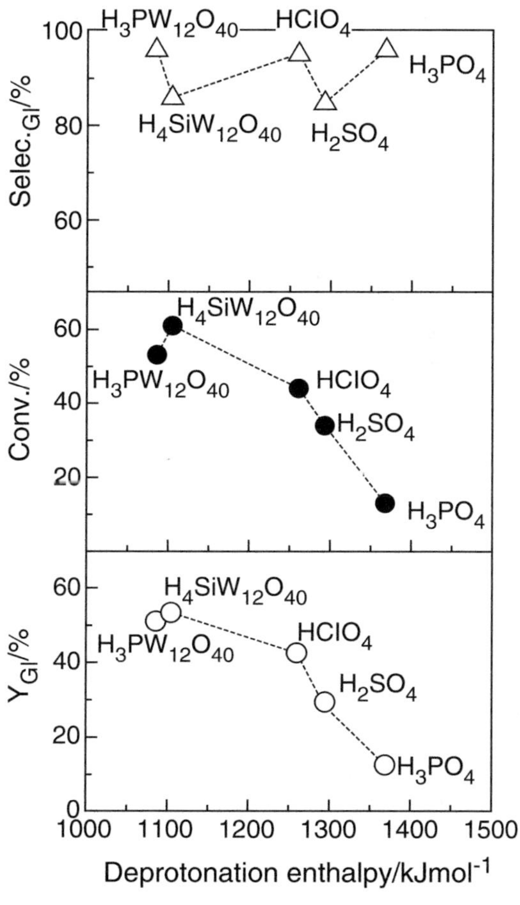

Fig. 2.30 Variation of glucose yield, cellobiose conversion, and glucose selectivity for cellobiose substrate with acid catalyst deprotonation enthalpy (DPE) (Reprinted with permission from Ref. [105], Copyright © 2009 The Royal Society of Chemistry)

material catalyzes the hydrolysis of cellulose to oligosaccharides while Ru promotes the conversion of oligosaccharides into glucose.

In summary, for efficient conversion of cellulose to glucose, solid acid catalysts with strong and accessible Brønsted acid sites and good adsorption ability of β-1,4 glucanare needed. The design of acids catalysts from the aspect of the catalyst structure may provide another approach, in which the acidic properties are varied within the catalyst structure.

2.3.4 Lewis Acid Sites Promote 5-hydroxymethylfurfural Formation

The biomass-derived platform compound 5-hydroxymethylfurfural (HMF) is highly versatile as a feedstock and can be converted into many types of chemicals and fuels. For example, HMF can be hydrogenated into 2,5-dimethylfuran and 2,5-dimethyltetrahydrofuran that are important fuel additives [118, 119]. HMF can also undergo hydrogenolysis to form polyols and oxidized into furan-2,5-dicarboxylic acid which are important polymer precursors [120]. HMF yields higher than 90% can be obtained from fructose or glucose as starting materials [121]. However, direct transformation of cellulose to HMF would be of high technological interest for future biorefineries, because cellulose represents a major constituent of lignocellulosic biomass. The reaction pathway for the direct transformation of cellulose into HMF consists of a series of reaction steps (Fig. 2.31). First, cellulose

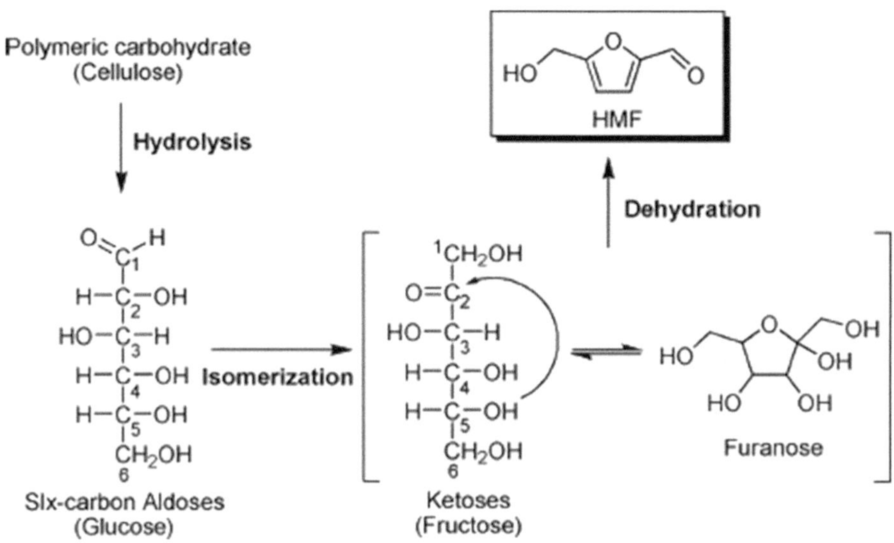

Fig. 2.31 Chemical pathway of direct transformation of cellulose into HMF (Reprinted with permission from Ref. [125], Copyright © 2011 The Royal Society of Chemistry)

undergoes hydrolysis into oligomers and six-carbon aldoses such as glucose that is known as saccharification. The generated aldose-type sugars are then isomerized into ketose-type sugars, which allow the formation of five-membered rings. Finally, these ketose-type sugars are dehydrated into HMF by removal of three water molecules per sugar molecule. This series of reactions could be effectively promoted with catalysts having Brønsted acid and Lewis acid sites, as the isomerization step is promoted by Lewis acid catalysts (e.g., $CrCl_3$, $AlCl_3$, Sn-containing zeolite) [122–124].

In 2002, Seri et al. reported that $LaCl_3$ catalyzes the transformation of microcrystalline cellulose into HMF in water at 250 °C although the yields (ca. 19%) were relatively low. It is important to note that metal salts can maintain their Lewis acid sites even in aqueous solutions and that they show good activity [126] so that their use has many possible applications.

Acidic metal salts-ionic liquids systems have good prospects for direct conversion of cellulose into HMF due to the excellent dissolution characteristics that some ionic liquids have for cellulose [127–129]. Zhao et al. [130] reported a Brønsted-Lewis-surfactant-combined $Cr[(DS)H_2PW_{12}O_{40}]_3$ (DS represents $OSO_3C_{12}H_{25}$ dodecyl sulfate) heteropolyacid for the conversion of cellulose into HMF in water. The catalyst gave a conversion of 77.1% and a HMF yield of 52.7%. The reason of high yield of HMF can be attributed to the assembly of the catalyst as micelles in the aqueous solution. Through interaction of cellulose with the surface of the micelles and its penetration into the micelles, the concentration of cellulose is concentrated around the catalyst, and so the catalyst activity is improved. Moreover, these micelles droplets are sufficient to provide a hydrophobic environment for protecting HMF from further decomposition.

Biphasic solvent systems consist of two immiscible liquid phases for which one phase contains the reactants and the other phase extracts the reaction product. Biphasic solvent systems can be used effectively for the direct conversion of cellulose into HMF. For example, in a biphasic solvent system, bimodal micro/mesoporous HZSM-5 with desilication gave cellulose conversions of 67% and 46% yields of HMF when using microcrystalline as material [131]. Shi et al. [132] developed a cost-effective biphasic solvent system with THF and concentrated $NaHSO_4$-$ZnSO_4$ aqueous solution for production of HMF from microcrystalline cellulose and obtained HMF yields of 53%.

Glucose is an intermediate that is formed when cellulose undergoes hydrolysis. The presence of a Lewis acid or a Brønsted acid promotes conversion of cellulose to glucose and then the reaction of glucose proceeds through a tandem reaction pathway involving the isomerization of glucose to fructose and then the dehydration of fructose to HMF. A catalyst that contains both Brønsted acid sites and Lewis acid sites could be effective for synthesis of HMF. Design of suitable catalysts for HMF will require consideration of catalyst structure and a proper balance of Brønsted acid and Lewis acid sites.

2.4 Supported Metal Catalysts for Cellulose Conversion

2.4.1 Supported Metal Catalysts

Supported metal-based catalysts are widely used in ammonia synthesis, Fischer-Tropsch synthesis and CO_2 hydrogenation reactions. A metal-based catalyst is composed of active transition metals and a support. The metals promote targeted reaction pathways and are closely associated with adsorption and activation mechanisms on the metal surfaces. The catalytic performance of a metal-based catalyst is highly-dependent on the combination of active metals and support.

2.4.1.1 Active Metals

Bare atoms on the metallic surface are coordinatively unsaturated and act as active sites for adsorption and activation of reactant molecules. The atoms of metals can be stacked in different arrangements, such as face-centered cubic (fcc-), body-centered cubic (bcc-) and hexagonal close-packed (hcp) crystal lattice. Exposure of different crystal faces to the substrate greatly affects catalytic performance. For example, N_2 dissociation readily occurs on the (111) face of Fe and demonstrates structural sensitivity [133].

The group 8-11metals such as Fe, Co, Ni, Cu, Ag, Au, Pt, Ru, Pd, and Ru are highly active for hydrogenation reactions. Catalytic performance depends on the occupancy of the metal (d) orbitals. For the transition metals, the outer s or d orbitals are not filled. For IB group metals (Cu, Ag, Au), the d orbitals are filled, while the outer s orbitals are not filled. Electron transition occurs between d and s orbitals that form energy levels with unpaired electrons. Interactions between reactants and the unpaired electrons in d orbitals produces adsorbed species and results in their activation and the promotion of their transformation. In contrast, filled d orbitals are relatively inactive for interaction and bonding. The activity of transition metals as catalysts can be explained by the d-band hole value (Table 2.3).

As shown in Table 2.3, values of the d-band hole decrease with an increase in d-band occupancy. In general, the adsorption of reactants is determined by the choices of metals with appropriate d-band hole values. Strong adsorption does not

Table 2.3 Summary of d-band hole values and percentage d character for selected transition metals

Elements	Fe	Co	Ni	Cu
group	8	9	10	11
atom	$3d^6\ 4s^2$	$3d^7\ 4s^2$	$3d^8\ 4s^2$	$3d^{10}\ 4s^1$
Energy band	$3d^{7.8}\ 4s^{0.2}$	$3d^{8.3}\ 4s^{0.7}$	$3d^{9.4}\ 4s^{0.6}$	$3d^{10}\ 4s^1$
d-band hole value	2.2	1.7	0.6	0
percentage d character	39.5	39.7	40	36

allow easy desorption and product removal and usually causes deactivation of the catalytic site. Nevertheless, the activation of reactants does not readily occur when molecules are weakly adsorbed onto the metal surfaces.

The percentage d character (d%) is also one way to assess the possible catalytic performance of a metal. The d% is the percentage of d orbitals in the hybridized orbitals (e.g., sp.-d hybridization) derived from valence bond theory. Usually, high values of d% imply low values d-band hole. Both d-band hole values and d% are parameters reflecting the electronic states of metals.

The combination of different metals is an effective way for modifying catalytic activity [134]. The metal particle size, shape and precursor are all important factors that influence the metal crystal face and electronic properties and that must be considered in designing an efficient catalyst [135].

2.4.1.2 Supports

Supports for active metals are mainly used for dispersion of the sites and for increasing the specific surface area. Thus, one of the most important criterions for active metal supports is high surface area. High surface area materials used for active metals include SiO_2, Al_2O_3, ZrO_2, CeO_2, carbon materials and zeolites. The determination of the dispersion of an active metal in a support is necessary to characterize the performance of an active metal catalyst, as discussed in Sect. 2.2.5. The support should also have high mechanical strength and high thermal conductivity to facilitate the removal of heat.

Besides physical dispersion, the support plays an important role in influencing the activity and selectivity of a catalyst due its interface with the substrate and catalytic sites that is called metal-support interaction. Metal-support interactions are divided into strong, medium and weak categories. Strong metal-support interactions are characteristic of metals supported on reducible oxides, such as TiO_2 [72, 76]. Weak metal-support interactions are characteristic of metals supported on non-reducible oxides, such as SiO_2 [136]. Metal support interactions stabilize the metal, form an interface and modify the electronic properties of the metal.

Hydrogenolysis reactions are typically "electron-sensitive" and the turnover frequency (TOF) may vary by several orders-of-magnitude among different supports.

A change in the electronic properties of the metal (electron-rich and electron poor) by the support has a large effect on catalyst activity and selectivity for different reactions [137]. For example, electron-poor metal nanoparticles showed higher activity than electron-rich metal nanoparticles in alkane hydrogenolysis. In contrast, electron-rich metals cause an increase in selectivity to alcohol products in C=O hydrogenation, while electron-poor metals result in a decrease in C=O hydrogenation. TiO_2 supported catalysts are known as materials with strong metal-support interactions. Baker et al. [138] probed that the Pt/TiO_2 interface provides a new and faster reaction pathway for furfural hydrogenation to furfuryl alcohol through the use of strong metal-support interactions.

Supports can provide additional active sites (e.g., oxygen vacancy sites, acid sites, and base sites) or show synergistic effects for some specific reactions. Mironenko et al. [139] demonstrated that oxygen vacancies in RuO_2 promote C–O bond scission, which is the most crucial step for catalytic transfer hydrogenation of furfural with 2-propanol. Basic sites in CaO/Al_2O_3 act as a CO_2 sorbent for the reforming reaction between CH_4 and CO_2 molecules over $Fe_2O_3/MgAl_2O_4$-CaO/Al_2O_3 composites [140]. Acidic sites on supports effectively promote the hydrogenation and hydrogenolysis reactions of carbon-oxygen bonds, by polarizing and activating carbon-oxygen bonds [141]. Since carbon-oxygen bonds widely occur in biomass resources, the importance of metal-acid bifunctional catalysts for oxygen removal reactions are discussed in the next section.

2.4.2 Supported Active Metals for Syngas and Hydrogen Production

Cellulose gasification to syngas and hydrogen offers substantial advantages for biomass conversion. In this case, the catalyst promotes the reforming and combustion reactions of pyrolyzed products of cellulose [142]. The reaction is environmentally clean and sustainable for both power generation and synthesis of chemicals and fuels (e.g., methanol, dimethyl ether, dimethyl oxalate, olefins and diesels). Fluidized bed gasification of cellulose has been found to be advantageous. Prior to the 1990s, dolomite, alkali metals and nickel catalysts were to be highly active for many reactions. However, the catalytic gasification of cellulose is challenging due to the formation of tar and char that occurs especially at temperatures around 500–600 °C. Sudden deactivation of catalysts was often observed due to the carbon deposition on the catalyst surface [143].

Tomishige and co-workers have made great progress in the gasification of biomass-derived cellulose [144, 145]. A series of supported Rh catalysts (Rh/CeO_2, Rh/ZrO_2, Rh/Al_2O_3, Rh/TiO_2, Rh/MgO and Rh/SiO_2) were studied for gasification of pressed microcrystalline cellulose (250–350 μm) and found to have good performance at moderate temperatures (450–550 °C) using air as the gasifying agent in fluidized-bed reactor [144]. Among these catalysts, Rh/CeO_2 catalyst showed the best performance and gave total conversion to gas. CeO_2 was demonstrated to promote the Rh/SiO_2 catalyst with improved stability and activity due to the interaction between Rh species and CeO_2. Through characterization studies on the catalysts, it was found that the oxidation and reduction of Rh species was promoted by CeO_2 [145] and that deactivation caused by carbon deposition was suppressed. Support effects (SiO_2, Al_2O_3 and ZrO_2) were further investigated for cellulose gasification in a fix-bed reactor [146]. The Rh/CeO_2/SiO_2 catalyst was applied to the gasification of jute and rice straw raw biomass resources. A dual bed reactor for separating the char and ash from volatile tar was also developed [147].

Although Rh-based catalysts are efficient in gasification of cellulose, the high cost and limited availability of Rh greatly hinders its industrial application. Supported nickel catalysts (Ni/Al_2O_3, Ni/ZrO_2, Ni/TiO_2, Ni/CeO_2 and Ni/MgO) on the other hand, are economic and have been investigated for cellulose gasification [148] in which it was found that the supports mainly act to disperse nickel species, while the conversion of cellulose is mainly controlled by nickel species. The CeO_2 support was found to decrease coke deposition on the catalyst surface. Xiao et al. [149] investigated the gasification of biomass using Ni-loaded brown coal char catalysts. The catalysts were found to have good resistance ability of coke formation at 650 °C, compared with Ni/Al_2O_3 catalysts. However, they did not compare the reactivity of the Ni/Al_2O_3 catalyst with Ni/CeO_2 catalysts [150, 151].

Promotion effects of noble metals (Pd, Pd, Rh and Ru) on supported Ni-CeO_2 catalysts have been studied, among which the metal Pt was found to be most effective [152, 153]. The promotion effect of Pt species is attributed to the formation of Pt-Ni alloy and the enhancement of reducibility of Ni species [152]. The promotion effect of metal Co has been studied [154]. The bimetallic Ni-Co/Al_2O_3 catalyst was found to have much more activity than monometallic Ni and Co species and the bimetallic catalyst resisted coke formation and exhibited good stability. The favorable performance of the Ni-Co/Al_2O_3 catalyst is related to the formation of Ni-Co alloy and the synergistic effect between Ni and Co atoms on the Ni–Co alloy surface. The good performance of bimetallic Ni-Co catalyst has been also demonstrated in the catalytic steam reforming of the aqueous fraction of bio-oil [155]. The NiCo/AlMg catalyst lowers the amount of coke deposits on the catalyst surface compared with Ni/AlMg, thus improving catalyst stability.

Gasification of cellulose and biomass resources offers renewable production of syngas and hydrogen and is compatible with traditional chemical and power industries [156]. Supported Rh catalysts and Ni catalysts have good reactivity and high resistance to coke deposition for biomass gasification. Further investigations are necessary to understand reaction mechanisms and to elucidate support effects exhibited by non-noble metal based catalysts. Development of catalysts for industrial-scale gasification with actual biomass resources is one goal of the future biorefinery.

2.5 Metal-Acid Bifunctional Catalysts for Cellulose Conversion

2.5.1 Metal-Acid Bifunctional Catalysts

The combination of active sites (e.g., metal-acid, metal-base, acid-base, etc.) forms bifunctional or multifunctional catalysts. Among them, metal-acid bifunctional catalysts are attracting attention due to their successful application in hydrocracking and hydrogenolysis reactions. For example, metal-acid bifunctional catalysts have

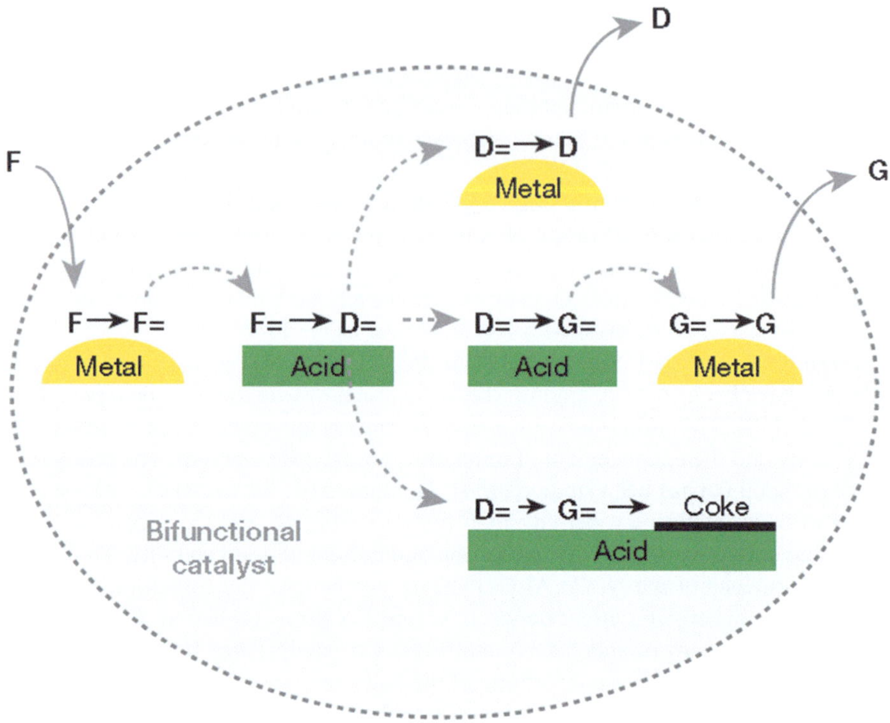

Fig. 2.32 Scheme of bifunctional catalysts for tandem reactions. Feed normal alkane molecules (F) are dehydrogenated on a metal surface, producing alkene intermediates (F=). The alkenes diffuse to acid sites, on which they undergo acid-catalyzed skeletal isomerization, which can be followed by one (D=) or more (G=) cracking events, sometimes leading to coke formation. Isomerized (D=) and/or cracked (D=, G=) alkene intermediates diffuse to the metal site and are hydrogenated to form isomerized or cracked products D (diesel) and G (gas) (Reprinted with permission from Ref. [157], Copyright © 2015 Nature)

been proposed for hydrocracking of fossil and renewable hydrocarbon sources to provide high-quality gasoline/diesel fuel [157]. The metal sites catalyze (de)hydrogenation reactions, while the acid sites catalyze the isomerization and cracking reactions for the conversion of alkanes (Fig. 2.32).

The intimacy criterion is one of the most important concepts in design of metal-acid bifunctional catalysts. The intimacy criterion requires a proper distance between metal and acid sites, at which the catalytic performance is optimal. Song et al. [158] demonstrated the improvement of proximal acid sites for the C–O hydrogenolysis reactions in phenol, catechol and guaiacol through a synergistic effect. Zecevic et al. [157] developed a bifunctional catalyst by combining a mixture of zeolite Y and alumina binder with Pt metal and controllably depositing the Pt on either the zeolite or on the binder (Fig. 2.33). By adjusting the distance between metal and acidic sites, that work demonstrates that nanoscale intimacy between metal and acid sites rather than closest intimacy will benefit the hydrocracking of alkanes.

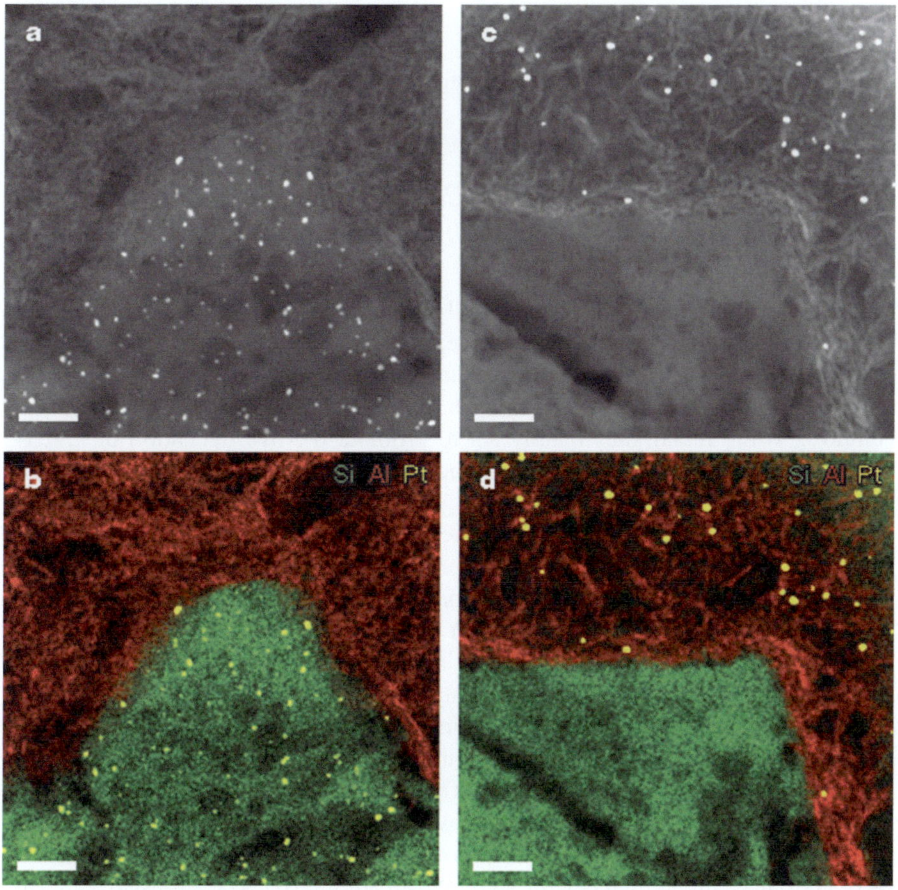

Fig. 2.33 Controlled deposition of Pt on either the zeolite Y or the alumina component of Y/A extrudates [157]. (**a**), (**b**), HAADF-STEM image and EDX map for Pt-Y/A sample; (**c**), (**d**), HAADF-STEM image and EDX map for Pt-A/Y sample (Reprinted with permission from Ref. [157], Copyright © 2015 Nature)

Acid sites activate C–O/C=O bonds and catalyze dehydration reactions, while metal sites catalyze hydrogenation reactions. Thus, metal-acid bifunctional catalysts are highly effective for promoting reactions leading to oxygen removal in biomass materials. The synergy between metal and acid allows realization of the hydrogenation/hydrogenolysis reactions. Lercher's group explored the C–O cleavage over monofunctional Ni/SiO_2 and bifunctional Ni/HZSM-5 catalysts, respectively, in the aqueous phase and demonstrated the promotion effect of proximal acid sites for C–O bonds [158]. Nickel phyllosilicate derived Ni/SiO_2 catalysts realized the C=O/C–O hydrogenolysis at relatively mild conditions, which was ascribed to the presence of proximal acid sites from the phyllosilicate structure [53].

In the design of bifunctional catalysts, the intimacy between the two active sites is one of the keys to achieving catalytic performance. By adjusting the intimacy between the active sites, the catalytic performance can be optimized.

2.5.2 Metal-Acid Bifunctional Catalysts for Sugar Alcohols Production

Metal-acid bifunctional catalysts are well known materials for promoting the conversion of cellulose into sugar alcohols (e.g., sorbitol). The conversion of cellulose into sorbitol is a cascade reaction with metastable glucose as a reaction intermediate. Both Brønsted acids and Lewis acids can efficiently catalyze the hydrolysis of cellulose, while metal sites mainly act to promote the hydrogenation of glucose (Fig. 2.34). The metal sites can also promote the hydrolysis of cellulose via heterolytic dissociation of H_2 [13].

Metals supported on acidic supports are good candidates for sorbitol production. Fukuoka et al. [159] investigated metals on a series of supports for microcrystalline cellulose conversion to sorbitol and demonstrated that solid acidic supports such as the H form of ultrastable Y zeolite (HUSY), γ-Al_2O_3, and SiO_2-Al_2O_3 were effective for sorbitol formation. Ru/SiO_2–SO_3H is a bifunctional catalyst with Brønsted acidic sites and metal sites (Ru). The catalyst was investigated for cellulose conversion to sorbitol. Compared with mechanical mixtures of silica-supported Brønsted acid (SiO_2–SO_3H) and the Ru/SiO_2 catalysts, the bifunctional catalyst exhibited much higher yields (61.2%) of sorbitol for relatively mild conditions (150 °C for 10 h reaction time) [160] and demonstrates the importance of the intimacy criterion. Ru nanoparticles loaded onto a Keggin-type polyoxometalate ($Cs_3PW_{12}O_{40}$) are reported to have high efficiency for sorbitol formation [161]. The Brønsted acid sites generated *in situ* from H_2 play a key role in the formation of sorbitol. Ru nanopar-

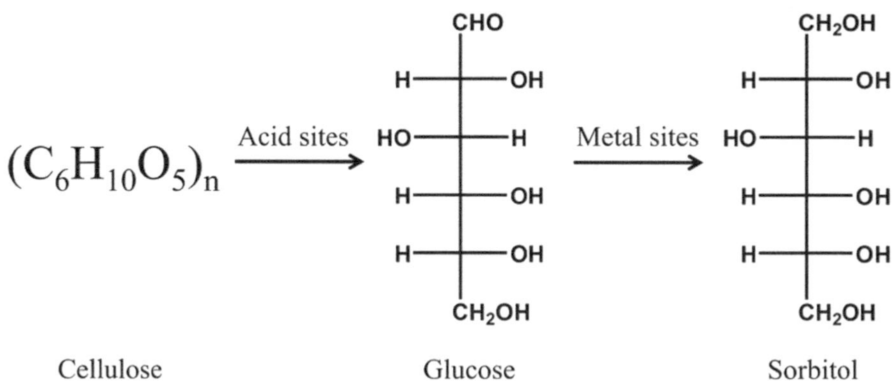

Fig. 2.34 Pathway for the selective transformation of cellulose to sorbitol over bifunctional catalysts

ticles supported on carbon nanotubes (CNTs) bearing acidic groups could also catalyze the sorbitol formation from cellulose [70]. Moreover, a combination of strong mineral H_2SO_4 or heteropolyacid with Ru/C affords excellent yields of sorbitol from ball-milled cellulose [162, 163].

Further development of metal-acid bifunctional catalysts is still needed to produce sorbitol on a practical scale. The balance and intimacy of the two sites are key design parameters for efficient bifunctional catalysts.

2.5.3 Metal-Acid Bifunctional Catalysts for Alkanes Production

The transformation of cellulose into alkanes is a necessary strategy for future society as it transitions away from fossil fuels. As discussed above, metal-acid bifunctional catalysts are good candidates for oxygen removal. There are some elegant examples in the literature describing the production of biofuels from sugars, sugar alcohols [164] or other platform molecules such as HMF [165] and levulinic acid [166]. However, research on the direct route from low cost cellulose to alkanes is still in its infancy with the main challenge being to selectively break C–O bonds in presence of C–C bonds.

Hexane can be formed by two different reaction pathways through cellulose (Fig. 2.35) [167]. As demonstrated in many works, alkanes can be obtained by hydrogenolysis of sorbitol [164]. Thus, the conversion of cellulose to hexane through sorbitol is one possible route. For sorbitol hydrogenolysis to alkanes, both Brønsted acid site and metal sites are needed. Liu et al. [168] reported on the selective one-pot conversion of cellulose pretreated using a ball-mill to n-hexane over Ir-Re/SiO_2 catalyst combined with HZSM-5 with ~80% yield in a biphasic reaction system. First, protons produced from HZSM-5 or hot water catalyze the hydrolysis of cellulose to water-soluble oligosaccharides, which is the rate-determining step. Oligosaccharides are further hydrolyzed to glucose. Then, glucose is hydrogenated to sorbitol over the Ir-ReO_x/SiO_2 catalyst. Finally, hydrogenolysis of sorbitol over Ir-ReO_x/SiO_2 and HZSM-5 produces the end product of n-hexane (~80% yield).

Alkane can also be obtained through HMF, which is different from the currently accepted sorbitol-to-alkane route. De Beeck et al. [167] proposed that the dominant route to the liquid alkanes proceeds via HMF over tungstosilicic acid (TSA) modified Ru/C catalyst in a biphasic reaction system. By modifying TSA on Ru/C, the glucose hydrogenation ability was suppressed and HMF hydrogenation was favored. The reaction pathway was therefore changed.

Products with other carbon number can be obtained through cellulose. For example, levulinic acid is also an important platform chemical that can be obtained from cellulose. Levulinic acid can be catalytically processed through the intermediate formation of γ-valerolactone to 5-nonanone [170]. The group of Dumesic proposed a catalytic process to convert cellulose cellulose into liquid hydrocarbons with levu-

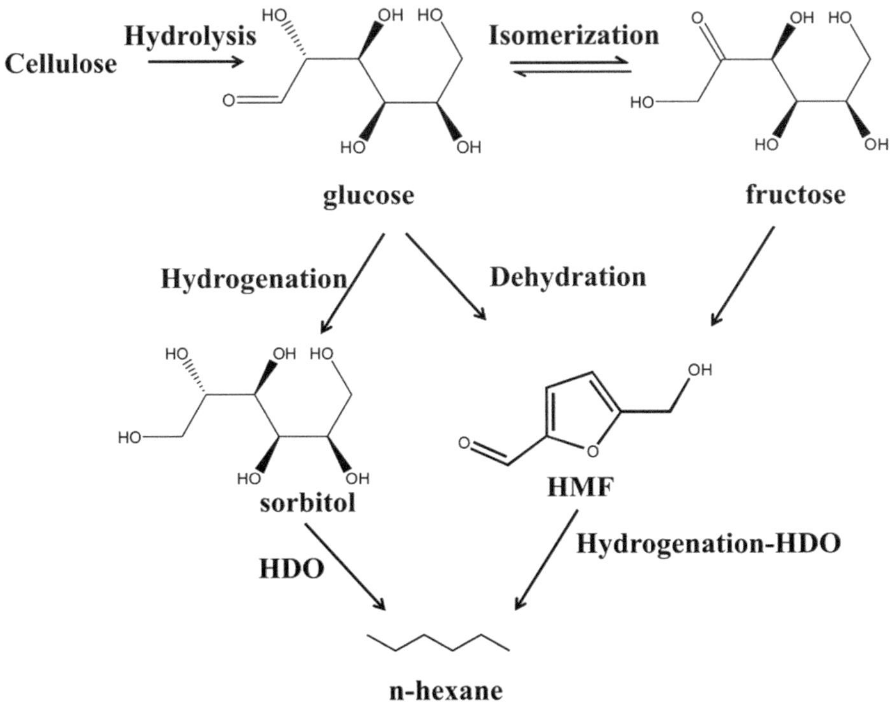

Fig. 2.35 Two different pathways for the selective one-pot conversion of cellulose to *n*-hexane. HDO, hydrodeoxygenation

linic acid as an intermediate (Fig. 2.36) [169]. The process starts with the deconstruction of cellulose in aqueous solution yielding a mixture of levulinic acid and formic acid. The formic acid is used to reduce levulinic acid to γ-valerolactone over Ru/C catalyst. The product of γ-valerolactone is then upgraded to 5-nonanone with high yields by using a dual catalyst bed of Pd/Nb_2O_5 and ceria-zirconia. The 5-nonanone product possesses a functional group (C=O) that can be used to achieve hydrocarbon fuel components. In particular, 5-nonanone can be converted, by means of hydrogenation/dehydration cycles over bifunctional Pt/Nb_2O_5 to linear nonane. Additionally, branched C_9 hydrocarbons for use as gasoline components can be obtained by dehydration/isomerization of 5-nonanol (hydrogenation product of 5-nonanone) over a zeolite such as USY.

Light hydrocarbons ($C_{n(<5)}$ alkanes) can be synthesized from cellulose directly over molecular sieves by hydrocracking reactions. By successive hydrocracking and condensation reactions of biomass-derived materials, C_2-C_9 alkanes are formed directly over Pt/H-ZSM-5 catalyst [171]. Kato et al. [172] reported that a Pt/H-beta zeolite catalyst prepared by ion-exchanges catalyzes conversion of ball-milled cellulose to C_3 and C_4 hydrocarbons.

Despite the importance of directly converting cellulose into alkanes, the catalytic efficiency and the selectivity are still far from what is needed on a practical scale.

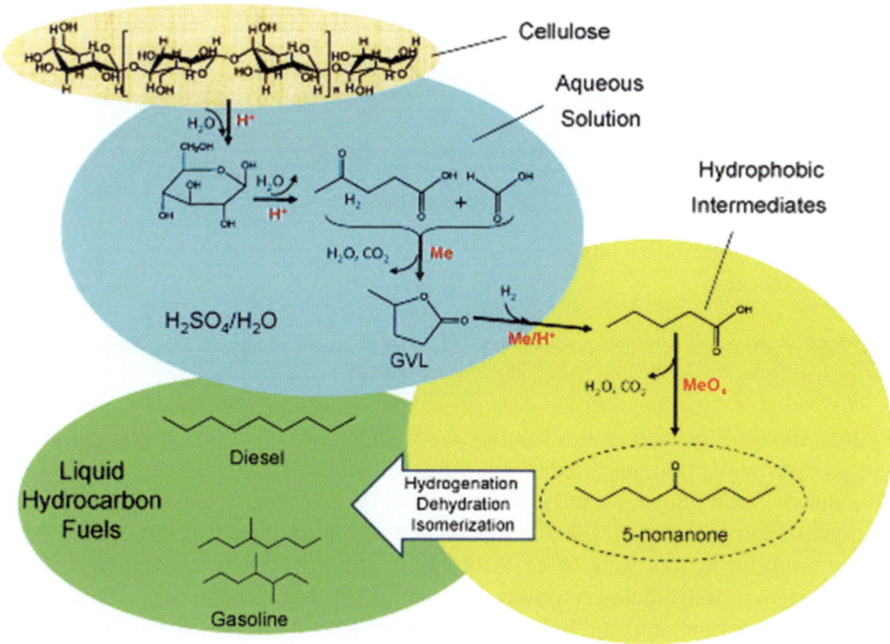

Fig. 2.36 Strategy for conversion of solid cellulose to liquid hydrocarbon fuels. H^+: acid sites; Me: metal sites; MeO_x: metal oxide sites (Reprinted with permission from Ref. [169], Copyright © 2015 Elsevier)

The conversion of cellulose to alkanes is composed of many successive reactions that need in-depth understanding on the relevant catalytic active sites and their kinetics. Moreover, the structural and surface evolution as an assembly of multifunctional components needs to be investigated.

2.5.4 *Metal-Acid Bifunctional Catalysts for Ethylene Glycol Production*

Ethylene glycol (EG) is an important bulk chemical with an annual consumption of 20 million tons per year and is widely applied to polymer precursors and as anti-freeze [13, 173]. It is now produced commercially from the hydration of petroleum-based ethylene oxide and could be synthesized from syngas via dimethyl oxalate hydrogenation [173, 174]. The one-pot catalytic conversion of cellulose to EG is an emerging area, which is firstly discovered by the group of Zhang [175]. This is a cascade reaction, containing at least three steps: the hydrolysis of cellulose, C–C bond cleavage and subsequent hydrogenation to EG, as illustrated in Fig. 2.37. The reaction needs the synergy of multifunctional active catalytic sites. First, the

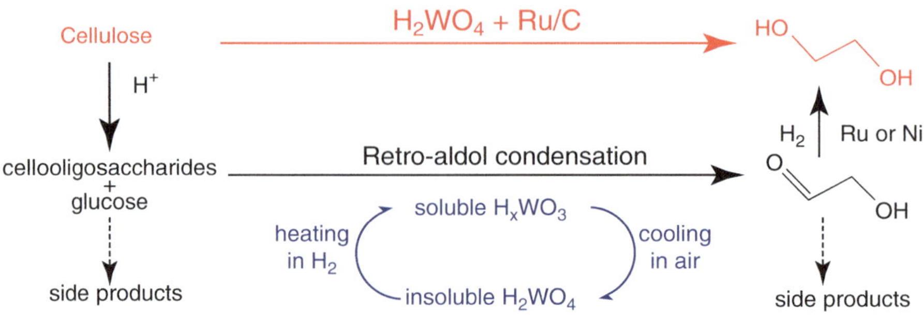

Fig. 2.37 Pathway of cellulose conversion to EG over W-based metal catalysts (Reprinted with permission from Ref. [13], Copyright © 2013 American Chemical Society)

hydrolysis of cellulose to water soluble oligosaccharides and glucose proceeds over the acid; secondly, the intermediates undergo chemical transformation to form glycolaldehyde with tungsten catalysts; finally, EG is generated through hydrogenation of glycolaldehyde by the transition metal catalysts. This route is greener than some other catalytic utilizations of cellulose (e.g., hydrolysis by acids), from both environmental and efficiency points of view [176]. Metal-acid sites catalyze the hydrolysis reactions and the hydrogenation reactions. Different from sorbitol and alkane formation, the EG synthesis needs an additional retro-aldol condensation reaction [177]. To date, tungsten-based compounds are the most efficient catalysts to promote the C–C bond cleavage reaction (i.e., retro-aldol condensation reaction) for the cellulose-to-EG conversion. This catalyst could also be classified as that with multifunctional active sites.

Carbon-supported tungsten carbide can effectively catalyze the conversion of cellulose to ethylene glycol. The W_2C/AC catalyst completely depolymerizes cellulose and produces EG in yields of 27%. Interestingly, the W_2C/AC catalyst modified with a small amount of Ni gives EG yield as high as 61%, indicating a synergistic effect between W_2C and Ni metals [175]. The performance of W/AC catalyst has been further elucidated for cellulose conversion [178]. The W/AC catalyst gives a cellulose conversion of 100% and an EG yield of 2% and negligible other polyols, which is lower than that obtained for the W_2C/AC catalyst (27%). With an increase in Ni/W, the conversion of cellulose is constant at 100% while the yield of EG decreases, which shows that neither W nor W_2C are efficient for depolymerization of cellulose (e.g., the C–C cracking reactions). The difference of EG yields between W/AC and W_2C/AC catalysts is attributed to the inactivity of W metal for hydrogenation of unsaturated intermediates [178] and the strong noble-metallic properties of W_2C compound [179].

To promote the reactivity of Ni-WC_x catalysts, Zhang et al. [176] proposed 3D mesoporous carbon supported Ni-WC_x to catalyze the conversion of cellulose to EG and obtained EG yields as high as 72.9%. Because of the superior depolymerization performance of W metal and W_2C, the authors thus further used transition metals to compensate for the hydrogenation ability and discovered a series of supported transition metal-tungsten catalysts for cellulose conversion to EG, including Ni-W, Pd-W, Pt-W, Ru-W and Ir-W bimetallic systems [178]. By assembling transition metals with W and changing the metal/W ratios, hydrogenation and degradation pathways promoted by the catalysts could be finely tuned, and thus the EG yield could be effectively modulated. The promotion effect of Ni on the hydrogenation processes was confirmed by Cao et al. [180].

Although great progress for cellulose-to-EG has been achieved by the former researchers, future work is still needed, including (1) the mechanistic understanding of the tungsten components on the molecular level; (2) the development of efficient, stable and cheap catalysts based on present understanding of reaction mechanisms, active site intimacy and interdisciplinary approaches with material scientists.

2.6 Conclusions and Future Outlook

A wide range of catalysts and catalytic methods are needed for efficient conversion of cellulose to target products, such as glucose, 5-hydroxymethylfurfural and platform chemicals. In this chapter, an overview of solid acid, supported metal and metal-acid bifunctional catalysts has been given with some examples for cellulose conversion. Firstly, an overview of characterization methods for catalysts was discussed to elucidate the nature of active sites and further correlate their catalytic performance, including XRD, N_2 adsorption, XPS, H_2-TPR, chemisorption, TPD, TPO, TPSR, TGA, IR, Raman, TEM, SEM, STEM, AFM and XAS. Based on the characterization, a clear correlation between catalyst performance and product formation can be seen.

Acid catalysts catalyze cellulose conversion to glucose and to HMF, however, for glucose formation, efficient solid acid catalysts with strong and accessible Brønsted acid sites and that are robust are still needed. For HMF formation, catalysts with Lewis acid sites could help to promote the process. To achieve a high yield of products, the modulation of catalyst acidity and type is crucial, which is based on the structure of the acid catalyst.

Supported metal catalysts are able to catalyze cellulose conversion to syngas and hydrogen. Metal electronic properties and metal support interactions are prerequisites for the design of highly-efficient catalysts.

Metal-acid bifunctional catalysts are able to catalyze cellulose conversion to sorbitol, alkane and ethylene glycol. Besides the above discussed properties for metal and acid sites, the intimacy between the metal and acid sites is an important factor for bifunctional catalysts, which could improve catalytic performance.

Catalytic conversion of cellulose has many special requirements related to the design of catalysts. The design of solid catalysts requires an understanding of the structure and surface changes. Thus, more comprehensive characterization methods that can be applied to the catalysts under actual reaction conditions would help to accelerate catalyst design and lead to improved catalytic performance. In this chapter, the structure of catalysts for cellulose reactions has been reviewed.

Catalytic research that improves substrate access to active sites via biomimetic and other approaches will be attractive for further study in the design of cellulose conversion processes.

References

1. Geboers JA, Van de Vyver S, Ooms R, Op de Beeck B, Jacobs PA, Sels BF. Chemocatalytic conversion of cellulose: opportunities, advances and pitfalls. Catal Sci Technol. 2011;1:714–26.
2. Sun Y, Cheng J. Hydrolysis of lignocellulosic materials for ethanol production: a review. Bioresour Technol. 2002;83:1–11.
3. Bhanja P, Bhaumik A. Porous nanomaterials as green catalyst for the conversion of biomass to bioenergy. Fuel. 2016;185:432–41.
4. Zheng X, Gu X, Ren Y, Zhi Z, Lu X. Production of 5-hydroxymethyl furfural and levulinic acid from lignocellulose in aqueous solution and different solvents. Biofuels Bioprod Biorefin. 2016;10:917–31.
5. Werpy T, Petersen G, Aden A, Bozell J, Holladay J, White J, Manheim A, Eliot D, Lasure L, Jones S, Top value added chemicals from biomass. Volume 1-Results of screening for potential candidates from sugars and synthesis gas, NREL/TP-510-35523; National Renewable Energy Laboratory, 2004.
6. Rinaldi R, Schüth F. Design of solid catalysts for the conversion of biomass. Energy Environ Sci. 2009;2:610–26.
7. Okuhara T. Water-tolerant solid acid catalysts. Chem Rev. 2002;102:3641–66.
8. van Putten R-J, van der Waal JC, de Jong E, Rasrendra CB, Heeres HJ, de Vries JG. Hydroxymethylfurfural, a versatile platform chemical made from renewable resources. Chem Rev. 2013;113:1499–597.
9. Climent MJ, Corma A, Iborra S. Conversion of biomass platform molecules into fuel additives and liquid hydrocarbon fuels. Green Chem. 2014;16:516–47.
10. Lu B, Xu A, Wang J. Cation does matter: how cationic structure affects the dissolution of cellulose in ionic liquids. Green Chem. 2014;16:1326–35.
11. O'Sullivan A. Cellulose: the structure slowly unravels. Cellulose. 1997;4:173–207.
12. Klemm D, Heublein B, Fink HP, Bohn A. Cellulose: fascinating biopolymer and sustainable raw material. Angew Chem Int Ed. 2005;44:3358–93.
13. Wang A, Zhang T. One-pot conversion of cellulose to Ethylene Glycol with multifunctional tungsten-based catalysts. Acc Chem Res. 2013;46:1377–86.
14. Onda A, Ochi T, Yanagisawa K. Selective hydrolysis of cellulose into glucose over solid acid catalysts. Green Chem. 2008;10:1033–7.
15. Bridgwater AV. Review of fast pyrolysis of biomass and product upgrading. Biomass Bioenergy. 2012;38:68–94.
16. Ertl G, Knözinger H, Weitkamp J. Handbook of heterogeneous catalysis. Weinheim: Wiley–VCH; 1997.
17. Zhu Y, Zhu Y, Ding G, Zhu S, Zheng H, Li Y. Highly selective synthesis of ethylene glycol and ethanol via hydrogenation of dimethyl oxalate on Cu catalysts: influence of support. Appl Catal A Gen. 2013;468:296–304.
18. Sharafutdinov I, Elkjær CF, Pereira de Carvalho HW, Gardini D, Chiarello GL, Damsgaard CD, Wagner JB, Grunwaldt J-D, Dahl S, Chorkendorff I. Intermetallic compounds of Ni and Ga as catalysts for the synthesis of methanol. J Catal. 2014;320:77–88.
19. Storck S, Bretinger H, Maier WF. Characterization of micro- and mesoporous solids by physisorption methods and pore-size analysis. Appl Catal A Gen. 1998;174:137–46.
20. Sing KSW. The use of gas adsorption for the characterization of porous solids. Colloids Surf. 1989;38:113–24.

21. Rinaldi R, Palkovits R, Schüth F. Depolymerization of cellulose using solid catalysts in ionic liquids. Angew Chem Int Ed. 2008;47:8047–50.
22. Osswald J, Kovnir K, Armbruster M, Giedigkeit R, Jentoft R, Wild U, Grin Y, Schlögl R. Palladium–gallium intermetallic compounds for the selective hydrogenation of acetylene: Part II: surface characterization and catalytic performance. J Catal. 2008;258:219–27.
23. Biesinger MC, Lau LW, Gerson AR, Smart RSC. Resolving surface chemical states in XPS analysis of first row transition metals, oxides and hydroxides: Sc, Ti, V, Cu and Zn. Appl Surf Sci. 2010;257:887–98.
24. Dupin J-C, Gonbeau D, Vinatier P, Levasseur A. Systematic XPS studies of metal oxides, hydroxides and peroxides. Phys Chem Chem Phys. 2000;2:1319–24.
25. Linsmeier C. Auger electron spectroscopy. Vacuum. 1994;45:673–90.
26. Antonides E, Janse EC, Sawatzky GA. LMM Auger spectra of Cu, Zn, Ga, and Ge. I. Transition probabilities, term splittings, and effective Coulomb interaction. Phys Rev B. 1977;15:1669–79.
27. Zhu Y, Kong X, Li X, Ding G, Zhu Y, Li Y-W. Cu Nanoparticles Inlaid Mesoporous Al_2O_3 As a high-performance bifunctional catalyst for ethanol synthesis via Dimethyl Oxalate Hydrogenation. ACS Catal. 2014;4:3612–20.
28. Sitthisa S, Pham T, Prasomsri T, Sooknoi T, Mallinson RG, Resasco DE. Conversion of furfural and 2-methylpentanal on Pd/SiO_2 and Pd–Cu/SiO_2 catalysts. J Catal. 2011;280:17–27.
29. Nie R, Liang D, Shen L, Gao J, Chen P, Hou Z. Selective oxidation of glycerol with oxygen in base-free solution over MWCNTs supported PtSb alloy nanoparticles. Appl Catal B Environ. 2012;127:212–20.
30. Takeguchi T, Manabe S, Kikuchi R, Eguchi K, Kanazawa T, Matsumoto S, Ueda W. Determination of dispersion of precious metals on CeO_2-containing supports. Appl Catal A Gen. 2005;293:91–6.
31. Perrichon V, Retailleau L, Bazin P, Daturi M, Lavalley JC. Metal dispersion of CeO_2–ZrO_2 supported platinum catalysts measured by H_2 or CO chemisorption. Appl Catal A Gen. 2004;260:1–8.
32. Sun S, Tsubaki N, Fujimoto K. The reaction performances and characterization of Fischer–Tropsch synthesis Co/SiO_2 catalysts prepared from mixed cobalt salts. Appl Catal A Gen. 2000;202:121–31.
33. Pannell RB, Chung KS, Bartholomew CH. The stoichiometry and poisoning by sulfur of hydrogen, oxygen and carbon monoxide chemisorption on unsupported nickel. J Catal. 1977;46:340–7.
34. Reuel RC, Bartholomew CH. The stoichiometries of H_2 and CO adsorptions on cobalt: effects of support and preparation. J Catal. 1984;85:63–77.
35. Wang X, Gorte RJ, Wagner JP. Deactivation mechanisms for Pd/Ceria during the water–gas-shift reaction. J Catal. 2002;212:225–30.
36. Zhang J, Wang H, Dalai AK. Effects of metal content on activity and stability of Ni-Co bimetallic catalysts for CO_2 reforming of CH_4. Appl Catal A Gen. 2008;339:121–9.
37. Gervasini A, Bennici S. Dispersion and surface states of copper catalysts by temperature-programmed-reduction of oxidized surfaces (s-TPR). Appl Catal A Gen. 2005;281:199–205.
38. Chinchen GC, Hay CM, Vandervell HD, Waugh KC. The measurement of copper surface areas by reactive frontal chromatography. J Catal. 1987;103:79–86.
39. Gong J, Yue H, Zhao Y, Zhao S, Zhao L, Lv J, Wang S, Ma X. Synthesis of ethanol via syngas on Cu/SiO_2 catalysts with balanced Cu^0-Cu^+ sites. J Am Chem Soc. 2012;134:13922–5.
40. Liu S, Ren J, Zhang H, Lv E, Yang Y, Li Y-W. Synthesis, characterization and isomerization performance of micro/mesoporous materials based on H-ZSM-22 zeolite. J Catal. 2016;335:11–23.
41. Silva CCCM, Ribeiro NFP, Souza MMVM, Aranda DAG. Biodiesel production from soybean oil and methanol using hydrotalcites as catalyst. Fuel Process Technol. 2010;91:205–10.

42. Watanabe M, Aizawa Y, Iida T, Nishimura R, Inomata H. Catalytic glucose and fructose conversions with TiO_2 and ZrO_2 in water at 473 K: relationship between reactivity and acid–base property determined by TPD measurement. Appl Catal A Gen. 2005;295:150–6.
43. Hou Z, Yokota O, Tanaka T, Yashima T. Investigation of CH_4 reforming with CO_2 on mesoporous Al_2O_3-supported Ni catalyst. Catal Lett. 2003;89:121–7.
44. Si R, Liu J, Yang K, Chen X, Dai W, Fu X. Temperature-programed surface reaction study of CO oxidation over Au/TiO_2 at low temperature: an insight into nature of the reaction process. J Catal. 2014;311:71–9.
45. Zhang W, Zhu Y, Niu S, Li Y. A study of furfural decarbonylation on K-doped Pd/Al_2O_3 catalysts. J Mol Catal A Chem. 2011;335:71–81.
46. Shi D, Vohs JM. Deoxygenation of biomass-derived oxygenates: reaction of furfural on Zn-Modified Pt(111). ACS Catal. 2015;5:2177–83.
47. Dong F, Zhu Y, Zheng H, Zhu Y, Li X, Li Y. Cr-free Cu-catalysts for the selective hydrogenation of biomass-derived furfural to 2-methylfuran: the synergistic effect of metal and acid sites. J Mol Catal A Chem. 2015;398:140–8.
48. Kong X, Zheng R, Zhu Y, Ding G, Zhu Y, Li Y-W. Rational design of Ni-based catalysts derived from hydrotalcite for selective hydrogenation of 5-hydroxymethylfurfural. Green Chem. 2015;17:2504–14.
49. Yuan Z, Wang L, Wang J, Xia S, Chen P, Hou Z, Zheng X. Hydrogenolysis of glycerol over homogenously dispersed copper on solid base catalysts. Appl Catal B Environ. 2011;101:431–40.
50. Jitianu M, Bălăsoiu M, Marchidan R, Zaharescu M, Crisan D, Craiu M. Thermal behaviour of hydrotalcite-like compounds: study of the resulting oxidic forms. Int J Inorg Mater. 2000;2:287–300.
51. Furusawa T, Sato T, Saito M, Ishiyama Y, Sato M, Itoh N, Suzuki N. The evaluation of the stability of Ni/MgO catalysts for the gasification of lignin in supercritical water. Appl Catal A Gen. 2007;327:300–10.
52. Chen L, Guo P, Qiao M, Yan S, Li H, Shen W, Xu H, Fan K. Cu/SiO_2 catalysts prepared by the ammonia-evaporation method: Texture, structure, and catalytic performance in hydrogenation of dimethyl oxalate to ethylene glycol. J Catal. 2008;257:172–80.
53. Kong X, Zhu Y, Zheng H, Li X, Zhu Y, Li Y-W. Ni nanoparticles inlaid nickel phyllosilicate as a metal-acid bifunctional catalyst for low-temperature hydrogenolysis reactions. ACS Catal. 2015;5:5914–20.
54. Zhang C, Yue H, Huang Z, Li S, Wu G, Ma X, Gong J. Hydrogen production via steam reforming of ethanol on phyllosilicate-derived Ni/SiO_2: enhanced metal–support interaction and catalytic stability. ACS Sustain Chem Eng. 2012;1:161–73.
55. Zhang C, Zhu W, Li S, Wu G, Ma X, Wang X, Gong J. Sintering-resistant Ni-based reforming catalysts obtained via the nanoconfinement effect. Chem Commun. 2013;49:9383–5.
56. Torii K, Onodera Y, Iwasaki T, Shirai M, Arai M, Nishiyama Y. Hydrothermal synthesis of novel smectite-like mesoporous materials. J Porous Mater. 1997;4:261–8.
57. Van Der Grift CJG, Elberse PA, Mulder A, Geus JW. Preparation of silica-supported copper catalysts by means of deposition-precipitation. Appl Catal. 1990;59:275–89.
58. Qiao B, Wang A, Yang X, Allard LF, Jiang Z, Cui Y, Liu J, Li J, Zhang T. Single-atom catalysis of CO oxidation using Pt_1/FeO_x. Nat Chem. 2011;3:634–41.
59. Ammari F, Lamotte J, Touroude R. An emergent catalytic material: Pt/ZnO catalyst for selective hydrogenation of crotonaldehyde. J Catal. 2004;221:32–42.
60. Holzapfel HH, Wolfbeisser A, Rameshan C, Weilach C, Rupprechter G. PdZn surface alloys as models of methanol steam reforming catalysts: molecular studies by LEED, XPS, TPD and PM-IRAS. Top Catal. 2014;57:1218–28.
61. Freund HJ, Meijer G, Scheffler M, Schlögl R, Wolf M. CO oxidation as a prototypical reaction for heterogeneous processes. Angew Chem Int Ed. 2011;50:10064–94.
62. Layman KA, Bussell ME. Infrared spectroscopic investigation of CO adsorption on silica-supported nickel phosphide catalysts. J Phys Chem B. 2004;108:10930–41.

63. Wöll C. The chemistry and physics of zinc oxide surfaces. Prog Surf Sci. 2007;82:55–120.
64. Pang SH, Schoenbaum CA, Schwartz DK, Medlin JW. Effects of thiol modifiers on the kinetics of furfural hydrogenation over Pd catalysts. ACS Catal. 2014;4:3123–31.
65. Ferrari AC. Raman spectroscopy of graphene and graphite: disorder, electron–phonon coupling, doping and nonadiabatic effects. Solid State Commun. 2007;143:47–57.
66. Iqbal M, Singh AK, Iqbal M, Eom J. Raman fingerprint of doping due to metal adsorbates on graphene. J Phys Condens Matter. 2012;24:335301.
67. Dresselhaus MS, Dresselhaus G, Saito R, Jorio A. Raman spectroscopy of carbon nanotubes. Phys Rep. 2005;409:47–99.
68. Hao Y, Wang Y, Wang L, Ni Z, Wang Z, Wang R, Koo CK, Shen Z, Thong JTL. Probing layer number and stacking order of few-layer graphene by Raman spectroscopy. Small. 2010;6:195–200.
69. Bergeret G, Gallezot P. Particle size and dispersion measurements. In: Ertl G, Knoezinger H, Schueth F, Weitkamp J, editors. Handbook of heterogeneous catalysis, vol. 2. 2nd ed. Weinheim: Wiley-VCH Verlag GmbH; 2008. p. 738–65.
70. Deng W, Liu M, Tan X, Zhang Q, Wang Y. Conversion of cellobiose into sorbitol in neutral water medium over carbon nanotube-supported ruthenium catalysts. J Catal. 2010;271:22–32.
71. Liu X, Liu M-H, Luo Y-C, Mou C-Y, Lin SD, Cheng H, Chen J-M, Lee J-F, Lin T-S. Strong metal–support interactions between gold nanoparticles and ZnO Nanorods in CO oxidation. J Am Chem Soc. 2012;134:10251–8.
72. Tauster S, Fung S, Garten RL. Strong metal-support interactions. Group 8 noble metals supported on titanium dioxide. J Am Chem Soc. 1978;100:170–5.
73. Lunkenbein T, Schumann J, Behrens M, Schlögl R, Willinger MG. Formation of a ZnO Overlayer in industrial Cu/ZnO/Al_2O_3 catalysts induced by strong metal–support interactions. Angew Chem Int Ed. 2015;127:4627–31.
74. Bowker M, Stone P, Morrall P, Smith R, Bennett R, Perkins N, Kvon R, Pang C, Fourre E, Hall M. Model catalyst studies of the strong metal-support interaction: surface structure identified by STM on Pd nanoparticles on TiO_2(110). J Catal. 2005;234:172–81.
75. Willinger MG, Zhang W, Bondarchuk O, Shaikhutdinov S, Freund H-J, Schlögl R. A case of strong metal-support interactions: combining advanced microscopy and model systems to elucidate the atomic structure of interfaces. Angew Chem Int Ed. 2014;53:5998–6001.
76. Lee J, Burt SP, Carrero CA, Alba-Rubio AC, Ro I, O'Neill BJ, Kim HJ, Jackson DHK, Kuech TF, Hermans I, Dumesic JA, Huber GW. Stabilizing cobalt catalysts for aqueous-phase reactions by strong metal-support interaction. J Catal. 2015;330:19–27.
77. Cao X, Shi Y, Shi W, Lu G, Huang X, Yan Q, Zhang Q, Zhang H. Preparation of Novel 3D Graphene networks for supercapacitor applications. Small. 2011;7:3163–8.
78. Sugimoto Y, Pou P, Abe M, Jelinek P, Pérez R, Morita S, Custance O. Chemical identification of individual surface atoms by atomic force microscopy. Nature. 2007;446:64–7.
79. Wastl DS, Weymouth AJ, Giessibl FJ. Atomically resolved graphitic surfaces in air by atomic force microscopy. ACS Nano. 2014;8:5233–9.
80. Müller DJ, Dufrene YF. Atomic force microscopy as a multifunctional molecular toolbox in nanobiotechnology. Nat Nanotechnol. 2008;3:261–9.
81. Ando T, Uchihashi T, Scheuring S. Filming biomolecular processes by high-speed atomic force microscopy. Chem Rev. 2014;114:3120–88.
82. Peng Y, Wang Z, Li C. Study of nanotribological properties of multilayer graphene by calibrated atomic force microscopy. Nanotechnology. 2014;25:305701–9.
83. Iwata K, Yamazaki S, Mutombo P, Hapala P, Ondráček M, Jelínek P, Sugimoto Y. Chemical structure imaging of a single molecule by atomic force microscopy at room temperature. Nat Commun. 2015;6:7766–73.
84. Albrecht F, Pavliček N, Herranz-Lancho C, Ruben M, Repp J. Characterization of a surface reaction by means of atomic force microscopy. J Am Chem Soc. 2015;137:7424–8.
85. Qin C, Clarke K, Li K. Interactive forces between lignin and cellulase as determined by atomic force microscopy. Biotechnol Biofuels. 2014;7:65–73.

86. van Oversteeg CHM, Doan HQ, de Groot FMF, Cuk T. In situ X-ray absorption spectroscopy of transition metal based water oxidation catalysts. Chem Soc Rev. 2017;46:102–25.
87. Koso S, Watanabe H, Okumura K, Nakagawa Y, Tomishige K. Comparative study of Rh–MoO_x and Rh–ReO_x supported on SiO_2 for the hydrogenolysis of ethers and polyols. Appl Catal B Environ. 2012;111:27–37.
88. Tupy SA, Karim AM, Bagia C, Deng W, Huang Y, Vlachos DG, Chen JG. Correlating ethylene glycol reforming activity with In Situ EXAFS detection of Ni segregation in supported NiPt bimetallic catalysts. ACS Catal. 2012;2:2290–6.
89. Gonzalez-delaCruz VM, Pereñiguez R, Ternero F, Holgado JP, Caballero A. In Situ XAS study of synergic effects on Ni–Co/ZrO_2 methane reforming catalysts. J Phys Chem C. 2012;116:2919–26.
90. Spanjers CS, Sim RS, Sturgis NP, Kabius B, Rioux RM. In Situ spectroscopic characterization of $Ni_{1-x}Zn_x$/ZnO catalysts and their selectivity for acetylene semihydrogenation in excess ethylene. ACS Catal. 2015;5:3304–15.
91. Giordanino F, Borfecchia E, Lomachenko KA, Lazzarini A, Agostini G, Gallo E, Soldatov AV, Beato P, Bordiga S, Lamberti C. Interaction of NH3 with Cu-SSZ-13 catalyst: a complementary FTIR, XANES, and XES study. J Phys Chem Lett. 2014;5:1552–9.
92. Bordiga S, Groppo E, Agostini G, van Bokhoven JA, Lamberti C. Reactivity of surface species in heterogeneous catalysts probed by in situ X-ray absorption techniques. Chem Rev. 2013;113:1736–850.
93. Singh JP, Gautam S, Lim WC, Asokan K, Singh BB, Raju M, Chaudhary S, Kabiraj D, Kanjilal D, Lee J-M. Electronic structure of magnetic Fe/MgO/Fe/Co multilayer structure by NEXAFS spectroscopy. Vacuum. 2017;138:48–54.
94. Singh J, Lamberti C, van Bokhoven JA. Advanced X-ray absorption and emission spectroscopy: in situ catalytic studies. Chem Soc Rev. 2010;39:4754–66.
95. Bui L, Luo H, Gunther WR, Román-Leshkov Y. Domino reaction catalyzed by zeolites with Brønsted and lewis acid sites for the production of γ-valerolactone from furfural. Angew Chem Int Ed. 2013;52:1–5.
96. Zhu Z, Chen Q, Xie Z, Yang W, Li C. The roles of acidity and structure of zeolite for catalyzing toluene alkylation with methanol to xylene. Microporous Mesoporous Mater. 2006;88:16–21.
97. Chai S-H, Wang H-P, Liang Y, Xu B-Q. Sustainable production of acrolein: investigation of solid acid-base catalysts for gas-phase dehydration of glycerol. Green Chem. 2007;9:1130–6.
98. Chen J, Liang T, Li J, Wang S, Qin Z, Wang P, Huang L, Fan W, Wang J. Regulation of framework aluminum siting and acid distribution in H-MCM-22 by boron incorporation and its effect on the catalytic performance in methanol to hydrocarbons. ACS Catal. 2016;6:2299–313.
99. Lucas N, Bordoloi A, Amrute AP, Kasinathan P, Vinu A, Bohringer W, Fletcher JCQ, Halligudi SB. A comparative study on liquid phase alkylation of 2-methylnaphthalene with long chain olefins using different solid acid catalysts. Appl Catal A Gen. 2009;352:74–80.
100. Chang CD, Chu CTW, Socha RF. Methanol conversion to olefins over ZSM-5. J Catal. 1984;86:289–96.
101. Ozaki A, Kimura K. The effective site on acid catalysts revealed in n-butene isomerization. J Catal. 1964;3:395–405.
102. Bondeson D, Mathew A, Oksman K. Optimization of the isolation of nanocrystals from microcrystalline celluloseby acid hydrolysis. Cellulose. 2006;13:171–80.
103. Mosier N, Wyman C, Dale B, Elander R, Lee Y, Holtzapple M, Ladisch M. Features of promising technologies for pretreatment of lignocellulosic biomass. Bioresour Technol. 2005;96:673–86.
104. Gandarias I, Arias P, Requies J, Güemez M, Fierro J. Hydrogenolysis of glycerol to propanediols over a Pt/ASA catalyst: The role of acid and metal sites on product selectivity and the reaction mechanism. Appl Catal B Environ. 2010;97:248–56.

105. Shimizu K, Furukawa H, Kobayashi N, Itaya Y, Satsuma A. Effects of Bronsted and Lewis acidities on activity and selectivity of heteropolyacid-based catalysts for hydrolysis of cellobiose and cellulose. Green Chem. 2009;11:1627–32.
106. Kozhevnikov IV, Kloetstra KR, Sinnema A, Zandbergen HW, van Bekkum H. Study of catalysts comprising heteropoly acid $H_3PW_{12}O_{40}$ supported on MCM-41 molecular sieve and amorphous silica. J Mol Catal A Chem. 1996;114:287–98.
107. Narasimharao K, Brown DR, Lee AF, Newman AD, Siril PF, Tavener SJ, Wilson K. Structure–activity relations in Cs-doped heteropolyacid catalysts for biodiesel production. J Catal. 2007;248:226–34.
108. Harmer MA, Sun Q. Solid acid catalysis using ion-exchange resins. Appl Catal A Gen. 2001;221:45–62.
109. Song X, Sayari A. Sulfated zirconia-based strong solid-acid catalysts: recent progress. Catal Rev. 1996;38:329–412.
110. Suganuma S, Nakajima K, Kitano M, Yamaguchi D, Kato H, Hayashi S, Hara M. Hydrolysis of cellulose by amorphous carbon bearing SO_3H, COOH, and OH groups. J Am Chem Soc. 2008;130:12787–93.
111. Pang J, Wang A, Zheng M, Zhang T. Hydrolysis of cellulose into glucose over carbons sulfonated at elevated temperatures. Chem Commun. 2010;46:6935–7.
112. Kobayashi H, Komanoya T, Hara K, Fukuoka A. Water-tolerant mesoporous-carbon-supported ruthenium catalysts for the hydrolysis of cellulose to glucose. ChemSusChem. 2010;3:440–3.
113. Onda A, Ochi T, Yanagisawa K. Hydrolysis of cellulose selectively into glucose over sulfonated activated-carbon catalyst under hydrothermal conditions. Top Catal. 2009;52:801–7.
114. Van de Vyver S, Peng L, Geboers J, Schepers H, de Clippel F, Gommes CJ, Goderis B, Jacobs PA, Sels BF. Sulfonated silica/carbon nanocomposites as novel catalysts for hydrolysis of cellulose to glucose. Green Chem. 2010;12:1560.
115. Lai DM, Deng L, Li J, Liao B, Guo QX, Fu Y. Hydrolysis of cellulose into glucose by magnetic solid acid. ChemSusChem. 2011;4:55–8.
116. Tian J, Wang J, Zhao S, Jiang C, Zhang X, Wang X. Hydrolysis of cellulose by the heteropoly acid $H_3PW_{12}O_{40}$. Cellulose. 2010;17:587–94.
117. Hegner J, Pereira KC, DeBoef B, Lucht BL. Conversion of cellulose to glucose and levulinic acid via solid-supported acid catalysis. Tetrahedron Lett. 2010;51:2356–8.
118. Yang W, Sen A. One-step catalytic transformation of carbohydrates and cellulosic biomass to 2,5-Dimethyltetrahydrofuran for liquid fuels. ChemSusChem. 2010;3:597–603.
119. Román-Leshkov Y, Barrett CJ, Liu ZY, Dumesic JA. Production of dimethylfuran for liquid fuels from biomass-derived carbohydrates. Nature. 2007;447:982–5.
120. Gupta NK, Nishimura S, Takagaki A, Ebitani K. Hydrotalcite-supported gold-nanoparticle-catalyzed highly efficient base-free aqueous oxidation of 5 hydroxymethylfurfural into 2,5-furandicarboxylic acid under atmospheric oxygen pressure. Green Chem. 2011;13:824–7.
121. Hu S, Zhang Z, Zhou Y, Han B, Fan H, Li W, Song J, Xie Y. Conversion of fructose to 5-hydroxymethylfurfural using ionic liquids prepared from renewable materials. Green Chem. 2008;10:1280–3.
122. Román-Leshkov Y, Moliner M, Labinger JA, Davis ME. Mechanism of glucose isomerization using a solid Lewis acid catalyst in water. Angew Chem Int Ed. 2010;49:8954–7.
123. Li G, Pidko EA, Hensen EJ. Synergy between Lewis acid sites and hydroxyl groups for the isomerization of glucose to fructose over Sn-containing zeolites: a theoretical perspective. Cat Sci Technol. 2014;4:2241–50.
124. Pagán-Torres YJ, Wang T, Gallo JMR, Shanks BH, Dumesic JA. Production of 5-Hydroxymethylfurfural from glucose using a combination of Lewis and Brønsted acid catalysts in water in a biphasic reactor with an alkylphenol solvent. ACS Catal. 2012;2:930–4.
125. Kim B, Jeong J, Lee D, Kim S, Yoon H-J, Lee Y-S, Cho JK. Direct transformation of cellulose into 5-hydroxymethyl-2-furfural using a combination of metal chlorides in imidazolium ionic liquid. Green Chem. 2011;13:1503–6.

126. Seri K, Sakaki T, Shibata M, Inoue Y, Ishida H. Lanthanum(III)-catalyzed degradation of cellulose at 250°C. Bioresour Technol. 2002;81:257–60.
127. Swatloski RP, Spear SK, Holbrey JD, Rogers RD. Dissolution of cellose with ionic liquids. J Am Chem Soc. 2002;124:4974–5.
128. Ohira K, Abe Y, Kawatsura M, Suzuki K, Mizuno M, Amano Y, Itoh T. Design of cellulose dissolving ionic liquids inspired by nature. ChemSusChem. 2012;5:388–91.
129. Su Y, Brown HM, Huang X, Zhou X-D, Amonette JE, Zhang ZC. Single-step conversion of cellulose to 5-hydroxymethylfurfural (HMF), a versatile platform chemical. Appl Catal A Gen. 2009;361:117–22.
130. Zhao S, Cheng M, Li J, Tian J, Wang X. One pot production of 5-hydroxymethylfurfural with high yield from cellulose by a Bronsted-Lewis-surfactant-combined heteropolyacid catalyst. Chem Commun. 2011;47:2176–8.
131. Nandiwale KY, Galande ND, Thakur P, Sawant SD, Zambre VP, Bokade VV. One-pot synthesis of 5-Hydroxymethylfurfural by cellulose hydrolysis over highly active bimodal micro/mesoporous H-ZSM-5 catalyst. ACS Sustain Chem Eng. 2014;2:1928–32.
132. Shi N, Liu Q, Zhang Q, Wang T, Ma L. High yield production of 5-hydroxymethylfurfural from cellulose by high concentration of sulfates in biphasic system. Green Chem. 2013;15:1967–74.
133. Tsai M-C, Ship U, Bassignana I, Küppers J, Ertl G. A vibrational spectroscopy study on the interaction of N_2 with clean and K-promoted Fe (111) surfaces: π-bonded dinitrogen as precursor for dissociation. Surf Sci. 1985;155:387–99.
134. Rodriguez JA, Goodman DW. Surface science studies of the electronic and chemical properties of bimetallic systems. J Phys Chem. 1991;95:4196–206.
135. Mäki-Arvela P, Hájek J, Salmi T, Murzin DY. Chemoselective hydrogenation of carbonyl compounds over heterogeneous catalysts. Appl Catal A Gen. 2005;292:1–49.
136. Kennedy G, Baker LR, Somorjai GA. Selective amplification of C–O bond hydrogenation on Pt/TiO_2: catalytic reaction and sum-frequency generation vibrational spectroscopy studies of crotonaldehyde hydrogenation. Angew Chem Int Ed. 2014;126:3473–6.
137. Ruppert AM, Weckhuysen BM. Metal-support interactions, in: handbook of heterogeneous catalysis. Weinheim: Wiley-VCH; 2008.
138. Baker LR, Kennedy G, Van Spronsen M, Hervier A, Cai X, Chen S, Wang LW, Somorjai GA. Furfuraldehyde hydrogenation on titanium oxide-supported platinum nanoparticles studied by sum frequency generation vibrational spectroscopy: acid-base catalysis explains the molecular origin of strong metal-support interactions. J Am Chem Soc. 2012;134:14208–16.
139. Mironenko AV, Vlachos DG. Conjugation-driven "reverse Mars-van Krevelen"-type radical mechanism for low-temperature C–O bond activation. J Am Chem Soc. 2016;138:8104–13.
140. Buelens LC, Galvita VV, Poelman H, Detavernier C, Marin GB. Super-dry reforming of methane intensifies CO_2 utilization via Le Chatelier's principle. Science. 2016;354:449–52.
141. Zhou C-HC, Beltramini JN, Fan Y-X, Lu GM. Chemoselective catalytic conversion of glycerol as a biorenewable source to valuable commodity chemicals. Chem Soc Rev. 2008;37:527–49.
142. Asadullah M, Ito S-i, Kunimori K, Yamada M, Tomishige K. Biomass gasification to hydrogen and syngas at low temperature: novel catalytic system using fluidized-bed reactor. J Catal. 2002;208:255–9.
143. Saeman JF. Kinetics of wood saccharification-hydrolysis of cellulose and decomposition of sugars in dilute acid at high temperature. Ind Eng Chem. 1945;37:43–52.
144. Asadullah M, Fujimoto K, Tomishige K. Catalytic performance of Rh/CeO_2 in the gasification of cellulose to synthesis gas at low temperature. Ind Eng Chem Res. 2001;40:5894–900.
145. Miyazawa T, Okumura K, Kunimori K, Tomishige K. Promotion of oxidation and reduction of Rh species by interaction of Rh and CeO_2 over Rh/CeO_2/SiO_2. J Phys Chem C. 2008;112:2574–83.
146. Asadullah M, Ito S-i, Kunimori K, Yamada M, Tomishige K. Energy efficient production of hydrogen and syngas from biomass: development of low-temperature catalytic process for cellulose gasification. Environ Sci Technol. 2002;36:4476–81.

147. Tomishige K, Asadullah M, Kunimori K. Syngas production by biomass gasification using Rh/CeO_2/SiO_2 catalysts and fluidized bed reactor. Catal Today. 2004;89:389–403.
148. Miyazawa T, Kimura T, Nishikawa J, Kado S, Kunimori K, Tomishige K. Catalytic performance of supported Ni catalysts in partial oxidation and steam reforming of tar derived from the pyrolysis of wood biomass. Catal Today. 2006;115:254–62.
149. Xiao X, Cao J, Meng X, Le DD, Li L, Ogawa Y, Sato K, Takarada T. Synthesis gas production from catalytic gasification of waste biomass using nickel-loaded brown coal char. Fuel. 2013;103:135–40.
150. Kimura T, Miyazawa T, Nishikawa J, Kado S, Okumura K, Miyao T, Naito S, Kunimori K, Tomishige K. Development of Ni catalysts for tar removal by steam gasification of biomass. Appl Catal B Environ. 2006;68:160–70.
151. Tomishige K, Kimura T, Nishikawa J, Miyazawa T, Kunimori K. Promoting effect of the interaction between Ni and CeO_2 on steam gasification of biomass. Catal Commun. 2007;8:1074–9.
152. Nishikawa J, Miyazawa T, Nakamura K, Asadullah M, Kunimori K, Tomishige K. Promoting effect of Pt addition to Ni/CeO_2/Al_2O_3 catalyst for steam gasification of biomass. Catal Commun. 2008;9:195–201.
153. Nishikawa J, Nakamura K, Asadullah M, Miyazawa T, Kunimori K, Tomishige K. Catalytic performance of Ni/CeO_2/Al_2O_3 modified with noble metals in steam gasification of biomass. Catal Today. 2008;131:146–55.
154. Wang L, Li D, Koike M, Watanabe H, Xu Y, Nakagawa Y, Tomishige K. Catalytic performance and characterization of Ni–Co catalysts for the steam reforming of biomass tar to synthesis gas. Fuel. 2013;112:654–61.
155. Remón J, Medrano JA, Bimbela F, García L, Arauzo J. Ni/Al–Mg–O solids modified with Co or Cu for the catalytic steam reforming of bio-oil. Appl Catal B Environ. 2013;132:433–44.
156. Kırtay E. Recent advances in production of hydrogen from biomass. Energy Convers Manag. 2011;52:1778–89.
157. Zecevic J, Vanbutsele G, de Jong KP, Martens JA. Nanoscale intimacy in bifunctional catalysts for selective conversion of hydrocarbons. Nature. 2015;528:245–8.
158. Song W, Liu Y, Baráth E, Zhao C, Lercher JA. Synergistic effects of Ni and acid sites for hydrogenation and C–O bond cleavage of substituted phenols. Green Chem. 2015;17:1204–18.
159. Fukuoka A, Dhepe PL. Catalytic conversion of cellulose into sugar alcohols. Angew Chem Int Ed. 2006;45:5161–3.
160. Zhu W, Yang H, Chen J, Chen C, Guo L, Gan H, Zhao X, Hou Z. Efficient hydrogenolysis of cellulose into sorbitol catalyzed by a bifunctional catalyst. Green Chem. 2014;16:1534–42.
161. Liu M, Deng W, Zhang Q, Wang Y, Wang Y. Polyoxometalate-supported ruthenium nanoparticles as bifunctional heterogeneous catalysts for the conversions of cellobiose and cellulose into sorbitol under mild conditions. Chem Commun. 2011;47:9717–9.
162. Palkovits R, Tajvidi K, Procelewska J, Rinaldi R, Ruppert A. Hydrogenolysis of cellulose combining mineral acids and hydrogenation catalysts. Green Chem. 2010;12:972–8.
163. Geboers J, Van de Vyver S, Carpentier K, de Blochouse K, Jacobs P, Sels B. Efficient catalytic conversion of concentrated cellulose feeds to hexitols with heteropoly acids and Ru on carbon. Chem Commun. 2010;46:3577–9.
164. Li N, Huber GW. Aqueous-phase hydrodeoxygenation of sorbitol with Pt/SiO_2-Al_2O_3: Identification of reaction intermediates. J Catal. 2010;270:48–59.
165. Chatterjee M, Matsushima K, Ikushima Y, Sato M, Yokoyama T, Kawanami H, Suzuki T. Production of linear alkane via hydrogenative ring opening of a furfural-derived compound in supercritical carbon dioxide. Green Chem. 2010;12:779–82.
166. Serrano-Ruiz JC, Wang D, Dumesic JA. Catalytic upgrading of levulinic acid to 5-nonanone. Green Chem. 2010;12:574–7.
167. de Beeck BO, Dusselier M, Geboers J, Holsbeek J, Morré E, Oswald S, Giebeler L, Sels BF. Direct catalytic conversion of cellulose to liquid straight-chain alkanes. Energy Environ Sci. 2014;8(1):230–40.

168. Liu S, Tamura M, Nakagawa Y, Tomishige K. One-pot conversion of cellulose into n-Hexane over the Ir-ReO_x/SiO_2 catalyst combined with HZSM-5. ACS Sustain Chem Eng. 2014;2:1819–27.
169. Serrano-Ruiz JC, Braden DJ, West RM, Dumesic JA. Conversion of cellulose to hydrocarbon fuels by progressive removal of oxygen. Appl Catal B Environ. 2010;100:184–9.
170. Alonso DM, Bond JQ, Serrano-Ruiz JC, Dumesic JA. Production of liquid hydrocarbon transportation fuels by oligomerization of biomass-derived C9 alkenes. Green Chem. 2010;12:992–9.
171. Murata K, Liu Y, Inaba M, Takahara I. Hydrocracking of biomass-derived materials into alkanes in the presence of platinum-based catalyst and hydrogen. Catal Lett. 2010;140:8–13.
172. Kato Y, Sekine Y. One pot direct catalytic conversion of cellulose to hydrocarbon by decarbonation using Pt/H-beta zeolite catalyst at low temperature. Catal Lett. 2013;143:418–23.
173. Yue H, Zhao Y, Ma X, Gong J. Ethylene glycol: properties, synthesis, and applications. Chem Soc Rev. 2012;41:4218–44.
174. Zhu Y, Kong X, Cao D-B, Cui J, Zhu Y, Li Y-W. The rise of calcination temperature enhances the performance of Cu catalysts: contributions of support. ACS Catal. 2014;4(10):3675–81.
175. Ji N, Zhang T, Zheng M, Wang A, Wang H, Wang X, Chen JG. Direct catalytic conversion of cellulose into ethylene glycol using nickel-promoted tungsten carbide catalysts. Angew Chem Int Ed. 2008;47:8510–3.
176. Zhang Y, Wang A, Zhang T. A new 3D mesoporous carbon replicated from commercial silica as a catalyst support for direct conversion of cellulose into ethylene glycol. Chem Commun. 2010;46:862–4.
177. Li C, Zheng M, Wang A, Zhang T. One-pot catalytic hydrocracking of raw woody biomass into chemicals over supported carbide catalysts: simultaneous conversion of cellulose, hemicellulose and lignin. Energy Environ Sci. 2012;5:6383–90.
178. Zheng MY, Wang AQ, Ji N, Pang JF, Wang XD, Zhang T. Transition metal-tungsten bimetallic catalysts for the conversion of cellulose into ethylene glycol. ChemSusChem. 2010;3:63–6.
179. Oyama S. Preparation and catalytic properties of transition metal carbides and nitrides. Catal Today. 1992;15:179–200.
180. Cao Y, Wang J, Kang M, Zhu Y. Efficient synthesis of ethylene glycol from cellulose over Ni–WO_3/SBA-15 catalysts. J Mol Catal A Chem. 2014;381:46–53.

Part II
Bifunctional Chemocatalysis

Chapter 3
Brønsted-Lewis Acids for Efficient Conversion of Renewables

Zichun Wang and Jun Huang

Abstract Acid-catalyzed conversion of renewables, e.g. biomass, is a major route for the sustainable production of fuels and chemicals through a "closed carbon balance" benign to the environment. Development of solid acids offer cost-effective and clean chemical processes are considerable preferred. The performance of solid acids depends on the type (Brønsted acid vs Lewis acid), density, strength, accessibility and local environment of the active sites. Most efforts focus on four key factors for efficient catalysis, for instance, preparing catalyst with higher density and strength with one type of acid sites that are constrained by further acidity enhancement. An alternative way is to combine Brønsted and Lewis acid sites in the local structure, introducing Brønsted-Lewis cooperativity that can have synergic effects. The Brønsted-Lewis synergy effects can enhance the acid strength of Brønsted acid sites, while Brønsted-Lewis cooperativity influence the molecular adsorption and activation, resulting in Brønsted-Lewis bifunctional acid-catalysis. Brønsted-Lewis bifunctional acid-catalysts that works cooperatively during reaction rises a new research direction. The chapter focuses on the application of Brønsted-Lewis bifunctional catalysts for efficient biomass conversion emerging sustainable biorefining processes.

3.1 Definition and Characterization of Acid Catalysts

Acid-catalyzed reactions are widely utilized in chemical processes for fine chemistry, oil refinery, petrochemical and pharmaceutical industries. Acid-catalyzed reactions are fundamental to a chemical process based on an acid. This section provides a brief introduction of acid catalysts and the characterization of their acidity as a key to understand their catalytic performance.

Z. Wang • J. Huang (✉)
Laboratory for Catalysis Engineering, University of Sydney, Sydney, NSW, Australia
e-mail: jun.huang@sydney.edu.au

Z. Fang et al. (eds.), *Production of Biofuels and Chemicals with Bifunctional Catalysts*, Biofuels and Biorefineries 8, https://doi.org/10.1007/978-981-10-5137-1_3

3.1.1 Definition of Acids

Protonic acids, such as H_2SO_4 and HCl, were defined by S. A. Arrhenius in 1903 as any hydrogen-containing species that is able to release protons and a base is any species able to release hydroxide ions in aqueous solution. This definition was modified by J. M. Brønsted [1] and T. M. Lowry [2] independently in 1923 to read as an acid is any species that can release proton(s) while a base is any species that is able to accept proton(s), to the extent that the use of acid-base catalysis is in aqueous solutions. This theory was used to explain reactions initiated via proton transfer, such as esterification and aldol reactions.

In the same year, G. N. Lewis proposed an alternative definition: as an acid is any species is able to accept an electron pair via its incomplete electronic grouping to form a dative or coordination bond, while a base is any species are capable to donate a nonbonding electron pair to form a dative or coordination bond [3]. Based on Lewis' definition, Brønsted-type acid and base are also Lewis-type, but not the converse. For example, alumina with the unoccupied orbital, is a Lewis-type but not Brønsted-type acid, since it can accept electron pair from NH_3 to form coordination bond. Here, NH_3 is a Lewis-type base, which is also a Brønsted-type basic molecule that can accept a proton. Therefore, the Lewis acid-base theory is more general and is able to explain various acid-base catalyzed reactions.

3.1.2 Classification of Acid Catalysts

3.1.2.1 Homogeneous Acids and Metal Complexes

Homogeneous acid catalysts, including protonic liquid acids and metal complexes, can be well mixed with reactants, and thus, provide high activity at relative low temperatures, as well as low investment cost and flexibility. Currently, H_2SO_4, HF and HCl/$AlCl_3$ are widely used in acid-catalyzed alkylation and acylation processes in the refinery industries [4]. The inherent disadvantages of liquid acids with respect of solid acids arise from (1) safety problems related to their toxic and corrosive nature, (2) difficult, expensive and energy-inefficient processes for catalyst separation and regeneration, as well as (3) issues connected with the environment and sustainability.

Most of the disadvantages associated to the homogeneous acid catalysts can be addressed by employing suitable solid acids [4–6], taking advantage of (1) less corrosion in the reactor and plant, (2) simple separation and regeneration, and (3) environmentally-friendly characteristics and (4) safety in handling. Therefore, solid acids have attracted great attention as alternative catalysts in various important acid-catalyzed processes for sustainable and green chemical and fuel production in the chemical, food, pharmaceutical and petrochemical industries.

3.1.2.2 Zeolite-Type Catalysts and Amorphous Silica-Alumina

Zeolite-type catalysts and amorphous silica-alumina (ASA) are the most widely used acid catalysts in hydrocarbon conversion industries [4, 5] or as support materials for multifunctional catalysts in many chemical processes, such as hydrocracking, isomerization, alkylation, as well as hydrogenation and oxidation reactions [7, 8]. In the last two decades, both non-siliceous zeolites [9] and the incorporation of metal component(s) (B, Fe, Ga, Ge, Mg, Nb, Sn, Ti, V, Zr and Zn) into the zeolite framework have been applied [10–16]. Over 218 framework types of zeolites have been discovered and characterized by XRD techniques. Crystalline aluminosilicate zeolites, featured with micropores (< 2 nm), high surface area and large pore volume, are the most widely used in heterogeneous catalysis, particularly in hydrocarbon reactions. In pure silica zeolites, the framework consists mainly of tetrahedrally coordinated silicon atoms shared by four neighboring oxygens (SiO_4). One of the most important modifications is the incorporation of Al into the zeolite framework to generate Brønsted acidity. When an Al^{3+} atom is substitute for a Si^{4+} atom in the framework (Fig. 3.1a), the compensation of a negative charge on framework oxygen by protons results in surface Brønsted acid sites (BAS) [17]. The density and strength of BAS can be flexibly tuned through adjusting Si/Al ratios in zeolites, dealumination and ion-exchange techniques. Moreover, both dealumination and ion-exchange techniques, as well as dehydroxylation and alkaline-assisted metalation [18, 19], can introduce Lewis acidic extra-framework metal cations, such as oxoaluminum Al cations (AlO^+, $Al(OH)_2^+$, $AlOH^{2+}$), AlOOH and $Al(OH)_3$ [20], and thus, are suitable techniques in the preparation of bifunctional Brønsted-Lewis acid catalyst.

Similarly as crystalline aluminosilicate zeolites, ASA is able to provide surface BAS, which can be generated via flexibly coordination or pseudo-bridging silanols between surface hydroxyl groups and neighboring aluminum centers (Fig. 3.1 c–d) [17, 21, 22]. ASA can be prepared by co-gelation [23], grafting [24], co-precipitation [23] and hydrolysis [25]. A key achievement in ASA synthesis was the development of mesoporous (2–50 nm) ASA materials with ordered structures, such as Al-doped siliceous MCM-41 [26, 27], providing Brønsted acidity. With these preparation techniques, the intrinsic homogeneity results in Al species tend to condense to form

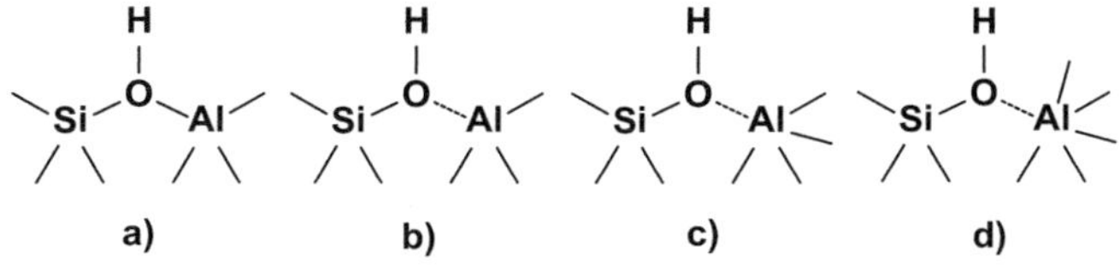

Fig. 3.1 Proposed models for BASs in silica-alumina catalysts. (**a**) BAS consisting of a bridging silanol site (Si(OH)Al) in zeolites [30]. (**b**) BAS consisting of the flexible coordination between silanol oxygen and neighboring tetra-coordinated Al (Al^{IV}) center [17] and BAS consisting of PBS interacting with (**c**) Al^{IV} site and (**d**) Al^{IV} site [21, 22]. In that later case, the dotted line does not denote a covalent bond but only the close proximity between O and Al atoms (Adapted with permission from Ref. [22], Copyright © Nature 2016)

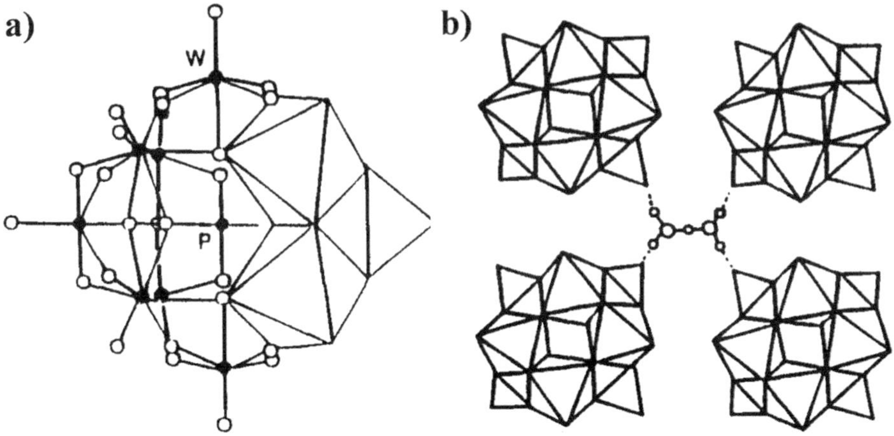

Fig. 3.2 Heteropoly anion with the Keggin structure, $PW_{12}O_{40}^{3-}$: (**a**) primary structure; (**b**) secondary structure ($H_3PW_{12}O_{40}{\cdot}6H_2O$) (Adapted with permission from Ref. [31], Copyright © 1934, Royal Society)

an alumina phase, particularly, at high Al/Si ratios. The LAS originate from alumina phase enables ASA as a bifunctional Brønsted-Lewis acid catalyst. With the improved diffusion properties of large molecules compared to that inside the microporous of zeolites [28, 29], these materials are promising for acid or bifunctional catalysis.

3.1.2.3 Heteropolyacids

Heteropolyacids, such as $H_3PW_{12}O_{40}$ of a common Keggin unit [31] as shown in Fig. 3.2, are generated by the condensation of more than two different types of oxoanions [5, 32]. The anion contains a central atom (e.g. P, Si, or Ge) shared by four neighboring oxygen atoms, connected to oxygen-linked octahedral secondary metal atoms, such as Mo, W, V, Nb and Ta. Heteropolyacids have mainly Brønsted acidity, which is generated by the protons compensating for the negative charge of anions, such as in $H_3PW_{12}O_{40}$.The strong acidity of heteropolyacids, which is even higher than that of the most acidic zeolite, H-ZSM-5, enable them to have application in various hydrocarbon conversion [33–35] and biomass conversion [36, 37] processes. However, the rapid deactivation of highly-active heteropolyacids limit their applications in industrial gas-phase reactions [5].

3.1.2.4 Mixed and Functionalized Oxides

Generally, pure oxides of multivalent metals, such as Al_2O_3, ZrO_2 and TiO_2, exhibit medium to strong Lewis acidity [4]. Their acidity is due to the ionic network formed between metal cations and oxygen. For example, the presence of Lewis acidic Zr^{4+} species in pure zirconia has been evidenced by CO adsorption experiments [38, 39].

Mixed oxides providing both Brønsted and Lewis acidity, (e.g. Al_2O_3-SiO_2, ZrO_2-SiO_2 and V_2O_5-Al_2O_3, can be prepared by co-gel [40], sol-gel [41], and coprecipitation [42] techniques. Lewis acidity originates from Lewis metal cations in the single-metal enriched phase, while Brønsted acidity are generated by protons to neutralize the electronegativity difference between different metal components [43]. The properties of mixed oxides can be flexibly controlled by the ratio between different components, pH value, calcination temperature and gel aging time [40, 41, 44], and thus, have been widely used in chemical reactions either directly or as support materials [42, 45–49].

Introducing quite strong Brønsted acidity to metal oxides (e.g. ZrO_2, TiO_2, SnO_2, Fe_2O_3, ZrO_2-TiO_2) can be achieved via acid treatment with H_2SO_4, $(NH_4)_2SO_4$, etc. [4]. On sulfated zirconia, the presence of BAS is characterized by IR with a band in the range of 3645–3630 cm^{-1} [50]. These BAS exhibit ^1H NMR chemical shift larger than those of zeolite protons [51], hinting a higher Brønsted acid strength. Several models were proposed to explain the formation of BAS on sulfated metal oxides [52–54], which suggested the acid strength of the SO_4^{2-}/MO_2 catalyst directly correlates to the electronegativity of the metal [55, 56].

3.1.3 *Characterization of Acidity*

The determination of acidity in acid catalysts is key for their catalytic application. Various methods have been utilized for this purpose, such as titration methods often used to determine the strength and density of liquid acid catalysts. For solid acids, their acidity depends on local structure related type, density, strength and accessibility of surface acid sites, which cannot always be obtained by titration methods. Up to now, temperature-programmed desorption (TPD), microcalorimetry, Fourier transform infrared spectroscopy (FTIR), and solid-state nuclear magnetic resonance (SSNMR) spectroscopy have been applied to characterize the acidity of solid acids.

3.1.3.1 Titration Methods

The strength of acids in water strongly depends on acid dissociation and can be determined by titration with bases (e.g. *n*-butylamine) as suitable color indicators [57]. In general, acids can convert bases into the conjugate acid with a color change, which indicates their acid strength. Hammett and Hirschler indicators are the most widely used and have been extensively reviewed [5, 58]. Some efforts are being made to improve the titration methods [59–61]. For example, arylcarbinols as indicators can detect mainly Brønsted acid sites, rather than both Brønsted acid sites and Lewis sites by conventional methods [59].

3.1.3.2 Temperature-Programmed Desorption

More detailed information of surface acidity (density and strength) of solid acids can be obtained by temperature-programmed desorption (TPD) of bases. In a typical TPD experiment, the pretreated sample with a gas (e.g. He, Ar or N_2) are loaded with adsorbates (e.g. NH_3 or pyridine). The amount of adsorbates is determined by gas chromatography or mass spectrometry, with a curve as a function of programmed desorption temperature [62, 63]. The desorption temperature can be used to evaluate the strength of acid sites, which depends on the adsorption energies between the adsorbed molecules and the surface sites [64]. For quantitative study, the area under the desorption curve gives the total number of acid sites, including both BAS and Lewis acid sites (LAS).

However, TPD cannot distinguish BAS and LAS, and the desorption curve are poorly resolved in most cases. The deconvolution of TPD curve to determine the intensity of each sites can be achieved by coupling with other spectroscopy techniques to determine the assignment of each signal when different type of acid sites occurred on the surface [65]. It should be noted that there are structure effects on the adsorbate desorption, such as the porous structure in zeolites, that affects the diffusion of desorbed molecules that can be readsorbed on their way out of the pores [66, 67], which is able to shift the desorption peak to higher temperatures.

3.1.3.3 FTIR Spectroscopy

FTIR spectroscopy is a frequently used technique in the characterization of surface acidity [4, 5, 30]. Taking advantages of dipole moment of OH groups, Brønsted acid sites can be studied directly through their excitation vibration states induced by electromagnetic radiation. Generally, different types of OH groups have different stretching vibration states (υ_{OH}), and therefore, give identical band in the spectrum. Such as for zeolite H-Y [68], external SiOH groups show a signal at ca. 3750 cm^{-1} while the signals at ca. 3650 and 3560 cm^{-1} are assigned to bridging OH groups in different local environments. The concentration of each OH groups can be determined from the intensity of its corresponding IR band. It should be noted that different types of OH groups contribute to different extinction coefficients, which must be determined prior to the quantitative study.

Alternatively, adsorption of base as probe molecule is a suitable way to characterize both the concentration and strength of acid sites as summarized in Table 3.1. Strong base molecules, such as pyridine and ammonia, are able to be protonated by Brønsted acidic OH groups, but coordinate on Lewis acid sites [69, 70]. For example, pyridines protonated on BAS exhibit a band at 1540 cm^{-1} while when coordinated to LAS results in a band at 1450 cm^{-1}, which can be used to distinguish between these surface sites [70]. The concentration of BAS and LAS can be determined by evaluating the intensities of corresponding bands. Loading weak bases (such as benzene, acetone, CO and ethane) on dehydrated samples can characterize the strength of acid sites [71]. The wavenumber shift ($\Delta\upsilon_{OH}$) of the stretching vibra-

Table 3.1 IR bands region of pyridine, CO and N_2, adsorbed on BAS and LAS on aluminosilicates

	BAS	LAS	References
Pyridine		1447–1460	[72–79]
	1485–1500	1488–1503	
	1540	1580	
	1640	1600–1633	
		2350	
CO	2173	2225	[80]
N_2	2334	2349–2357	[81]

tions of these probe molecules are proportional to the strength of BAS, e.g. a larger $\Delta\upsilon_{OH}$ indicates a higher strength.

3.1.3.4 Solid-State NMR Spectroscopy

SSNMR spectroscopy is a powerful technique that not only probes the type, concentration, strength and accessibility of surface sites, but also reveals the local structure of the surface sites [17, 82]. By applying magic-angle-spinning (MAS) techniques, it allows one to characterize surface OH groups by ^1H MAS NMR spectroscopy directly. Upon dehydration (e.g. at 723 K under vacuum), different hydroxyl groups can be identified by their ^1H chemical shifts (δ_{1H}). In zeolites, the signal of isolated SiOH groups is in the range of $\delta_{1H} = 1.2–2.2$ ppm, while the signal at $\delta_{1H} = 3.6–4.3$ has been attributed to protons from bridging OH groups (Si-OH-Al) acting as surface BAS. Taking advantage of the linear correlation between the signal intensity and spin concentration for ^1H nuclei, the concentration of hydroxyl groups in solid catalysts can be quantified by using an external standard.

Unlike zeolites, the ^1H MAS NMR spectra of the dehydrated ASA is often dominated by the non-acidic terminal silanol groups ($\delta_{1H} = 1.2–2.2$ ppm) and the direct characterization of BAS in ASA by ^1H MAS NMR spectroscopy is not possible [17]. Loading with strong base as probe molecules, e.g. pyridine, ammonia and trimethylphosphine, is a suitable method for probing these BAS. Upon ammonia adsorption, the formation of ammonium ions ($\delta_{1H} = 6.5–7$ ppm) on BAS in a quantitative manner can be used to probe the presence and density of BAS. Weak bases, such as acetone-2-^{13}C, acetonitrile-d_3, perchloroethylene and trimethylphosphine oxide, can be hydrogen-bonded to surface OH groups. The adsorbate-induced chemical shift ($\Delta\delta$) of functional groups in probe molecules (^{13}C, ^{15}N and ^{31}P) is a scale of OH groups with different acid strength.

The coordination of probe molecules on LAS, such as ^{15}N–pyridine, ^{13}CO, trimethylphosphine, trimethylphosphine oxide, ammonia and acetone-2-^{13}C, can be studied by SSNMR spectroscopy as well [17, 83]. As an example, the characterization of both BAS and LAS on zeolites and ASA by NMR spectroscopy with probe molecules has been summarized in Table 3.2. It enables one to obtain the BAS/LAS ratio from the same NMR experiment. With external standard applied, the densities of BAS and LAS can be determined.

Table 3.2 Probe molecules for characterizing both Brønsted and Lewis acid sites (BAS and LAS) on aluminosilicates by solid-state NMR spectroscopy

	Brønsted acid sites	Lewis acid sites	References
^{15}N–pyridine	pyridinium ions at $\delta_{15N} = 198$ ppm	$\delta_{15N} = 265$ ppm	[84–86]
^{13}C-2-acetone	bridging Si(OH)Al groups of zeolites at $\delta_{13C} = 216–225$ ppm, weak to strong BAS on ASA at $\delta_{13C} = 213–227$ ppm, strongly acidic OH groups of heteropoly acids at $\delta_{13C} = 235$ ppm	$\delta_{13C} = 233–245$ ppm for acetone at extra-framework aluminum species in zeolite and on pure or supported $AlCl_3$	[29, 87–96]
Trimethylphosphine	$(CH_3)_3PH^+$ at $\delta_{31P} = -3$ ppm	$\delta_{31P} = -32$ to -67 ppm	[86, 97–99]
Trimethylphosphine oxide	at strong BAS of zeolite H,Na-Y at $\delta_{31P} = 65$ ppm, zeolite H-Beta at $\delta_{31P} = 78$ ppm, zeolites H-MOR and H-ZSM-5 at 86 ppm	$\delta_{31P} = 37$ ppm at dealuminated zeolites Y and γ-Al_2O_3	[83, 100–104]
Ammonia	ammonium ions at $\delta_{1H} = 6.5–7$ ppm	ammonia adsorbed on aluminum species $\delta_{1H} = -0.5–4.6$ ppm	[29, 83, 87, 95–97, 105–107]

^1H and ^{13}C NMR shifts are referenced to tetramethylsilane ($\delta_{1H} = 0$ ppm, $\delta_{13C} = 0$ ppm), and ^{15}N and ^{31}P NMR shifts are referenced to liquid $^{15}NH_3$ at $\delta_{15N} = 0$ ppm and H_3PO_4 (85%) at $\delta_{31P} = 0$ ppm, respectively

3.2 Bifunctional BAS-LAS Cooperative Acid-Catalysis

Acid catalysts involved in various important biomass conversion chemical processes [108, 109]. Recent investigations focus on lignocellulosic biomass as renewable feedstocks for the production of biofuels and biochemical. In the presence of acid catalysts [110–123], lignocellulose can be converted into value-added platform chemicals, such as sugars. Acid catalysts are also significant in the biodiesel production directly from vegetable oils, animal fats and algae with high oil content. Acid-catalyzed transesterification of triglyceride (or other esters) with alcohols can produce alkyl esters (biodiesel) with glycerol as the major byproduct by using zeolites, heteropolyacids, sulfated and mixed metal oxides [124, 125]. The distribution of products varies and is closely associated with the type, density, strength and local environment of acid sites. For example, in acid-catalyzed hydrolysis/depolymerization of cellulose, $-SO_3H$ functionalized active carbon (sulfonation at 523 K) obtained a higher glucose yield (61%) than H_2SO_4 (0.01 M, 40.5%), while Lewis acidic γ-Al_2O_3 is inactive at the similar conditions [116, 119, 120]. Lewis acids is active in the isomerization of sugars, including glucose, to promote subsequent dehydration over both Brønsted and Lewis acids.

Extensive efforts are devoted to improve the catalytic performance through the proper control of the catalyst acidity (type, density, strength, and accessibility), for instance, preparing catalyst with higher density and strength for a single type of acid sites (e.g. Brønsted acidity in aluminosilicates), however, are often constrained in acidity enhancement. Alternatively, the integration of multiple catalytic steps together in one-pot procedure can remarkably reduce energy consumption, simplify the reaction system and improve the reaction efficiency. The cooperativity of acids (Brønsted and/or Lewis acid) with base or metal catalysts for bifunctional catalysis has been extensively studied and reviewed for hydrocarbon and biomass conversions [108, 126, 127].

Bifunctional BAS-LAS cooperative acid-catalysis has attracted much attention for both homogeneous and heterogeneous systems. Bifunctional catalysts have excellent catalytic performance in the transformation of biomass-derived platform molecules (e.g. glucose and glycerol) into high value-added chemicals compared with single-type acid-catalyzed chemical processes [128]. Unlike to combine both homogeneous Brønsted and Lewis catalysts together for bifunctional catalysis, various techniques has been devoted to the preparation of BAS-LAS bifunctional catalysts, such as impregnation, ion-exchange, acid treatment and dealumination processes. The tuning of surface BAS/LAS ratios can be easily achieved via either preparation methods and catalysts composition or post-treatment. In this section, we will briefly introduce the potential use of BAS-LAS bifunctional catalysts in the conversion of biomass into fuels and value-added chemicals.

3.2.1 5-Hydroxymethylfurfural Production from Glucose

As a versatile and important building block, 5-Hydroxymethylfurfural (HMF) is a potential candidate for the production of various value-added chemicals, such as liquid alkanes, biofuels, furan derivatives [129, 130]. Hexose (mainly glucose and fructose) dehydration to produce HMF is a promising route since it is abundant and easily derived from biomass depolymerization [116]. Compared with fructose, glucose is favored due to its wide availability and low cost. Catalytic conversion of glucose to HMF involves two reaction steps: (1) LAS for glucose isomerization to fructose; and (2) fructose dehydration over BAS [131]. Lewis acidic metal halides combined with protonic liquid acids have been applied for this purpose. Recent efforts focus on the development of suitable BAS-LAS bifunctional catalysts for efficient glucose dehydration to HMF. Various bifunctional catalysts, including zeolites, heteropolyacid, pure and supported Nb_2O_5, metal phosphates, sulphated zirconia and metal-organic framework (MOF), combined with ILs and/or water-organic biphasic systems have been investigated [131–135].

3.2.1.1 Combined BAS-LAS Bifunctional Catalysts

Metal halides combined with protonic liquid acids, such as $HCl/AlCl_3$, have been widely utilized on a large scale [4]. $HCl/AlCl_3$ can provide strong Brønsted and Lewis acidity. When combined with water/2-*sec*-butylphenol biphasic solvent, a high yield of HMF (62%) from glucose dehydration at 443 K after 40 min was reported, which is higher than that obtained for other metal halides (e.g. $SnCl_4$, VCl_3, $GaCl_3$, $InCl_3$, $YbCl_3$, $DyCl_3$ and $LaCl_3$) at the same conditions [136]. A BAS-LAS cooperativity reaction pathway has been suggested [136, 137], in which $[Al(OH)_2(aq)]^+$ species response for Lewis-acid-catalyzed glucose isomerization and HCl act as BAS for fructose dehydration reaction. The using of organic phase can simultaneously extract HMF (97%) and promote the reaction in the aqueous phase. The substitution of 2-*sec*-butylphenol with cyclopentyl methyl ether significant increases the HMF yield (54.4%) for much shorter reaction time under similar conditions [138], which has been explained by cyclopentyl methyl ether as a dipolar aprotic solvent with alkaline characteristics. Moreover, the addition of NaCl can improve the extraction of HMF into the organic phase through the salting-out effect. By optimizing these conditions, high HMF yield (30.5%) was achieved from plant biomass (corncob acid hydrolysis residues) in one-pot synthesis [138].

The use of $CrCl_3$ with inorganic acids for glucose dehydration to HMF has been reported [139, 140]. Lewis acidic $CrCl_3$ is active for glucose isomerization to fructose [139]. Kinetic results reveal the hydrolyzed Cr(III) complex $[Cr(H_2O)_5OH]^{2+}$ is the most active Cr species [139]. The strong interaction between the Cr center and glucose molecule for isomerization was probed by EXAFS and confirmed by molecular modelling studies [139]. Even without the addition of Brønsted acids, protons can be released from the hydrolysis of Cr(III) ($Cr(H_2O)_6{}^{3+}$) to form $(Cr(H_2O)_5OH)^{2+}$ [141]. This intrinsic Brønsted acidity is capable to drive the fructose dehydration reaction to yield HMF [139, 140, 142]. Combined with HCl as co-catalyst, 59% HMF yield is obtained in a biphasic water/THF solvent under 413 K after 3 h [139]. Further investigations show increasing reaction temperature and $CrCl_3$ concentration in $CrCl_3$/HCl system has minor effects on fructose and HMF yields [140]. However, adjusting the $CrCl_3$/HCl ratio has significant influence on HMF yield at constant glucose conversion [140]. The maximum HMF yield is obtained at moderate HCl concentration due to the balance of BAS and LAS for the reaction [139].

In contrast to $CrCl_3$/HCl, mainly levulinic acid (LA) (52%) is produced at a complete glucose conversion in glucose dehydration under 443 K for 4 h over $CrCl_3$/H_3PO_4 as BAS-LAS co-catalysts [143]. LA is a versatile building block for the synthesis of various organic commodity chemicals and transportation fuels. It can be generated by HMF rehydration [144]. The catalytic results suggest a coupling of $CrCl_3$ and H_3PO_4 as mixed catalyst exhibiting a positive effect on glucose conversion to LA compared with using $CrCl_3$ or H_3PO_4 solely [143, 145]. Possible bifunctional BAS-LAS cooperative acid-catalysis mechanisms have been proposed [145]. The activation energies for both glucose dehydration into HMF and HMF rehydration to LA were 65.4 and 60.6 kJ/mol, respectively, which are significantly lower than those in previous reports with single homogeneous Brønsted acid or Lewis acid catalysts [143]. Therefore, one-pot preparation of LA from biomass feed stock may be promising with BAS-LAS bifunctional catalysts just as it is for glucose conversion.

Acid-catalyzed sugar conversion have been investigated for both $CrCl_3$ and $AlCl_3$ in ILs [146–148]. For example, when dissolved in 1-alkyl-3-methylimidazolium chloride, [AMIM]Cl, a high HMF yield of 68–70% is achieved in the conversion of glucose with [EMIM]$CrCl_3$ (E = ethyl, $CrCl_2$ combined with [EMIM]CrCl) compared with only $CrCl_2$ (28%) [148]. The catalytic activity can be introduced by the BAS-LAS cooperativity from hydrolysis of Cr(III) [139, 140, 142]. Combining ILs with biphasic water/organic solvent for glucose conversion is interesting. In a 1-butyl-3-methylimidazolium/glycol dimethyl ether/water ternary biphasic system, $CrCl_3$ gives a HMF yield of 64.5% from a very high glucose concentration (80 wt.% with respect to ILs) at 381 K in 1 h reaction time [147]. The HMF yield is almost double of that obtained without water/organic solvent, which is attributed to multi-functional glycol dimethyl ether.

Up to date, a HMF yield of over 90% can be achieved from glucose by homogeneous catalysts [149]. However, the upgrading and purification of HMF and the environment friendly chemical process drives recent research focus on the development of solid acid catalysts providing BAS-LAS bifunctionality for efficient HMF production from sugars. Bifunctional BAS-LAS acids applied in the conversion of glucose into 5-HMF are summarized in Table 3.3.

3.2.1.2 Zeolite-Based BAS-LAS Bifunctional Catalysts

Zeolites are some of the most important solid catalysts in current refining and petrochemical industry, due to their relatively low cost, wide availability, and widespread baseline knowledge [4, 30]. The surface Brønsted and Lewis acidity on zeolites can be easily tuned through various post-treatment techniques, such as dealumination and ion-exchange with introducing metal cations and alkaline-assisted metalation [17–19, 94, 96, 190]. Therefore, they are of great interest in the conversion of glucose dehydration into HMF.

Tin-containing Beta (Sn-Beta) zeolite combined with Brønsted acids as co-catalysts has been tested for the production of HMF from biomass-derived feedstocks [164, 191, 192]. Sn-Beta zeolite, containing exclusive Lewis acidity, is highly active for the isomerization of glucose to fructose in water, but inactive towards HMF as the desired dehydration product [191, 192]. Mechanistic studies by ^{1}H and ^{13}C NMR spectroscopies on isotopically labeled glucose reveal that the isomerization of glucose to fructose over Sn-Beta zeolite via an intramolecular hydride shift rather than proton transfer [191, 192] as shown in Fig. 3.3. When combined with solid Brønsted acid resin, Amberlyst-70 (Amb-70) [164], the HMF yields obtained are 59, 55, 60 and 63% in the corresponding solvents, γ-valerolactone, γ-hexalactone, tetrahydrofuran:methyltetrahydrofuran (1:1 by weight) and tetrahydrofuran, respectively. Sn-Beta is stable in highly acidic aqueous environments, showing a potential combination with Brønsted acids for sequential hydrolysis/isomerization or isomerization/dehydration reactions, including the production of HMF from glucose or directly from plant biomass.

Sn-Mont catalysts, prepared from calcium montmorillonite (Ca-Mont) exchanged with Sn(VI) precursor, have been tested for glucose dehydration in a mono-phase solvent of THF/dimethyl sulfoxide (DMSO) [167]. The presence of both BAS and LAS are probed by NH_3-TPD and FTIR upon pyridine adsorption. The high HMF yield (53.5%) from glucose dehydration at 433 K for 3 h over Sn-Mont has been

Table 3.3 Bifunctional Brønsted–Lewis acid catalysts used in the conversion of glucose into 5-hydroxymethylfurfural with summary of reaction conditions and maximum catalytic activity

Entry	Substrate (wt%)	Catalyst[a]	Reaction conditions			Catalytic activity		References
			Solvent	Temp (°C)	Time	Conv. (%)	Yield (%)	
1	Glucose (10)	15 mol% $GeCl_4$	[BMIM]Cl	100	1 h	98	38	[150]
2	Glucose (10)	10 mol% $CrCl_3{\cdot}6H_2O$	Et-DBUBS	110	2 h	100	83	[151]
3	Glucose (3.3)	[CO_2H-PMIM]Cl	DMSO	120	7 h	–	12	[152]
4	Glucose (3.3)	0.2 M $ZrOCl_2$ + [CO_2H-PMIM] Cl	DMSO	120	7 h	–	51	[152]
5	Glucose (4)	10 mol% $CrCl_2$	DMF	120	3 h	–	50	[153]
6	Glucose (4)	10 mol% P[BVIM]Cl-$CrCl_2$	DMF	120	3 h	–	66	[153]
7	Glucose (4)	10 mol% P[BVIM]Cl-$AlCl_3$	DMF	120	3 h	–	49	[153]
8	Glucose (5)	30 wt% SO3H-PIL-PW	DMSO	150	2 h	–	18	[154]
9	Glucose (5)	30 wt% SO3H-PIL-PW/$CrCl_3$	DMSO	150	2 h	–	34	[154]
10	Glucose (10)	10 mg $CrCl_2$-Im-SBA-15	H_2O:DMSO/2-BuOH:MIBK	150	3 h	50	35	[155]
11	Fructose (30)	3.3 wt% Taa-SBA-15	H_2O/2-BuOH:MIBK	180	0.5 h	66	49	[156]
12	Glucose (6.7)	100 wt% H^+-D001-cc resin	[BMIM]Cl	110	0.5 h	–	10	[157]
13	Glucose (6.7)	100 wt% Cr^{3+}-D001-cc resin	[BMIM]Cl	110	0.5 h	–	61	[157]
14	Fructose (3)	26.7 wt% [HSO_3 + (ILs/ $CrCl_2$)]-MSN	DMSO	90	3 h	98	73	[158]
15	Glucose (5)	60 wt% Cr-HAP	[BMIM]Cl	Microwave 400 W	150 s	78	40	[159]
16	Glucose (10)	33.3 wt% Al-MCM-550	20 wt% NaCl-H_2O/MIBK (3:7, v/v)	195	0.5 h	98	63	[160]
17	Glucose (2.5)	83 wt% H-Beta (Si/Al: 15)-750	H_2O-DMSO (9:1)/THF (1:3)	180	3 h	78	43	[161]
18	Glucose (10)	40 wt% H-Beta (Si/Al: 25)	[BMIM]Cl	180	50 min	81	50	[162]
19	Glucose (10)	0.5 mol% Sn-Beta	1-Butanol + H_2O + 35 wt% NaCl	160	1.5 h	75	14	[163]

20	Glucose (10)	HCl ($pH = 1$) 0.5 mol% Sn-Beta	1-Butanol + H_2O + 35 wt% NaCl	160	1.5 h	75	41	[163]
21	Glucose (10)	HCl ($pH = 1$) 0.5 mol% Sn-Beta	H_2O	160	1.5 h	45	2.7	[163]
22	Glucose (10)	HCl ($pH = 1$) 0.5 mol% Sn-Beta	1-Butanol + H_2O	160	1.5 h	77	20	[163]
23	Glucose (10)	HCl ($pH = 1$) 0.5 mol% Sn-Beta	THF + H_2O + 35 wt% NaCl	160	1.5 h	79	57	[163]
24	Glucose (2)	3.3 wt% Amberlyst-70 3.3 wt% Sn-Beta	GVL/H_2O (9:1, w/w)	130	0.33 h	92	59	[164]
25	Glucose (2)	3.3 wt% Amberlyst-70 3.3 wt% Sn-SBA-15	GVL/H_2O (9:1, w/w)	130	0.25 h	90	46	[164]
26	Glucose (2)	3.3 wt% Amberlyst-70 3.3 wt% Sn-Beta	THF/H_2O (9:1, w/w)	130	0.5 h	90	63	[164]
27	Glucose (2)	3.3 wt% Amberlyst-70 3.3 wt% Sn-SBA-15	THF/H_2O (9:1, w/w)	130	0.33 h	90	36	[164]
28	Glucose (2)	3.3 wt% Amberlyst-70 3.3 wt% Sn-Beta	MTHF:THF/H_2O (9:1, w/w)	130	0.67 h	90	59	[164]
29	Glucose (3)	33 wt% H-ZSM-5	H_2O/MIBK (3/7, v/v)	195	150 min	57	1	[165]
30	Glucose (3)	33 wt% Fe-ZSM-5	H_2O/MIBK (3/7, v/v)	195	150 min	95	33	[165]
31	Glucose (3)	33 wt% Cu-ZSM-5	H_2O/MIBK (3/7, v/v)	195	150 min	100	30	[165]
32	Glucose (3)	33 wt% H-ZSM-5	H_2O/MIBK (3/7, v/v) + 35 wt% NaCl	195	150 min	80	43	[165]
33	Glucose (3)	33 wt% Fe-ZSM-5	H_2O/MIBK (3/7, v/v) + 35 wt% NaCl	195	150 min	45	25	[165]
34	Glucose (3)	33 wt% Cu-ZSM-5	H_2O/MIBK (3/7, v/v) + 35 wt% NaCl	195	150 min	80	50	[165]
35	Glucose (37.5)	10 wt% Mordenite (Si/Al = 11.2) treated with NH_4Cl (1 M)	[BMIM]Br	100	6 h	97	64	[166]
36	Glucose (6)	10 wt% Mordenite (Si/Al = 11.2) treated with NH_4Cl (1 M) + NH_4F (2.4 M)	Acetone/water/ethyl acetate (3:2:7.5)	65	9 h	98	50	[166]

(continued)

Table 3.3 (continued)

Entry	Substrate (wt%)	Catalyst[a]	Reaction conditions			Catalytic activity		References
			Solvent	Temp (°C)	Time	Conv. (%)	Yield (%)	
37	Glucose (10)	100% mesoporous AlSiO-20 (Si/Al = 18)	THF/H_2O (3:1, v/v)-NaCl	160	1.5 h	–	63.1	[132]
38	Glucose (5)	Sn-Mont	THF-DMSO (7:3, v/v)	160	3 h	98	54	[167]
39	Cellulose (5)	Sn-Mont	THF/H_2O-NaCl (5:1, v/v)	160	3 h	–	39	[167]
40	Glucose (5)	6 wt % mesoporous sulfated zirconia	DMSO	120	1 h	–	15	[168]
41	Fructose (5)	6 wt % mesoporous sulfated zirconia	DMSO	120	1 h	–	91.9	[168]
42	Fructose (5)	50 wt% PO_4/TiO_2	DMA-LiCl	140	5 min	–	44	[169]
43	Glucose (5)	20 wt% H_3PO_4/TiO_2	DMA-LiCl	140	5 min	–	22	[169]
44	Sugarcane bagasse (5)	20 wt% PO_4/TiO_2	DMA-LiCl	140	5 min	–	26	[169]
45	Glucose (5)	2 wt% Zr-Mo-SA	$[BMIM]Cl/H_2O$	150	3 h	–	27	[170]
46	Glucose (5)	50 wt% TiO_2 nanospheres	DMA-LiCl	130	2 min	–	30	[171]
47	Glucose (5)	50 wt% TiO_2 nanospheres	DMSO	140	5 min	–	36	[172]
48	Glucose (2)	30 wt% PO_4/TiO_2	H_2O/n-BuOH	175	3 h	97	81	[173]
49	Fructose (6)	100 wt% PO_4/Nb_2O_5	H_2O	130	0.5 h	58	45	[174]
50	Glucose (1)	100 wt% PO_4/Nb_2O_5	H_2O	120	3 h	92	48	[175]
51	Glucose (10)	33.3 wt% PO_4/Ta_2O_5	H_2O/MIBK (3:7, v/v)	175	1.5 h	69	23	[176]
52	Glucose (6)	8.3 wt% PO_4/Ta_2O_5 (TA-p)	H_2O/2-BuOH (2:3, v/v)	160	3 h	99	90	[177]
53	Glucose (10)	33.3 wt% PO_4/Ta_2O_5	H_2O/MIBK (3:7, v/v)	170	1 h	56	33	[178]
54	Glucose (16)	10 wt% Nb_2O_5-γ-Al_2O_3	DMSO	150	4 h	–	54	[179]
55	Glucose (1)	100 wt% Nb_2O_5	H_2O	180	3 h	–	13	[180]
56	Glucose (1)	100 wt% HNb_3O_8	H_2O	180	3 h	–	18	[180]
57	Glucose (1)	100 wt% hy-Nb-TEOA	H_2O	180	3 h	–	28	[180]

58	Glucose (1)	100 wt% Meso Nb_2O_5	H_2O	180	3 h	–	36	[180]
59	Glucose (1)	1000 wt% $Nb_{0.2}$–WO_3	H_2O	120	3 h	86	34	[181]
60	Glucose (18)	55.6 wt% Sn-W oxide	THF/H_2O (5/1, v/v)	120	18 h		48	[182]
61	Glucose (10)	500 wt% Sn-VPO	DMSO	110	6 h	100	74	[183]
62	Glucose (10)	10 wt% α-$Sr(PO_3)_2$	H_2O	220	5 min	60	21	[184]
63	Glucose (10)	10 wt% CaP_2O_6	H_2O	220	5 min	70	20	[184]
64	Glucose (10)	33.3 wt% SO_4/Zr-MCM-550	H_2O/MIBK (3:7, v/v)	175	2.5 h	82	23	[185]
65	Glucose (30)	3.3 wt% $Ag_3PW_{12}O_{40}$	H_2O/MIBK (1:2.25, v/v)	130	4 h	89	76	[186]
66	Fructose (30)	21.3 wt% $Cs_{2.5}H_{0.5}PW$	H_2O/MIBK (1:3, v/v)	115	1 h	95	74	[187]
67	Glucose (15)	5.9 mol% $C_{16}H_3PW_{11}CrO_{39}$	H_2O	130	2 h	84	35	[187]
68	Glucose (3)	60 wt% $Cs_2[Cr_3O(OOCC_2H_5)_6(H_2O)_3]_2[\alpha$-$SiW_{12}O_{40}]$	H_2O or DMSO	140	4 h	100	48	[188]
69	Glucose (15)	$(C_{16}H_{33}N(CH_3)_3)H_4PW_{11}TiO_{40}$	H_2O	130	2 h	88	51	[189]
70	Glucose (15)	$(C_{16}H_{33}N(CH_3)_3)H_4PW_{11}TiO_{40}$	H_2O + n-Butanol (1:3, v/v)	130	2 h	92	77	[189]

[a]Catalyst dosage relative to the substrate

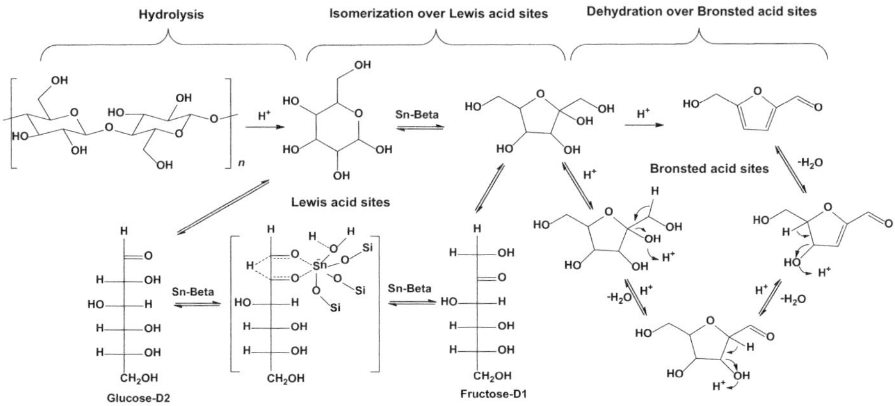

Fig. 3.3 Proposed scheme of glucose dehydration over Sn-Beta catalysts combined with Brønsted and Lewis acidity. (Modified from Ref. [141], Copyright 2014, American Chemical Society)

attributed to the BAS-LAS bifunctional catalysis. In direct conversion of glucose-based carbohydrates, including sucrose, cellobiose, inulin, starch and cellulose under the same conditions, Sn-Mont catalysts exhibit high activity in the conversion of these substrates to HMF with a yield of 39.1% from cellulose in a THF/H_2O-NaCl bi-phasic system.

Zeolites modified through ion-exchange technique have been widely investigated in the catalytic conversion of glucose into HMF [165, 166]. Introducing Lewis acidity on H-ZSM-5 zeolites through ion-exchange with Fe(II) and Cu(II) precursor has been achieved [165]. The formation of LAS is attributed to metal cations and EFAl species, while BAS is due to the parent H-ZSM-5 zeolite. The LAS/BAS ratios are 0.25, 0.7 and 2.5 for H-ZSM-5, Fe/H-ZSM-5, and Cu/H-ZSM-5, respectively, determined by NH_3-TPD coupled with pyridine adsorption FTIR. Increasing LAS/BAS ratio can significantly improve HMF yield from 8.5% (H-ZSM-5) to 26–30% for ion-exchanged H-ZSM-5 at 468 K for 150 min in water/ methyl isobutyl ketone biphasic system. By adding NaCl (20 wt%) into the biphasic water/MIBK system, it can remarkably reduce the reaction time with a HMF yield of 42% at 80% glucose conversion for 30 min reaction time. Mordenites featuring both Brønsted and Lewis acidity can be prepared by ion-exchange with NH_4Cl, NH_4Ac, NH_4F independently or combined [166]. With properly tuned the BAS/LAS ratios, the maximum HMF yield reaches 50% in water/ethyl acetate biphasic systems. These observations demonstrate that the density and strength of surface acid sites depend also on the reaction medium for obtaining the maximum HMF yield.

Compared with ion-exchanged zeolites, introducing LAS on H-form zeolites by dealumination is more convenient and economic. Tatsumi and co-workers prepared dealuminated H-Beta zeolites for glucose dehydration [161]. The H-form Beta zeolites are either calcined under temperature from 773 to 1023 K or treated with steam at 773 and 873 K. The formation of EFAl species as LAS is detected and evaluated by ^{27}Al MAS NMR and pyridine adsorption FTIR spectroscopies. The LAS/BAS ratio increases with increasing calcination or steam treatment temperature. These

BAS-LAS bifunctional catalysts are effective for glucose dehydration, where the decrease of BAS/LAS ratio (from 1.9 to 1.2) improves the HMF yield from 20.5 to 42.9% at 453 K for 3 h. Furthermore, these dealuminated Beta zeolites can promote the direct conversion of other hexoses and even cellulose into HMF [131], where a HMF yield of 42% from cellulose is obtained.

NaX zeolite fused with multiwalled carbon nanotubes (MWCNT-NaX) are able to catalyze glucose conversion to HMF [193]. The hybrid with MWCNT can significantly improve the stability of NaX, which is deactivated rapidly under harsh reaction conditions. The presence of LAS on NaX and BAS from sulfated MWCNT (MWCNT-SO_3H) have been identified. Both NaX and MWCNT-SO_3H are unable to convert glucose to HMF with high selectivity. However, by combination of both in hybrid MWCNT-NaX catalyst, fructose from the isomerization of glucose on LAS can be dehydrated rapidly over BAS. In this way, the BAS-LAS cooperativity results in a HMF yield of ca. 32% in a water (35 wt% NaCl)/THF biphasic system at 423 K with 600 rpm.

Mesoporous silica-alumina that has higher diffusion of reactant/product molecules than microporous zeolites have been studied for the production of HMF from glucose [132, 160]. The acidic properties are characterized by NH_3-TPD and pyridine adsorption FTIR techniques. Aluminum doped MCM-41 silica ([Al]MCM-41) with Si/Al of 5 and 10 were prepared [160]. With BAS-LAS bifunctional catalysis, a HMF yield of 36% at 87% glucose conversion are obtained in water/MIBK system at 468 K for 150 min reaction time. By using NaCl aqueous solutions (20 wt%) and MIBK, the partition coefficient between the organic and the aqueous phases increases up to 1.9, which improves the HMF yield (63%) at 98% glucose conversion for only 30 min. This indicates that [Al]MCM-41 catalysts are highly active, even compared with Cu-doped H-ZSM-5 zeolites that give a HMF yield of 42% at 80% glucose conversion at the same conditions after 30 min [165]. Mesoporous silica-alumina (AlSiO) has been tested in a THF/H_2O-NaCl biphasic system [132]. AlSiO-20 (Si/Al = 18) with medium BAS/LSA ratio (0.26) produced the highest HMF yield (63.1%) at 433 K for 1.5 h, which can be reused for 30 times. This confirms a suitable BAS/LAS ratio is one of the keys for glucose conversion to obtain HMF with high yields.

Zeolite-catalyzed conversion of glucose in ILs, e.g. 1-butyl-3-methylimidazolium chloride, has been investigated [162, 166]. Various zeolites (HY, H-Beta, H-ZSM-5) with BAS/LAS ratio in the range of 0.46–2.34 are studied [162]. Over H-ZSM-5, increasing BAS/LAS ratio from 0.98 to 1.25 can promote the formation of HMF (14.2–20.5%), but decreases to 8.3% with increasing the BAS/LAS ratio to 2.34. An HMF yield (50.3%) is obtained with H-Beta (Si/Al = 25, BAS/LAS = 1.03) at 80.6% glucose conversion at 423 K for 50 min reaction time. Often, the presence of strong LAS can promote HMF formation by efficiently initiating glucose isomerization [133]. However, strong adsorption of ILs, e.g. 1-butyl-3-methylimidazolium bromide, on adsorption sites are observed to limit the diffusion of reactant/product molecules over mordenites [166]. By optimizing the zeolite mesoporosity and BAS/LAS ratio, a HMF yield of 64% can be achieved at 97% glucose conversion over mordenites.

A mechanistic investigation has been carried out by combining experimental and computational studies [194, 195]. The model examined the multicomponent adsorption, homogeneous Brønsted acid catalyzed chemistry of fructose, HMF dehydration and humin formation. Using glucose as feedstock, BAS prefers to protonate the hydroxyl group at the meta-position of the anomeric carbon atom in the dehydration of fructose to HMF [194]. Further rehydration to levulinic acid (LA) is more difficult and results in the formation of humins as byproducts. On the other hand, the protonation of the hydroxyl group at the ortho-position is favored to result in humins and reversion products, rather than HMF or LA. The differences in the reactivity of glucose and fructose over BAS are ascribed by the regioselectivity in the initial protonation step [194]. Therefore, LAS, e.g. EFAl sites, are crucial for glucose isomerization to fructose as the initial step. Two reaction pathways are revealed based on surface BAS and LAS acid sites on BEA-25 zeolite [195]. With a certain amount of acid sites, the HMF production rate displays a maximum with increasing LAS/BAS ratio. On the other hand, with a fixed number of LAS, increasing the concentration of BAS leads to a linear increase at the beginning and then reaches a plateau. This kinetic model predicts the maximum HMF yield is 60% at 403 K with LAS/BAS ratio of 0.3, while it can be maximized when a higher LAS/BAS ratio is applied in a single aqueous phase [195].

3.2.1.3 Mixed and Functionalized Metal Oxides

Immobilization of inorganic acids on solid catalysts have been studied for converting glucose to HMF [168, 196]. Sulfated zirconia nanoparticles having both BAS and LAS were tested for hexose conversion [168]. The yield of HMF from fructose reached 72.8 and 91.9% under 383 K after 0.5 and 2 h in DMSO, respectively. However, using glucose instead of fructose dramatically decreased the HMF yield to 15.4% under 393 K after 1 h. The HMF yield from glucose may be improved through further optimization of BAS/LAS ratio and density.

Various Nb-based catalysts have been investigated for glucose dehydration to HMF [179–181, 197]. The catalytic conversion of sugars, e.g. glucose and sucrose, was studied over niobium oxides ranging from bulk to nanoparticles and multi- to monolayer [180, 197]. Mechanistic investigation of niobium oxides reveals both LAS (more efficient) and strong BAS can catalyze glucose isomerization to fructose, while weak BAS is more selective for fructose dehydration to HMF [180]. High temperature calcination induces nanoparticle sintering and significant loss of surface acid sites, and consequently, is nearly inactive for Nb_2O_5 calcined above 973 K [197]. When Nb_2O_5 nanoparticles are dispersed on SBA-15, only pure Brønsted acidity can be detected. The increase of Nb_2O_5 content (2–10%) on SBA-15 is proportional to an increase in fructose conversion (6–18%), however, the glucose is unable to be isomerized to fructose [197].

Niobium oxide mixed with other metal oxides, e.g. niobium-tungsten oxides, provides both Brønsted and Lewis acidity [181]. The LAS/BAS ratio increases with reducing Nb content. By optimizing the LAS/BAS ratio, complete glucose conversion with HMF yield of 38% is achieved in a THF/water mixture at 393 K after 3 h. Niobium-tungsten oxides is more active than both pure Nb_2O_5 and WO_3 under the same conditions. Moreover, the Nb_2O_5 modified γ-Al_2O_3 nanofibers is the most active among Nb-based catalysts [179]. HMF yield from glucose is in the range of 55.9–59% with a loading of 0.5–1 wt.% Nb_2O_5 at 423 K for 4 h in DMSO. This has been attributed to the modification of γ-Al_2O_3 nanofibers with Nb_2O_5 which provides $Nb^{\delta+}$ as strong LAS and intensive BAS. Increasing Nb_2O_5 loading up to 40 wt.% strongly influences the distribution and density of BAS and LAS on the surface due to the formation of two/three dimensional polymerized niobia species, and thus, the HMF yield decreases greatly but significantly increases when the substrate is fructose.

Niobium phosphate (NbPO) has been applied for producing HMF via sugar dehydration [134, 135]. The performance of various metal phosphates in glucose transformation is in the order of NbPO > ZrPO > TiPO > AlPO [134]. The highest HMF selectivity (56%) is obtained with NbPO having suitable BAS/LAS ratio ($\approx$ 1), while an excess of LAS results in undesired humins causing catalyst deactivation. The activity are suggested to originate from Lewis acidic $Nb^{\delta+}$ which can convert glucose to fructose for further dehydration on BAS from phosphate support [134]. NbPO modified by HCl (0.1, 1 and 10 M) was assessed for HMF production from cellobiose [135]. The BAS-LAS cooperativity can combine the hydrolysis of cellobiose to hexose with glucose isomerization and fructose dehydration together, which is useful for yielding HMF from polysaccharides in one-pot reaction systems.

Phosphate TiO_2 catalysts have been introduced for glucose dehydration to HMF [173, 198]. The incorporation of phosphorus into the TiO_2 framework can promote the formation of Ti-O-P bond, evidenced by XPS analysis [173]. Both Brønsted and Lewis acidity on surface are remarkably enhanced, and consequently, catalytic performance is improved. In a water-butanol biphasic system at 448 K after 3 h reaction, excellent glucose conversion (97%) and HMF yield (81%) are obtained for a glucose concentration of 2 wt.%, but decreased to 45% for a glucose concentration of 10 wt.%. Phosphate TiO_2 catalysts have been tested in the conversion of various mono- and di saccharides, including glucose, fructose, galactose, cellobiose and maltose, in water/THF biphasic solvent under 413 K for 3 h [198]. The presence of both BAS and LAS enables yields of HMF from these sugars (e.g. 37% from glucose), however, they are significantly lower than those from fructose (55% HMF yield), which has been attributed to the limitation of hydrolysis and isomerization of these sugars.

A Sn grafted-vanadium phosphate (Sn-VPO) has been prepared by a wetness impregnation technique [183] Various metal cations ($Sn^{\delta+}$ and $V^{\delta+}$) are detected by XPS analysis. The co-existence of BAS and LAS in high density on Sn-VPO (e.g. 7.35 mmol/g with 20 wt.% Sn loading) have been probed by NH_3-TPD and pyridine adsorption FTIR. The conversion of glucose to HMF was carried out under 383 K for 6 h in DMSO. HMF yield significantly increased from 38.1 to 73.8% with increasing Sn loading from 0 to 20 wt.%, which demonstrates by the increase of Lewis acidic Sn^{2+} cations.

3.2.1.4 Other Bifunctional BAS-LAS Acids

Heterogeneous metal halides, e.g. AlF_3 and MgF_2 have been used for glucose dehydration to obtain HMF [199, 200]. The co-existence of LAS and BAS in partially hydroxylated mesoporous AlF_3 is evidenced by FTIR with pyridine adsorption and NH_3-TPD [199]. The cooperativity of BAS and LAS are observed. Under optimum conditions, a HMF yield of 57.3% at complete glucose conversion is achieved under 413 K for 10 h reaction time. Over partially hydroxylated MgF_2, unsaturated Mg are able to isomerize glucose to fructose, followed by dehydration to HMF over BAS [200]. By further N_2-stripping of MgF_2, the HMF yield from glucose water solution can be increased from 21 to 33% under 453 K for 8 h.

Heteropoly acids (HPAs) prepared by introducing Lewis acid sites on their surface have been investigated for glucose conversion into HMF [188, 189]. $Cs_2[Cr_3O(OOCC_2H_5)_6(H_2O)_3]_2[\alpha\text{-}SiW_{12}O_{40}]$, which is a chromium-based HPA ionic crystal, can convert glucose into HMF with a yield of 48% under 413 K after 4 h in a mixture of water and 56% in DMSO. It has been observed that DMSO can stabilize HMF, while water leads to the formation of byproducts (levulinic and formic acids) [201]. The presence of both Brønsted and Lewis acidity on $Cs_2[Cr_3O(OOCC_2H_5)_6(H_2O)_3]_2[\alpha\text{-}SiW_{12}O_{40}]$ affords a higher activity compared with $H_4SiW_{12}O_{40}$ that mainly has Brønsted acidity.

Glucose dehydration over a sulfonic acid functionalized MOF MIL-101(Cr)-SO_3H has been studied in both batch and fixed-bed systems [196]. The Lewis acidic metal centers (e.g. Cr^{3+}) afford glucose isomerization to fructose, which is then dehydrated on Brønsted acidic $-SO_3H$ sites. A yield of HMF of 44.9% is obtained at a complete glucose conversion under 423 K for 4 h in a mixture of γ-valerolactone with 10 wt.% H_2O. Using γ-valerolactone can significantly reduce the activation energy from 114.0 kJ/mol in pure water to 100.9 kJ/mol. In a fixed-bed reaction system, a similar moderate HMF yield (35–45%) is achieved under similar conditions with a feed rate of 0.4 mL/min. In both batch and continuous flow systems, MIL-101(Cr)-SO_3H exhibit high stability and nearly no activity loss can be observed after 55 h and five recycle uses.

3.2.2 Synthesis of Furfural from Pentoses

Furfural and its derivatives are potential biofuel additives, green solvents, and key feedstocks [130]. Furfural can be synthesized from xylose via xylose isomerization at LAS to xylulose, and dehydration of xylulose to furfural is preferred at BAS over LAS [202–204]. At 145 °C in aqueous phase, the combination of Lewis ($CrCl_3$) and Brønsted (HCl) acids obtained a furfural yield (ca. 38%) much higher than that obtained with HCl only [205], as shown in Table 3.4 (entry 1–3). Metal halides, such as $AlCl_3$, $SnCl_2$ and MgF_2, exhibit a higher furfural yields of 71–80% in biphasic water/organic solution (Table 3.4 entry 4–6) [206, 207]. The hydrolysis of metal halides in water are able to release Lewis acidic metal species and protons [139,

Table 3.4 Bifunctional Brønsted–Lewis acid catalysts used in the conversion of xylose into furfural with summary of reaction conditions and maximum catalytic activity

Entry	Xylose substrate (wt%)	Catalyst[a]	Reaction conditions			Catalytic activity		References
			Solvent	Temp (°C)	Time	Conv. (%)	Yield (%)	
1	1	0.1 M HCl	H_2O	145	5 h	–	29	[205]
2	1	0.1 M HCl + 6 mM $CrCl_3$	H_2O	145	1.5 h	–	38	[205]
3	1	0.1 M HCl + 6 mM $CrCl_3$	Toluene/H_2O	140	2 h	96	76	[205]
4	10	50 wt% $AlCl_3$	H_2O/MIBK (1:5, v/v)	150	25 min	81	75	[206]
5	10	50 wt% $SnCl_2$	H_2O/MIBK (1:5, v/v)	150	25 min	82	80	[206]
6	20	20 wt% MgF_2-PF	Toluene/H_2O (1:1, v/v)	160	20 h	79	71	[207]
7	15	66.7 wt% acidic MCM-41	*n*-BuOH/H_2O (3:2, v/v)	170	3 h	97	44	[215]
8	10	100 wt% Nb/MCM-41	Toluene/H_2O (7:3, v/v)	160	6 h	99	39	[209]
9	10	100 wt% AM-11 niobium silicate	Toluene/H_2O (7:3, v/v)	160	6 h	90	50	[209]
10	10	33.3 wt% Nb_2O_5/MCM-41	Toluene/H_2O + NaCl (7:5, v/v)	170	3 h	98	60	[208, 209]
11	20 g/L	20 wt% SBA-12Nb	Toluene/H_2O (1:1, v/v)	160	24 h	84	93	[216]
12	3	66.7 wt% H-MCM-22 zeolite (Si/Al = 24)	Toluene/H_2O (7:3, v/v)	170	16 h	98	71	[212]
13	10	66.7 wt% Beta/TUD-1	Toluene/H_2O (7:3, v/v)	170	8 h	98	74	[210]
14	10	del-Nu-6(1)	Toluene/H_2O (7:3, v/v)	170	4 h	90	47	[213]
15	10	66.7 wt% SAPO-11 silicoaluminophosphate	Toluene/H_2O (7:3, v/v)	175	4 h	69	38	[214]
16	10	66.7 wt% eHTiNbO_5-MgO	Toluene/H_2O (7:3, v/v)	160	4 h	92	55	[217]
17	10	100 wt% TiO_2-ZrO_2	H_2O	250	5 min	80	35	[218]
18	10	66.7 wt% SO_4/ZrO_2-Al_2O_3	Toluene/H_2O (7:3, v/v)	160	4 h	>90	50	[219]
19	10	66.7 wt% ZrAlW-MP	Toluene/H_2O (7:3, v/v)	170	4 h	98	51	[220]
20	10	66.7 wt% $(VO)_2P_2O_7$	Toluene/H_2O (7:3, v/v)	170	4 h	91	53	[221]
21	10	66.7 wt% SO_4/ZrO_2-Al_2O_3/SBA-15	Toluene/H_2O (7:3, v/v)	160	4 h	99	53	[222]

[a]Catalyst dosage relative to the substrate

141], which catalyze xylose isomerization and dehydration to furfural. Recent works are focus on the development bifunctional solids to replace these corrosive acids for efficient conversion of pentoses into furfural.

3.2.2.1 Molecular Sieve-Based BAS-LAS Bifunctional Catalysts

Molecular sieves combining both Brønsted and Lewis acidity can significantly enhance their activity in the conversion of xylose into furfural. For example, Lewis acidic Nb_2O_5 provides a furfural yield of 17% after 100 min reaction time at 170 °C in toluene/H_2O (7:5, v/v) [208]. Introducing the cooperativity of Brønsted and Lewis acidity to molecular sieves can be achieved via incorporation of niobium into the framework and impregnation of Nb_2O_5 [208, 209]. At the same conditions, impregnation of Lewis acidic Nb_2O_5 on [Si]MCM-41 can enhance the catalytic activity of Nb_2O_5 with a furfural yield of 30%, which can be improved to 60% by adding NaCl due to salting out effect [208]. The BAS/LAS ratio can be flexibly tuned by Si/Nb ratios via controlling Nb incorporated into the framework of molecular sieves, such as Nb/MCM-41 [209]. Optimizing Si/Nb ratio provides a furfural yield of 40% after 6 h reaction time at 160 °C in Toluene/H_2O (7:3, v/v).

BAS-LAS bifunctional zeolites can be prepared by techniques, such as dealumination, desilication, impregnation or ion-exchange with other species (e.g. Al^{3+}, Mg^{2+}, Ca^{2+}, La^{3+}, Ce^{3+}, *etc.*) [18, 19]. Combining the intrinsic strong Brønsted acidity, bifunctional zeolites exhibit high catalytic activity in the conversion of xylose into furfural [210–214]. With the same Si/Al ratio and the similar total number of acid sites, increasing LAS/BAS ratio of H-MCM-22 from 0.4 to 1.0 results in a decrease of furfural yield from 70 to 66% after 16 h reaction time at 170 °C in Toluene/H_2O (7:3, v/v) [212]. Under the same conditions after 6 h reaction time, Beta zeolite (Si/Al = 12, LAS/BAS = 1.3) and Al-TUD-1 (Si/Al = 21, LAS/BAS = 2.3) provides furfural yields of 49% and 60%, respectively, while a furfural yield of 69% is obtained via incorporation of Beta into mesoporous TUD-1 [210, 211]. Therefore, the optimization of both the density of acid sites and the LAS/BAS ratio are important in obtaining high furfural yield.

3.2.2.2 Metal Oxides

Metal oxide-based BAS-LAS bifunctional catalysts are extensively investigated in the synthesis of furfural from xylose, including exfoliated nanosheets, TiO_2-ZrO_2, ZrAlW-MP, sulfate and phosphate metal oxides (Table 3.4 entry 16–21) [217–222]. At 160–170 °C in toluene/H_2O (7:5, v/v) after 4 h reaction time, similar furfural yields of 50–55% are obtained with xylose conversions of 92–99% over these metal oxides. Tuning the surface Brønsted and Lewis acidity of metal oxides is the key in obtaining furfural with high yield, e.g. eHTiNbO_5-MgO having medium total number of acid sites (0.382 mmol/g) and LAS/BAS ratio (1.4), obtained the highest furfural yield (55% vs. 40–50%) among exfoliated nanosheets (0.046–0.449 mmol/g, LAS/BAS = 1.2–5.5) at 160 °C in toluene/H_2O (7:3, v/v) after 6 h reaction time

[217]. The furfural yield obtained by metal oxides is comparable or even higher than molecular sieves at the similar conditions [208, 209, 213, 214]. Taking advantage of highly hydrothermal stable compared with zeolites, development of robust and stable BAS-LAS bifunctional metal oxides for the sugar conversion on large scale is desirable.

3.2.3 Glycerol and Derivatives Conversion Over BAS-LAS Bifunctional Catalysts

Glycerol is the major byproduct from biodiesel industry with an annual production of over a million tons and it is rapidly increasing with the growing market of renewable energy [223, 224]. Glycerol can be converted into value-added chemicals via oxidation, hydrolysis, pyrolysis, dehydration, transesterification, oligomerization, and carboxylation, such as acrolein [224–230].

3.2.3.1 Glycerol Dehydration to Acrolein

The dehydration of glycerol into acrolein is of great interest in producing acrylic acid esters, super absorber polymers or detergents [228], which is a promising area for applications with solid acids including zeolites, niobia, metal sulphates and phosphates, heteropolyacids, and metal oxides [231–236]. Two reaction pathways have been proposed based on the type of acid sites [225]: (a) A sequential dehydration of the secondary and the primary hydroxyl groups at BAS yields acrolein (Fig. 3.4a) and (b) LAS mainly catalyze the dehydration of primary hydroxyl groups of glycerol to generate acetol as byproduct (Fig. 3.4b) [224, 225].

Improving the acrolein yield from glycerol is limited by exclusively tuning Brønsted acidity [96]. Tailoring BAS-LAS bifunctional catalysts, such as Al-exchanged H-ZSM-5, exhibit extremely high acrolein yield (54.4%) compared to the parent H-ZSM-5 zeolite (35.9%) under the same conditions. This has been attributed to a two-step dehydration of glycerol to acrolein via the dehydration of the secondary hydroxyl group of glycerol at BAS, followed by the dehydration of the primary hydroxyl group at LAS to form acrolein as shown in Fig. 3.5.

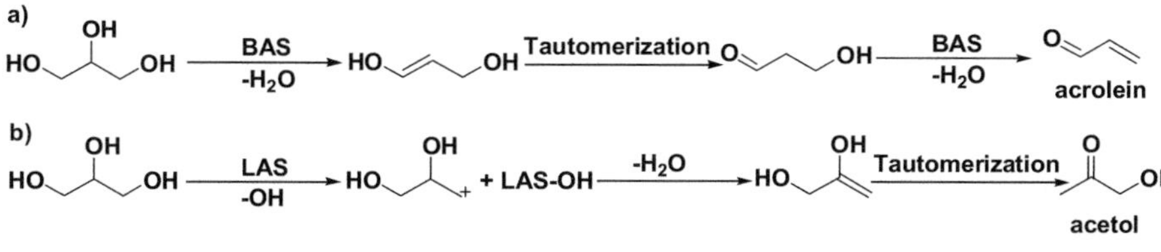

Fig. 3.4 Proposed mechanism for glycerol dehydration on single type acid sites: (**a**) Brønsted acid sites to yield desired acrolein, and (**b**) Lewis acid sites for acetol as the major byproduct (Adapted with permission from Ref. [96], Copyright © 2014, American Chemical Society)

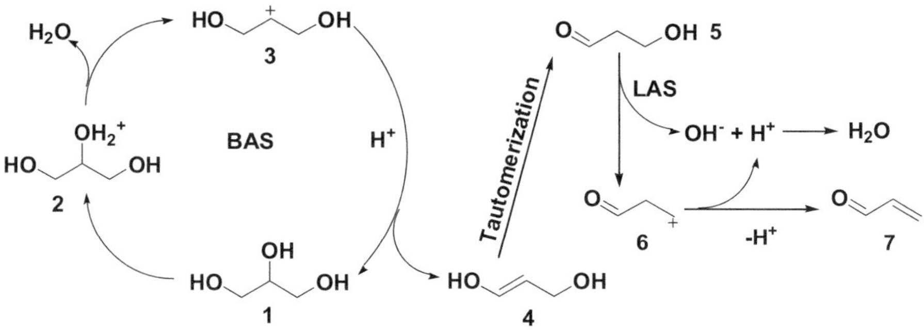

Fig. 3.5 The proposed mechanism for two-step dehydration of glycerol to acrolein via BAS-LAS cooperativity (Adapted with permission from Ref. [96], Copyright © 2014, American Chemical Society)

3.2.3.2 Glycerol Derived Trioses Conversion Over BAS-LAS Bifunctional Catalysts

Glycerol selective oxidation over supported metal catalysts can produce glyceraldehyde (GLY) and dihydroxyacetone (DHA) [228, 237]. These trioses have been applied in the one-pot synthesis of lactate esters [238, 239], which are used as green solvents and cosmetics, in healthcare formulations, and in the skin care and biopolymer industry [240–242]. The reaction proceeds via the dehydration of primary hydroxyl groups of trioses to form pyruvic aldehyde (PAL) on acid sites, and then LAS catalyze the conversion of PAL into lactic acid and derivatives via nucleophilic addition of a water or alcohol molecule to carbonyl carbon atom of aldehyde, followed by 1,2-hydride shift through an intramolecular Cannizzaro reaction mechanism [239] (Fig. 3.6).

BAS-LAS bifunctional catalysts, such as Sn(VI)-grafted carbon-silica catalysts (Sn-Si-CSM) [239], exhibit an extremely high ethyl lactate yield (100%), compared with those have exclusive BAS (23.4%) or LAS (1.25%) at the same conditions, where Sn(IV) species and carboxylic acids act as LAS and BAS, respectively. Mechanistic studies via isotope labelling experiments suggested mild BAS response for the dehydration of primary hydroxyl group of trioses, while LAS catalyze intramolecular Cannizzaro reaction between the α-keto aldehyde and alcohols to yield alkyl lactates [238].This demonstrates the existence of both BAS and LAS is a prerequisite to a sequential two-step reaction for fast and selective conversion of trioses into alkyl lactates and lactic acid. With this idea, mesoporous ZrSBA-15 silicate having both strong LAS and weak BAS can yield ethyl lactate directly from cellulose in a supercritical mixture of ethanol and water [243]. The BAS-LAS cooperativity can promote consecutive reactions, such as depolymerization, retro-aldol condensation, and esterification. Under the optimum reaction conditions, ca. 33% yield of ethyl lactate are produced from cellulose over Zr-SBA-15 catalyst at 533 K in supercritical 95:5 (w/w) ethanol/water.

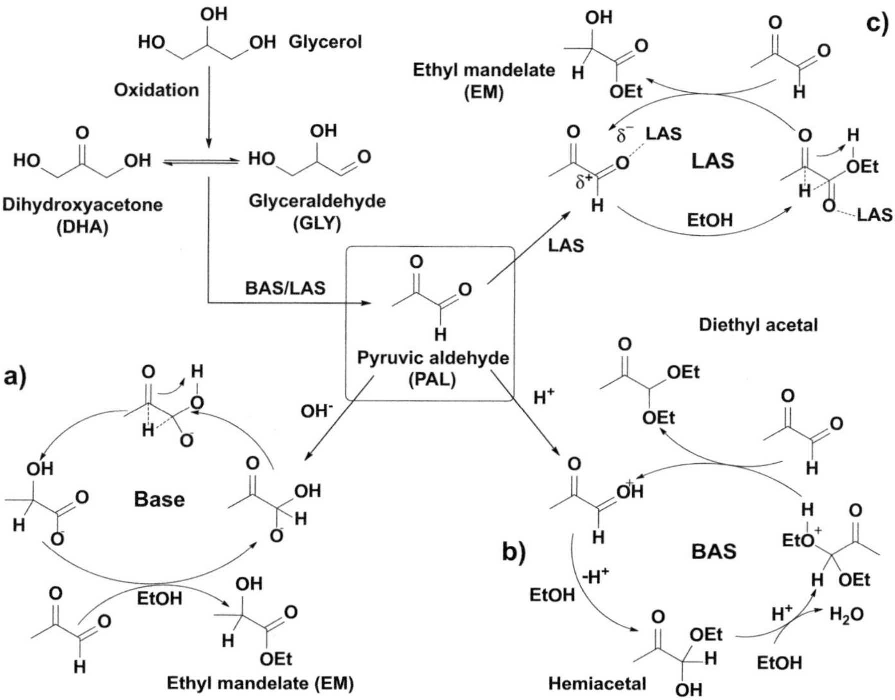

Fig. 3.6 Proposed Reaction Mechanisms of Catalytic Conversion of PG in EtOH: (**a**) Pathway in Homogeneous Base System; (**b**) Pathway in Homogeneous Brønsted Acid System [238]; (**c**) Pathway on Lewis Acid Sites (Adapted with permission from Ref. [239], Copyright © 2012, American Chemical Society)

3.3 Conclusion and Future Outlook

The utilization of Brønsted-Lewis acid bifunctional catalysts can combine both the properties of BAS and LAS for sequential reactions. The cooperativity of BAS and LAS at optimal ratio can significantly enhance the reaction efficiency and selectivity to target product(s) compared to reaction preformed with mainly BAS or LAS, such as for glucose dehydration reaction. Therefore, BAS-LAS bifunctional catalysis can greatly improves the catalytic performance and will shed new light on a reasonable design of efficient catalysts.

Future work for the development of BAS-LAS bifunctional catalysts are focus on two major directions: (1) the development of suitable preparation techniques to tailor BAS-LAS bifunctional catalysts with high stability and precious control of BAS/LAS ratio to optimize the cooperative effect of BAS and LAS. For instance, the current techniques to generate leaching of LAS generated via dealumination of zeolites often reduces catalyst performance in liquid phase reaction and limits their applications. (2) Moreover, the development of novel reaction pathway based on BAS-LAS cooperativity and understanding of the cooperative reaction mechanism is promising for efficient catalysis. It is not only focus on the cooperative action to integrate a sequential reactions based on BAS and LAS together for one-pot synthesis, but also tune the negative role of acid sites (e.g. LAS in glycerol dehydration) to a positive contribution to target product(s) in reactions via site-site cooperation.

References

1. Brönsted JN. Einige Bemerkungen über den Begriff der Säuren und Basen. Recueil des Travaux Chimiques des Pays-Bas. 1923;42(8):718–28.
2. Lowry TM. The uniqueness of hydrogen. J Soc Chem Ind. 1923;42(3):43–7.
3. Lewis GN. Valency and structure of atoms and molecules. New York: Wiley; 1923.
4. Busca G. Acid catalysts in industrial hydrocarbon chemistry. Chem Rev. 2007;107(11):5366–410.
5. Corma A. Inorganic solid acids and their use in acid-catalyzed hydrocarbon reactions. Chem Rev. 1995;95(3):559–614.
6. Chai SH, Wang HP, Liang Y, Xu BQ. Sustainable production of acrolein: gas-phase dehydration of glycerol over 12-tungstophosphoric acid supported on ZrO_2 and SiO_2. Green Chem. 2008;10(10):1087–93.
7. Maki-Arvela P, Hajek J, Salmi T, Murzin DY. Chemoselective hydrogenation of carbonyl compounds over heterogeneous catalysts. Appl Catal A-Gen. 2005;292:1–49.
8. Ennaert T, Van Aelst J, Dijkmans J, De Clercq R, Schutyser W, Dusselier M, Verboekend D, Sels BF. Potential and challenges of zeolite chemistry in the catalytic conversion of biomass. Chem Soc Rev. 2016;45(3):584–611.
9. Wilson ST, Lok BM, Messina CA, Cannan TR, Flanigen EM. Aluminophosphate molecular sieves – a new class of microporous crystalline inorganic solids. J Am Chem Soc. 1982;104(4):1146–7.
10. Wang YY, Mu Y, Sun YJ, Li JY. Two new four-connected zeolite-like magnesium aluminophosphates with intersecting 8-ring channels. RSC Adv. 2014;4(99):56288–93.
11. Millini R, Montanari L, Bellussi G. Synthesis and characterization of a potassium borosilicate with ANA framework type-structure. Microporous Mater. 1993;1(1):9–15.
12. Kumar R, Thangaraj A, Bhat RN, Ratnasamy P. Synthesis of iron-silicate analogs of zeolite-beta. Zeolites. 1990;10(2):85–9.
13. Camblor MA, Perezpariente J, Fornes V. Synthesis and characterization of gallosilicates and galloaluminosilicates isomorphous to zeolite-beta. Zeolites. 1992;12(3):280–6.
14. Kornatowski J, Wichterlova B, Jirkovsky J, Loffler E, Pilz W. Spectroscopic studies of vanadium-substituted zeolitic silicates of MFI topology. J Chem Soc-Faraday Trans. 1996;92(6):1067–78.
15. Ausavasukhi A, Sooknoi T. Tunable activity of Ga-HZSM-5 with H_2 treatment: ethane dehydrogenation. Catal Commun. 2014;45:63–8.
16. Li YP, Head-Gordon M, Bell AT. Analysis of the reaction mechanism and catalytic activity of metal-substituted Beta zeolite for the isomerization of glucose to fructose. ACS Catal. 2014;4(5):1537–45.
17. Jiang Y, Huang J, Dai W, Hunger M. Solid-state nuclear magnetic resonance investigations of the nature, property, and activity of acid sites on solid catalysts. Solid State Nucl Magn Reson. 2011;39(3–4):116–41.
18. Guth JL, Kessler H. In: Weitkamp J, Puppe L, editors. Catalysis and Zeolites-fundamentals and applications. 1st ed. Berlin/Heidelberg/New York: Springer; 1999.

19. Dapsens PY, Mondelli C, Perez-Ramirez J. Highly selective Lewis acid sites in desilicated MFI zeolites for dihydroxyacetone isomerization to lactic acid. ChemSusChem. 2013;6(5):831–9.
20. Shannon RD, Gardner KH, Staley RH, Bergeret G, Gallezot P, Auroux A. The nature of the nonframework aluminum species formed during the dehydroxylation of H-Y. J Phys Chem. 1985;89(22):4778–88.
21. Chizallet C, Raybaud P. Pseudo-Bridging silanols as versatile Brønsted acid sites of amorphous aluminosilicate surfaces. Angew Chem Int Ed. 2009;48(16):2891–3.
22. Wang ZC, Jiang YJ, Lafon O, Trebosc J, Kim KD, Stampfl C, Baiker A, Amoureux JP, Huang J. Brønsted acid sites based on penta-coordinated aluminum species. Nat Commun. 2016. https://doi.org/10.1038/ncomms13820.
23. Dewitte BM, Grobet PJ, Uytterhoeven JB. Pentacoordinated aluminum in noncalcined amorphous aluminosilicates, prepared in alkaline and acid-medium. J Phys Chem. 1995;99(18):6961–5.
24. de Boer JH. Constitution and properties of silica-alumina-catalysts. Discuss Faraday Soc. 1971;52:109–12.
25. Peri JB. Infrared study of adsorption of carbon dioxide, hydrogen chloride, and other molecules on "acid" sites on dry silica—alumina and γ-alumina. J Phys Chem. 1966;70(10):3168–79.
26. Beck JS, Vartuli JC, Roth WJ, Leonowicz ME, Kresge CT, Schmitt KD, Chu CTW, Olson DH, Sheppard EW, McCullen SB, Higgins JB, Schlenker JL. A new family of mesoporous molecular-sieves prepared with liquid-crystal templates. J Am Chem Soc. 1992;114(27):10834–43.
27. Kresge CT, Leonowicz ME, Roth WJ, Vartuli JC, Beck JS. Ordered mesoporous molecular-sieves synthesized by a liquid-crystal template mechanism. Nature. 1992;359(6397):710–2.
28. Corma A, Fornes V, Navarro MT, Perezpariente J. Acidity and stability of MCM-41 crystalline aluminosilicates. J Catal. 1994;148(2):569–74.
29. Wang Z, Jiang Y, Rachwalik R, Liu Z, Shi J, Hunger M, Huang J. One-step room-temperature synthesis of Al MCM-41 materials for the catalytic conversion of phenylglyoxal to ethylmandelate. ChemCatChem. 2013;5(12):3889–96.
30. Weitkamp J, Hunger M. Acid and base catalysis on zeolites. In: Ejka J, van Bekkum H, Corma A, Schüth F, editors. Studies in surface science and catalysis. Amsterdam: Elsevier; 2007. p. 787–835.
31. Keggin JF. The structure and formula of 12-phosphotungstic acid. Proc R Soc Lond A. 1934;144(A851):75–100.
32. Marcì G, García-López EI, Palmisano L. Heteropolyacid-based materials as heterogeneous photocatalysts. Eur J Inorg Chem. 2014;2014(1):21–35.
33. Micek-Ilnicka A. The role of water in the catalysis on solid heteropolyacids. J Mol Catal A Chem. 2009;308(1–2):1–14.
34. Beltrame P, Zuretti G. Kinetic studies for processes of liquid-phase alkylation of aromatics over solid acid catalysts. Green Chem. 2004;6(1):7–13.
35. Baronetti G, Thomas H, Querini CA. Wells-Dawson heteropolyacid supported on silica: isobutane alkylation with C4 olefins. Appl Catal A Gen. 2001;217(1–2):131–41.
36. Su F, Guo YH. Advancements in solid acid catalysts for biodiesel production. Green Chem. 2014;16(6):2934–57.
37. Shimizu KI, Satsuma A. Toward a rational control of solid acid catalysis for green synthesis and biomass conversion. Energy Environ Sci. 2011;4(9):3140–53.
38. Bolis V, Cerrato G, Magnacca G, Morterra C. Surface acidity of metal oxides. Combined microcalorimetric and IR-spectroscopic studies of variously dehydrated systems. Thermochim Acta. 1998;312(1–2):63–77.
39. Pokrovski K, Jung KT, Bell AT. Investigation of CO and CO_2 adsorption on tetragonal and monoclinic zirconia. Langmuir. 2001;17(14):4297–303.
40. Soled S, McVicker GB. Acidity of silica-substituted zirconia. Catal Today. 1992;14(2):189–94.

41. Miller JB, Rankin SE, Ko EI. Strategies in controlling the homogeneity of zirconia silica aerogels – effect of preparation on textural and catalytic properties. J Catal. 148(2):673–82.
42. Bosman HJM, Kruissink EC, Vanderspoel J, Vandenbrink F. Characterization of the acid strength of SiO_2-ZrO_2 mixed oxides. J Catal. 1994;148(2):660–72.
43. Shibata K, Kiyoura T, Kitagawa J, Sumiyosh T, Tanabe K. Acidic properties of binary metal-oxides. Bull Chem Soc Jpn. 1973;46(10):2985–8.
44. Toba M, Mizukami F, Niwa S, Sano T, Maeda K, Annila A, Komppa V. The effect of preparation methods on the properties of zirconia silicas. J Mol Catal. 1994;94(1):85–96.
45. Ramirez J, Cedeno L, Busca G. The role of titania support in Mo-based hydrodesulfurization catalysts. J Catal. 1999;184(1):59–67.
46. Maity SK, Ancheyta J, Rana MS, Rayo P. Alumina-titania mixed oxide used as support for hydrotreating catalysts of Maya heavy crude-effect of support preparation methods. Energy Fuel. 2006;20(2):427–31.
47. Xu J, Zheng AM, Yang J, Su YC, Wang JQ, Zeng DL, Zhang MJ, Ye CH, Deng F. Acidity of mesoporous MoO_x/ZrO_2 and WO_x/ZrO_2 materials: a combined solid-state NMR and theoretical calculation study. J Phys Chem B. 2006;110(22):10662–71.
48. Vartuli JC, Santiesteban JG, Traverso P, Cardona-Martinez N, Chang CD, Stevenson SA. Characterization of the acid properties of tungsten/zirconia catalysts using adsorption microcalorimetry and n-pentane isomerization activity. J Catal. 1999;187(1):131–8.
49. Miller JB, Ko EI. Acidic properties of silica-containing mixed oxide aerogels: preparation and characterization of zirconia-silica and comparison to titania-silica. J Catal. 1996;159(1):58–68.
50. Manoilova OV, Olindo R, Arean CO, Lercher JA. Variable temperature FTIR study on the surface acidity of variously treated sulfated zirconias. Catal Commun. 2007;8:865–70.
51. Semmer V, Batamack P, Doremieux-Morin C, Fraissard J. NMR studies of the Brønsted acidity of solids: application to superacidic solids. Top Catal. 1998;6:119–25.
52. Arata K, Hino M. Solid catalyst treated with anion. 18. benzoylation of toluene with benzoyl chloride and benzoic anhydride catalyzed by solid superacid of sulfate-supported alumina. Appl Catal. 1990;59:197–204.
53. Clearfield A, Serrette GPD, Khazi-Syed AH. Nature of hydrous zirconia and sulfated hydrous zirconia. Catal Today. 1994;20:295–312.
54. Yamaguchi T. Recent progress in solid superacid. Appl Catal. 1990;61:1–25.
55. Matsuhashi H, Hino M, Arata K. Solid catalyst treated with anion. Appl Catal. 1990;59:205–12.
56. Waqif M, Bachelier J, Saur O, Lavalley JC. Acidic properties and stability of sulfate-promoted metal-oxides. J Mol Catal. 1992;72:127–38.
57. Benesi HA. Acidity of catalyst surfaces. 1. Acid strength from colors of adsorbed indicators. J Am Chem Soc. 1956;78(21):5490–4.
58. Hunger M. Catalytically active sites: generation and characterization, in zeolites and catalysis. In: Zeolites and catalysis. Weinheim: Wiley-VCH Verlag GmbH & Co. KGaA; 2010. p. 493–546.
59. Hirschler AE. The measurement of catalyst acidity using indicators forming stable surface carbonium ions. J Catal. 1963;2(5):428–39.
60. Chai SH, Wang HP, Liang Y, Xu BQ. Sustainable production of acrolein: investigation of solid acid-base catalysts for gas-phase dehydration of glycerol. Green Chem. 2007;9(10):1130–6.
61. Leftin HP, Hobson MC. Application of spectrophotometry to the study of catalytic systems. Adv Catal. 1963;14:115–201.
62. Cvetanov RJ, Amenomiy Y. Temperature programmed desorption technique for investigation of practical catalysts. Catal Rev. 1972;6(1):21–48.
63. Parker LM, Bibby DM, Meinhold RH. An evaluation by TD MS of the use of weak bases (ammonia, primary amines and pyridine) as probes for the study of zeolite acid sites. Zeolites. 1985;5(6):384–8.
64. Yang L, Aizhen Y, Qinhua X. Acidity, diffusion and catalytic properties of the silicoaluminophosphate SAPO-11. Appl Catal. 1991;67(2):169–77.

65. Weingarten R, Tompsett GA, Conner WC Jr, Huber GW. Design of solid acid catalysts for aqueous-phase dehydration of carbohydrates: the role of Lewis and Brønsted acid sites. J Catal. 2011;279(1):174–82.
66. Gorte RJ. Design parameters for temperature programmed desorption from porous catalysts. J Catal. 1982;75(1):164–74.
67. Demmin RA, Gorte RJ. Design parameters for temperature-programmed desorption from a packed-bed. J Catal. 1984;90(1):32–9.
68. Qin GL, Zheng L, Xie YM, Wu CC. On the framework hydroxyl-groups of H-ZSM-5 zeolites. J Catal. 1985;95(2):609–12.
69. Parker WON Jr, Wegner S. Aluminum in mesoporous silica-alumina. Microporous Mesoporous Mater. 2012;158:235–40.
70. Stebbins JF, Xu Z. NMR evidence for excess non-bridging oxygen in an aluminosilicate glass. Nature. 1997;390(6655):60–2.
71. Beebe TP, Gelin P, Yates JT. Infrared spectroscopic observations of surface bonding in physical adsorption – the physical adsorption of CO on SiO_2 surfaces. Surf Sci. 148(2–3):526–50.
72. Parry EP. An infrared study of pyridine adsorbed on acidic solids characterization of surface acidity. J Catal. 1963;2:371–9.
73. Basila MR, Rhee KH, Kantner TR. Nature of acidic sites on silica-alumina. Characterization by infrared spectroscopic studies of trimethylamine + pyridine chemisorption. J Phys Chem. 1964;68:3197–207.
74. Hughes TR, White HM. A study of surface structure of decationized Y zeolite by quantitative infrared spectroscopy. J Phys Chem. 1967;71:2192–201.
75. Occelli ML, Biz S, Auroux A, Ray GJ. Effects of the nature of the aluminum source on the acidic properties of some mesostructured materials. Microporous Mesoporous Mater. 1998;26:193–213.
76. Fan Y, Bao XJ, Lin XY, Shi G, Liu HY. Acidity adjustment of HZSM-5 zeolites by dealumination and realumination with steaming and citric acid treatments. J Phys Chem B. 2006;110:15411–6.
77. Machado V, Rocha J, Carvalho AP, Martins A. A modification of MCM-22 zeolite through sequential post-synthesis treatments. Implications on the acidic and catalytic behavior. Appl Catal A Gen. 2012;445:329–38.
78. Guisnet M, Ayrault P, Coutanceau C, Alvarez MF, Datka J. Acid properties of dealuminated beta zeolites studied by IR spectroscopy. J Chem Soc Faraday Trans. 1997;93:1661–5.
79. Marques JP, Gener I, Ayrault P, Bordado JC, Lopes JM, Ribeiro FR, Guisnet M. Infrared spectroscopic study of the acid properties of dealuminated BEA zeolites. Microporous Mesoporous Mater. 2003;60:251–62.
80. Bordiga S, Lamberti C, Geobaldo F, Zecchina A, Palomino GT, Arean CO. Fourier-transform infrared study of CO adsorbed at 77 K on H-mordenite and alkali-metal-exchanged mordenites. Langmuir. 1995;11:527–33.
81. Geobaldo F, Lamberti C, Ricchiardi G, Bordiga S, Zecchina A, Palomino GT, Arean CO. N_2 adsorption at 77 K on H-mordenite and alkali-metal-exchanged mordenites – an IR study. J Phys Chem. 1995;99:11167–77.
82. Zheng A, Li SH, Liu SB, Deng F. Acidic properties and structure-activity correlations of solid acid catalysts revealed by solid-state NMR spectroscopy. Acc Chem Res. 2016;49(4):655–63.
83. Lang S, Benz M, Obenaus U, Himmelmann R, Hunger M. Novel approach for the characterization of Lewis acidic solid catalysts by solid-state NMR spectroscopy. ChemCatChem. 2016;8:2031–6.
84. Haw JF, Chuang IS, Hawkins BL, Maciel GE. Surface titration of silica-alumina monitored by N-15 NMR with cross polarization and magic-angle spinning. J Am Chem Soc. 1983;105:7206–7.
85. Ripmeester JA. Surface acid site characterization by means of CP/MAS N-15 NMR. J Am Chem Soc. 1983;105:2925–7.

86. Haw JF, Zhang JH, Shimizu K, Venkatraman TN, Luigi DP, Song WG, Barich DH, Nicholas JB. NMR and theoretical study of acidity probes on sulfated zirconia catalysts. J Am Chem Soc. 2000;122:12561–70.
87. Huang J, van Vegten N, Jiang Y, Hunger M, Baiker A. Increasing the Brønsted acidity of flame-derived silica/alumina up to zeolitic strength. Angew Chem Int Ed. 2010;49:7776–81.
88. Filek U, Bressel A, Sulikowski B, Hunger M. Structural stability and Brønsted acidity of thermally treated $AlPW_{12}O_{40}$ in Comparison with $H_3PW_{12}O_{40}$. J Phys Chem C. 2008;112:19470–6.
89. Biaglow AI, Gorte RJ, Kokotailo GT, White D. A probe of Brønsted site acidity in zeolites – C-13 chemical-shift of acetone. J Catal. 1994;148:779–86.
90. Biaglow AI, Gorte RJ, White D. C-13 NMR-studies of acetone in dealuminated faujasites – a probe for nonframework alumina. J Catal. 1994;150:221–4.
91. Biaglow AI, Sepa J, Gorte RJ, White D. A C-13 NMR-study of the condensation chemistry of acetone and acetaldehyde adsorbed at the Brønsted acid sites in H-ZSM-5. J Catal. 1995;151:373–84.
92. Song WG, Nicholas JB, Haw JF. A persistent carbenium ion on the methanol-to-olefin catalyst HSAPO-34: Acetone shows the way. J Phys Chem B. 2001;105:4317–23.
93. Xu M, Wang W, Seiler M, Buchholz A, Hunger M. Improved Brønsted acidity of mesoporous [Al]MCM-41 material treated with ammonium fluoride. J Phys Chem B. 2002;106:3202–8.
94. Li S, Zheng A, Su Y, Zhang H, Chen L, Yang J, Ye C, Deng F. Brønsted/Lewis acid synergy in dealuminated HY zeolite: a combined solid-state NMR and theoretical calculation study. J Am Chem Soc. 2007;129:11161–71.
95. Wang Z, Pokhrel S, Chen M, Hunger M, Mädler L, Huang J. Palladium-doped silica–alumina catalysts obtained from double-flame FSP for chemoselective hydrogenation of the model aromatic ketone acetophenone. J Catal. 2013;302:10–9.
96. Wang Z, Wang L, Jiang Y, Hunger M, Huang J. Cooperativity of Brønsted and Lewis acid sites on zeolite for glycerol dehydration. ACS Catal. 2014;4:1144–7.
97. Ma D, Han XW, Xie SJ, Bao XH, Hu HB, Au-Yeung SCF. An investigation of the roles of surface aluminum and acid sites in the zeolite MCM-22. Chem Eur J. 2002;8:162–70.
98. Lunsford JH, Tutunjian PN, Chu PJ, Yeh EB, Zalewski DJ. Solid-state NMR study using trimethylphosphine as a probe of acid sites in normal and dealuminated zeolite Y. J Phys Chem. 1989;93(6):2590–5.
99. Zhao BY, Pan HJ, Lunsford JH. Characterization of $(CH_3)_3P\text{-}H^+$ complexes in normal H-Y, dealuminated H-Y, and H-ZSM-5 zeolites using P-31 solid-state NMR spectroscopy. Langmuir. 1999;15(8):2761–5.
100. Kao HM, Yu CY, Yeh MC. Detection of the inhomogeneity of Brønsted acidity in H-mordenite and H-P zeolites: a comparative NMR study using trimethylphosphine and trimethylphosphine oxide as P-31 NMR probes. Microporous Mesoporous Mater. 2002;53(1–3):1–12.
101. Alonso B, Klur I, Massiot D. Studies of surfaces through molecular probe adsorption and solid-state NMR. Chem Commun. 2002;8:804–5.
102. Sutovich KJ, Peters AW, Rakiewicz EF, Wormsbecher RF, Mattingly SM, Mueller KT. Simultaneous quantification of Brønsted- and Lewis-acid sites in a USY zeolite. J Catal. 1999;183(1):155–8.
103. Karra MD, Sutovich KJ, Mueller KT. NMR characterization of Brønsted acid sites in faujasitic zeolites with use of perdeuterated trimethylphosphine oxide. J Am Chem Soc. 2002;124(6):902–3.
104. Rakiewicz EF, Peters AW, Wormsbecher F, Sutovich KJ, Mueller KT. Characterization of acid sites in zeolitic and other inorganic systems using solid-state P-31 NMR of the probe molecule trimethylphosphine oxide. J Phys Chem B. 1998;102(16):2890–6.
105. Pfeifer H, Freude D, Hunger M. Nuclear magnetic-resonance studies on the acidity of zeolites and related catalysts. Zeolites. 1985;5(5):274–86.
106. Jacobs WPJH, de Haan JW, van de Ven LJM, van Santen RA. Interaction of ammonia with Brønsted acid sites in different cages of zeolite Y as studied by proton MAS NMR. J Phys Chem. 1993;97(40):10394–402.

107. Yin F, Blumenfeld AL, Gruver V, Fripiat JJ. NH_3 as a Probe Molecule for NMR and IR study of zeolite catalyst acidity. J Phys Chem B. 1997;101(10):1824–30.
108. Huber GW, Iborra S, Corma A. Synthesis of transportation fuels from biomass: chemistry, catalysts, and engineering. Chem Rev. 2006;106(9):4044–98.
109. Besson M, Gallezot P, Pinel C. Conversion of biomass into chemicals over metal catalysts. Chem Rev. 2014;114(3):1827–70.
110. Emmel A, Mathias AL, Wypych F, Ramos LP. Fractionation of eucalyptus grandis chips by dilute acid-catalysed steam explosion. Bio Technol. 2003;86(2):105–15.
111. Hu R, Lin L, Liu T, Liu S. Dilute sulfuric acid hydrolysis of sugar maple wood extract at atmospheric pressure. Bioresour Technol. 2010;101(10):3586–94.
112. Kawamoto H, Saito S, Hatanaka W, Saka S. Catalytic pyrolysis of cellulose in sulfolane with some acidic catalysts. J Wood Sci. 2007;53(2):127–33.
113. Liu A, Park Y, Huang Z, Wang B, Ankumah RO, Biswas PK. Product identification and distribution from hydrothermal conversion of walnut shells. Energy Fuel. 2006;20(2):446–54.
114. Binder JB, Raines RT. Fermentable sugars by chemical hydrolysis of biomass. Proc Natl Acad Sci U S A. 2010;107(10):4516–21.
115. Lenihan P, Orozco A, O'Neill E, Ahmad MNM, Rooney DW, Walker GM. Dilute acid hydrolysis of lignocellulosic biomass. Chem Eng J. 2010;156(2):395–403.
116. Onda A, Ochi T, Yanagisawa K. Selective hydrolysis of cellulose into glucose over solid acid catalysts. Green Chem. 2008;10(10):1033–7.
117. Yamaguchi D, Kitano M, Suganuma S, Nakajima K, Kato H, Hara M. Hydrolysis of cellulose by a solid acid catalyst under optimal reaction conditions. J Phys Chem C. 2009;113(8):3181–8.
118. Wu Y, Fu Z, Yin D, Xu Q, Liu F, Lu C, Mao L. Microwave-assisted hydrolysis of crystalline cellulose catalyzed by biomass char sulfonic acids. Green Chem. 2010;12(4):696–700.
119. Suganuma S, Nakajima K, Kitano M, Yamaguchi D, Kato H, Hayashi S, Hara M. Hydrolysis of cellulose by amorphous carbon bearing SO_3H, COOH, and OH groups. J Am Chem Soc. 2008;130(38):12787–93.
120. Pang J, Wang A, Zheng M, Zhang T. Hydrolysis of cellulose into glucose over carbons sulfonated at elevated temperatures. Chem Commun. 2010;46(37):6935–7.
121. Van de Vyver S, Peng L, Geboers J, Schepers H, de Clippel F, Gommes CJ, Goderis B, Jacobs PA, Sels BF. Sulfonated silica/carbon nanocomposites as novel catalysts for hydrolysis of cellulose to glucose. Green Chem. 2010;12(9):1560–3.
122. Tian J, Wang J, Zhao S, Jiang C, Zhang X, Wang X. Hydrolysis of cellulose by the heteropoly acid $H_3PW_{12}O_{40}$. Cellulose. 2010;17(3):587–94.
123. Deng W, Liu M, Zhang Q, Tan X, Wang Y. Acid-catalysed direct transformation of cellulose into methyl glucosides in methanol at moderate temperatures. Chem Commun. 2010;46(15):2668–70.
124. Bray RG (2004) Biodiesel production. SRI Consulting.
125. Koh TS, Chung KH. Production of biodiesel from waste frying oil by transesterification on zeolite catalysts with different acidity. Kor Ind Eng Chem. 2008;19(2):214–21.
126. Hara M, Nakajima K, Kamata K. Recent progress in the development of solid catalysts for biomass conversion into high value-added chemicals. Sci Technol Adv Mater. 2015;16(3):034903.
127. Zhou CH, Xia X, Lin CX, Tong DS, Beltramini J. Catalytic conversion of lignocellulosic biomass to fine chemicals and fuels. Chem Soc Rev. 2011;40(11):5588–617.
128. Li H, Fang Z, Smith RL, Yang S. Efficient valorization of biomass to biofuels with bifunctional solid catalytic materials. Prog Energy Combust Sci. 2016;55:98–194.
129. Werpy T, Petersen G, Aden A, Bozell J, Holladay J, White J, Manheim A, Eliot D, Lasure L, Jones S (2004) Top value added chemicals from biomass. Volume 1-Results of screening for potential candidates from sugars and synthesis gas. DTIC document.
130. Corma A, Iborra S, Velty A. Chemical routes for the transformation of biomass into chemicals. Chem Rev. 2007;107(6):2411–502.

131. Otomo R, Tatsumi T, Yokoi T. Beta zeolite: a universally applicable catalyst for the conversion of various types of saccharides into furfurals. Catal Sci Technol. 5(8):4001–7.
132. Li XC, Xia QN, Nguyen VC, Peng KH, Liu XH, Essayem N, Wang YQ. High yield production of HMF from carbohydrates over silica-alumina composite catalysts. Catal Sci Technol. 2016;6(20):7586–96.
133. Wang T, Nolte MW, Shanks BH. Catalytic dehydration of C-6 carbohydrates for the production of hydroxymethylfurfural (HMF) as a versatile platform chemical. Green Chem. 2014;16(2):548–72.
134. Ordomsky VV, Sushkevich VL, Schouten JC, van der Schaaf J, Nijhuis TA. Glucose dehydration to 5-hydroxymethylfurfural over phosphate catalysts. J Catal. 2013;300:37–46.
135. Carniti P, Gervasini A, Bossola F, Dal Santo V. Cooperative action of Brønsted and Lewis acid sites of niobium phosphate catalysts for cellobiose conversion in water. Appl Catal B Environ. 2016;193:93–102.
136. Pagan-Torres YJ, Wang TF, Gallo JMR, Shanks BH, Dumesic JA. Production of 5-Hydroxymethylfurfural from glucose using a combination of Lewis and Brønsted acid catalysts in water in a biphasic reactor with an alkylphenol solvent. ACS Catal. 2012;2(6):930–4.
137. Tang JQ, Zhu LF, Fu X, Dai JH, Guo XW, Hu CW. Insights into the kinetics and reaction network of aluminum chloride-catalyzed conversion of glucose in NaCl-H_2O/THF biphasic system. ACS Catal. 2017;7(1):256–66.
138. Wang C, Zhang LM, Zhou T, Chen JC, Xu F. Synergy of Lewis and Brønsted acids on catalytic hydrothermal decomposition of carbohydrates and corncob acid hydrolysis residues to 5-hydroxymethylfurfural. Sci Rep. 2017;7:40908.
139. Choudhary V, Mushrif SH, Ho C, Anderko A, Nikolakis V, Marinkovic NS, Frenkel AI, Sandler SI, Vlachos DG. Insights into the interplay of Lewis and Brønsted acid catalysts in glucose and fructose conversion to 5-(hydroxymethyl)furfural and levulinic acid in aqueous media. J Am Chem Soc. 2013;135(10):3997–4006.
140. Dallas Swift T, Nguyen H, Anderko A, Nikolakis V, Vlachos DG. Tandem Lewis/Brønsted homogeneous acid catalysis: conversion of glucose to 5-hydoxymethylfurfural in an aqueous chromium(iii) chloride and hydrochloric acid solution. Green Chem. 2015;17(10):4725–35.
141. Caratzoulas S, Davis ME, Gorte RJ, Gounder R, Lobo RF, Nikolakis V, Sandler SI, Snyder MA, Tsapatsis M, Vlachos DG. Challenges of and insights into acid-catalyzed transformations of sugars. J Phys Chem C. 2014;118(40):22815–33.
142. Wrigstedt P, Keskiväli J, Leskelä M, Repo T. The role of salts and Brønsted acids in Lewis acid-catalyzed aqueous-phase glucose dehydration to 5-hydroxymethylfurfural. ChemCatChem. 2015;7(3):501–7.
143. Wei W, Wu S. Experimental and kinetic study of glucose conversion to levulinic acid catalyzed by synergy of Lewis and Brønsted acids. Chem Eng J. 2017;307:389–98.
144. Zhang J, Weitz E. An in situ NMR study of the mechanism for the catalytic conversion of fructose to 5-hydroxymethylfurfural and then to levulinic acid using C-13 labeled D-fructose. ACS Catal. 2012;2(6):1211–8.
145. Yang F, Fu J, Mo J, Lu X. Synergy of Lewis- and Brønsted acids on catalytic hydrothermal decomposition of hexose to levulinic acid. Energy Fuel. 2013;27(11):6973–8.
146. Yang Y, Liu WT, Wang NN, Wang HJ, Li W, Song ZX. Effect of different ionic liquids on 5-hydroxymethylfurfural preparation from glucose in DMA over $AlCl_3$: experimental and theoretical study. Chin J Chem. 2015;33(5):583–8.
147. Zhou JX, Xia Z, Huang TY, Yan PF, WJ X, ZW X, Wang JJ, Zhang ZC. An ionic liquid-organics-water ternary biphasic system enhances the 5-hydroxymethylfurfural yield in catalytic conversion of glucose at high concentrations. Green Chem. 2015;17(8):4206–16.
148. Zhao HB, Holladay JE, Brown H, Zhang ZC. Metal chlorides in ionic liquid solvents convert sugars to 5-hydroxymethylfurfural. Science. 2007;316(5831):1597–600.
149. Zakrzewska ME, Bogel-Lukasik E, Bogel-Lukasik R. Ionic liquid-mediated formation of 5-hydroxymethylfurfural-a promising biomass-derived building block. Chem Rev. 2011;111(2):397–417.

150. Zhang ZH, Wang QA, Xie HB, Liu WJ, Zhao ZB. Catalytic conversion of carbohydrates into 5-hydroxymethylfurfural by germanium(IV) chloride in ionic liquids. ChemSusChem. 2011;4(1):131–8.
151. Wu L, Song J, Zhang B, Zhou B, Zhou H, Fan H, Yang Y, Han B. Very efficient conversion of glucose to 5-hydroxymethylfurfural in DBU-based ionic liquids with benzenesulfonate anion. Green Chem. 2014;16(8):3935–41.
152. Hu Z, Liu B, Zhang ZH, Chen LQ. Conversion of carbohydrates into 5-hydroxymethylfurfural catalyzed by acidic ionic liquids in dimethyl sulfoxide. Ind Crop Prod. 2013;50:264–9.
153. Liu DJ, Chen EYX. Polymeric ionic liquid (PIL)-supported recyclable catalysts for biomass conversion into HMF. Biomass Bioenergy. 2013;48:181–90.
154. Li H, Zhang QY, Liu XF, Chang F, Zhang YP, Xue W, Yang S. Immobilizing Cr^{3+} with SO_3H-functionalized solid polymeric ionic liquids as efficient and reusable catalysts for selective transformation of carbohydrates into 5-hydroxymethylfurfural. Bioresource Technol. 2013;144:21–7.
155. Degirmenci V, Pidko EA, Magusin P, Hensen EJM. Towards a selective heterogeneous catalyst for glucose dehydration to 5-hydroxymethylfurfural in water: $CrCl_2$ catalysis in a thin immobilized ionic liquid layer. ChemCatChem. 2011;3(6):969–72.
156. Crisci AJ, Tucker MH, Dumesic JA, Scott SL. Bifunctional solid catalysts for the selective conversion of fructose to 5-hydroxymethylfurfural. Top Catal. 2010;53(15–18):1185–92.
157. Liu H, Wang H, Li Y, Yang W, Song C, Li H, Zhu W, Jiang W. Glucose dehydration to 5-hydroxymethylfurfural in ionic liquid over Cr^{3+}-modified ion exchange resin. RSC Adv. 2015;5(12):9290–7.
158. Lee YY, KCW W. Conversion and kinetics study of fructose-to-5-hydroxymethylfurfural (HMF) using sulfonic and ionic liquid groups bi-functionalized mesoporous silica nanoparticles as recyclable solid catalysts in DMSO systems. Phys Chem Chem Phys. 2012;14(40):13914–7.
159. Zhang ZH, Zhao ZB. Production of 5-hydroxymethylfurfural from glucose catalyzed by hydroxyapatite supported chromium chloride. Bioresource Technol. 2011;102(4):3970–2.
160. Jimenez-Morales I, Moreno-Recio M, Santamaria-Gonzalez J, Maireles-Torres P, Jimenez-Lopez A. Production of 5-hydroxymethylfurfural from glucose using aluminium doped MCM-41 silica as acid catalyst. Appl Catal B Environ. 2015;164:70–6.
161. Otomo R, Yokoi T, Kondo JN, Tatsumi T. Dealuminated Beta zeolite as effective bifunctional catalyst for direct transformation of glucose to 5-hydroxymethylfurfural. Appl Catal A Gen. 2014;470:318–26.
162. Hu L, Wu Z, Xu J, Sun Y, Lin L, Liu S. Zeolite-promoted transformation of glucose into 5-hydroxymethylfurfural in ionic liquid. Chem Eng J. 2014;244:137–44.
163. Nikolla E, Roman-Leshkov Y, Moliner M, Davis ME. "One-Pot" synthesis of 5-(hydroxymethyl)furfural from carbohydrates using tin Beta zeolite. ACS Catal. 2011;1(4):408–10.
164. Gallo JMR, Alonso DM, Mellmer MA, Dumesic JA. Production and upgrading of 5-hydroxymethylfurfural using heterogeneous catalysts and biomass-derived solvents. Green Chem. 2013;15(1):85–90.
165. Moreno-Recio M, Santamaria-Gonzalez J, Maireles-Torres P. Brønsted and Lewis acid ZSM-5 zeolites for the catalytic dehydration of glucose into 5-hydroxymethylfurfural. Chem Eng J. 2016;303:22–30.
166. Mamo W, Chebude Y, Marquez-Alvarez C, Diaz I, Sastre E. Comparison of glucose conversion to 5-HMF using different modified mordenites in ionic liquid and biphasic media. Catal Sci Technol. 2016;6(8):2766–74.
167. Wang J, Ren J, Liu X, Xi J, Xia Q, Zu Y, Lu G, Wang Y. Direct conversion of carbohydrates to 5-hydroxymethylfurfural using Sn-Mont catalyst. Green Chem. 2012;14(9):2506–12.
168. Wang N, Yao Y, Li W, Yang Y, Song Z, Liu W, Wang H, Xia XF, Gao H. Catalytic dehydration of fructose to 5-hydroxymethylfurfural over a mesoscopically assembled sulfated zirconia nanoparticle catalyst in organic solvent. RSC Adv. 2014;4(100):57164–72.

169. Dutta A, Patra AK, Dutta S, Saha B, Bhaumik A. Hierarchically porous titanium phosphate nanoparticles: an efficient solid acid catalyst for microwave assisted conversion of biomass and carbohydrates into 5-hydroxymethylfurfural. J Mater Chem. 2012;22(28):14094–100.
170. Li H, Zhang Q, Liu J, Liu X, Chang F, Liu Y, Xue W, Yang S. Selective transformation of carbohydrates into HMF promoted by carboxylic acids modified ZrMo mixed oxides. Biomass Conv Biorefinery. 2014;4(1):59–66.
171. De S, Dutta S, Patra AK, Bhaumik A, Saha B. Self-assembly of mesoporous TiO_2 nanospheres via aspartic acid templating pathway and its catalytic application for 5-hydroxymethyl-furfural synthesis. J Mater Chem. 2011;21(43):17505–10.
172. Dutta S, De S, Patra AK, Sasidharan M, Bhaumik A, Saha B. Microwave assisted rapid conversion of carbohydrates into 5-hydroxymethylfurfural catalyzed by mesoporous TiO_2 nanoparticles. Appl Catal A Gen. 2011;409:133–9.
173. Atanda L, Mukundan S, Shrotri A, Ma Q, Beltramini J. Catalytic conversion of glucose to 5-hydroxymethyl-furfural with a phosphated TiO_2 catalyst. ChemCatChem. 2015;7(5):781–90.
174. Zhang Y, Wang JJ, Ren JW, Liu XH, Li XC, Xia YJ, GZ L, Wang YQ. Mesoporous niobium phosphate: an excellent solid acid for the dehydration of fructose to 5-hydroxymethylfurfural in water. Catal Sci Technol. 2012;2(12):2485–91.
175. Nakajima K, Baba Y, Noma R, Kitano M, Kondo JN, Hayashi S, Hara M. Nb_2O_5 center dot nH_2O as a heterogeneous catalyst with water-tolerant Lewis acid sites. J Am Chem Soc. 2011;133(12):4224–7.
176. Jiménez-Morales I, Moreno-Recio M, Santamaría-González J, Maireles-Torres P, Jiménez-López A. Mesoporous tantalum oxide as catalyst for dehydration of glucose to 5-hydroxymethylfurfural. Appl Catal B Environ. 2014;154–155:190–6.
177. Yang FL, Liu QS, Yue M, Bai XF, YG D. Tantalum compounds as heterogeneous catalysts for saccharide dehydration to 5-hydroxymethylfurfural. Chem Commun. 2011;47(15):4469–71.
178. Jiménez-Morales I, Teckchandani-Ortiz A, Santamaría-González J, Maireles-Torres P, Jiménez-López A. Selective dehydration of glucose to 5-hydroxymethylfurfural on acidic mesoporous tantalum phosphate. Appl Catal B Environ. 2014;144:22–8.
179. Jiao HF, Zhao XL, Lv CX, Wang YJ, Yang DJ, Li ZH, Yao XD. Nb_2O_5-gamma-Al_2O_3 nanofibers as heterogeneous catalysts for efficient conversion of glucose to 5-hydroxymethylfurfural. Sci Rep. 2016;6:9.
180. Kreissl HT, Nakagawa K, Peng YK, Koito Y, Zheng J, Tsang SCE. Niobium oxides: correlation of acidity with structure and catalytic performance in sucrose conversion to 5-hydroxymethylfurfural. J Catal. 2016;338:329–39.
181. Yue C, Li G, Pidko EA, Wiesfeld JJ, Rigutto M, Hensen EJM. Dehydration of glucose to 5-hydroxymethylfurfural using Nb-doped Tungstite. ChemSusChem. 2016;9(17):2421–9.
182. Yamaguchi K, Sakurada T, Ogasawara Y, Mizuno N. Tin-Tungsten mixed oxide as efficient heterogeneous catalyst for conversion of saccharides to furan derivatives. Chem Lett. 2011;40(5):542–3.
183. Behera GC, Parida KM. One-pot synthesis of 5-hydroxymethylfurfural: a significant biomass conversion over tin-promoted vanadium phosphate (Sn-VPO) catalyst. Catal Sci Technol. 2013;3(12):3278–85.
184. Daorattanachai P, Khemthong P, Viriya-Empikul N, Laosiripojana N, Faungnawakij K. Conversion of fructose, glucose, and cellulose to 5-hydroxymethylfurfural by alkaline earth phosphate catalysts in hot compressed water. Carbohydr Res. 2012;363:58–61.
185. Jimenez-Morales I, Santamaria-Gonzalez J, Jimenez-Lopez A, Maireles-Torres P. Glucose dehydration to 5-hydroxymethylfurfural on zirconium containing mesoporous MCM-41 silica catalysts. Fuel. 2014;118:265–71.
186. Fan CY, Guan HY, Zhang H, Wang JH, Wang ST, Wang XH. Conversion of fructose and glucose into 5-hydroxymethylfurfural catalyzed by a solid heteropolyacid salt. Biomass Bioenergy. 2011;35(7):2659–65.

187. Zhao QA, Wang L, Zhao S, Wang XH, Wang ST. High selective production of 5-hydroymethylfurfural from fructose by a solid heteropolyacid catalyst. Fuel. 2011;90(6):2289–93.
188. Yi XH, Delidovich I, Sun Z, Wang ST, Wang XH, Palkovits R. A heteropoly acid ionic crystal containing Cr as an active catalyst for dehydration of monosaccharides to produce 5-HMF in water. Cat Sci Technol. 2015;5(4):2496–502.
189. Lu Y, Sun Z, Huo M. Fabrication of a micellar heteropolyacid with Lewis-Brønsted acid sites and application for the production of 5-hydroxymethylfurfural from saccharides in water. RSC Adv. 2015;5(39):30869–76.
190. Huang J, Jiang Y, Marthala VRR, Thomas B, Romanova E, Hunger M. Characterization and acidic properties of aluminum-exchanged zeolites X and Y. J Phys Chem C. 2008;112(10):3811–8.
191. Roman-Leshkov Y, Moliner M, Labinger JA, Davis ME. Mechanism of glucose isomerization using a solid Lewis acid catalyst in water. Angew Chem Int Ed. 2010;49(47):8954–7.
192. Moliner M, Roman-Leshkov Y, Davis ME. Tin-containing zeolites are highly active catalysts for the isomerization of glucose in water. Proc Natl Acad Sci U S A. 2010;107(14):6164–8.
193. Faria J, Pilar Ruiz M, Resasco DE. Carbon nanotube/zeolite hybrid catalysts for glucose conversion in water/oil emulsions. ACS Catal. 2015;5(8):4761–71.
194. Yang G, Pidko EA, Hensen EJM. Mechanism of Brønsted acid-catalyzed conversion of carbohydrates. J Catal. 2012;295:122–32.
195. Swift TD, Nguyen H, Erdman Z, Kruger JS, Nikolakis V, Vlachos DG. Tandem Lewis acid/Brønsted acid-catalyzed conversion of carbohydrates to 5-hydroxymethylfurfural using zeolite beta. J Catal. 2016;333:149–61.
196. Su Y, Chang GG, Zhang ZG, Xing HB, BG S, Yang QW, Ren QL, Yang YW, Bao ZB. Catalytic dehydration of glucose to 5-hydroxymethylfurfural with a bifunctional metal-organic framework. AICHE J. 2016;62(12):4403–17.
197. Reche MT, Osatiashtiani A, Durndell LJ, Isaacs MA, Silva A, Lee AF, Wilson K. Niobic acid nanoparticle catalysts for the aqueous phase transformation of glucose and fructose to 5-hydroxymethylfurfural. Catal Sci Technol. 2016;6(19):7334–41.
198. Alam MI, De S, Singh B, Saha B, Abu-Omar MM. Titanium hydrogenphosphate: An efficient dual acidic catalyst for 5-hydroxymethylfurfural (HMF) production. Appl Catal A Gen. 2014;486:42–8.
199. YM L, Li H, He J, Liu YX, ZB W, DY H, Yang S. Efficient conversion of glucose to 5-hydroxymethylfurfural using bifunctional partially hydroxylated AlF_3. RSC Adv. 2016;6(16):12782–7.
200. Agirrezabal-Telleria I, Guo Y, Hemmann F, Arias PL, Kemnitz E. Dehydration of xylose and glucose to furan derivatives using bifunctional partially hydroxylated MgF_2 catalysts and N_2-stripping. Catal Sci Technol. 2014;4(5):1357–68.
201. Rosatella AA, Simeonov SP, Frade RFM, Afonso CAM. 5 Hydroxymethylfurfural (HMF) as a building block platform: biological properties, synthesis and synthetic applications. Green Chem. 2011;13(4):754–93.
202. Antal MJ, Leesomboon T, Mok WS, Richards GN. Kinetic-studies of the reactions of ketoses and aldoses in water at high-temperature.3. Mechanism of formation of 2-furaldehyde from d-xylose. Carbohydr Res. 1991;217:71–85.
203. Rasmussen H, Sorensen HR, Meyer AS. Formation of degradation compounds from lignocellulosic biomass in the biorefinery: sugar reaction mechanisms. Carbohydr Res. 2014;385:45–57.
204. Choudhary V, Pinar AB, Sandler SI, Vlachos DG, Lobo RF. Xylose isomerization to xylulose and its dehydration to furfural in aqueous media. ACS Catal. 2011;1(12):1724–8.
205. Choudhary V, Sandler SI, Vlachos DG. Conversion of xylose to furfural using Lewis and Brønsted acid catalysts in aqueous media. ACS Catal. 2012;2(9):2022–8.
206. Tao FR, Song HL, Chou LJ. Efficient process for the conversion of xylose to furfural with acidic ionic liquid. Can J Chem Revue. 2011;89(1):83–7.

207. Agirrezabal-Telleria I, Hemmann F, Jager C, Arias PL, Kemnitz E. Functionalized partially hydroxylated MgF_2 as catalysts for the dehydration of D-xylose to furfural. J Catal. 2013;305:81–91.
208. Garcia-Sancho C, Sadaba I, Moreno-Tost R, Merida-Robles J, Santamaria-Gonzalez J, Lopez-Granados M, Maireles-Torres P. Dehydration of xylose to furfural over MCM-41-supported niobium-oxide catalysts. ChemSusChem. 2013;6(4):635–42.
209. Dias AS, Lima S, Brandao P, Pillinger M, Rocha J, Valente AA. Liquid-phase dehydration of D-xylose over microporous and mesoporous niobium silicates. Catal Lett. 2006;108(3–4):179–86.
210. Lima S, Antunes MM, Fernandes A, Pillinger M, Ribeiro MF, Valente AA. Catalytic cyclodehydration of xylose to furfural in the presence of zeolite H-Beta and a micro/mesoporous Beta/TUD-1 composite material. Appl Catal A Gen. 2010;388(1–2):141–8.
211. Lima S, Antunes MM, Fernandes A, Pillinger M, Ribeiro MF, Valente AA. Acid-catalysed conversion of saccharides into furanic aldehydes in the presence of three-dimensional mesoporous Al-TUD-1. Molecules. 2010;15(6):3863–77.
212. Antunes MM, Lima S, Fernandes A, Pillinger M, Ribeiro MF, Valente AA. Aqueous-phase dehydration of xylose to furfural in the presence of MCM-22 and ITQ-2 solid acid catalysts. Appl Catal A Gen. 2012;417:243–52.
213. Lima S, Pillinger M, Valente AA. Dehydration of D-xylose into furfural catalysed by solid acids derived from the layered zeolite Nu-6(1). Catal Commun. 2008;9(11–12):2144–8.
214. Lima S, Fernandes A, Antunes MM, Pillinger M, Ribeiro MF, Valente AA. Dehydration of xylose into furfural in the presence of crystalline microporous silicoaluminophosphates. Catal Lett. 2010;135(1–2):41–7.
215. Zhang JH, Zhuang JP, Lin L, Liu SJ, Zhang Z. Conversion of D-xylose into furfural with mesoporous molecular sieve MCM-41 as catalyst and butanol as the extraction phase. Biomass Bioenergy. 2012;39:73–7.
216. Garcia-Sancho C, Agirrezabal-Telleria I, Guemez MB, Maireles-Torres P. Dehydration of D-xylose to furfural using different supported niobia catalysts. Appl Catal B Environ. 2014;152:1–10.
217. Dias AS, Lima S, Carriazo D, Rives V, Pillinger M, Valente AA. Exfoliated titanate, niobate and titanoniobate nanosheets as solid acid catalysts for the liquid-phase dehydration of D-xylose into furfural. J Catal. 2006;244(2):230–7.
218. Chareonlimkun A, Champreda V, Shotipruk A, Laosiripojana N. Catalytic conversion of sugarcane bagasse, rice husk and corncob in the presence of TiO_2, ZrO_2 and mixed-oxide TiO_2-ZrO_2 under hot compressed water (HCW) condition. Bioresource Technol. 2010;101(11):4179–86.
219. Dias AS, Lima S, Pillinger M, Valente AA. Modified versions of sulfated zirconia as catalysts for the conversion of xylose to furfural. Catal Lett. 2007;114(3–4):151–60.
220. Antunes MM, Lima S, Fernandes A, Candeias J, Pillinger M, Rocha SM, Ribeiro MF, Valente AA. Catalytic dehydration of D-xylose to 2-furfuraldehyde in the presence of Zr-(W,Al) mixed oxides. Tracing by-products using two-dimensional gas chromatography-time-of-flight mass spectrometry. Catal Today. 2012;195(1):127–35.
221. Sadaba I, Lima S, Valente AA, Granados ML. Catalytic dehydration of xylose to furfural: vanadyl pyrophosphate as source of active soluble species. Carbohydr Res. 2011;346(17):2785–91.
222. Shi XJ, YL W, Li PP, Yi HF, Yang MD, Wang GH. Catalytic conversion of xylose to furfural over the solid acid SO_4^{2-}/ZrO_2-Al_2O_3/SBA-15 catalysts. Carbohydr Res. 2011;346(4):480–7.
223. Biofuels for Transport. An international perspective. Paris: Int Energy Agency; 2005. p. 169.
224. Katryniok B, Paul S, Capron M, Dumeignil F. Towards the sustainable production of acrolein by glycerol dehydration. ChemSusChem. 2009;2(8):719–30.
225. Alhanash A, Kozhevnikova EF, Kozhevnikov IV. Gas-phase dehydration of glycerol to acrolein catalysed by caesium heteropoly salt. Appl Catal A Gen. 2010;378(1):11–8.

226. Atia H, Armbruster U, Martin A. Dehydration of glycerol in gas phase using heteropolyacid catalysts as active compounds. J Catal. 2008;258(1):71–82.
227. Clacens JM, Pouilloux Y, Barrault J. Selective etherification of glycerol to polyglycerols over impregnated basic MCM-41 type mesoporous catalysts. Appl Catal A Gen. 2002;227(1–2):181–90.
228. Pagliaro M, Ciriminna R, Kimura H, Rossi M, Della Pina C. From glycerol to value-added products. Angew Chem Int Ed. 2007;46(24):4434–40.
229. Zhou CH, Beltramini JN, Fan YX, GQ L. Chemoselective catalytic conversion of glycerol as a biorenewable source to valuable commodity chemicals. Chem Soc Rev. 2008;37(3):527–49.
230. Brett GL, He Q, Hammond C, Miedziak PJ, Dimitratos N, Sankar M, Herzing AA, Conte M, Lopez-Sanchez JA, Kiely CJ, Knight DW, Taylor SH, Hutchings GJ. Selective oxidation of glycerol by highly active bimetallic catalysts at ambient temperature under base-free conditions. Angew Chem Int Ed. 2011;50(43):10136–9.
231. Gu Y, Liu S, Li C, Cui Q. Selective conversion of glycerol to acrolein over supported nickel sulfate catalysts. J Catal. 2013;301(0):93–102.
232. Possato LG, Diniz RN, Garetto T, Pulcinelli SH, Santilli CV, Martins L. A comparative study of glycerol dehydration catalyzed by micro/mesoporous MFI zeolites. J Catal. 2013;300(0):102–12.
233. Massa M, Andersson A, Finocchio E, Busca G, Lenrick F, Wallenberg LR. Performance of ZrO_2-supported Nb- and W-oxide in the gas-phase dehydration of glycerol to acrolein. J Catal. 2013;297(0):93–109.
234. Martin A, Armbruster U, Atia H. Recent developments in dehydration of glycerol toward acrolein over heteropolyacids. Eur J Lipid Sci Technol. 2012;114(1):10–23.
235. Chai SH, Wang HP, Liang Y, BQ X. Sustainable production of acrolein: gas-phase dehydration of glycerol over Nb_2O_5 catalyst. J Catal. 2007;250(2):342–9.
236. Cavani F, Guidetti S, Marinelli L, Piccinini M, Ghedini E, Signoretto M. The control of selectivity in gas-phase glycerol dehydration to acrolein catalysed by sulfated zirconia. Appl Catal B Environ. 2010;100(1–2):197–204.
237. Ning X, Li Y, Yu H, Peng F, Wang H, Yang Y. Promoting role of bismuth and antimony on Pt catalysts for the selective oxidation of glycerol to dihydroxyacetone. J Catal. 2016;335:95–104.
238. Pescarmona PP, Janssen KPF, Delaet C, Stroobants C, Houthoofd K, Philippaerts A, De Jonghe C, Paul JS, Jacobs PA, Sels BF. Zeolite-catalysed conversion of C(3) sugars to alkyl lactates. Green Chem. 2010;12(6):1083–9.
239. de Clippel F, Dusselier M, Van Rompaey R, Vanelderen P, Dijkmans J, Makshina E, Giebeler L, Oswald S, Baron GV, Denayer JFM, Pescarmona PP, Jacobs PA, Sels BF. Fast and selective sugar conversion to alkyl lactate and lactic acid with bifunctional carbon-silica catalysts. J Am Chem Soc. 2012;134(24):10089–101
240. Coppola GM, Schuster HF. α-Hydroxy Acids in enantioselective synthesis. Weinheim: VCH; 1997.
241. Pereira CSM, Silva VMTM, Rodrigues AE. Ethyl lactate as a solvent: properties, applications and production processes – a review. Green Chem. 2011;13(10):2658–71.
242. Datta R, Henry M. Lactic acid: recent advances in products, processes and technologies — a review. J Chem Technol Biotechnol. 2006;81(7):1119–29.
243. Yang L, Yang X, Tian E, Lin H. Direct conversion of cellulose into ethyl lactate in supercritical ethanol–water solutions. ChemSusChem. 2016;9(1):36–41.

Chapter 4
Design of Bifunctional Solid Catalysts for Conversion of Biomass-Derived Syngas into Biofuels

Hao Wang, Yan Pei, Minghua Qiao, and Baoning Zong

Abstract Biomass resources can be thought of as an important source of syngas ($CO + H_2$). The conversion of the biomass-derived syngas (bio-syngas) into biofuels represents a carbon dioxide-neutral route for the production of substitute of the petroleum-derived fuels. Fischer–Tropsch synthesis (FTS) is by far the most effective approach to convert syngas into biofuels. However, FTS produces unselectively normal aliphatic hydrocarbons with a broad distribution of carbon numbers. To selectively produce biofuels, bifunctional FTS catalysts composed of an FTS-active metal and an acidic zeolite can be used. Such catalysts are capable of narrowing down the distribution of the products into the liquid fuel fraction by means of hydrocracking, isomerization, and hydrogenolysis. At the beginning of this chapter, the pathways of biofuel production from lignocellulosic biomass and the processes involved in transforming lignocellulosic biomass to bio-syngas are introduced. Then, research activities on converting syngas into biofuels over bifunctional FTS catalysts are described and ideas for catalyst engineering, catalyst structure, and catalytic outcome are highlighted. This chapter describes some of the selectivity control strategies that can be adopted with bifunctional catalysts so that restrictions given by classical Anderson–Schulz–Flory distribution can be overcome and industrialization can be realized.

H. Wang • Y. Pei • M. Qiao (✉)
Collaborative Innovation Center of Chemistry for Energy Materials, Department of Chemistry and Shanghai Key Laboratory of Molecular Catalysis and Innovative Materials, Fudan University, Shanghai, People's Republic of China
e-mail: mhqiao@fudan.edu.cn

B. Zong (✉)
State Key Laboratory of Catalytic Materials and Chemical Engineering, Research Institute of Petroleum Processing, SINOPEC, Beijing, People's Republic of China
e-mail: zongbn.ripp@sinopec.com

Z. Fang et al. (eds.), *Production of Biofuels and Chemicals with Bifunctional Catalysts*, Biofuels and Biorefineries 8, https://doi.org/10.1007/978-981-10-5137-1_4

4.1 Introduction

The growing concern of petroleum depletion has spurred worldwide interest in exploring alternative ways to sustainably produce chemical commodities and liquid fuels. In particular, the ever-increasing demand of transportation fuels [1] has prompted enormous efforts in producing biofuels from biomass, which simultaneously reduces greenhouse-gas emissions.

In general, biofuels produced from biomass can be classified as being first generation or second generation. The first generation biofuels include biodiesel catalytically transformed from vegetable oils and bioethanol fermented from grain (rice, wheat), sugarcane, potatoes, and corns, i.e., edible biomass as feedstocks. It is apparent that feedstocks for first-generation biofuels are not abundant [2]. For the sake of high expenditure for the feedstocks, first generation biofuels can only satisfy a small portion of the consumption of transportation fuels [1], and their further development is uncertain unless efficacious fermentation strategies are developed for the utilization of the more widely available, inedible lignocellulosic biomass as the feedstock.

The second generation biofuels are devoid of the shortcoming of feedstock deficiency faced by first generation biofuels. They can be produced from lignocellulosic biomass ranging from agriculture, forestry, energy crop planting, and food residues and byproducts. Hence, second generation biofuels are free from competition with food production and have good prospects for industrial application [3, 4].

There are two major routes to transform lignocellulosic biomass into biofuels that are classified as biochemical or thermochemical route (Fig. 4.1) [4]. Biochemical transformation of lignocellulosic biomass into biofuels has low efficiency [5], and direct pyrolysis/liquefaction of lignocellulosic biomass to biofuels has low selectivity. The third route, biomass-to-liquids (BTL), is very promising since lignocellulosic biomass is gasified to small molecules and then the bio-syngas is used as the feedstock [5]. BTL is constituted by processes of gasification, gas cleaning and conditioning, and Fischer–Tropsch synthesis (FTS) (Fig. 4.2), which is analogous to the well-established industrial coal-to-liquids (CTL) process or the natural gas-to-liquids (GTL) process [6]. The H_2/CO volume ratio in bio-syngas is close to 1, making it a suitable feedstock for the production of light olefins. To produce liquid fuels, this H_2/CO ratio can be improved by implementing a water–gas shift (WGS) reactor. Alternatively, one can adopt a Fe-containing FTS catalyst that is also active in catalyzing the WGS reaction.

As the first step of BTL, biomass gasification is used to transform the lignocellulosic biomass into a gas-phase mixture under high operating temperatures in the presence of a gasifying agent. The gasification agent usually contains oxygen, air, steam, or their mixture. The gaseous product is a mixture of H_2, CO, CO_2, CH_4, C_2H_4, and N_2, in addition to several kinds of byproducts, including tars (benzene and other aromatic hydrocarbons), nitrogen-containing compounds (primarily NH_3 and HCN), hydrogen sulfide (H_2S), hydrogen chloride (HCl), and other alkali metals in biomass (K, Na, Ca, and Mg) [4, 7]. The quality of bio-syngas can be improved by adjusting the gasification conditions, such as gasifying agent, gasifying temperature, and gasification reactors (gasifiers) [8].

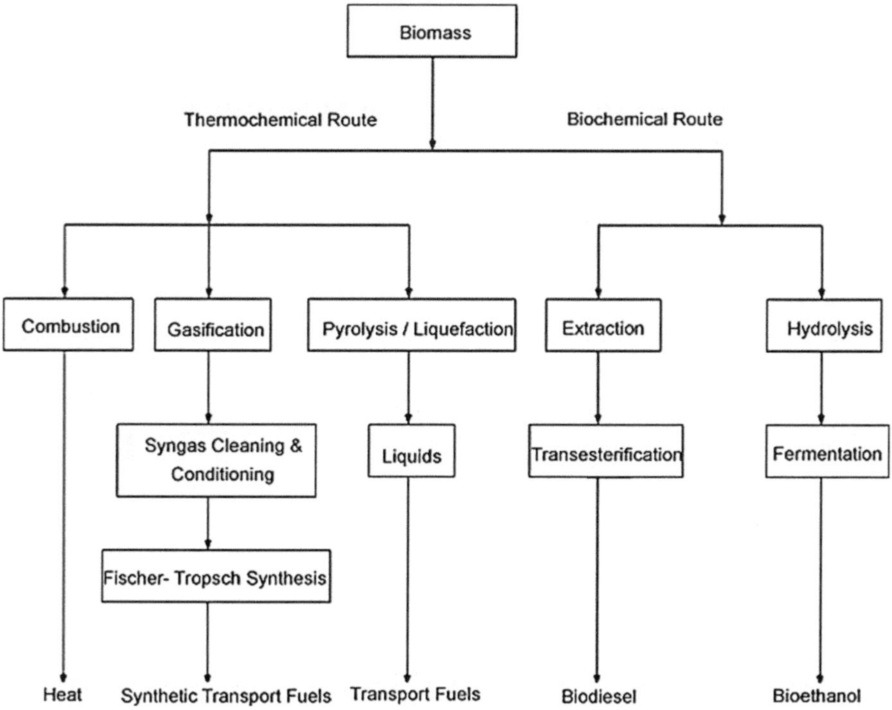

Fig. 4.1 Schematic illustration of major routes for conversion of biomass into biofuels (Reprinted with permission from [4], Copyright © 2011 Elsevier)

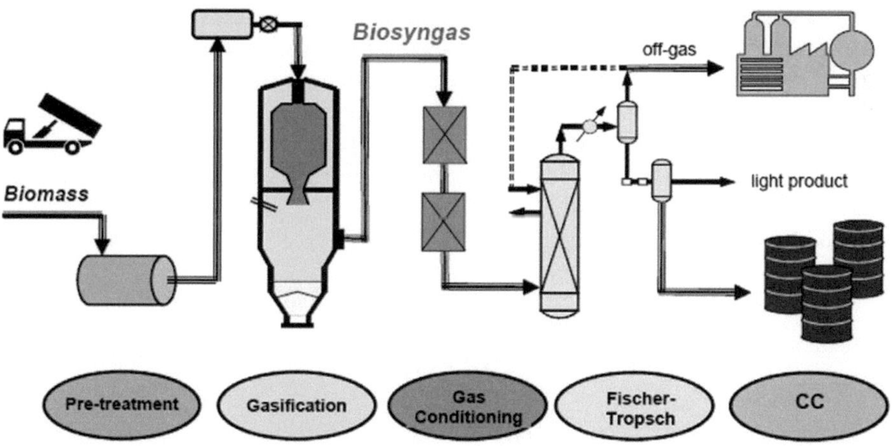

Fig. 4.2 Schematic diagram for thermochemically converting biomass into biofuels (Reprinted with permission from [9], Copyright © 2009 John Wiley & Sons, Ltd.)

There are mainly two types of gasifiers for biomass gasification, that is, fluidized-bed gasifier and fixed-bed gasifier [10, 11]. The former is often operated at a comparatively low temperature of about 1173 K. The bio-syngas mixture from this gasifier is mainly composed of H_2 and CO, along with unwanted products such as methane, CO_2, and higher hydrocarbons. The latter requires temperature as high as 1373 K. By means of fixed-bed gasification, all biomass can be completely transformed into bio-syngas containing only H_2 and CO, which is chemically equivalent to and can displace syngas derived from fossil sources.

Gas cleaning and conditioning is an essential step to remove the impurities from bio-syngas, such as particulate matter, tars, and sulfur- and nitrogen-containing compounds. During FTS for converting bio-syngas into biofuels, the particulate matter may damage the gas-handling equipment. The sulfur- and nitrogen-containing compounds may poison the catalyst while the tars may cause equipment fouling. Gas cleaning and conditioning can be achieved in a filtration system downstream from the gasifier for the capture of the particulate matter. Sorbents are added to remove the sulfur- and nitrogen-containing compounds. Furthermore, catalytic thermal cracking or steam reforming is implemented to diminish the tars [4].

4.2 Fischer–Tropsch Synthesis

FTS is capable of converting syngas into paraffins, olefins, alcohols, and aldehydes [4]. In the 1920s, Franz Fischer and Hans Tropsch firstly reported on the conversion of syngas to long-chain hydrocarbons [12]. FTS was commercialized in Germany in the 1930s for CTL to reduce the petroleum dependence of industry. Plants for GTL have also been successfully established [13]. Synthetic biofuels produced by FTS have several advantages over conventional petroleum-derived fuels in that they are almost free of aromatics, sulfur, and nitrogen, which means that fewer particles and pollutants would be released after combustion. Synthetic diesel also has very good ignition properties with a high cetane number of around 70. Therefore, much research effort has been dedicated to commercializing the BTL process using biomass as the feedstock [14, 15].

The FTS reaction can be represented by Eqs. (4.1) and (4.2):

$$n\mathrm{CO} + 2n\mathrm{H}_2 \rightarrow \mathrm{C}_n\mathrm{H}_{2n} + n\mathrm{H}_2\mathrm{O} \tag{4.1}$$

$$n\mathrm{CO} + (2n+1)\mathrm{H}_2 \rightarrow \mathrm{C}_n\mathrm{H}_{2n+2} + n\mathrm{H}_2\mathrm{O} \tag{4.2}$$

Many reviews have been published on the mechanism [16–21] and kinetics of FTS [18, 22, 23]. Generally, CO undergoes direct dissociation or hydrogen-assisted dissociation and hydrogenation to generate the monomeric CH_x species ($x = 0$–3). Then, the monomeric CH_x species polymerize into straight-chain species through C–C coupling and yield C_nH_m intermediates. Finally, the C_nH_m intermediates undergo hydrogenation, dehydrogenation, or carbonylation to produce paraffins, olefins, or alcohols, respectively [24, 25].

Table 4.1 Influence of reaction parameters on product selectivity in Fischer–Tropsch synthesis

	Carbon chain			Selectivity		
Increase in parameter	Length	Branching	Carbon deposition	Olefin	Alcohol	Methane
Temperature	↓	↑	↑	C	↓	↑
Pressure	↑	↓	C	C	↑	↓
H_2/CO ratio	↓	↑	↓	↓	↓	↑
Conversion	C	C	↑	↓	↓	↑
Space velocity	C	C	C	↑	↑	↓

Adapted with permission from [18], Copyright © 1999 Taylor& Francis
Note: ↑: increase; ↓: decrease; C: complex

The distribution of the FTS products follows the Anderson–Schulz–Flory (ASF) kinetic model derived from the polymerization mechanism and can be depicted by eq. (4.3):

$$M_n = (1-\alpha)\alpha^{n-1} \tag{4.3}$$

In this equation, M_n is the molar fraction of each FTS product with carbon number n. α is the chain-growth probability, defined as a function of the propagation and termination rates of the hydrocarbon chain. Larger values of α indicate high selectivity for long-chain hydrocarbons. By plotting $\log(M_n/n)$ against n, the ASF plot should ideally display a straight line, with the slope being α. The product distribution sometimes deviates from the ASF distribution caused by the readsorption of short α-olefins and the occurrence of secondary reactions [26].

Factors affecting the α value include temperature, pressure, syngas composition (H_2/CO ratio) [27], catalyst type, catalyst composition, and reactor type [19]. Table 4.1 summarizes the influence of operating parameters on the chain length as well as other properties of the products [18, 28]. In general, the α value increases when increasing pressure or decreasing temperature. In industry, FTS is conducted at temperatures between 463 K and 623 K at moderate pressures of 15–40 bar [15]. Depending on the products of interest, FTS is operated in two different temperature regions: low-temperature FTS (LTFTS) at 463–513 K and high-temperature FTS (HTFTS) at 573–623 K. Because a larger α value is expected during LTFTS, while a smaller α value is expected during HTFTS [29], the former is mainly for the synthesis of diesel and waxes, whereas the latter is mainly for gasoline or light olefins.

Due to the restriction imposed by the ASF distribution, the exclusive production of a specific hydrocarbon fraction by FTS is not feasible with conventional FTS catalysts, except for methane ($\alpha = 0$) and an infinite chain-length wax ($\alpha = 1$). As illustrated in Fig. 4.3 [30], biofuels cannot be produced with high selectivity. For instance, the maximum selectivity of the gasoline-range fuels (C_5–C_{11}) is *ca.* 45% ($\alpha = 0.76$) and that of the diesel-range fuels (C_{12}–C_{20}) is only *ca.* 30% ($\alpha = 0.89$) [31]. As a compromise, the FTS industry adopts an indirect two-stage approach to improve the selectivity for liquid fuels. In the first stage, the FTS catalyst with an α value larger than 0.9 is used to minimize the production of the short-chain

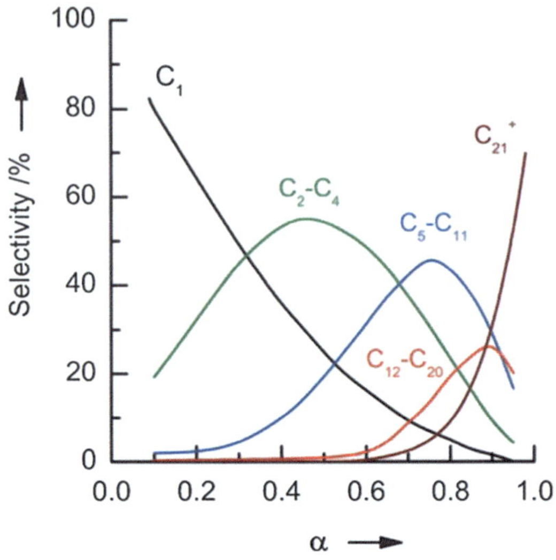

Fig. 4.3 Distribution of Fischer–Tropsch synthesis (FTS) products predicted by the Anderson–Schulz–Flory (ASF) law (Reprinted with permission from [30], © 2014 WILEY-VCH Verlag GmbH & Co. KGaA, Weinheim)

hydrocarbons. In the second stage, the waxes are subjected to hydrotreatment (hydrocracking, isomerization) in another reactor to maximize the production of the desired fuels that are in most cases, diesel [6, 32]. Aside from the need of an extra hydrotreatment reactor, this two-stage approach also requires an extra high-pressure H_2 plant for hydrotreatment, which greatly adds to the investment cost and increases the safety risk. Therefore, the development of catalysts that can afford high selectivity to liquid fuels in one step is a central issue in the realm of FTS [31, 33, 34].

Conceptually, the development of such catalytic systems can be achieved by combining one catalyst component for CO hydrogenation with another one for hydrocracking or isomerization of long-chain hydrocarbons to liquid fuels to obtain FTS catalysts with the desired bifunctionality [25, 30, 35, 36]. In practice, the former is usually borrowed from conventional FTS catalysts. The latter is based on solid acid catalysts, which are generally chosen as in most cases, acidic zeolites. Many bifunctional FTS catalysts with different structures and purposes have been designed and evaluated, which showed attractive results in liquid fuel production. Understanding the characteristics and functions of each catalyst component is a prerequisite to the rational formulation of more effective bifunctional FTS catalysts for converting bio-syngas directly into biofuels.

4.3 Bifunctional FTS Catalyst

4.3.1 FTS-Active Metal

All group VIII metals are able to catalyze CO hydrogenation to hydrocarbons. The average molecular weight of hydrocarbons produced descends in the order of Ru > Fe > Co > Rh > Ni > Ir > Pt > Pd. But only four of them, Ru, Fe, Co, and Ni, show appreciable activity in FTS. Ru is the most active one and can be used even below 463 K. Moreover, the large α value provided by Ru makes it an ideal candidate as the CO hydrogenation component of bifunctional FTS catalysts. However, Ru is scarce and has a high price [31, 37]. Ni is not applicable due to its high selectivity to low-value methane characteristics [37, 38]. Hence, despite numerous reports on new FTS catalysts, the Fe- and Co-based catalysts retain their dominant position in the FTS industry in terms of both catalytic performance and economy.

The Fe-based catalysts can be used over a broad temperature range and are therefore suitable for both LTFTS and HTFTS [39]. Aside from the low cost, high versatility in product control, and relatively high tolerance to sulfur impurities in the syngas, the Fe-based catalysts also display a unique activity in the WGS reaction, thus favorably compensating the deficiency of H_2 in syngas derived from coal and biomass. There are two types of commercial Fe-based FTS catalysts, i.e., fused Fe and precipitated Fe [40], which are mainly employed in HTFTS and LTFTS, respectively. The major shortcoming of the Fe-based catalysts is the relatively short lifetime. Systematic investigations have found that factors underlying Fe-based FTS catalyst deactivation include sintering, surface carbon deposition, mechanical attrition, oxidation, and poisoning [39, 41, 42].

The Co-based catalysts are more costly, but are more selective for long-chain linear hydrocarbons than the Fe-based catalysts. The Co-based catalysts are therefore preferred in LTFTS over the Fe-based catalysts to produce diesel and waxes. However, the Co-based catalysts are sensitive to changes in the operating temperature and pressure. A drastic and unwanted increase in methane formation will occur when temperature is increased [41, 43]. On the other hand, the Co-based catalysts are more resistant to deactivation caused by water than the Fe-based catalysts. Hence, although the Co-based and Fe-based catalysts are comparable in activity at low CO conversions, for high CO conversions, the Co-based catalysts are superior to the Fe-based catalysts [42, 43]. However, since the Co-based catalysts display negligible WGS activity, improving the H_2/CO ratio in bio-syngas is necessary prior to FTS. Therefore, it is anticipated that the Fe–Co bimetallic catalysts, with Fe possessing high WGS activity and Co possessing good water-resistance, are promising candidates for BTL. Sonal et al. have studied a series of Fe–Co bimetallic catalysts for FTS at low H_2/CO ratio [44, 45]. The addition of Fe increased the WGS activity, CO conversion, and C_{5+} hydrocarbon selectivity as compared to the monometallic Fe or Co catalyst.

4.3.2 *Zeolite*

Zeolites are microporous crystalline aluminosilicate materials with uniform pore sizes, cavities, and regularly arranged channels. The unique pore structure can control the diffusion and reaction of the reactants and the reaction intermediates [46]. Along with the shape selectivity and the acidic property, the zeolites have been widely used to tailor the product selectivity in the catalytic conversion of biomass [47]. For bifunctional FTS catalysts, the acidity, topology, and porosity of the zeolites are the main factors that affect catalytic performance.

4.3.2.1 Acidity

Zeolites have been used as a constituent in bifunctional FTS catalysts to alter product selectivity mainly by taking advantage of their controllable acidity. Goldwasser et al. prepared a series of the Fe/H-ZSM-5 catalysts with the Si/Al ratios in H-ZSM-5 between 19 and 114 [48]. The results showed that the acidity of H-ZSM-5 significantly influenced the product distribution. Cracking was not prominent at low density or strength of the acidic sites, thus long-chain hydrocarbons (C_{5+}) were dominant. When increasing the density or strength of the acidic sites, more long-chain hydrocarbons were cracked into lower molecular-weight paraffins and olefins [48]. Similar results were reported by Botes and Böhringer [49]. On the catalyst containing alkali-promoted Fe and a high-acidity H-ZSM-5 zeolite (Si/Al ratio = 30), high activity towards aromatics and light paraffins was observed at the beginning of the reaction at 603 K and 20 bar. However, rapid deactivation of the zeolite occurred during the reaction, which affected negatively on the selectivity to the gasoline-range fuels. In contrast, despite the catalyst containing a low-acidity H-ZSM-5 zeolite (Si/Al ratio = 280) was much less active than the one containing a high-acidity H-ZSM-5 zeolite, it deactivated less significantly and afforded a 25–35% increase in the selectivity to the gasoline-range fuels.

Martínez and López physically mixed a K/Fe/Co catalyst with H-ZSM-5 having Si/Al ratios of 15, 25, 40, and 140 [50]. At 583 K, 20 bar, and for a H_2/CO ratio of 1, the addition of zeolites shifted the selectivity to high-octane C_5–C_{11} *iso*-paraffins by transforming the primary hydrocarbons on the acid sites through cracking of long-chain hydrocarbons to the short ones, isomerization of *n*-paraffins, or aromatization of light α-olefins. The probability of the occurrence of these secondary reactions is related to the crystallite size and the Si/Al ratio of the zeolite. A higher density of the Brønsted acid sites (at Si/Al ratios of 15–25) favored the aromatization of light α-olefins. The olefin to paraffin (O/P) ratio in the C_2–C_4 range decreased with a decrease in the crystal size and the amount of the Brønsted acid sites of H-ZSM-5. Small zeolite sizes facilitated diffusion of the aromatics and hence avoided their condensation into carbonaceous deposits, so that the deactivation of H-ZSM-5 was suppressed [50].

4.3.2.2 Topology

The catalytic performances of zeolites in bifunctional FTS catalysts are related to their topological properties. In general, selectivity for the desired FTS products can be maximized by choosing a zeolite with optimal topology [51–53]. Martínez et al. reported on the effect of zeolites with different topologies (Beta, USY, Mordenite, and ZSM-5) in combination with a Co/SiO_2 catalyst in cracking long-chain hydrocarbons into gasoline-range fuels with FTS [51]. It was shown that the deactivation rate and the amount of carbonaceous deposits increased in the order of H-ZSM-5 < H-Mordenite < H-Beta < USY. Large-pore H-Mordenite with monodirectional pore systems produced smaller amounts of aromatic-type coke and were more stable than the large-pore H-Beta and USY zeolites with tri-directional interconnected channel systems, while the medium-pore H-ZSM-5 with bi-directional interconnected channel system showed the lowest deactivation rate. The formation of aromatic-type coke was impeded on the H-ZSM-5 zeolite owing to the narrow 10-MR (membered ring) pore openings, thus leaving more acid sites to crack and isomerize heavy hydrocarbons than the 12-MR zeolites (H-Beta and USY).

The authors further investigated the catalytic performances of medium-pore (10- or 12-MR) zeolites (ZSM-5, MCM-22, IM-5, and ITQ-2) in conjugation with a Co/SiO_2 catalyst to produce high-quality gasoline-range fuels directly from syngas under the reaction conditions of 523 K, H_2/CO ratio of 2, and 20 bar [52]. The activity decreased in the order of ITQ-2 > MCM-22 > IM-5 > ZSM-5 after 15 h on stream mainly attributed to coking during the reaction. It was observed that for ITQ-2, the majority of the coke was located on the external surface of the zeolite, which provided an unrestricted space for coke accumulation. In the case of MCM-22, most of the coke resided inside its large 12-MR supercage, which restrained its escape through the narrower 10-MR channels. For IM-5, most of the coke preferentially resided within the 10-MR channels, while for ZSM-5 the coke mainly resided on the exterior of the 10-MR pore openings.

It was reported that incorporating Fe into ZSM-11/12, beta, omega, or LTL zeolite by the impregnation, ion exchange, or precipitation method could increase the production of olefins and *iso*-paraffins and the resistance to coking [54]. Large-pore (12-MR) LTL is usually commercially used to catalyze hydrocarbon conversions such as reforming [55, 56]. For isomorphously Fe-substituted LTL, low CO conversions and improved selectivity for methane were reported, which was attributed to the low content of iron in the isomorphously substituted material [56]. Cagnoli et al. impregnated iron on the potassium-form of LTL (K-LTL). The as-reduced Fe/K-LTL catalyst had two different types of microcrystals of Fe^0 distinguished by their location inside the structure or on the exterior of K-LTL. After reaction, the former remained intact, while the latter was carburized [57]. The higher selectivity for light olefins on the Fe/K-LTL catalyst compared with the Fe/Al_2O_3, Fe/C, and Fe/SiO_2 catalysts was attributed to the promotion effect of potassium and the high percentage of the Fe^0 microcrystals inside K-LTL. On the other hand, the carburized Fe^0 on the exterior of K-LTL might be responsible for the improved selectivity for the diesel-range fuels.

4.3.2.3 Porosity

The intrinsic pore size of the zeolites is smaller than 2 nm, which falls in the range of microporosity. Microporous zeolites allow the occurrence of secondary reactions of primary FTS products on the acid sites inside channels or cavities, thus enhancing the selectivity of a certain product spectrum through pore size screening [58]. In this connection, the formation of long-chain hydrocarbons could be diminished and the product distribution could be shifted towards liquid fuels on microporous zeolites [59].

Li et al. synthesized ZSM-5 microspheres embedded with well-dispersed uniform Fe_3O_4 nanoparticles (Fe_3O_4@ZSM-5) using an *in situ* crystallization route [60]. For the synthesis of this catalyst, a Fe_3O_4@SiO_2 composite with Fe_3O_4 nanoparticles of ~10 nm in diameter confined in the pores of SiO_2 was used as the precursor, which was subjected to hydrothermal treatment in a weak basic solution. The resulting zeolite microspheres with uniform sizes between 6 μm and 9 μm were composed of concentric nanorod-like ZSM-5 crystals and Fe_3O_4 nanoparticles embedded between the nanorods. For the bifunctional Fe_3O_4@ZSM-5 catalyst, the selectivity for gasoline-range fuels was as high as 44.6% after 110 h on stream. In contrast, the selectivity of the Fe_3O_4@ZSM-5 catalyst for diesel-range fuels was low (3.2%), and no C_{21+} products were detected.

Although microporous zeolite-based bifunctional FTS catalysts are capable of producing the gasoline-range fuels, incorporating metal into the microporous zeolites sometimes has adverse effects on the activity or selectivity for the following reasons: (1) reducibility of the metal inside the zeolite structure may be lowered due to the strong metal–support interactions [61, 62], and (2) mass transport limitations in microporous zeolite may increase the H_2/CO ratio with respective to the original ratio, thus improving the selectivity for methane and decreasing the olefin/paraffin ratio. Microporous zeolites may also cause over-cracking of the FTS products due to prolonged residence time on the acid sites [63, 64].

Imparting mesoporosity to microporous zeolites has attracted great attention in recent years [65–72]. Mesoporous zeolites having bimodal porous structures, which are also referred to as hierarchical zeolites [73, 74], possess both the advantages of traditional microporous zeolites and high efficiency of mesoporous materials in mass transport. Mesoporous zeolites have shown enhanced performance in acid-catalyzed reactions in the petrochemical industry, such as in hydrocracking and alkylation [75]. Rangel et al. developed a mesostructured Beta zeolite (MB) supported Co catalyst for FTS [76]. The MB zeolite exhibited a lower micropore/mesopore volume ratio (20–24%) than the pristine Beta zeolite. Owing to the mesoporous structure, the Co/MB catalyst showed not only higher activity, but also lower selectivity for methane and higher selectivity for C_{6+} hydrocarbons than the Co/Beta catalyst.

Wang and co-workers synthesized mesoporous ZSM-5 (meso-ZSM-5) and Beta (meso-Beta) zeolites by treating H-ZSM-5 and Beta zeolites with NaOH aqueous solutions to construct bifunctional FTS catalysts [63, 77, 78]. The hierarchical meso-ZSM-5 had micropores of *ca.* 0.55 nm in size and mesopores with sizes of *ca.* 3.8–7.2 nm depending on the concentration of NaOH. As compared to pristine H-ZSM-5, the meso-ZSM-5 supported Ru nanoparticles had markedly decreased selectivities for methane and C_2–C_4 hydrocarbons, while increased selectivity for gasoline-range fuels (Fig. 4.4). The selectivity on the Ru/meso-ZSM-5 catalyst was closely related to the concentration of NaOH during the treatment of ZSM-5. For meso-ZSM-5 treated by 1.0 M NaOH, the selectivity to CH_4 decreased to about 5%, while the selectivity to the gasoline-range fuels increased to about 80% (Fig. 4.4). The CO conversion also increased with an increase in the concentration of NaOH

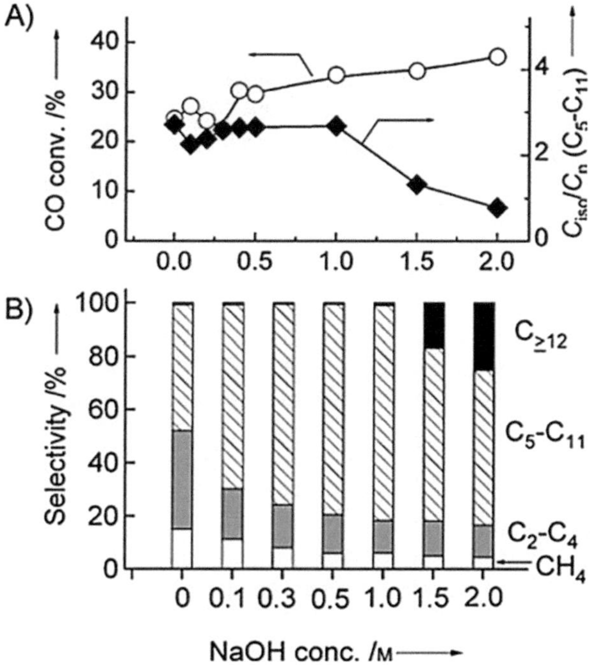

Fig. 4.4 Fischer–Tropsch synthesis results over the Ru/meso-ZSM-5 catalysts prepared by treating H-ZSM-5 with NaOH of different concentrations. (**a**) CO conversion and C_{iso}/C_n ratio; (**b**) Product selectivity. Reaction conditions: 0.5 g catalyst, $H_2/CO = 1.0$, temperature of 533 K, pressure of 20 bar, total flow rate of 20 mL min^{-1}, and time on stream of 12 h (Reprinted with permission from [63], Copyright © 2011 WILEY-VCH Verlag GmbH & Co. KGaA, Weinheim)

probably due to enhanced mass transport (Fig. 4.4). The *iso*-paraffin to normal paraffin (C_{iso}/C_n) ratio in the C_5–C_{11} range was approximately 2.5–2.7 at the NaOH concentration of ≤1.0 M (Fig. 4.4), showing promise of the Ru/meso-ZSM-5 catalyst for the production of high-quality gasoline.

Sartipi et al. prepared two meso-H-ZSM-5 supports by treating ZSM-5 with either NaOH aqueous solution or tetrapropylammonium hydroxide (TPAOH) organic base [64]. Under similar treatment conditions, TPAOH gave mild desilication of ZSM-5, thus generating mesostructures with pores in the range of 4–8 nm. In contrast, NaOH led to a severe desilication, creating mesostructures with pores in the range of 15–32 nm. It was found that the CO conversion was about 15% higher on the Co/meso-H-ZSM-5 catalyst treated with TPAOH than on that treated with NaOH. Both Co/meso-H-ZSM-5 catalysts exhibited higher CO conversions than the untreated Co/H-ZSM-5 catalyst [64, 79]. Imparting mesoporosity to H-ZSM-5 could create mesoporous surfaces that facilitate the dispersion of the Co nanoparticles and increase the amount of the Co nanoparticles at the vicinity of the acid sites [79]. In the preparation of the Co/meso-H-ZSM-5 catalyst, it was observed that most Co nanoparticles were well distributed on the external surface, and the rest were located inside the mesopores. However, the Co nanoparticles were inhomogeneously dispersed on the external surface of the untreated H-ZSM-5. The conversion over the Co/meso-H-ZSM-5 catalyst was at least 28% higher than the one over the Co/H-ZSM-5 catalyst after 30 h on stream. The homogeneously dispersed Co nanoparticles over meso-H-ZSM-5 might account for its enhanced activity and stability [64].

The authors [80] further studied three mesoporous zeolites (meso-H-ITQ-2, meso-H-USY, and meso-H-ZSM-5) with different topologies as supports for cobalt. The catalytic activities over the Co/meso-H-USY and Co/meso-H-ZSM-5 catalysts were similar, but they were higher than those over the Co/ITQ-2 and Co/SiO_2 catalysts. The difference between the FTS activities of the different catalysts was attributed to the dispersion of Co in the materials. The Co/meso-H-USY and Co/meso-H-ZSM-5 catalysts were identified to have smaller Co particles that were well dispersed.

Aside from enhancing the selectivity for the gasoline-range fuels, mesoporous zeolites are also capable of enhancing the selectivity for diesel-range fuels. The diesel fuels produced by FTS have merits of reduced NO_x and particulate matter (PM) emissions [81]. Wang and co-workers discovered that a sodium-form mesoporous zeolite Y (Na-meso-Y) supported Co nanoparticles were highly selective for diesel-range fuels with a value being ~60%, along with a desirably low C_{iso}/C_n ratio (0.3) and a low methane selectivity (5%) [82]. During 515 h on stream, the Co/Na-meso-Y catalyst remained stable, which might be due to alkali promotion that impeded coke formation. Surprisingly, the Co/Na-meso-Y catalyst did not have any Brønsted acid sites, indicating that the acid-catalyzed hydrocracking/isomerization mechanism is not responsible for the enhanced diesel production. Instead, the authors proposed that the hydrogenolysis mechanism, which has long been regarded as an undesired secondary reaction that forms excessive methane, gave rise to the high selectivity to diesel. This peculiar mechanism of selectivity control was substantiated by the conversion of *n*-hexadecane model molecule on the Co/meso-Y catalysts with and without Brønsted acid sites. It was found that the sizes of the Co nanoparticles and the mesopores were two critical factors affecting the catalytic performance. The Co nanoparticles and mesopores with medium sizes could afford higher selectivity of the catalyst for diesel-range fuels.

4.3.3 Type of Bifunctional FTS Catalyst

On the basis of the catalyst structure, bifunctional FTS catalysts can be divided into three categories: (1) physically mixed catalysts [50, 51, 83–85], (2) zeolite-supported catalysts [61, 64, 77–80, 86–89], and (3) zeolite-coated catalysts [90–98].

4.3.3.1 Physically Mixed Catalyst

For physically-mixed bifunctional FTS catalysts, there are two packing configurations. One is the dual-layer packing configuration, with a separate zeolite layer being packed downstream of the FTS catalyst layer [25, 35, 49]. The other is the single-layer packing configuration, with the two catalyst components being homogeneously mixed [25, 35, 51, 85, 99].

Both Fe- and Co-based bifunctional FTS catalysts have been investigated in these two packing configurations. In general, the single-layer configuration leads to more *iso*-paraffins [49, 84, 100–102] due to an enhanced contribution of the acid-catalyzed reactions. However, severe alkali migration from a fused Fe catalyst to H-ZSM-5 during the course of reaction took place for the catalysts packed in the single-layer configuration [49]. As a result, the activity declined, and the selectivity shifted considerably towards low-value light paraffins including methane [49]. This shortcoming could be amended by substitution of La for alkali as the promoter. La with larger atomic number is less mobile than alkali and has been reported as an excellent promoter to enhance the product selectivity and lifetime of the catalysts [103, 104]. Pour et al. found that over a La-promoted Fe catalyst, higher selectivity for the gasoline fuels was obtained in the single-layer configuration [103].

Analogous to Fe-based catalysts, Co-based bifunctional FTS catalysts showed increased selectivity to the gasoline-range fuels in the single-layer configuration [25, 30, 35, 36, 84, 100, 101]. Schaub and co-workers revealed that under identical reaction conditions, both packing configurations of the Co-based catalysts led to similar fractions of liquid fuels (C_5–C_{20}), however, the gasoline fraction was higher over the catalyst in the single-layer configuration than in the dual-layer configuration [100].

Many studies reported that the methane selectivity is enhanced when the packing configuration is changed from the dual-layer configuration to the single-layer configuration [84, 100, 103, 105, 106]. To have an insight into this difference, Gormley et al. added inert SiO_2 to a homogeneously mixed Co/SiO_2 and H-ZSM-5 catalyst and found that the methane selectivity was reduced by one half, while it did not change without the zeolite [106]. Moreover, fewer percentages of aromatics in the liquid products were detected after dilution by SiO_2. More recently, Co/SiO_2 were homogeneously mixed with H-USY, H-Beta, H-Mordenite, or H-ZSM-5 and diluted with the highly thermoconductive SiC [51, 52]. It was found that the presence of zeolites did not add up to methane production. These observations show that the acidic zeolites do not lead to excessive methane production. Therefore, as far as the reaction heat can be dissipated in time by proper catalyst and reactor engineering, it is promising to take full advantage of the catalyst packed in the single-layer configuration without producing excessive methane.

4.3.3.2 Supported Catalyst

For supported bifunctional FTS catalysts, there are two important parameters that determine the product selectivity, i.e., acidity and pore structure. Bessell found that the selectivity to *iso*-paraffins depended strongly on the acidity of the support, while the FTS products followed the classic ASF distribution by using low-acidity supports. Compared with Y and H-Mordenite zeolites supported catalysts, the ZSM-5 supported Co catalyst displayed the highest selectivity for *iso*-paraffins owing to its strongest acidity [86].

To disclose the effect of the pore structure of the zeolite on the catalytic performance, Bessell prepared a series of H-ZSM-5, H-ZSM-11, H-ZSM-12, and H-ZSM-34 supported Co catalysts [107]. Catalytic evaluation indicated that the activity increased with the pore size of the zeolite. The 12-MR ZSM-12 supported Co catalyst exhibited the highest activity, followed by the two 10-MR ZSM-5 and ZSM-11 supported Co catalysts, with the 8-MR ZSM-34 supported Co catalyst showing the lowest activity.

4.3.3.3 Zeolite-Coated Catalyst

The zeolite-coated bifunctional FTS catalysts are also called core–shell catalysts or capsule catalysts [95, 97, 98, 108–110]. Tsubaki and co-workers developed a series of core–shell-structured bifunctional FTS catalysts by a hydrothermal synthesis method [90, 91, 95, 108, 111–113]. Namely, the Co-based catalyst pellets were added into the zeolite precursor solution, which was then subjected to hydrothermal treatment. In some cases, the pretreatment of the Co-based catalyst may be necessary to improve the affinity of the catalyst surface to zeolites such as H-Beta [91]. The coating of the Co/SiO_2 catalyst with a H-ZSM-5 membrane improved the selectivity to the gasoline-range fuels and suppressed the formation of C_{11+} hydrocarbons [90, 95]. The C_{iso}/C_n ratio increased from 0.37 to 1.88 with the increase in the amount of the zeolite coating from 11.5 to 24.3%. However, the methane selectivity increased concomitantly.

Huang et al. synthesized a Co/ZrO_2@H-ZSM-5 catalyst by the hydrothermal synthesis method [92]. The Co/ZrO_2 catalyst itself displayed higher selectivity to C_{5+} hydrocarbons [113]. The core–shell catalyst exhibited a higher selectivity to the gasoline-range *iso*-paraffins (23–25%) than the homogeneously mixed Co/ZrO_2 and H-ZSM-5 catalyst (17%), though the CO conversion on the former was lower.

Xu et al. prepared a skeletal Co@H-ZSM-5 catalyst by coating skeletal Co with H-ZSM-5 using the hydrothermal method [114]. The core–shell catalyst exhibited a higher efficiency than the homogeneously mixed one in cracking long-chain hydrocarbons, thus giving rise to a higher selectivity to the gasoline-range fuels. At an optimized thickness of the zeolite shell, which was tuned by adjusting the time of hydrothermal crystallization, the long-chain hydrocarbons were thoroughly cracked at appreciably high FTS activity, thus leading to a high selectivity to the gasoline-range fuels. Elevating the reaction temperature resulted in higher FTS activity and cracking activity, but the product distribution shifted to short-chain hydrocarbons.

The selectivity to the gasoline-range fuels amounted to as high as 79% over the optimal skeletal Co@H-ZSM-5 catalyst at 523 K, 20 bar, and H_2/CO ratio of 2.

Aside from the Co-based core–shell bifunctional catalysts, Tsubaki and co-workers synthesized the core–shell FTS catalyst with fused Fe as the core and H-ZSM-5 as the shell by the hydrothermal method [115]. The hydrothermal treatment was repeated twice so as to obtain a compact zeolite shell of *ca.* 3 μm in thickness. Consistent with the Co-based bifunctional catalysts, the selectivity for long-chain hydrocarbons was drastically reduced on the fused-Fe@H-ZSM-5 catalyst. The selectivity of the catalyst for the gasoline-range fuels was about 50%, and the C_{iso}/C_n ratio (in C_{4+}) was 4.17 over this core–shell catalyst. In comparison, fewer *iso*-paraffins were formed on the fused Fe catalyst and the homogeneously mixed fused Fe and H-ZSM-5 catalyst.

Sun et al. prepared a Raney Fe@HZSM-5 bifunctional catalyst through a one-step hydrothermal approach using $Fe_{50}Al_{50}$ alloy as the precursor for the core [94]. TPAOH acted not only as a template for the hydrothermal synthesis of the H-ZSM-5 shell, but also as a base for the de-alumination of the $Fe_{50}Al_{50}$ alloy. The leached Al species was harnessed exclusively as the Al source for the synthesis of zeolite. By means of this facile preparation strategy, a Raney-Fe@HZSM-5 catalyst with the iron core being coated with the H-ZSM-5 shell (Si/Al ratio about 63) of *ca.* 4.1 μm in thickness was obtained. This core–shell catalyst displayed a higher CO conversion than Raney Fe alone and the homogeneously mixed catalyst (Raney-Fe–HZSM-5). At 543 K, 20 bar, and H_2/CO ratio of 2, a selectivity to the gasoline-range fuels of 71% and a C_{iso}/C_n ratio of 1.9 were obtained over the core–shell catalyst, which were much higher than those over the Raney-Fe and Raney-Fe–HZSM-5 catalysts. The methane selectivity over the core–shell catalyst was substantially lower than over other two catalysts (Fig. 4.5).

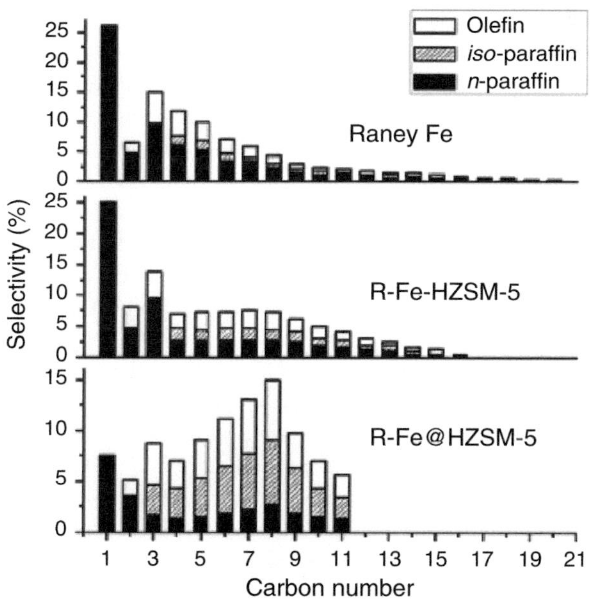

Fig. 4.5 Product selectivity over the Raney-Fe@HZSM-5, Raney-Fe–HZSM-5, and Raney Fe catalysts after 150 h on stream. Reaction conditions: $T = 543$ K, $P = 20$ bar, H_2/CO = 2, and $W_{Fe}/F_{(CO+H2)} = 10$ g h mol^{-1} (Reprinted with permission from [94], Copyright ©2012, Royal Society of Chemistry)

It is unusual that coating H-ZSM-5 on the Raney-Fe core decreased simultaneously the selectivity for long-chain hydrocarbons (C_{12+}) and the selectivities for methane and C_2–C_4 hydrocarbons. Consequently, the selectivity for the gasoline-range fuels on the core–shell catalyst exceeded that to C_{5+} hydrocarbons over the Raney-Fe catalyst (Fig. 4.5), which suggests that the function of the H-ZSM-5 shell is not limited to the cracking and isomerization of long-chain hydrocarbons formed on the core. ^{57}Fe Mossbauer absorption spectroscopy revealed that the content of Hägg carbide (Fe_5C_2), which is the well-acknowledged active phase for FTS, decreased in the order of Raney-Fe@HZSM-5 > Raney-Fe–HZSM-5 > Raney-Fe. It is plausible that the hydrophilicity of the H-ZSM-5 shell is conducive to a less oxidizing chemical environment around the iron core during FTS by preferentially interacting with water, thus retarding the oxidation of the active phase.

4.4 Conclusions and Future Outlook

Bifunctional FTS catalysts are usually composed of a conventional FTS catalyst for CO hydrogenation and an acidic zeolite. One-step conversion of bio-syngas to biofuels over bifunctional FTS catalysts is highly promising to bridge the gap between lignocellulosic biomass and biofuels, as it will enable a cost-effective, energy-efficient, and fuel-oriented BTL process.

Extensive investigations on the effect of the structure of bifunctional FTS catalysts reveal that the zeolite-supported catalysts and zeolite-coated catalysts are more selective to liquid fuels than the physically mixed ones, manifesting that an intimate contact between the two catalyst components is essential. Ideally, the FTS core component should be coated by a pinhole-free zeolite shell to avoid product flow bypass. However, existing synthetic strategies to realize such a perfect core–shell structure are tedious and costly. Hence, concise and economic synthetic strategies need to be developed in future work. Moreover, due to mass transport limitations imposed by the microporous zeolite shell, in some cases the core–shell bifunctional FTS catalysts are inferior to the uncoated ones in activity. Imparting mesoporosity to the microporous zeolite shell by means of hard templating, soft templating, or post-treatment [72] will be effective to solve this problem, while the advantages of the core–shell structure are well retained. Since the core catalyst is exposed to a hydrothermal environment in the presence of zeolite precursor solution with structure-directing agent, it should be noted that the phase transformation of the support of the core catalyst sometimes takes place [89]. This transformation will alter the catalytic performance of the pristine core catalyst and complicate the understanding of the synergy between the core and shell components.

Although acidic zeolites are dominant in bifunctional FTS catalysts, work using surface-acidified carbon nanotubes (CNTs) that exhibit high selectivity for the diesel-range fuels break the limitations imposed by the ASF distribution [33]. More interestingly, Wang et al. demonstrated that on a nanosized Ru/P25 catalyst, by simply switching between the nickel sulfate and potassium sulfate additives during aqueous-phase FTS, the selectivities for the gasoline-range fuels and diesel-range fuels could be reversibly modulated to 60% and 50%, respectively, which exceed the values predicted by the ASF distribution [116]. Metal sulfates modify the stability of the endogenous α-olefins, which are the reaction intermediates, thus changing the product distribution. This new strategy imparts unprecedented versatility to FTS in converting syngas into liquid fuels. We anticipate that by further extending the scope of the acidic components and exploring new selectivity control strategies, novel bifunctional FTS catalysts with exciting catalytic performances can be developed, which will propel the industrialization of the environmentally significant BTL process.

Acknowledgement The support from the National Basic Research Program of China (2016YFB0301602), the National Science Foundation of China (21373055), the International Joint Laboratory on Resource Chemistry (IJLRC), and the Science and Technology Commission of Shanghai Municipality (08DZ2270500) are cordially acknowledged.

References

1. Abelló S, Montané D. Exploring iron-based multifunctional catalysts for Fischer–Tropsch synthesis: a review. ChemSusChem. 2011;4(11):1538–56.
2. Sanchez OJ, Cardona CA. Trends in biotechnological production of fuel ethanol from different feedstocks. Bioresour Technol. 2008;99(13):5270–95.
3. Suurs RAA, Hekkert MP. Competition between first and second generation technologies: lessons from the formation of a biofuels innovation system in the Netherlands. Energy. 2009;34(5):669–79.
4. Damartzis T, Zabaniotou A. Thermochemical conversion of biomass to second generation biofuels through integrated process design—a review. Renew Sust Energ Rev. 2011;15(1):366–78.
5. Sims REH, Mabee W, Saddler JN, Taylor M. An overview of second generation biofuel technologies. Bioresour Technol. 2010;101(6):1570–80.
6. de Klerk A. Fischer–Tropsch refining: technology selection to match molecules. Green Chem. 2008;10(12):1249–79.
7. Torres W, Pansare SS, Goodwin JG Jr. Hot gas removal of tars, ammonia, and hydrogen sulfide from biomass gasification gas. Catal Rev. 2007;49(4):407–56.
8. Claude V, Courson C, Köhler M, et al. Overview and essentials of biomass gasification technologies and their catalytic cleaning methods. Energy Fuel. 2016;30(11):8791–814.
9. Zwart R, van Ree R. Bio-based Fischer–Tropsch diesel production technologies. In: Biofuels, Chichester, UK: John Wiley & Sons, Ltd; 2009; pp. 95–116.
10. McKendry P. Energy production from biomass (part 3): gasification technologies. Bioresour Technol. 2002;83(1):55–63.
11. Boerrigter H, Calis H P, Slort D J, Bodenstaff H. Gas cleaning for integrated biomass gasification (BG) and Fischer–Tropsch (FT) systems; experimental demonstration of two BG-FT systems. The 2nd world conference and technology exhibition on biomass for energy, industry and climate protection. 2004; pp. 51–56.
12. Fischer F, Tropsch H. Über die synthese höherer glieder der aliphatischen reihe aus kohlenoxyd. Eur J Inorg Chem. 1923;56(11):2428–43.
13. Steynberg AP, Dry ME, Davis BH, Breman BB. Fischer–Tropsch reactors. Stud Surf Sci Catal. 2004;152:64–195.

14. Evans G, Smith C. Biomass to liquids technology. In: Comprehensive renewable energy. Oxford: Elsevier; 2012. p. 155–204.
15. Hu J, Yu F, Lu Y. Application of Fischer–Tropsch synthesis in biomass to liquid conversion. Catalysts. 2012;2(2):303–26.
16. Maitlis PM, Zanotti V. The role of electrophilic species in the Fischer–Tropsch reaction. Chem Commun. 2009;45(13):1619–34.
17. Adesina AA. Hydrocarbon synthesis via Fischer–Tropsch reaction: travails and triumphs. Appl Catal A. 1996;138(2):345–67.
18. Van Der Laan GP, Beenackers A. Kinetics and selectivity of the Fischer–Tropsch synthesis: a literature review. Catal Rev. 1999;41(3–4):255–318.
19. Bell AT. Catalytic synthesis of hydrocarbons over group VIII metals. A discussion of the reaction mechanism. Catal Rev. 1981;23(1–2):203–32.
20. Dry ME. Practical and theoretical aspects of the catalytic Fischer–Tropsch process. Appl Catal A. 1996;138(2):319–44.
21. Davis BH. Fischer–Tropsch synthesis: current mechanism and futuristic needs. Fuel Process Technol. 2001;71(1):157–66.
22. Zimmerman WH, Bukur DB. Reaction kinetics over iron catalysts used for the Fischer–Tropsch synthesis. Can J Chem Eng. 1990;68(2):292–301.
23. Botes FG. The effects of water and CO_2 on the reaction kinetics in the iron-based low-temperature Fischer–Tropsch synthesis: a literature review. Catal Rev. 2008;50(4):471–91.
24. Dry ME. High quality diesel via the Fischer–Tropsch process–a review. J Chem Technol Biotechnol. 2002;77(1):43–50.
25. Martínez A, Prieto G. The application of zeolites and periodic mesoporous silicas in the catalytic conversion of synthesis gas. Top Catal. 2009;52(1–2):75.
26. Kuipers EW, Vinkenburg IH, Oosterbeek H. Chain length dependence of α-olefin readsorption in Fischer–Tropsch synthesis. J Catal. 1995;152(1):137–46.
27. Lu Y, Lee T. Influence of the feed gas composition on the Fischer–Tropsch synthesis in commercial operations. J Nat Gas Chem. 2007;16(4):329–41.
28. Hamelinck CN, Faaij APC, den Uil H, Boerrigter H. Production of FT transportation fuels from biomass; technical options, process analysis and optimisation, and development potential. Energy. 2004;29(11):1743–71.
29. De Klerk A, Furimsky E. Catalysis in the refining of Fischer–Tropsch syncrude. Cambridge: Royal Society of Chemistry; 2010.
30. Zhang QH, Cheng K, Kang JC, Deng W, Wang Y. Fischer–Tropsch catalysts for the production of hydrocarbon fuels with high selectivity. ChemSusChem. 2014;7(5):1251–64.
31. Zhang QH, Kang JC, Wang Y. Development of novel catalysts for Fischer–Tropsch synthesis: tuning the product selectivity. ChemCatChem. 2010;2(9):1030–58.
32. de Klerk A. Fischer–Tropsch fuels refinery design. Energy Environ Sci. 2011;4(4):1177–205.
33. Kang JC, Zhang S, Zhang QH, Wang Y. Ruthenium nanoparticles supported on carbon nanotubes as efficient catalysts for selective conversion of synthesis gas to diesel fuel. Angew Chem Int Ed. 2009;121(14):2603–6.
34. Yu GB, Sun B, Pei Y, Xie SH, Yan SR, Qiao MH, Fan KN, Zhang XX, Zong BN. Fe_xO_y@C spheres as an excellent catalyst for Fischer–Tropsch synthesis. J Am Chem Soc. 2010;132(3):935–7.
35. Sun B, Qiao MH, Fan KN, Ulrich J, Tao F. Fischer–Tropsch synthesis over molecular sieve supported catalysts. ChemCatChem. 2011;3(3):542–50.
36. Sartipi S, Makkee M, Kapteijn F, Gascon J. Catalysis engineering of bifunctional solids for the one-step synthesis of liquid fuels from syngas: a review. Cat Sci Technol. 2014;4(4):893–907.
37. Schulz H. Short history and present trends of Fischer–Tropsch synthesis. Appl Catal A. 1999;186(1):3–12.
38. Enger BC, Holmen A. Nickel and Fischer–Tropsch synthesis. Catal Rev. 2012;54(4):437–88.
39. Luque R, de la Osa AR, Campelo JM, Romero AA, Valverde JL, Sanchez P. Design and development of catalysts for Biomass-To-Liquid-Fischer–Tropsch (BTL-FT) processes for biofuels production. Energy Environ Sci. 2012;5(1):5186–202.

40. Dry M, Steynberg A. Fischer–Tropsch technology. Stud Surf Sci Catal. Elsevier Science & Technology. 2004.
41. Dry ME. The Fischer–Tropsch process: 1950–2000. Catal Today. 2002;71(3):227–41.
42. van Steen E, Claeys M. Fischer–Tropsch catalysts for the Biomass-To-Liquid (BTL)-Process. Chem Eng Technol. 2008;31(5):655–66.
43. Khodakov AY, Chu W, Fongarland P. Advances in the development of novel cobalt Fischer–Tropsch catalysts for synthesis of long-chain hydrocarbons and clean fuels. Chem Rev. 2007;107(5):1692–744.
44. Sonal PKK, Upadhyayula S. Synthesis of C_{5+} hydrocarbons from low H_2/CO ratio syngas over silica supported bimetallic Fe–Co catalyst. Catal Today. 2017;291:133–45.
45. Sonal KK, Pant KK, Upadhyayula S. Synergistic effect of Fe–Co bimetallic catalyst on FTS and WGS activity in the Fischer–Tropsch process: a kinetic study. Ind Eng Chem Res. 2017;56(16):4659–71.
46. De Vos RM, Verweij H. High-selectivity, high-flux silica membranes for gas separation. Science. 1998;279(5357):1710–1.
47. Ennaert T, Van Aelst J, Dijkmans J, De Clercq R, Schutyser W, Dusselier M, Verboekend D, Sels BF. Potential and challenges of zeolite chemistry in the catalytic conversion of biomass. Chem Soc Rev. 2016;45(3):584–611.
48. Goldwasser MR, Navas F, Zurita MJP, Cubeiro ML, Lujano E, Franco C, Jiménez FG, Jaimes E, Moronta D. Iron-pentasil molecular sieves: characterisation and catalytic behaviour in syngas conversion. Appl Catal A. 1993;100(1):85–95.
49. Botes FG, Böhringer W. The addition of HZSM-5 to the Fischer–Tropsch process for improved gasoline production. Appl Catal A. 2004;267(1):217–25.
50. Martínez A, López C. The influence of ZSM-5 zeolite composition and crystal size on the in situ conversion of Fischer–Tropsch products over hybrid catalysts. Appl Catal A. 2005;294(2):251–9.
51. Martínez A, Rollán J, Arribas MA, Cerqueira HS, Costa AF, Aguiar EFS. A detailed study of the activity and deactivation of zeolites in hybrid Co/SiO_2-zeolite Fischer–Tropsch catalysts. J Catal. 2007;249(2):162–73.
52. Martínez A, Valencia S, Murciano R, Cerqueira HS, Costa AF, Aguiar EFS. Catalytic behavior of hybrid Co/SiO_2-(medium-pore) zeolite catalysts during the one-stage conversion of syngas to gasoline. Appl Catal A. 2008;346(1):117–25.
53. Fujimoto K, Adachi M, Tominaga H. Direct synthesis of isoparaffins from synthesis gas. Chem Lett. 1985;14(6):783–6.
54. Hammer H, Joisten M, Lüngen S, Winkler D. New zeolites in Fischer–Tropsch synthesis. Int J Energy Res. 1994;18(2):223–31.
55. Marchetti SG, Cagnoli MV, Alvarez AM, Bengoa JF, Mercader RC, Yeramián AA. Study of the Fe/zeolite-L system: part I: Characterization of iron species and their structural properties. Appl Surf Sci. 2000;165(2):91–9.
56. Latham K, Round CI, Williams CD. Synthesis, further characterisation and catalytic activity of iron-substituted zeolite LTL, prepared using tetrahedral oxo-anion species. Microporous Mesoporous Mater. 2000;38(2):333–44.
57. Cagnoli MV, Gallegos NG, Alvarez AM, Bengoa JF, Yeramián AA, Schmal M, Marchetti SG. Catalytic CO hydrogenation on potassic Fe/zeolite LTL. Appl Catal A. 2002;230(1):169–76.
58. Guczi L, Kiricsi I. Zeolite supported mono-and bimetallic systems: structure and performance as CO hydrogenation catalysts. Appl Catal A. 1999;186(1):375–94.
59. Pour AN, Zamani Y, Tavasoli A, Shahri SMK, Taheri SA. Study on products distribution of iron and iron–zeolite catalysts in Fischer–Tropsch synthesis. Fuel. 2008;87(10):2004–12.
60. Li B, Sun B, Qian X, Li W, Wu Z, Sun Z, Qiao M, Duke M, Zhao D. In-situ crystallization route to nanorod-aggregated functional ZSM-5 microspheres. J Am Chem Soc. 2013;135(4):1181–4.

61. Sartipi S, Alberts M, Santos VP, Nasalevich M, Gascon J, Kapteijn F. Insights into the catalytic performance of mesoporous H-ZSM-5-supported cobalt in Fischer–Tropsch synthesis. ChemCatChem. 2014;6(1):142–51.
62. Ngamcharussrivichai C, Imyim A, Li X, Fujimoto K. Active and selective bifunctional catalyst for gasoline production through a slurry-phase Fischer–Tropsch synthesis. Ind Eng Chem Res. 2007;46(21):6883–90.
63. Kang J, Cheng K, Zhang L, Zhang Q, Ding J, Hua W, Lou Y, Zhai Q, Wang Y. Mesoporous zeolite-supported ruthenium nanoparticles as highly selective Fischer–Tropsch catalysts for the production of C_5–C_{11} isoparaffins. Angew Chem Int Ed. 2011;123(22):5306–9.
64. Sartipi S, Parashar K, Makkee M, Gascon J, Kapteijn F. Breaking the Fischer–Tropsch synthesis selectivity: direct conversion of syngas to gasoline over hierarchical Co/H-ZSM-5 catalysts. Cat Sci Technol. 2013;3(3):572–5.
65. Van Donk S, Janssen AH, Bitter JH, de Jong KP. Generation, characterization, and impact of mesopores in zeolite catalysts. Catal Rev. 2003;45(2):297–319.
66. Hartmann M. Hierarchical zeolites: a proven strategy to combine shape selectivity with efficient mass transport. Angew Chem Int Ed. 2004;43(44):5880–2.
67. Groen JC, Moulijn JA, Pérez-Ramírez J. Desilication: on the controlled generation of mesoporosity in MFI zeolites. J Mater Chem. 2006;16(22):2121–31.
68. Schmidt W. Solid catalysts on the nanoscale: design of complex morphologies and pore structures. ChemCatChem. 2009;1(1):53–67.
69. Meng X, Nawaz F, Xiao F-S. Templating route for synthesizing mesoporous zeolites with improved catalytic properties. Nano Today. 2009;4(4):292–301.
70. Chal R, Gerardin C, Bulut M, Van Donk S. Overview and industrial assessment of synthesis strategies towards zeolites with mesopores. ChemCatChem. 2011;3(1):67–81.
71. Verboekend D, Pérez-Ramírez J. Design of hierarchical zeolite catalysts by desilication. Cat Sci Technol. 2011;1(6):879–90.
72. Chen L-H, Li X-Y, Rooke JC, Zhang Y-H, Yang X-Y, Tang Y, Xiao F-S, Su B-L. Hierarchically structured zeolites: synthesis, mass transport properties and applications. J Mater Chem. 2012;22(34):17381–1740373.
73. Hartmann M, Machoke AG, Schwieger W. Catalytic test reactions for the evalution of hierarchical zeolites. Chem Soc Rev. 2016;45(12):3313–30.
74. Valtchev V, Mintova S. Hierarchical zeolites. MRS Bull. 2016;41(9):689–93.
75. de Jong KP, Zečević J, Friedrich H, de Jongh PE, Bulut M, Van Donk S, Kenmogne R, Finiels A, Hulea V, Fajula F. Zeolite Y crystals with trimodal porosity as ideal hydrocracking catalysts. Angew Chem Int Ed. 2010;122(52):10272–6.
76. Pereira ALC, González-Carballo JM, Pérez-Alonso FJ, Rojas S, Fierro JLG, do Carmo Rangel M. Effect of the mesostructuration of the beta zeolite support on the properties of cobalt catalysts for Fischer–Tropsch synthesis. Top Catal. 2011;54(1–4):179–89.
77. Cheng K, Kang J, Huang S, You Z, Zhang Q, Ding J, Hua W, Lou Y, Deng W, Wang Y. Mesoporous beta zeolite-supported ruthenium nanoparticles for selective conversion of synthesis gas to C_5–C_{11} isoparaffins. ACS Catal. 2012;2(3):441–9.
78. Cheng K, Zhang L, Kang J, Peng X, Zhang Q, Wang Y. Selective transformation of syngas into gasoline-range hydrocarbons over mesoporous H-ZSM-5-supported cobalt nanoparticles. Chem Eur J. 2015;21(5):1928–37.
79. Sartipi S, Parashar K, Valero-Romero MJ, Santos VP, van der Linden B, Makkee M, Kapteijn F, Gascon J. Hierarchical H-ZSM-5-supported cobalt for the direct synthesis of gasoline-range hydrocarbons from syngas: advantages, limitations, and mechanistic insight. J Catal. 2013;305:179–90.
80. Sartipi S, Alberts M, Meijerink MJ, Keller TC, Pérez-Ramírez J, Gascon J, Kapteijn F. Towards liquid fuels from bio-syngas: effect of zeolite structure in hierarchical-zeolite-supported cobalt catalysts. ChemSusChem. 2013;6(9):1646–50.
81. Gill SS, Tsolakis A, Dearn KD, Rodríguez-Fernández J. Combustion characteristics and emissions of Fischer–Tropsch diesel fuels in IC engines. Prog Energy Combust. 2011;37(4):503–23.

82. Peng X, Cheng K, Kang J, Gu B, Yu X, Zhang Q, Wang Y. Impact of hydrogenolysis on the selectivity of the Fischer–Tropsch synthesis: diesel fuel production over mesoporous zeolite-Y-supported cobalt nanoparticles. Angew Chem Int Ed. 2015;54(15):4553–6.
83. Pabst K, Kraushaar-Czarnetzki B, Schaub G. Combination of Fischer–Tropsch synthesis and hydroprocessing in a single-stage reactor. Part II. Effect of catalyst combinations. Ind Eng Chem Res. 2013;52(26):8988–95.
84. Freitez A, Pabst K, Kraushaar-Czarnetzki B, Schaub G. Single-stage Fischer–Tropsch synthesis and hydroprocessing: the hydroprocessing performance of Ni/ZSM-5/γ-Al_2O_3 under Fischer–Tropsch conditions. Ind Eng Chem Res. 2011;50(24):13732–41.
85. Tsubaki N, Yoneyama Y, Michiki K, Fujimoto K. Three-component hybrid catalyst for direct synthesis of isoparaffin via modified Fischer–Tropsch synthesis. Catal Commun. 2003;4(3):108–11.
86. Bessell S. Support effects in cobalt-based Fischer–Tropsch catalysis. Appl Catal A. 1993;96(2):253–68.
87. Calleja G, de Lucas A, van Grieken R. Co/HZSM-5 catalyst for syngas conversion: Influence of process variables. Fuel. 1995;74(3):445–51.
88. Kang S-H, Ryu J-H, Kim J-H, Prasad PSS, Bae JW, Cheon J-Y, Jun K-W. ZSM-5 supported cobalt catalyst for the direct production of gasoline range hydrocarbons by Fischer–Tropsch synthesis. Catal Lett. 2011;141(10):1464.
89. Sartipi S, van Dijk JE, Gascon J, Kapteijn F. Toward bifunctional catalysts for the direct conversion of syngas to gasoline range hydrocarbons: H-ZSM-5 coated Co versus H-ZSM-5 supported Co. Appl Catal A. 2013;456:11–22.
90. He J, Liu Z, Yoneyama Y, Nishiyama N, Tsubaki N. Multiple-functional capsule catalysts: a tailor-made confined reaction environment for the direct synthesis of middle isoparaffins from syngas. Chem-Eur J. 2006;12(32):8296–304.
91. Li X, He J, Meng M, Yoneyama Y, Tsubaki N. One-step synthesis of H-β zeolite-enwrapped Co/Al_2O_3 Fischer–Tropsch catalyst with high spatial selectivity. J Catal. 2009;265(1):26–34.
92. Huang X, Hou B, Wang J, Li D, Jia L, Chen J, Sun Y. CoZr/H-ZSM-5 hybrid catalysts for synthesis of gasoline-range isoparaffins from syngas. Appl Catal A. 2011;408(1):38–46.
93. Li C, Xu H, Kido Y, Yoneyama Y, Suehiro Y, Tsubaki N. A capsule catalyst with a zeolite membrane prepared by direct liquid membrane crystallization. ChemSusChem. 2012;5(5):862–6.
94. Sun B, Yu G, Lin J, Xu K, Pei Y, Yan S, Qiao M, Fan K, Zhang X, Zong B. A highly selective Raney Fe@ HZSM-5 Fischer–Tropsch synthesis catalyst for gasoline production: one-pot synthesis and unexpected effect of zeolites. Cat Sci Technol. 2012;2(8):1625–9.
95. He J, Yoneyama Y, Xu B, Nishiyama N, Tsubaki N. Designing a capsule catalyst and its application for direct synthesis of middle isoparaffins. Langmuir. 2005;21(5):1699–702.
96. Jin Y, Yang R, Mori Y, Sun J, Taguchi A, Yoneyama Y, Abe T, Tsubaki N. Preparation and performance of Co based capsule catalyst with the zeolite shell sputtered by Pd for direct isoparaffin synthesis from syngas. Appl Catal A. 2013;456:75–81.
97. Lin Q, Zhang Q, Yang G, Chen Q, Li J, Wei Q, Tan Y, Wan H, Tsubaki N. Insights into the promotional roles of palladium in structure and performance of cobalt-based zeolite capsule catalyst for direct synthesis of C_5–C_{11} iso-paraffins from syngas. J Catal. 2016;344:378–88.
98. Yamane N, Wang Y, Li J, He Y, Zhang P, Nguyen L, Tan L, Ai P, Peng X, Wang Y, Yang G, Tsubaki N. Building premium secondary reaction field with a miniaturized capsule catalyst to realize efficient synthesis of a liquid fuel directly from syngas. Cat Sci Technol. 2017;7(10):1996–2000.
99. Zhao T-S, Chang J, Yoneyama Y, Tsubaki N. Selective synthesis of middle isoparaffins via a two-stage Fischer–Tropsch reaction: activity investigation for a hybrid catalyst. Ind Eng Chem Res. 2005;44(4):769–75.
100. Subiranas AM, Schaub G. Combining Fischer–Tropsch (FT) and hydrocarbon reactions under FT reaction conditions: model compound and combined-catalyst studies. Int J Chem React Eng. 2009;7:A31.
101. Jothimurugesan K, Gangwal SK. Titania-supported bimetallic catalysts combined with HZSM-5 for Fischer–Tropsch synthesis. Ind Eng Chem Res. 1998;37(4):1181–8.

102. Karre AV, Kababji A, Kugler EL, Dadyburjor DB. Effect of addition of zeolite to iron-based activated-carbon-supported catalyst for Fischer–Tropsch synthesis in separate beds and mixed beds. Catal Today. 2012;198(1):280–8.
103. Pour AN, Housaindokht MR, Tayyari SF, Zarkesh J, Alaei MR. Kinetic studies of the Fischer–Tropsch synthesis over La, Mg and Ca promoted nano-structured iron catalyst. J Nat Gas Sci Eng. 2010;2(2):61–8.
104. He L, Teng B, Zhang Y, Fan M. Development of composited rare-earth promoted cobalt-based Fischer–Tropsch synthesis catalysts with high activity and selectivity. Appl Catal A. 2015;505:276–83.
105. Udaya V, Rao S, Gormley RJ. Bifunctional catalysis in syngas conversions. Catal Today. 1990;6(3):207–34.
106. Gormley RJ, Rao VUS, Anderson RR, Schehl RR, Chi RDH. Secondary reactions on metal-zeolite catalysts used in synthesis gas conversion. J Catal. 1988;113(1):193–205.
107. Bessell S. Investigation of bifunctional zeolite supported cobalt Fischer–Tropsch catalysts. Appl Catal A. 1995;126(2):235–44.
108. Bao J, He J, Zhang Y, Yoneyama Y, Tsubaki N. A core/shell catalyst produces a spatially confined effect and shape selectivity in a consecutive reaction. Angew Chem Int Ed. 2008;120(2):359–62.
109. Yang G, Tsubaki N, Shamoto J, Yoneyama Y, Zhang Y. Confinement effect and synergistic function of H-ZSM-5/Cu-ZnO-Al_2O_3 capsule catalyst for one-step controlled synthesis. J Am Chem Soc. 2010;132(23):8129–36.
110. He J, Xu B, Yoneyama Y, Nishiyama N, Tsubaki N. Designing a new kind of capsule catalyst and its application for direct synthesis of middle isoparaffins from synthesis gas. Chem Lett. 2004;34(2):148–9.
111. Yang G, He J, Yoneyama Y, Tan Y, Han Y, Tsubaki N. Preparation, characterization and reaction performance of H-ZSM-5/cobalt/silica capsule catalysts with different sizes for direct synthesis of isoparaffins. Appl Catal A. 2007;329:99–105.
112. Yang G, He J, Zhang Y, Yoneyama Y, Tan Y, Han Y, Vitidsant T, Tsubaki N. Design and modification of zeolite capsule catalyst, a confined reaction field, and its application in one-step isoparaffin synthesis from syngas. Energy Fuel. 2008;22(3):1463–8.
113. Enache DI, Rebours B, Roy-Auberger M, Revel R. In situ XRD study of the influence of thermal treatment on the characteristics and the catalytic properties of cobalt-based Fischer–Tropsch catalysts. J Catal. 2002;205(2):346–53.
114. Xu K, Cheng Y, Sun B, Pei Y, Yan S-R, Qiao M-H, Zhang X-X, Zong B-N. Fischer–Tropsch synthesis over skeletal Co@HZSM-5 core–shell catalysts. Acta Phys Chim Sin. 2015;31(6):1137–44.
115. Bao J, Yang G, Okada C, Yoneyama Y, Tsubaki N. H-type zeolite coated iron-based multiple-functional catalyst for direct synthesis of middle isoparaffins from syngas. Appl Catal A. 2011;397(1):195–200.
116. Wang H, Xu K, Cheng Y, Pei Y, Yan S, Qiao M, Li ZH, Zong B. Reversible selectivity modulation of gasoline and diesel by a facile metal-salt-modified Fischer–Tropsch synthesis strategy. ChemCatChem. 2016;24(8):3701–5.

Chapter 5
Reductive Conversion of 5-Hydroxymethylfurfural in Aqueous Solutions by Furan Ring Opening and Rearrangement

Junya Ohyama and Atsushi Satsuma

Abstract Biomass-derived 5-hydroxymethylfurfural (HMF) can be converted into chemicals that are useful for polymers, fine chemicals, and fuels. The upgrading of HMF by reductive conversion using supported metal catalysts has been intensively studied. From a practical viewpoint, conversion in water or in an aqueous organic solution is useful because HMF is expected to be supplied as an aqueous solution via the hydrolysis of cellulose. In the reductive conversion of HMF in water, acid catalysis due to water can cause hydrolytic furan ring opening or rearrangement. For instance, diketones can be generated by the hydrolytic ring opening of HMF, and cyclopentanones can be generated by a hydrogenative ring-rearrangement reaction. On the other hand, when the furan ring is not opened or rearranged, furfuryl alcohols, tetrahydrofurfuryl alcohols, and polyols are produced. The product selectivity of the reductive conversion in water is determined by the hydrogenation activity and acid–base properties of the catalytic system. In this chapter, the effect of catalyst functions on product selectivity in the reductive conversion of furfurals in aqueous solutions using heterogeneous catalysts is shown along with some studies on homogeneous catalysts.

Nomenclature

HMF:	5-hydroxymethylfurfural
BHF:	2,5-bis(hydroxymethyl)furan
BHTHF:	2,5-bis(hydroxymethyl)tetrahydrofuran
HHD:	1-hydroxyhexane-2,5-dione
HHED:	1-hydroxy-3-hexene-2,5-dione
HHCPEN:	4-hydroxy-4-hydroxymethyl-2-cyclopentenone

J. Ohyama (✉) • A. Satsuma
Materials Chemistry, Graduate School of Engineering, Nagoya University, Nagoya, Japan
e-mail: ohyama@apchem.nagoya-u.ac.jp; satsuma@chembio.nagoya-u.ac.jp

Z. Fang et al. (eds.), *Production of Biofuels and Chemicals with Bifunctional Catalysts*, Biofuels and Biorefineries 8, https://doi.org/10.1007/978-981-10-5137-1_5

HCPEN:	3-hydroxymethyl-2-cyclopentenone
HCPN:	3-hydroxymethylcyclopentanone
HCPO:	3-hydroxymethylcyclopentanol
HDN:	2,5-hexanedione
1,2,5-HT:	1,2,5-hexanetriol
1,2,6-HT:	1,2,6-hexanetriol
1,2,5,6-HT:	1,2,5,6-hexanetetraol
MHMF:	2-methyl-5-hydroxymethylfuran
DMF:	2,5-dimethylfuran
1,6-HD:	1,6-hexanediol
3-MCPEN:	3-methyl-2-cyclopenten-1-one
3-MCPN:	3-methylcyclopentanone
FA:	furfural
FOL:	furfuryl alcohol
CPN:	cyclopentanone
CPO:	cyclopentanol
4-HCPEN:	4-hydroxycyclopent-2-enone
THFOL:	tetrahydrofurfuryl alcohol
2-MTHF:	2-methyltetrahydrofuran

5.1 Introduction

Biorefinery technology enables the production of chemicals and fuels from biomass resources. The advantages of such technology are (1) the potential for replacing conventional petrochemistry and thus overcoming the problem of oil resource depletion and (2) the carbon neutrality of biomass resources, which prevents an increase in CO_2 emissions [1–4]. One of the platform chemicals in biorefinery processes is 5-hydroxymethylfurfural (HMF), which can be produced from cellulose and hemicellulose by hydrolysis, isomerization, and dehydration [5–8]. Various upgrading reactions of HMF, including hydrogenation, hydrogenolysis, hydrodeoxygenation, oxidation, aldol reaction, amination, decarbonylation, and acetalization, have been proposed to produce useful compounds for fuels, polymers, and fine chemicals. Among these reactions, reductive conversions of HMF, including hydrogenation, hydrogenolysis, and hydrodeoxygenation, have been the most intensively studied. Indeed, there are excellent reviews dealing with the reductive conversion of HMF to valuable compounds [1–4, 7, 8].

The upgrading of HMF in water is of practical importance because HMF is expected to be supplied as an aqueous solution in a biorefinery process. Using water as the solvent has benefits from the viewpoint of green chemistry [9]. Compared with the upgrading reactions of HMF via hydrogenation in organic solvents, those in water often show different products because the water causes furan ring opening or rearrangement. As shown in Scheme 5.1, hydrolytic ring opening and rearrange-

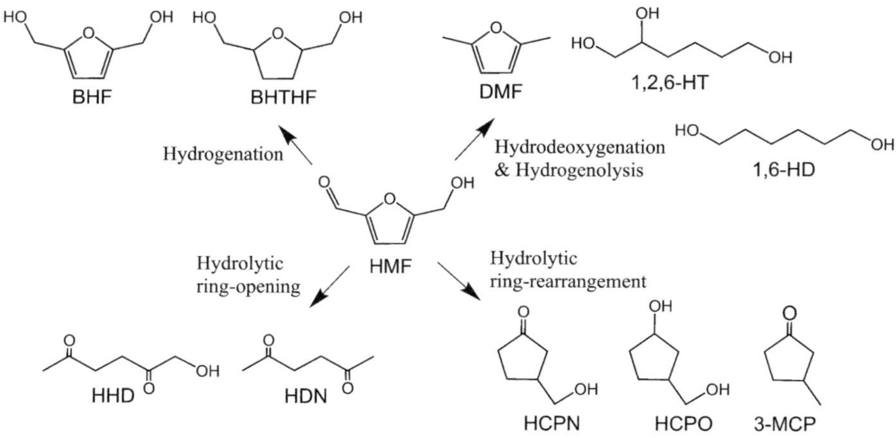

Scheme 5.1 Various products obtained by reaction of 5-hydroxymethylfurfural (HMF) with hydrogen using catalysts in water

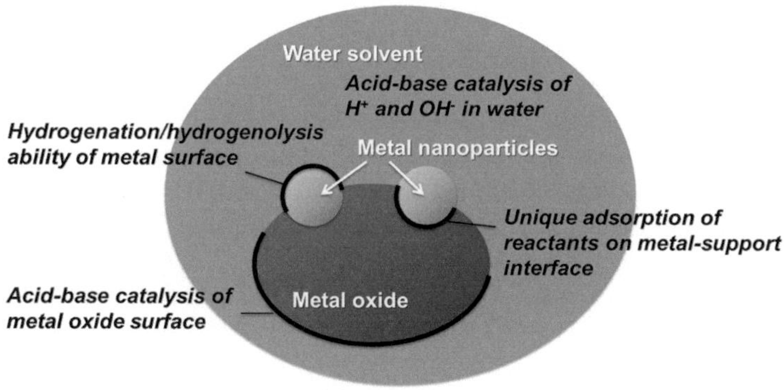

Fig. 5.1 Catalytic functions determining the selectivity of products in reaction systems for the hydrogenation of 5-hydroxymethylfurfural (HMF) using supported metal catalysts in water

ment of a furan ring can occur in water in addition to hydrogenation, hydrogenolysis, and hydrodeoxygenation. Thus, catalytic reaction systems involving water have additional functions compared with those without water. Figure 5.1 shows a simple diagram of the functional sites determining the selectivity of products in a catalytic reaction system for the hydrogenation of HMF using supported metal catalysts in water, where the following four functional sites can contribute to the reaction: (1) the metal surface, for hydrogenation or hydrogenolysis; (2) the support surface, for acid–base catalysis; (3) metal-support interface, for the unique adsorption of reactants; and (4) acid–base catalysis of the H^+ and OH^- of water. For the efficient production of desired compounds, it is necessary to understand how these functional sites affect the resulting products.

In this chapter, the focus of discussion is on the catalytic reaction of HMF with hydrogen in water. Based on selected papers, the functions and conditions of the catalyst systems effective for the following reactions of HMF in water will be shown: (1) hydrogenation without ring opening, (2) hydrolytic ring opening, (3) ring rearrangement, and (4) other reactions involving hydrodeoxygenation or hydrolysis.

5.2 Hydrogenation Without Ring Opening or Rearrangement

In this section, the focus is on the hydrogenation of HMF to 2,5-bis(hydroxymethyl) furan (BHF) and 2,5-bis(hydroxymethyl)tetrahydrofuran (BHTHF) using supported metal catalysts (Scheme 5.2). Hydrogenation of the C = O group of HMF forms BHF, and further hydrogenation of C = C of the furan ring generates BHTHF. These are useful compounds for applications as resin additives, monomers, and solvents.

Efficient catalyst systems for BHF and BHTHF are summarized in a previous review [10, 11]. Pt/MCM-41, Au/Al_2O_3, and Ir–ReOx/SiO_2 have been reported as efficient catalysts for conversion of HMF to BHF with ca. 100% yield [12–14]. On the other hand, Raney Ni and Ni – Pd/SiO_2 afford BHTHF from HMF in ca. 100% yield [15, 16]. In this section, the focus is on the functions of the catalyst systems required for selective hydrogenation of HMF to BHF and BHTHF in water based on several selected reports.

Figure 5.2 shows our results for the hydrogenation of HMF over various metal-oxide-supported Au catalysts [13]. The γ-Al_2O_3, La_2O_3, and CeO_2 are basic metal oxides, and TiO_2, Ta_2O_5, TiO_2-SiO_2, and sulfated zirconia (SO_4^{2-}/ZrO_2) are acidic metal oxides. The basic metal-oxide-supported Au catalysts exhibit high selectivity for BHF. On the other hand, acidic metal-oxide-supported Au catalysts show low selectivity or almost no yield of BHF and instead give the ring-opening product, 1-hydroxyhexane-2,5-dione (HHD). As described below, HHD is produced via the acid-catalyzed hydrolytic ring opening of BHF. Thus, the basicity of the metal oxide supports suppresses the ring opening to give BHF selectively.

The selectivity for BHTHF is also affected by the acid–base properties of the catalytic reaction system. Table 5.1 presents results for the hydrogenation of HMF on various supported Ru catalysts, taken from the report by Dumesic et al. [17]. When CeO_x, magnesia-zirconia (Mg-Zr), and γ-Al_2O_3, which each have an isoelectric point >7, were used as Ru supports, a high yield of BHTHF (ca. 90%) was

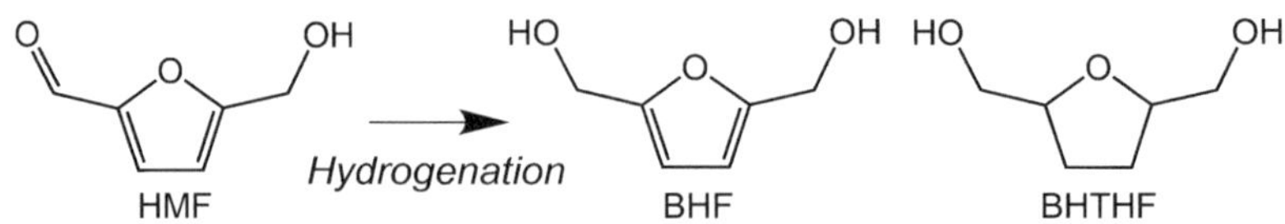

Scheme 5.2 Hydrogenation of HMF to BHF and BHTHF

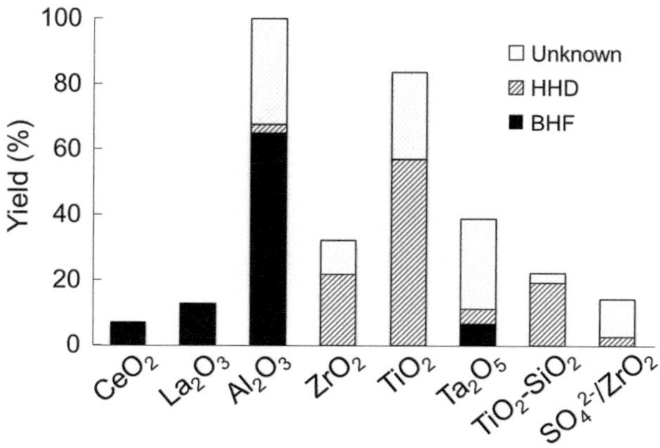

Fig. 5.2 Results for the hydrogenation of 5-hydroxymethylfurfural (HMF) over various metal-oxide-supported Au catalysts [13]. *BHF* 2,5-bis(hydroxymethyl)furan, *HHD* 1-hydroxyhexane-2,5-dione (Reprinted with permission from [13]. Copyright © 2013, The Royal Society of Chemistry)

Table 5.1 Product selectivity for 5-hydroxymethylfurfural (HMF) hydrogenation on various supports reported in [17].[a] Reprinted with permission from [17]. Copyright © 2012, The Royal Society of Chemistry

Support	Time (h)	HMF conv. (%)	Selectivity (%)					
			BHF[b]	BHTHF[c]	1,2,6-HT[d]	1,2,5-HT[e]	1,2,5,6-HT[f]	Total
CeOx	2	100	81	4	0	1	2	87
	12	100	0	91	3	1	5	100
Mg-Zr	2	99	94	2	0	0	2	99
	20	100	0	88	2	0	1	92
γ-Al_2O_3	2	92	81	5	0	0	2	88
	12	100	0	89	2	3	5	100
Vulcan carbon	2	100	0	50	2	9	28	89
Ru black	2	100	0	46	13	13	11	84
Silica	1	80	64	0	0	0	1	65
	2	100	0	53	3	13	20	89

[a]Reaction conditions: 5 wt% HMF in 2:1 bi-phasic 1-butanol–water solvent (24 g), 200 mg of catalyst, 403 K, H_2 2.8 MPa
[b]2,5-bis(hydroxymethyl)furan
[c]2,5-bis(hydroxymethyl)tetrahydrofuran
[d]1,2,6-hexanetriol
[e]1,2,5-hexanetriol
[f]1,2,5,6-hexanetetraol

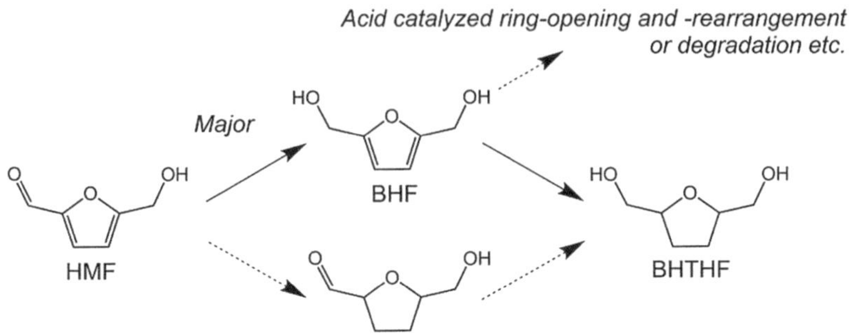

Scheme 5.3 The proposed reaction path for 5-hydroxymethylfurfural (HMF) hydrogenation to 2,5-bis(hydroxymethyl)furan (BHF) and 2,5-bis(hydroxymethyl)tetrahydrofuran (BHTHF) [15]

obtained. It should also be noted that such supported Ru catalysts also gave high yields of BHF by shortening the reaction time. On the other hand, when using SiO_2, which has an isoelectric point <7, as a support, BHTHF yield decreased; simultaneously, ring-opened compounds such as triols and tetraol increased. Based on the reaction results, the basicity of the supporting materials has a strong influence on the production of BHTHF. The effect of the acid–base properties of metal oxides was also investigated by adding metal oxides to the hydrogenation of HMF using Ru black. Compared with the reaction using only Ru black, the addition of basic metal oxides such as Al_2O_3 and MgO enhanced BHTHF selectivity. On the contrary, the addition of acidic SiO_2 lowered BHTHF selectivity and increased polyols by the ring opening of furan. The presence of homogeneous acids of H_2SO_4 and levulinic acid in the reaction using Ru black also caused a reduction in BHTHF yield. These results suggest that the pH of the aqueous reaction solution has a strong influence on the selectivity of BHTHF: a high pH reaction solution adjusted by basic metal oxides is effective for the selective synthesis of BHTHF, and a low pH induced by acidic oxides causes ring opening. The same can be said about the synthesis of BHF using supported metal catalysts.

Scheme 5.3 presents the proposed reaction path for the hydrogenation of HMF to BHF and BHTHF. BHTHF is mainly formed through BHF but not through 5-hydroxymethyl-tetrahydro-2-furaldehyde [15, 18–20]. BHF can suffer from conversion to other products via acid-catalyzed reactions such as a hydrolytic ring-opening reaction. Under basic conditions, the acid-catalyzed ring-opening reaction is suppressed and BHF and/or BHTHF can be obtained. Metals having C = O hydrogenation activity but low C = C hydrogenation, such as Au, are effective for BHF production, and those with high hydrogenation activity for both C = O and C = C, such as Pt group metals, can give both BHF and BHTHF.

In addition to the acid–base properties of a reaction system, the hydrogenation rate also has an influence on the yield or selectivity of BHF and BHTHF because BHF is relatively unstable under hydrothermal reaction conditions. Consequently, in the case of the hydrogenation of HMF using Au/γ-Al_2O_3, there is a certain reaction time that gives the maximum yield of BHF because BHF is gradually converted into other products such as the ring-opened or rearranged products (Sects. 2 and 3) under the reaction conditions. The maximum yield of BHF can be improved by increasing the hydrogenation rate of HMF. In other words, enhancement of the relative rate of HMF hydrogenation to BHF ring opening can improve BHF yield. Actually, an increase in the hydrogenation rate by raising H_2 pressure improves BHF yield and selectivity [13]. Control of Au nanoparticle size is also effective for the enhancement of the hydrogenation rate, leading to improvement in BHF yield [13, 21]. As for the hydrogenation of HMF to BHTHF using Pt group metal catalysts, the relative hydrogenation rate of C = C of the furan ring has an effect on the selectivity of BHTHF [17]. Ru/C has lower activity for HMF hydrogenation than does Pd/C, but it has higher activity for the hydrogenation of C = C of the furan ring, which causes higher BHTHF yield on Ru/C than on Pd/C. Considering the reaction path (Scheme 5.3), the key to selective production of BHTHF is hydrogenation of C = C of the furan ring before the acid-catalyzed degradation of BHF occurs in water.

Enhancement of the hydrogenation activity also allows one to lower the reaction temperature, which can suppress the degradation of BHF. For instance, small Au nanoparticles supported on FeO_x/Al_2O_3 gave 96% yield of BHF with 100% selectivity via the hydrogenation of HMF at 3 MPa of H_2 for 2 h at 80 °C, whereas relatively large Au nanoparticles supported on Al_2O_3 gave 62% yield of BHF with 66% selectivity via reaction at 3.8 MPa of H_2 for 2 h at 160 °C [21]. Tamura et al. performed hydrogenation of HMF on Ir-RhO_x at 30 °C at 1 MPa of H_2 for a reaction time of 6 h, and obtained a 100% yield of BHF without degradation [14]. Thus, hydrogenation using highly active catalysts at close to ambient temperatures is effective for the selective production of BHF. The same can apply to the production of BHTHF.

Based on the above results, it can be concluded that the selective hydrogenation of HMF to BHF and BHTHF in water requires basic conditions for the suppression of the acid-catalyzed ring opening of BHF, and it also requires high hydrogenation activity to complete the reaction before the degradation of BHF occurs.

5.3 Hydrolytic Ring Opening

The hydrogenation of HMF in water under acidic conditions can afford the ring-opening product 1-hydroxyhexane-2,5-dione (HHD) (Scheme 5.4), which is a useful diketone derivative for the preparation of valuable chemicals such as alcohols, amines, and cycloketones. Several catalytic systems for producing HHD have been reported, as listed in Table 5.2.

Hydrogenation of HMF using Pd/C in water under CO_2 offers HHD with up to ca. 80% yield (Table 5.2, entry 1) [22]. A combination catalyst system of Pd/C and the acidic resin Amberlyst-15 allows conversion of HMF to HHD with ca. 80% yield in THF containing a maximum water volume of 10% (Table 5.2, entry 2), although HHD is not formed by the reaction using either Pd/C or Amberlyst-15 [23]. It is known that protons generated by the dissolution of CO_2 in water and –

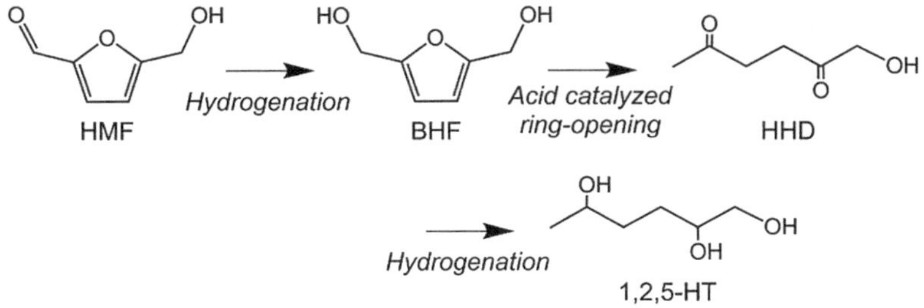

Scheme 5.4 Hydrogenative ring opening of 5-hydroxymethylfurfural (HMF) to 1-hydroxyhexane-2,5-dione (HHD)

Table 5.2 Catalytic reaction systems for producing 1-hydroxyhexane-2,5-dione (HHD) form 5-hydroxymethylfurfural (HMF) in water or a water-containing solution

Entry	Catalyst	Catalyst/HMF/ solvent (g/mmol/mL)	Temp. (°C)	H_2 pressure (MPa)	Time (h)	Conv. (%)	HHD yield (%)	Refs.
1	5 wt% Pd/C + 3 MPa CO_2	0.011/1.19/5	120	1	15	100	77	[22]
2[a]	5 wt% Pd/C + Amberlyst-15	0.008(Pd/C)0.03(A mberlyst)/1.19/5	80	2.5	15	100	77	[23]
3	Rh–ReO_x/ SiO_2 + Nafion SAC13	0.025(Rh–ReOx) 0.015(SAC13)/0.8/2	120	1 MPa for 1 h, 8 MPa for 17 h	18		81	[24]
4	1 wt% au/TiO_2 (with 4 wt% SO_3)	0.01/0.2/3	140	3.8	4	84	57	[13]
5	1 wt% au/ Nb_2O_5 + H_3PO_4 (8.5 mM)	0.01/0.2/3	140	8	12	81	60	[25]
6	[Ru]-1	1/1/10 (mo%l/ mmol/mL)	80	Formic acid 12 eq.	48	99	52	[26]
7	Cp*-Ir **A**	0.26/0.95/4 (mo%l/ mmol/mL)	120	0.7	2	99	86	[27]
8	Cp*-Ir **B**	0.0025/1/4 (mo%l/ mmol/mL)	120	1	2	43	30	[28]
9	Cp*-Ir **C**	0.01/100/500 (mo%l/mmol/mL)	130	Formic acid 5 eq.	2	–	92	[29]

[a]Solvent: tetrahydrofuran (THF) 5 mL containing a maximum of 0.5 mL water

SO_3H groups of Amberlyst-15 can work as Brønsted acid catalysts. The Rh–ReO_x catalyst gives HHD with 81% yield in combination with Nafion SAC-13 having –SO_3H groups acting as a Brønsted acid catalyst (Table 5.2, entry 3) [24]. Thus, the combination of metal catalysts with Brønsted acid catalysts is effective for the synthesis of HHD from HMF in water. These bifunctional catalysts also offer HHD from fructose and inulin by one-pot reaction since HMF can be synthesized by the acid-catalyzed dehydration of fructose or by the hydrolysis and dehydration of inulin. When fructose and inulin are used as starting substrates, 55% and 27% yields of HHD are given using Pd/C + Amberlyst-15, respectively [23].

Hydrogenation of HMF using Au catalysts gives HHD when they are combined with Brønsted acids, although Au catalysts are not effective for hydrogenation of a furan ring [13]. For instance, by using a TiO_2 support containing 3.64 wt.% of SO_3 (JRC-TIO-1), which should work as a Brønsted acid catalyst, the supported Au catalyst affords 57% yield of HHD (Table 5.2, entry 4). The combination of Au/Nb_2O_5 with H_3PO_4 as a Brønsted acid gives HHD with 60% yield (Table 5.2, entry 5) [25]. Thus, hydrogenation of the furan ring is not necessary for the synthesis of HHD. In other words, HHD synthesis from HMF requires a hydrogenation ability for C = O of HMF and Brønsted acid catalysis for the ring opening.

Further hydrogenation of HHD on metal catalysts can produce 1,2,5-HT, which has potential application in the field of polymers and solvents. It is expected that 1,2,5-HT is obtained using catalytic reaction systems similar to those used for the synthesis of HHD. However, there are few reports on the production of 1,2,5-HT from HMF [17, 30]. It has been reported that 1,2,5-HT can be mainly obtained by hydrogenation of HMF in water at 100 °C using Ru clusters immobilized in mesoporous zirconium silica (MSN-Zr) having acidic sites on their surface although the yield and selectivity were not provided [30]. The product distributions using Ru/MSN-Zr are strongly dependent on the reaction temperature, with BHTHF mainly being obtained by reaction at 25 °C. It is probable that the aqueous conversion of HMF at high temperatures promotes the ring opening of BHF, rather than hydrogenation of the furan ring. The effect of temperature on the product distribution is consistent with that represented in the synthesis of BHF and BHTHF (Sect. 2).

In addition to the above described reaction systems using heterogeneous catalysts, homogeneous catalytic systems have been reported for the synthesis of HHD from HMF. The 8-aminoquinoline coordinated arene-ruthenium(II) complex ([Ru]-**1**) offers 52% yield of HHD by hydrogenation of HMF in water using formic acid as a hydrogen source (Table 5.2, entry 6) [26]. The reaction system is also effective for the hydrogenative ring opening of furfural to levulinic acid. In these catalytic reactions using the Ru complex, the Ru complex catalyzes the formic acid driven transfer hydrogenation of furfurals to furfuryl alcohols, and H^+ ions formed from formic acid promote the hydrolytic ring opening of the furfuryl alcohols.

The 2,2′-bipyridine (bpy) coordinated Cp*-iridium(III) complex (Cp*, 1,2,3,4,5-pentamethylcyclopenta-1,3-diene) efficiently catalyzes the transformation of HMF to HHD using H_2 as a hydrogen source (Table 5.2, entries 7 and 8). The Cp*-Ir(bpy) complex offers H_2 heterolytic dissociation to form hydride on Ir, which causes high catalytic activity for the hydrogenation of C = O of HMF [27]. The simultaneously generated proton from H_2 heterolytic dissociation can serve as a catalyst for the hydrolytic ring opening. Consequently, the catalyst system using the Cp*Ir(bpy) complex works without the addition of other Brønsted acid catalysts. The catalytic activity of the Cp*-Ir(bpy) complex can be further improved by using an *ortho*-hydroxyl (*o*-OH) group functionalized bpy ligand with an NMe_2 group,

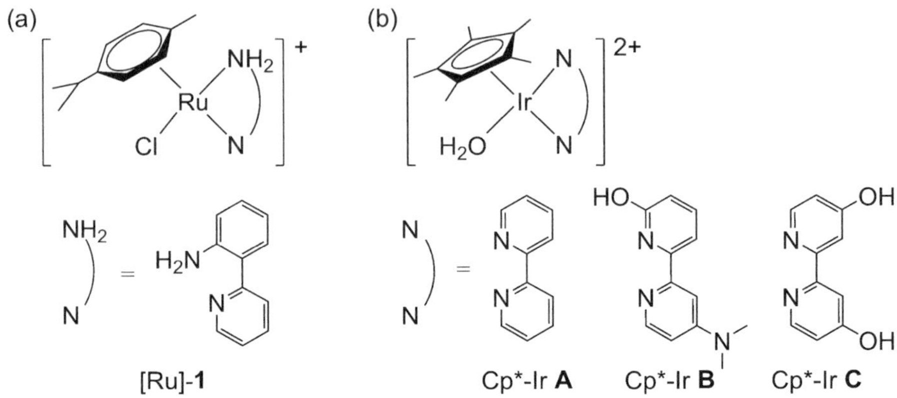

Fig. 5.3 Homogeneous metal complex catalysts for 1-hydroxyhexane-2,5-dione (HHD) production. (**a**) 8-Aminoquinoline coordinated arene-ruthenium(II) complex ([Ru]-**1**) [26]. (**b**) 2,2′-bpy coordinated Cp*-iridium(III) complex (Cp*-Ir **A**) [27], *ortho*-hydroxyl group functionalized bpy coordinated complex (Cp*-Ir **B**) [28], and 4,4′-dihydroxy-substituted bpy coordinated complex (Cp*-Ir **C**) [29]

i.e., Cp*-Ir **B** shown in Fig. 5.3 (Table 5.2, entry 8) [28]. The electron-donating NMe_2 group affords an *o*-oxyanion structure, which facilitates H_2 heterolytic dissociation to enhance hydrogenation activity. The NMe_2 group also allows the *o*-oxyanion structure under acidic conditions (pH = ca. 3), which promotes hydrolytic ring opening of BHF. Therefore, modification of the bpy ligand of Cp*-Ir complexes enhances both hydrogenation and hydrolytic ring opening to achieve an efficient synthesis of HHD (TOF of 31,560 h^{-1}, more than ca. 200 times higher than conventional catalysts).

Scheme 5.5 shows the proposed mechanism for the production of HHD from HMF using Cp*-Ir **A** [27]. HHD is formed through the hydrogenation of C = O of HMF, hydrolytic ring opening of BHF, and hydrogenation of C = C of 1-hydroxyhex-3-ene-2,5-dione (HHED). The hydrogenation of C = C of HHED is considered to be much easier than furan ring hydrogenation because the Au catalysts afford HHD, although they are not active for furan ring hydrogenation (Table 5.2, entries 4 and 5). Regarding the hydrolytic ring opening of BHF, two paths have been proposed. The first (path A) is a two-step process consisting of water addition to the double bond of BHF and subsequent H^+ and OH- elimination. The second (path B) is a hydroxymethyl group promoted one-step hydrolysis process. A mechanistic study for the reaction using the Cp*-Ir(bpy) complex demonstrated that the hydrolytic ring opening mainly proceeds via a hydroxymethyl group promoted hydrolysis process.

The effect of pH of reaction solution on HHD production from HMF has been studied by Fu et al. using the Cp*-Ir complex shown in Fig. 5.3c, which afford hydrogenation of HMF to HHD using H_2 as well as transfer hydrogenation using formic acid (Table 5.2 entry 9). It has been demonstrated that HHD production is strongly dependent on the pH of reaction solution, and a maximum yield of HHD

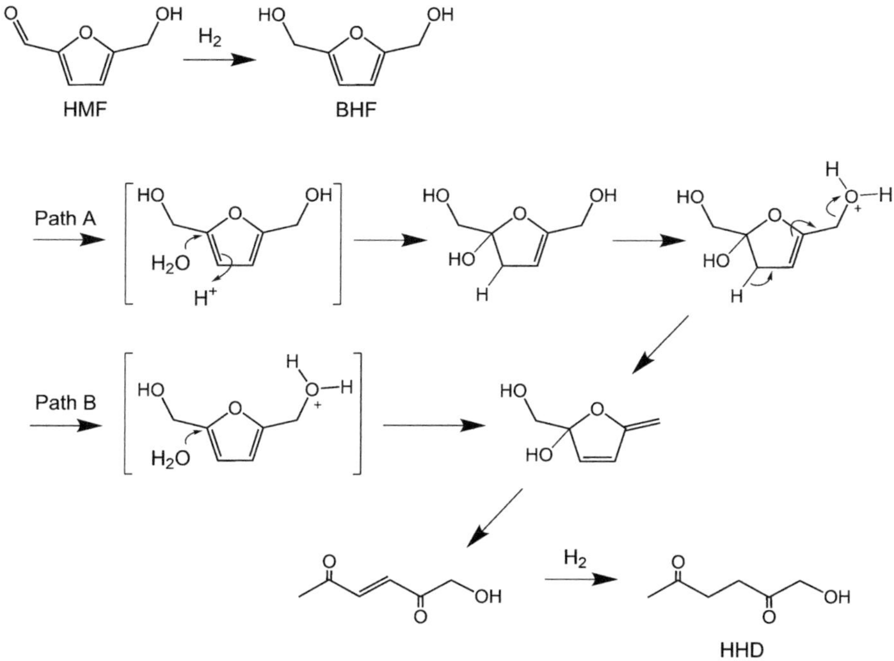

Scheme 5.5 Two potential reaction mechanisms for the conversion of 5-hydroxymethylfurfural (HMF) to 1-hydroxyhexane-2,5-dione (HHD) using Cp*-Ir **A**. The hydrolytic ring opening mainly proceeds via path B [27]

can be obtained at pH 2.5. HHD yield decreases at lower pH < 2 due to degradation of BHF, and at higher pH > 4 due to hindering hydrolytic ring-opening of BHF. Therefore, the weak acidity of reaction solution is effective for ring-opening as well as for avoiding the degradation of BHF.

To summarize this section, HHD and 1,2,5-HT can be synthesized using bifunctional catalysts composed of metals having hydrogenation activity toward C − O of HMF and Brønsted acid (H^+) catalysis for the hydrolytic ring opening, although too strong acidity causes degradation of HMF and the reaction intermediates [10].

5.4 Ring Rearrangement

5.4.1 Ring Rearrangement of Furfural (FA)

The hydrogenative ring rearrangement of biomass-derived furfural (FA) to cyclopentanone (CPN) and cyclopentanol (CPO) (Scheme 5.6) using solid catalysts has been developed [31, 32]. CPN and CPO are valuable chemicals used for solvents, perfumes, insecticides, pharmaceutical intermediates, and polymer raw materials.

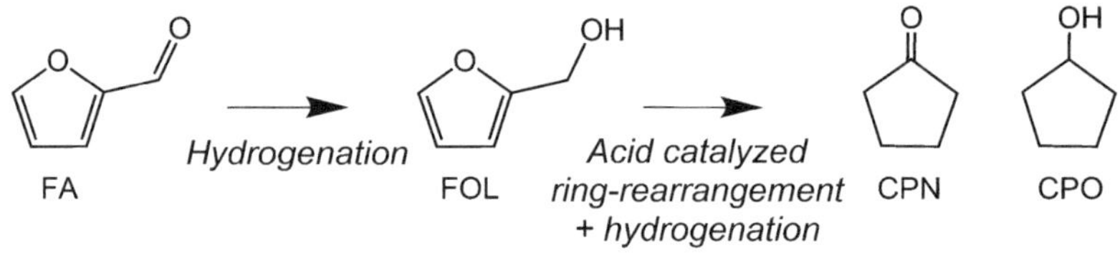

Scheme 5.6 Ring rearrangement of furfural (FA) to cyclopentanone (CPN) and cyclopentanol (CPO)

Table 5.3 Catalytic reaction systems for conversion of furfural (FA) to cyclopentanone (CPN) or cyclopentanol (CPO)

Catalyst	Catalyst/FA/ water (g)	Temp. (°C)	H_2 pressure (MPa)	Time (h)	Conv. (%)	CPN yield (%)	CPO yield (%)	Refs.
5 wt% Pt/C	0.05/1/20	160	8	0.5	100	77	5	[31]
5 wt% Pd-10 wt% Cu/C	0.01/1/20	160	3	1	98	92	< 1	[34]
Ru/MIL-101 (Ru 0.28 mol%)	0.015/0.5/5	160	4	2.5	100	96	–	[35]
6 wt% Ru/CNTs	0.02/0.45/40	160	1	1	99	91	4	[36]
10 wt% Ni/SiO_2	0.25–1/1/20	160	2.5	–	–	> 95	–	[37]
20 wt% Ni/HY	0.075/5/95	150	4	9	97	87	–	[38]
Ni-Cu/SBA	0.2/6/10	160	4	4	>99	62	3	[32]
5 wt% Cu–Co	0.05/0.02/10	170	2	1	–	–	68	[39]
Cu-Ni-Al hydrotalcites based catalyst	1.5/5.8/95	140	4	8	100	96	–	[40]
Cu/Zn/Al hydrotalcite-based catalyst	0.012/5.8/95	150	4	10	100	10	84	[41]
CuZnAl	0.2/5/8/15	150	4	6	98	60	3	[42]
0.1 wt% Au/TiO_2	0.5/4.8/100	160	4	14	99	99	–	[43]

Conventionally, CPN and CPO have been synthesized using petrochemicals, for instance, by the decarboxylation of adipic acid, the gas phase cyclization of 1,6-hexanediol or adipic ester, and liquid phase oxidation of cyclopentene using N_2O as oxidant. Therefore, the synthesis of CPN and CPO from biomass-derived FA is useful from the viewpoint of petroleum substitution.

Table 5.3 summarizes the reported results for the ring rearrangement of FA using solid catalysts. Hydrogenative ring rearrangement of FA using solid catalysts was first reported by Hronec et al. [31, 33]. Several research groups have reported catalytic conversion of FA to CPN or CPO using heterogeneous catalysts including Pt group metal catalysts, 3d transition metal (Ni, Co, Cu)-containing catalysts, and Au catalysts. Some of the catalyst systems gave high yields of CPN and CPO (>80%).

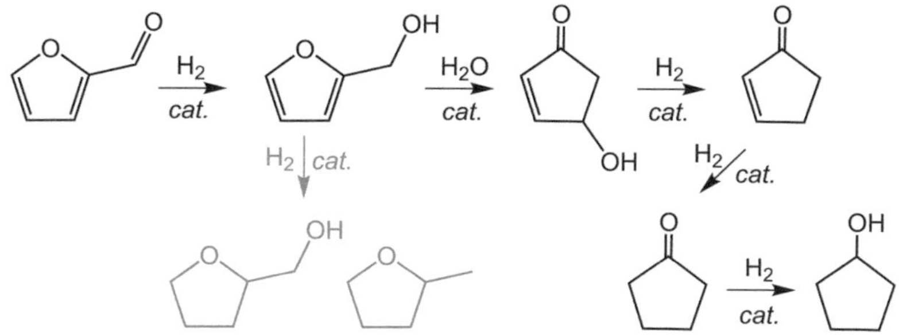

Scheme 5.7 The proposed reaction mechanism of ring rearrangement of furfural (FA) to cyclopentanone (CPN) and cyclopentanol (CPO) with a side reaction of furfuryl alcohol (FOL) to tetrahydrofurfuryl alcohol (THFOL) and 2-methyltetrahydrofuran (2-MTHF) [32]

Table 5.4 Effect of reaction solvent on conversion of furfural (FA) to cyclopentanone (CPN) or cyclopentanol (CPO) [31].[a] Reprinted with permission from [31]. Copyright © 2012, Elsevier

		Yield (%)				
Solvent	Conv. (%)	CPN	CPO	FOL	THFOL	2-MF[b]
Water	99	40	36	0	0	0
n-butanol/water (1/1 v/v)	99	10	2	0	7	31
n-butanol	99	0	0	48	6	40
THF	99	0	0	19	5	0

[a]Reaction conditions: 0.1 g 5 wt% Pt/C, 1 g FA, 20 mL solvent, 8 MPa H_2, 175 °C, 0.5 h
[b]2-methylfuran

Scheme 5.7 shows a reaction route for the hydrogenative ring rearrangement of FA, as proposed by Yang et al. [32]. They performed the hydrogenative ring rearrangement of FA to CPN and identified furfuryl alcohol, 4-hydroxy-2-cyclopentenone, and 2-cyclopentenone as reaction intermediates. Based on these intermediates, the reaction is considered to proceed as follows: hydrogenation of C = O of FA to FOL, ring rearrangement of FA to 4-hydroxy-2-cyclopentenone, hydrogenation and dehydration of 4-hydroxy-2-cyclopentenone to 2-cyclopentenone, and hydrogenation of 2-cyclopentenone to CPN. For the ring-rearrangement reaction, water as solvent is essential because the H^+ of water is important in this reaction [31]. In fact, as presented in Table 5.4, the ring-rearrangement reaction does not proceed when organic solvents (such as n-butanol and THF) are used [32]. Instead, hydrogenated products such as tetrahydrofurfuryl alcohol (THFOL) are produced. Yang et al. demonstrated that the acidity of the aqueous solution largely influences the selectivity of the ring-rearrangement reaction [32]. As presented in Table 5.5, when acidic or basic substances (Na_2CO_3, Na_2HPO_4, NaH_2PO_4, acetic acid, and H_3PO_4) were added to the reaction solution, the selectivity of CPN or CPO was decreased by the undesired resinification of FA or reaction intermediates under acidic conditions (pH = 1.4–4.8) and the formation of THFOL without furan ring rearrangement under basic conditions (pH = 9.2–11.0). Accordingly, the selectivity

Table 5.5 Effect of pH on conversion of furfural (FA) to cyclopentanone (CPN) or cyclopentanol (CPO) [32][a]

			Selectivity (%)				
Additives	pH	Conv. (%)	CPN	CPO	FOL	THFOL	2-MTHF[c]
Na_2CO_3	11.0	>99			40	36	n.d.
Na_2HPO_4	9.2	>99	2	1	36	28	n.d.
None	7.0	>99	62	3	n.d.	n.d.	17
NaH_2PO_4	4.8	>99	40	1	n.d.	n.d.	6
AcOH	3.3	94	40	2		n.d.	3
H_3PO_4[b]	1.4	–	–	–	–	–	–

[a]Reaction conditions: 10 g FA aq. (FA: 5.2 mmol), 0.2 g of NiCu-50/SBA-15 as catalyst, additive (0.1 mmol), 4 MPa H_2, 160 °C, 4 h
[b]1 mmol H_3PO_4. Black solid was observed
[c]2-methyltetrahydrofuran

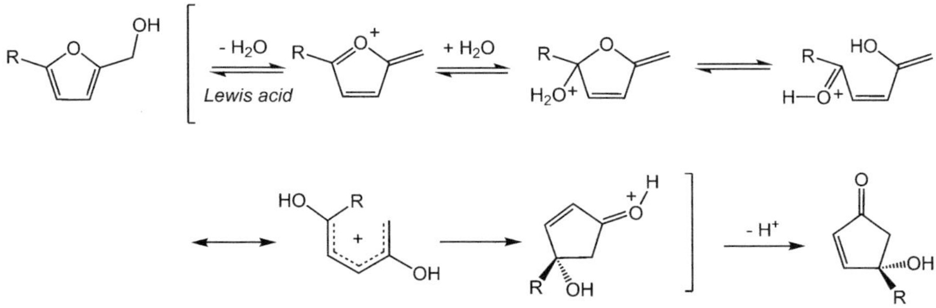

Scheme 5.8 Piancatelli reaction of furfural derivatives [45–49]. R = H: furfuryl alcohol (FOL) to 4-hydroxycyclopent-2-enone, R = CH_2OH: 2,5-bis(hydroxymethyl)furan (BHF) to 4-hydroxy-4-hydroxymethylcyclopent-2-enone (HHCPEN)

of CPN was the highest at pH 7 without the acid–base additives because side reactions proceeded under strongly acidic or basic conditions. It should be noted that the acid–base properties of metal-oxide supports can influence the yield of the ring-rearranged products [33].

The ring rearrangement of FOL to 4-hydroxycyclopent-2-enone proceeds by heating in water without a solid catalyst (4-HCPEN yield >50%) through the Piancatelli reaction [32, 44]. As shown in Scheme 5.8, the Piancatelli reaction is a cascade type reaction that proceeds by dehydration by H^+ and/or a Lewis acid, hydration of reaction intermediates, and 4π electron cyclic reaction of the pentadienyl cation [45–49]. Lewis acid catalysts are sometimes required for the Piancatelli reaction of furfural derivatives (e.g., 5-position-substituted furfurals such as BHF) where homogeneous rare earth metal triflate Lewis acid catalysts such as $Dy(OTf)_3$ and $Sc(OTf)_3$ have been used [47–49]. Fang et al. reported the ring rearrangement of FA over supported Ru catalysts on an acidic metal-organic framework material, MIL-101 (MIL: Matériaux de l'Institut Lavoisier), having Lewis acidic Cr^{3+} sites [35]. The Ru/MIL catalyst showed high catalytic performance for CPN production, which might be attributable to the Lewis acid sites for the ring-rearrangement step and the dehydration step of 4-HCPEN.

As described above, hydrogenation of FA in water affords CPN and CPO through the hydrogenation of FA on a metal catalyst and ring rearrangement by a proton and/or Lewis acid. In particular, the acidity of water has a strong effect on the selectivity of the ring-rearranged products: the selectivity decreases when the acidity of the water is too high or too low; accordingly, weak acidity (neutral water) is most suitable for the ring-rearrangement reaction.

5.4.2 Ring Rearrangement of HMF

The hydrogenative ring-rearrangement reaction of HMF affords 3-hydroxymethylcyclopentanone (HCPN) and 3-hydroxymethylcyclopentanol (HCPO) (Scheme 5.9) that are useful for many chemical products. As described above, in the current petrochemical process, cyclopentanone and its derivatives are produced from adipic acid; however, β-position-substituted cyclopentanones and cyclopentanols such as HCPN and HCPO are difficult to synthesize from cyclopentanone. Consequently, HCPN and HCPO may be unique and nonconventional building blocks for new materials derived from biomass.

Although Ni-Cu/SBA and Ru/MIL show good activity for hydrogenative ring rearrangement of FA to CPN or CPO (Table 5.3) and have been applied to the conversion of HMF to HCPN or HCPO, these catalysts do not give HCPN or HCPO [32, 35]. On the other hand, efficient conversion of HMF to HCPN with >80% yield has been achieved by the authors, particularly when using Au/Nb_2O_5 and a combination catalyst system of $Pt/SiO_2 + Ta_2O_5$ via the acid–base catalysis of metal oxides together with hydrogenation catalysis of metal nanoparticles [25, 50]. The ring-rearrangement reaction of HMF to HCPN has been reported by Rosseinsky et al. [51]. Using a Ni/Al_2O_3 catalyst prepared from layered double hydroxide, HCPN was obtained with 81% yield. Table 5.6 lists the catalytic reaction systems affording HCPN in high yields.

Scheme 5.10 presents a proposed reaction path for the hydrogenative ring rearrangement of HMF to HCPN based on the reaction intermediates, which are BHF, HHD, HHCPEN, and 4-hydroxymethyl-2-cyclopentenone (HCPEN), as identified in the HCPN synthesis using Au/Nb_2O_5 and $Pt/SiO_2 + Ta_2O_5$ catalysts [25, 50]. The

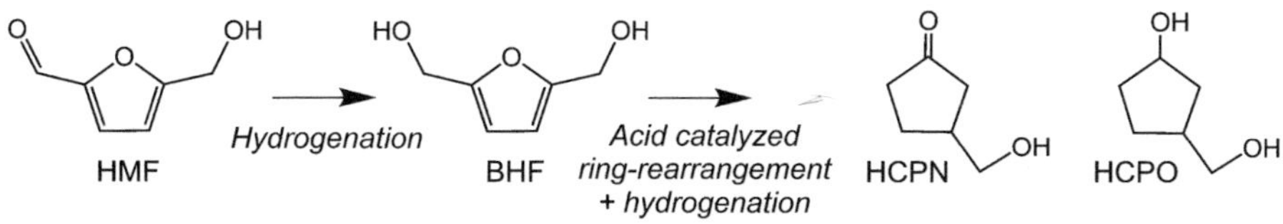

Scheme 5.9 Hydrogenative ring-rearrangement of 5-hydroxymethylfurfural (HMF) to 3-hydroxymethylcyclopentanone (HCPN) and 3-hydroxymethylcyclopentanol (HCPO)

Table 5.6 Catalytic reaction systems for conversion of 5-hydroxymethylfurfural (HMF) to 3-hydroxymethylcyclopentanone (HCPN) in high yields

Catalyst	Catalyst/HMF/water (g/g/g)	Temp. (°C)	H_2 pressure (MPa)	Time (h)	Conv. (%)	Yield (%)			Refs.
						HCPN	HHD	BHTHF	
Au/Nb_2O_5	0.01/0.025/3	140	8	12	100	86	7	0	[25]
$Pt/SiO_2 + Ta_2O_5$	0.02/0.025/3	140	3	12	100	82	3	0	[50]
Ni/Al_2O_3	0.035/0.27/45	140	2	6	100	81	11	8	[51]

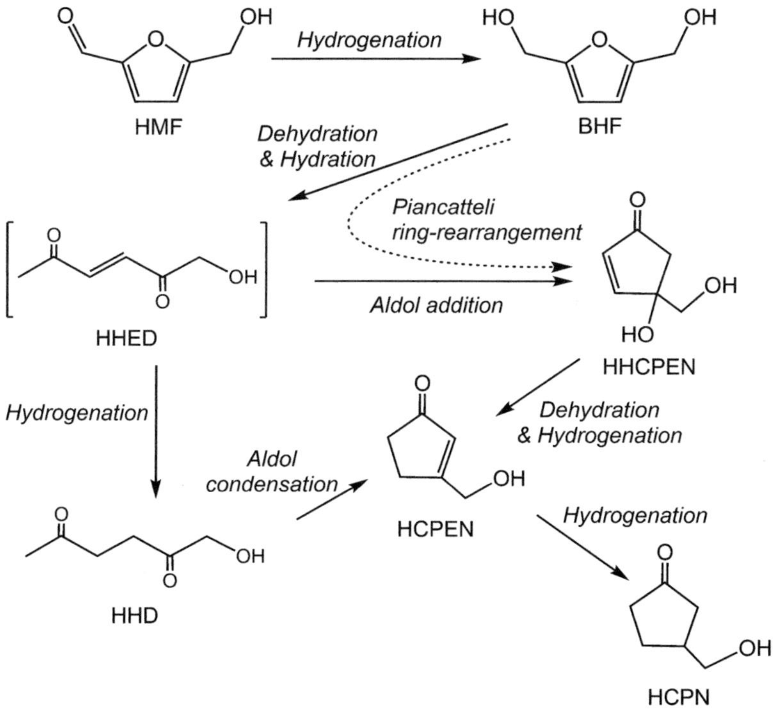

Scheme 5.10 Proposed reaction path for 5-hydroxymethylfurfural (HMF) conversion to 3-hydroxymethylcyclopentanone (HCPN) [25, 50]

reaction path is as follows: selective hydrogenation of HMF produces BHF; the furan ring of BHF opens to provide HHED (not detected, represented in parentheses in Scheme 5.10); selective hydrogenation of HMF produces BHF; the furan ring of BHF opens to provide 1-hydroxy-3-hexene-2,5-dione (HHED) (not detected, represented in parentheses in Scheme 5.10); intramolecular aldol reaction of HHED gives HHCPEN; HHCPEN is dehydrated and hydrogenated to produce HCPEN; and HCPEN is hydrogenated to form HCPN. HCPN can also be generated through the following two routes: intramolecular aldol condensation of HHD with the subsequent hydrogenation producing HCPN and Piancatelli reaction of BHF to HCPEN via HHCPEN. In the reaction scheme, the metal species (e.g., Au and Pt) facilitate the hydrogenation reaction, and the metal oxides (e.g., Nb_2O_5 and Ta_2O_5) are considered to facilitate the steps involving acid–base catalysis by H^+ and/or Lewis acid, which are hydration, aldol reaction, dehydration, and/or Piancatelli reaction.

On the basis of the reaction path in Scheme 5.10, it can be concluded that catalyst systems having high activity for furan ring hydrogenation and hydrogenolysis are not suitable for HCPN synthesis. Therefore, Au and Pt group metals with reduced hydrogenation ability are effective as catalyst components for HCPN synthesis. In fact, the authors have examined the effect of the hydrogenation ability of Pt group metals on HCPN yield [52]. The hydrogenation ability of Pt group metals was

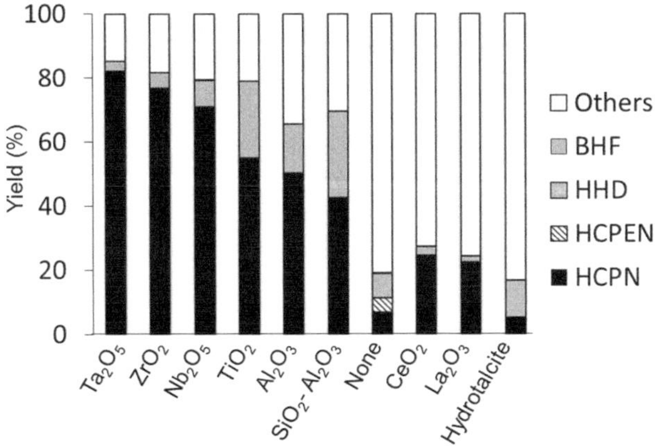

Fig. 5.4 Product distributions obtained by the reductive conversion of 5-hydroxymethylfurfural (HMF) in water using 1 wt.% Pt/SiO_2 with/without the various metal oxides [50] (Reprinted with permission from [43]. Copyright © 2016, The Royal Society of Chemistry)

adjusted by oxygen pretreatment. When using a Pt/Nb_2O_5 catalyst, HCPN yield was improved from 4% to 27% by oxygen treatment after hydrogen treatment. It should be noted that oxygen treatment was also applied to Pt/SiO_2 + Ta_2O_5 in Table 5.6.

The acid–base properties of metal oxides (as supports or additives) have a marked influence on ring rearrangement. Figure 5.4 shows the product distributions obtained by the reaction using 1 wt.% Pt/SiO_2 with/without the various metal oxides. The reaction using only Pt/SiO_2 (without metal oxides) produced a low yield of HCPN (7%). In the presence of CeO_2, La_2O_3, and hydrotalcite, which are basic metal oxides, the yield of HCPN did not largely increase (less than 25%). On the other hand, in the presence of Ta_2O_5, ZrO_2, Nb_2O_5, TiO_2, Al_2O_3, and SiO_2–Al_2O_3, which are known to be more acidic than CeO_2, La_2O_3, and hydrotalcite [53–56], the HCPN yield increased markedly. In particular, Ta_2O_5, ZrO_2, and Nb_2O_5 significantly enhanced the HCPN yield (72–82%). The effect of the acid–base properties of the metal oxides was also examined in HCPN synthesis using Au catalysts. Various metal oxides, including Nb_2O_5, Al_2O_3, ZrO_2, TiO_2, SiO_2–Al_2O_3, La_2O_3, CeO_2, and hydrotalcite, were used as supports for the Au catalysts. As a result, when the basicity or acidity of the metal oxide was too strong, the yield of HCPN decreased. The following two functions of the metal oxides are conceivable. One is pH adjustment of the aqueous solution, in which the reaction between the surface of the metal oxide and water changes the pH in the aqueous solution. Although the effect of the pH of the aqueous solvent in HCPN or HCPO synthesis has not yet been sufficiently studied, weakly acidic conditions are considered to be effective based on studies of the ring rearrangement of FA to CPN [32]. The other important function of metal oxides is Lewis acid catalysis on the surface. Regarding the intramolecular Piancatelli type rearrangement of furans, it is known that reactivity changes depending on whether or not the furans have a substituent at the 5-position [57]. The

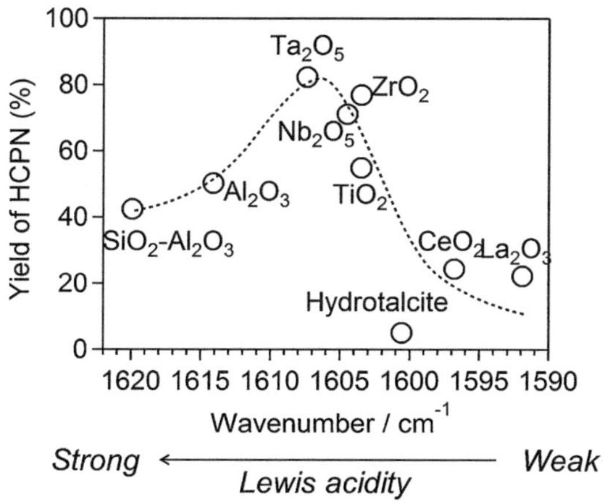

Fig. 5.5 Effect of Lewis acidity of metal oxides (MO_x) on production of 3-hydroxymethylcyclopentanone (HCPN) from 5-hydroxymethylfurfural (HMF) using Pt/SiO_2 + MO_x catalysts [50] (Reprinted with permission from [43]. Copyright © 2016, The Royal Society of Chemistry)

ring-rearrangement reaction of a furan having no substituent at the 5-position (e.g., furfuryl alcohol) proceeds only with the H^+ of water, whereas the ring-rearrangement reaction of a furan having a substituent at the 5-position (e.g., BHF) does not proceed efficiently with only H^+; a Lewis acid catalyst is required. Therefore, the Lewis acid catalysis of metal oxides becomes important for the hydrogenative ring rearrangement of HMF.

The authors studied the effect of the Lewis acid properties of metal oxides on HCPN synthesis using Pt/SiO_2 + MO_x based on pyridine adsorption FT-IR spectroscopy [50]. All of the IR spectra of MO_x showed bands around 1445 cm^{-1} and 1610 cm^{-1}, assignable to the ν_{19b} and ν_{8a} mode of pyridine adsorbed on Lewis acid sites. In Fig. 5.5, the HCPN yields are plotted against the peak wavenumbers of adsorbed pyridine (ν_{8a}). The higher wavenumber suggests a stronger Lewis acid of metal oxides [53, 56, 58]. The HCPN yield has a maximum in peak wavenumber of adsorbed pyridine, which indicates that a Lewis acid with moderate strength leads to a high yield of HCPN. It can be assumed that when the Lewis acid is too weak, the ring-rearrangement reaction does not proceed and the HCPN yield decreases; conversely, when the Lewis acid is too strong, the HCPN yield decreases due to acid-catalyzed side reactions such as polymerization of HMF and/or the intermediates. Therefore, a moderate Lewis acid is most effective for HCPN production. On the basis of the above, catalytic reaction systems using a catalyst with moderate Lewis acid together with a selective hydrogenation ability for the HMF aldehyde group and a weak pH condition are effective for the hydrogenative ring-rearrangement reaction of HMF in water.

5.5 Other Reactions Involving Hydrodeoxygenation or Hydrolysis

5.5.1 *Hydrodeoxygenation*

Hydrodeoxygenation of HMF to 2,5-dimethylfuran (DMF) (Scheme 5.11) has attracted intense attention in relation to the production of an alternative oil-based fuel from biomass. Selective synthesis of DMF has been achieved in organic solvents such as THF and 1-butanol. On the other hand, the hydrodeoxygenation of HMF is usually difficult to produce in the presence of water. In fact, according to the report by Dumesic et al., in the absence of water, HMF is converted primarily into 2-methyl-5-hydroxymethylfuran (MHMF) and DMF by hydrogenolysis, whereas reaction in THF-water mixtures (THF/H_2O = 95: 5 w/w) prevents the hydrogenolysis of HMF and leads to the selective production of BHTHF by hydrogenation [59]. However, there are a few reports on the efficient synthesis of DMF from HMF in water. Chatterjee et al. achieved a very high yield of DMF (ca. 100%) by hydrogenation of HMF using Pd/C in supercritical CO_2-water [60]. Bordoloi et al. reported that Ni nanoparticles supported on mesoporous nitrogen-rich carbon material (CN_x) afforded DMF with >98% yield by hydrogenation of HMF in water [61]. It has been proposed that synergy between metals and CN_x plays a crucial role in the hydrodeoxygenation of HMF to DMF in water. Further studies on the role of supercritical CO_2-water and CN_x support are desired for the development of DMF production systems as well as for synthesis of DMF-derived products by hydrolysis in water, as described below.

Sequential conversion of HMF by hydrodeoxygenation and ring opening (Scheme 5.11) gives 2,5-hexanedione (HDN), which can be used for the production alcohols, amines, cycloketones, and pyrrolidine. The reported yields of HDN are not yet very high when using HMF as a feedstock and H_2 as a hydrogenation source. Resasco et al. reported that 48% yield of HDN was obtained by the hydrogen of HMF in a water/decalin (1: 1) emulsion using a Ru nanoparticle catalyst anchored on hydrophobic MWCNT fused to aluminum oxide (Ru/MWNT/Al_2O_3) [62]. Two research groups, Jérôme et al. and Li et al., reported that 20–30% yield of HDN can be obtained by using combination catalysts of Pd/C and Brønsted acids including CO_2 and Amberlyst resins [22, 63].

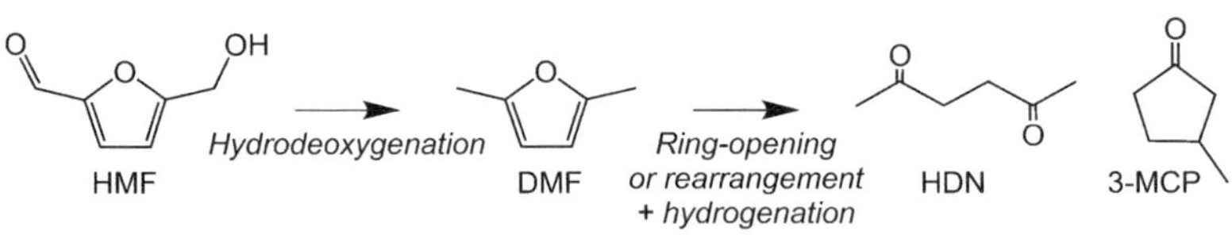

Scheme 5.11 Hydrodeoxygenation conversion of 5-hydroxymethylfurfural (HMF). 2,5-dimethylfuran (DMF) produced by only hydrodeoxygenation; 2,5-hexanedione (HDN) by hydrodeoxygenation and ring opening; 3-methyl-2-cyclopenten-1-one (3-MCPEN) by ring-rearrangement; 3-methylcyclopentanone (3-MCPN) by ring rearrangement and hydrodeoxygenation

HDN is considered to be formed through metal-catalyzed hydrodeoxygenation of HMF to MHMF or DMF and the subsequent acid-catalyzed hydrolytic ring opening. Jérôme et al. investigated HDN synthesis using MHMF and DMF as starting compounds. In the case of the reaction starting from MHMF, 28% yield of HDN was obtained using the combination catalyst of Pd/C and CO_2 [22]. On the other hand, when started from DMF, an almost equivalent amount of HDN was obtained under CO_2 in water without Pd/C. Based on these results, it is expected that efficient catalytic systems for DMF production can afford HDN synthesis by combination with acid catalysts.

Ring rearrangement of DMF, or the intermolecular aldol condensation of HDN, forms 3-methyl-2-cyclopenten-1-one (3-MCPEN), which is useful for synthesis of natural products such as jasmonoids and prostanoids. Huo et al. reported the production of 3-MCPEN (31% yield) and HDN (27% yield) from HMF by water splitting using zinc metal powder [64]. When starting from 5-methylfurfural, Au/TiO_2 efficiently produces 3-methylcyclopentanone (3-MCPN) (Scheme 5.11) [57]. It is expected that catalysts for the ring rearrangement of HMF, in particular metal oxides with Lewis acidity, are effective components for the synthesis of 3-MCPN and its derivatives. Bifunctional catalysts having hydrodeoxygenation and ring-rearrangement activity are considered to be useful for 3-MCPN production from HMF, although as yet there has been no efficient catalyst for 3-MCPN production from HMF.

5.5.2 *Hydrogenolytic Ring Opening*

The hydrogenolytic ring-opening reactions of HMF can form 1,2,6-hexanetriol (1,2,6-HT) (Scheme 5.12), which is used in the plastics, cosmetics, and pharmaceutical industries. Mu et al. reported that hydrogenation of HMF using Ni–Co–Al mixed oxide catalysts in methanol gave 1,2,6-HT with 65% yield [65]. In an analogous reaction, Mizugaki et al. reported the hydrogenolytic ring opening of furfural to 1,2-pentanediol in 2-propanol using Pt/HT [66]. The reaction scheme is similar

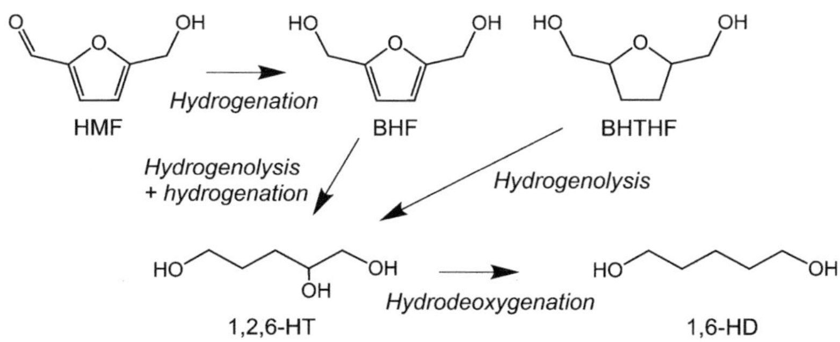

Scheme 5.12 Hydrogenolytic ring-opening conversion of 5-hydroxymethylfurfural (HMF)

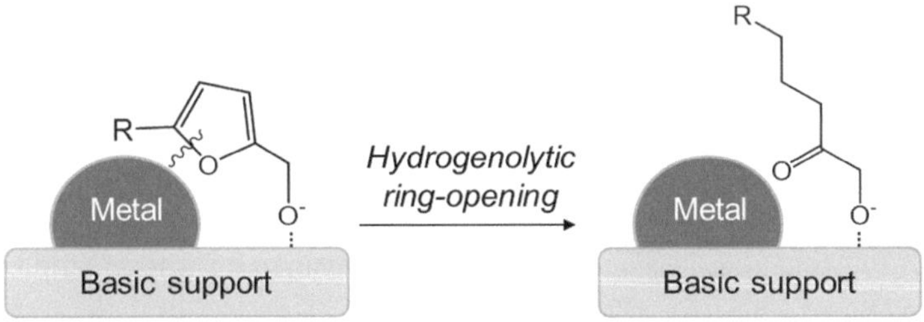

Fig. 5.6 Cooperative behavior of hydrogenation-active metal species and surface basic sites for selective hydrogenolytic ring opening of furfuryl alcohol intermediates [65, 66]. Adapted with permission from [65, 66]. Copyright © 2014, The American Chemical Society. R = CH_2OH: 2,5-bis(hydroxymethyl)furfural (BHF), R = H: furfuryl alcohol (FOL)

to 1,2,6-HT synthesis in that formation of furfuryl alcohol by hydrogenation of the aldehyde group of FA is followed by hydrogenolytic ring opening. Based on these reports, the cooperative behavior of hydrogenation-active metal species and surface basic sites adsorbing furfuryl alcohol intermediates is the key to selective hydrogenolytic ring-opening reactions (Fig. 5.6). The authors have tested the reductive conversion of HMF in water (0.2 M of HMF, 3 mL, 3 MPa H_2, 135 °C, 24 h) using 5 wt.% Pt/CeO_2 and Pt/La_2O_3 (10 mg). As a result, 1,2,6-HT was obtained with ca. 20% yield. In contrast, almost no 1,2,6-HT was obtained when the reaction was conducted using combination catalyst systems of 5 wt.% Pt/SiO_2 + CeO_2 and Pt/SiO_2 + La_2O_3 under the same reaction conditions. This result suggests that the interface between the metal nanoparticles and the basic carrier plays an important role. Tuning the interface between the basic supports and metal nanoparticles may further improve the efficiency in 1,2,6-HT synthesis.

The synthesis of 1,2,6-HT via BHTHF has been proposed [67]. Buntara et al. demonstrated catalytic conversion of BHTHF using Rh-Re/SiO_2 to 1,2,6-HT with high selectivity of 97% at 21% conversion of BHTHF [24]. Catalysts comprising ReO_x and Rh or Ir are effective for the hydrogenolysis of C–O in the furan ring of BHTHF to selectively give 1,2,6-HT owing to the bifunctionality of the catalysts: the acidic Re-OH species adsorbing furan and the hydrogenolysis activity of the noble metals [10, 68, 69]. In the synthesis of 1,2,6-HT from BHTHF, water as solvent is effective, possibly due to tuning of the Brønsted acidity of the catalysts through the reaction between ReO_x species and water on the surfaces [67, 68].

The combination of hydrogenolytic ring opening and hydrodeoxygenation can afford 1,6-hexanediol (1,6-HD) (Scheme 5.12), which is an important precursor for polyesters, polyurethanes, and adhesives. One-pot synthesis of 1,6-hexanediol (1,6-HD) from HMF has been reported, but to date, the yield of 1,6-HD is not very high (ca. 50%) [70]. Therefore, a two-step process for 1,6-HD synthesis from HMF has been proposed [24, 71]. Studies on the two-step synthesis have demonstrated that the Rh–and Ir–ReO_x catalysts, which catalyze the hydrogenolytic ring opening of BHTHF to 1,2,6-HT as mentioned above, are also effective for the hydrodeoxygenation of 1,2,6-HT to 1,6-HD. By using this knowledge, Zhang et al. performed 1,6-

HD synthesis from HMF in a fixed-bed reactor using double-layered catalysts of Pd/ SiO_2 and Ir–ReO_x/SiO_2, which are effective for total hydrogenation of HMF to BHTHF and hydrodeoxygenation of 1,2,6-HT to 1,6-HD, respectively [68]. As a result, they obtained ca. 60% yield of 1,6-HD. This result suggests that the combination of functional catalysts can improve the biorefinery process. With further studies on catalyst functions in various biomass conversion reactions and their combinations, a more efficient catalyst system for 1,6-HD synthesis can be developed.

5.6 Conclusions and Future Outlook

The acid catalysis of H^+ due to water can be involved in the reductive conversion of HMF in aqueous solution, in addition to reductive reactions on the metal surface, the metal-oxide interface, and acid–base reactions on metal-oxide supports.

When HMF is hydrogenated under suppression of Brønsted acid catalysis of H^+ in aqueous solution or on catalysts, BHF and BHTHF can be obtained in high yield. On the other hand, when Brønsted acid acts on BHF, HHD and HCPN can be obtained via the hydrolytic ring opening and the ring rearrangement, respectively. Strong Brønsted acids however cause degradation of HMF and BHF. Therefore, controlling reaction conditions affecting the Brønsted acid catalysis including reaction solution pH and temperature is important to obtain the desired products selectively. In the case of HCPN synthesis, Lewis acid on catalysts can promote ring-rearrangement of BHF in cooperation with Brønsted acid. Since hydrogenation catalysis, e.g., hydrogenation activity and selectivity for C = C and C = O groups of HMF and reaction intermediates, of course, have an influence on products, both hydrogenation and acid catalysis of reaction systems needs to be tuned for synthesis of the desired products. Previous studies have developed catalytic reaction systems for BHF, BHTHF, HHD, and HCPN, and one can obtain these four compounds from HMF in relatively high yield. Future detailed or systematic studies on the bifunctionality of catalytic reaction systems will further improve the production of the four compounds. Besides, efficient conversion of HMF to 1,2,5-HT and HCPO, which are the hydrogenation products of HHD and HCPN, respectively, is expected to be achieved by modification of catalytic reaction systems for HHD and HCPN synthesis or improving catalytic performance for hydrogenation of their C=O groups [72].

HMF conversion involving hydrodeoxygenation and hydrogenolysis offers valuable compounds such as DMF, HDN, 1,2,6-HT, and 1,6-HD. Although catalytic reaction systems showing hydrodeoxygenation and hydrogenolysis of HMF have been reported, the yields or selectivity of the products are still low. Understanding and control of adsorption of HMF or reaction intermediates on catalysts will be the key to future development of HMF conversion by hydrodeoxygenation and hydrogenolysis. More specifically, in the case of supported metal catalysts, functions of metal-support interface, which can serve as unique adsorption and reaction sites, will play an important role in selective hydrodeoxygenation or hydrogenolysis of

HMF. Further development of catalyst materials and elucidation of functions of metal-support interface are expected to achieve efficient reaction systems for HMF conversion involving hydrodeoxygenation and hydrogenolysis.

References

1. Corma A, Iborra S, Velty A. Chemical routes for the transformation of biomass into chemicals. Chem Rev. 2007;107(6):2411–502.
2. Besson M, Gallezot P, Pinel C. Conversion of biomass into chemicals over metal catalysts. Chem Rev. 2014;114(3):1827–70.
3. Delidovich I, Hausoul PJC, Deng L, Pfutzenreuter R, Rose M, Palkovits R. Alternative monomers based on lignocellulose and their use for polymer production. Chem Rev. 2016;116(3):1540–99.
4. Ragauskas AJ, Williams CK, Davison BH, Britovsek G, Cairney J, Eckert CA, Frederick WJ, Hallett JP, Leak DJ, Liotta CL, Mielenz JR, Murphy R, Templer R, Tschaplinski T. The path forward for biofuels and biomaterials. Science. 2006;311(5760):484–9.
5. Moliner M, Román-Leshkov Y, Davis ME. Tin-containing zeolites are highly active catalysts for the isomerization of glucose in water. Proc Natl Acad Sci U S A. 2010;107(14):6164–8.
6. Nakajima K, Baba Y, Noma R, Kitano MN, Kondo J, Hayashi S, Hara M. $Nb_2O_5nH_2O$ as a heterogeneous catalyst with water-tolerant Lewis acid sites. J Am Chem Soc. 2011;133(12):4224–7.
7. Rosatella AA, Simeonov SP, Frade RFM, Afonso CAM. 5-Hydroxymethylfurfural (HMF) as a building block platform: biological properties, synthesis and synthetic applications. Green Chem. 2011;13(4):754–93.
8. van Putten RJ, van der Waal JC, de Jong E, Rasrendra CB, Heeres HJ, de Vries JG. Hydroxymethylfurfural, a versatile platform chemical made from renewable resources. Chem Rev. 2013;113(3):1499–597.
9. Simon M-O, Li C-J. Green chemistry oriented organic synthesis in water. Chem Soc Rev. 2012;41(4):1415–27.
10. Nakagawa Y, Tamura M, Tomishige K. Catalytic reduction of biomass-derived Furanic compounds with hydrogen. ACS Catal. 2013;3(12):2655–68.
11. Li H, Fang Z, Smith RL, Yang S. Efficient valorization of biomass to biofuels with bifunctional solid catalytic materials. Prog Energy Combust Sci. 2016;55:98–194.
12. Chatterjee M, Ishizaka T, Kawanami H. Selective hydrogenation of 5-hydroxymethylfurfural to 2,5-bis-(hydroxymethyl)furan using Pt/MCM-41 in an aqueous medium: a simple approach. Green Chem. 2014;16(11):4734–9.
13. Ohyama J, Esaki A, Yamamoto Y, Arai S, Satsuma A. Selective hydrogenation of 2-hydroxymethyl-5-furfural to 2,5-bis(hydroxymethyl)furan over gold sub-nano clusters. RSC Adv. 2013;3(4):1033–6.
14. Tamura M, Tokonami K, Nakagawa Y, Tomishige K. Rapid synthesis of unsaturated alcohols under mild conditions by highly selective hydrogenation. Chem Commun. 2013;49(63):7034–6.
15. Nakagawa Y, Tomishige K. Total hydrogenation of furan derivatives over silica-supported Ni-Pd alloy catalyst. Catal Commun. 2010;12(3):154–6.
16. Connolly TJ, Considine JL, Ding Z, Forsatz B, Jennings MN, MacEwan MF, McCoy KM, Place DW, Sharma A, Sutherland K. Efficient synthesis of 8-Oxa-3-aza-bicyclo[3.2.1]octane Hydrochloride. Org Process Res Dev. 2010;14(2):459–65.
17. Alamillo R, Tucker M, Chia M, Pagan-Torres Y, Dumesic J. The selective hydrogenation of biomass-derived 5-hydroxymethylfurfural using heterogeneous catalysts. Green Chem. 2012;14(5):1413–9.
18. Nakagawa Y, Takada K, Tamura M, Tomishige K. Total hydrogenation of Furfural and 5-Hydroxymethylfurfural over supported Pd–Ir alloy catalyst. ACS Catal. 2014;4(8):2718–26.

19. Huber GW, Chheda JN, Barrett CJ, Dumesic JA. Production of liquid alkanes by aqueous-phase processing of biomass-derived carbohydrates. Science. 2005;308(5727):1446–50.
20. Wang X, Rinaldi R. Exploiting H-transfer reactions with RANEY[registered sign] Ni for upgrade of phenolic and aromatic biorefinery feeds under unusual, low-severity conditions. Energy Environ Sci. 2012;5(8):8244–60.
21. Ohyama J, Hayashi Y, Ueda K, Yamamoto Y, Arai S, Satsuma A. Effect of FeOx modification of Al_2O_3 on its supported Au catalyst for hydrogenation of 5-Hydroxymethylfurfural. J Phys Chem C. 2016;120(28):15129–36.
22. Liu F, Audemar M, Vigier KDO, Clacens J-M, De Campo F, Jerome F. Palladium/carbon dioxide cooperative catalysis for the production of Diketone derivatives from carbohydrates. ChemSusChem. 2014;7(8):2089–93.
23. liu F, Audemar M, De Oliveira Vigier K, Clacens J-M, De Campo F, Jerome F. Combination of Pd/C and Amberlyst-15 in a single reactor for the acid/hydrogenating catalytic conversion of carbohydrates to 5-hydroxy-2,5-hexanedione. Green Chem. 2014;16(9):4110–4.
24. Buntara T, Noel S, Phua PH, Melián-Cabrera I, de-Vries JG, Heeres HJ. Caprolactam from renewable resources: catalytic conversion of 5-Hydroxymethylfurfural into Caprolactone. Angew Chem Int Ed. 2011;50(31):7083–7.
25. Yamamoto Y, Arai S, Esaki A, Ohyama J, Satsuma A, Tanaka N. Statistical distribution of single atoms and clusters of supported Au catalyst analyzed by global high-resolution HAADF-STEM observation with morphological image-processing operation. Microscopy. 2014;63(3):209–18.
26. Gupta K, Tyagi D, Dwivedi AD, Mobin SM, Singh SK. Catalytic transformation of bio-derived furans to valuable ketoacids and diketones by water-soluble ruthenium catalysts. Green Chem. 2015;17(9):4618–27.
27. Xu Z, Yan P, Xu W, Liu X, Xia Z, Chung B, Jia S, Zhang ZC. Hydrogenation/Hydrolytic ring opening of 5-HMF by Cp*-Iridium(III) half-sandwich complexes for bioketones synthesis. ACS Catal. 2015;5(2):788–92.
28. Xu Z, Yan P, Li H, Liu K, Liu X, Jia S, Zhang ZC. Active Cp*Iridium(III) complex with ortho-Hydroxyl group functionalized Bipyridine ligand containing an electron-donating group for the production of Diketone from 5-HMF. ACS Catal. 2016;6(6):3784–8.
29. WP W, YJ X, Zhu R, Cui MS, Li XL, Deng J, Fu Y. Selective conversion of 5-Hydroxymethylfuraldehyde using Cp*Ir catalysts in Aqueous Formate Buffer solution. ChemSusChem. 2016;9(10):1209–15.
30. Chen J, Lu F, Zhang J, Yu W, Wang F, Gao J, Xu J. Immobilized Ru clusters in nanosized Mesoporous Zirconium Silica for the aqueous hydrogenation of Furan derivatives at room temperature. ChemCatChem. 2013;5(10):2822–6.
31. Hronec M, Fulajtarova K. Selective transformation of furfural to cyclopentanone. Catal Commun. 2012;24:100–4.
32. Yang Y, Du Z, Huang Y, Lu F, Wang F, Gao J, Xu J. Conversion of furfural into cyclopentanone over Ni-Cu bimetallic catalysts. Green Chem. 2013;15(7):1932–40.
33. Hronec M, Fulajtarova K, Liptaj T. Effect of catalyst and solvent on the furan ring rearrangement to cyclopentanone. Appl Catal A. 2012;437:104–11.
34. Hronec M, Fulajtarova K, Vavra I, Sotak T, Dobrocka E, Micusik M. Carbon supported Pd-Cu catalysts for highly selective rearrangement of furfural to cyclopentanone. Appl Catal B. 2016;181:210–9.
35. Fang RQ, Liu HL, Luque R, Li YW. Efficient and selective hydrogenation of biomass-derived furfural to cyclopentanone using Ru catalysts. Green Chem. 2015;17(8):4183–8.
36. Liu YH, Chen ZH, Wang XF, Liang Y, Yang XM, Wang ZC. Highly selective and efficient rearrangement of biomass-derived Furfural to Cyclopentanone over interface-active Ru/Carbon nanotubes catalyst in water. ACS Sustain Chem Eng. 2017;5(1):744–51.
37. Hronec M, Fulajtárova K, Soták T. Highly selective rearrangement of furfuryl alcohol to cyclopentanone. Appl Catal B. 2014;154–155:294–300.

38. Liu CY, Wei RP, Geng GL, Zhou MH, Gao LJ, Xiao GM. Aqueous-phase catalytic hydrogenation of furfural over Ni-bearing hierarchical Y zeolite catalysts synthesized by a facile route. Fuel Process Technol. 2015;134:168–74.
39. Li XL, Deng J, Shi J, Pan T, CG Y, HJ X, Fu Y. Selective conversion of furfural to cyclopentanone or cyclopentanol using different preparation methods of Cu-Co catalysts. Green Chem. 2015;17(2):1038–46.
40. Zhu HY, Zhou MH, Zeng Z, Xiao GM, Xiao R. Selective hydrogenation of furfural to cyclopentanone over Cu-Ni-Al hydrotalcite-based catalysts. Korean J Chem Eng. 2014;31(4):593–7.
41. Wang Y, Zhou MH, Wang TZ, Xiao GM. Conversion of Furfural to Cyclopentanol on Cu/Zn/Al catalysts derived from Hydrotalcite-like materials. Catal Lett. 2015;145(8):1557–65.
42. Guo J, Xu G, Han Z, Zhang Y, Fu Y, Guo Q. Selective conversion of Furfural to Cyclopentanone with CuZnAl catalysts. ACS Sustain Chem Eng. 2014;2(10):2259–66.
43. Zhang GS, Zhu MM, Zhang Q, Liu YM, He HY, Cao Y. Towards quantitative and scalable transformation of furfural to cyclopentanone with supported gold catalysts. Green Chem. 2016;18(7):2155–64.
44. Hronec M, Fulajtárová K, Soták T. Kinetics of high temperature conversion of furfuryl alcohol in water. J Ind Eng Chem. 2014;20(2):650–5.
45. Piancatelli G, Scettri A. Heterocycle steroids: synthesis of a new steroid alkaloid: 27 nor-23,26-imino-5α-cholesten-23(N)-3β-ol. Tetrahedron. 1976;32(14):1745–9.
46. Piancatelli G, Scettri A, Barbadoro S. A useful preparation of 4-substituted 5-hydroxy-3-oxocyclopentene. Tetrahedron Lett. 1976;17(39):3555–8.
47. Li S-W, Batey RA. Mild lanthanide(iii) catalyzed formation of 4,5-diaminocyclopent-2-enones from 2-furaldehyde and secondary amines: a domino condensation/ring-opening/electrocyclization process. Chem Commun. 2007;36:3759–61.
48. Veits GK, Wenz DR, de Read AJ. Versatile method for the synthesis of 4-Aminocyclopentenones: Dysprosium(III) Triflate catalyzed Aza-Piancatelli rearrangement. Angew Chem Int Ed. 2010;49(49):9484–7.
49. Bredihhin A, Luiga S, Vares L. Application of 5-Ethoxymethylfurfural (EMF) for the production of Cyclopentenones. Synthesis. 2016;48(23):4181–8.
50. Ohyama J, Kanao R, Ohira Y, Satsuma A. The effect of heterogeneous acid-base catalysis on conversion of 5-hydroxymethylfurfural into a cyclopentanone derivative. Green Chem. 2016;18(3):676–80.
51. Perret N, Grigoropoulos A, Zanella M, Manning TD, Claridge JB, Rosseinsky MJ. Catalytic response and stability of Nickel/Alumina for the hydrogenation of 5-Hydroxymethylfurfural in water. ChemSusChem. 2016;9(5):521–31.
52. Ohyama J, Kanao R, Esaki A, Satsuma A. Conversion of 5-hydroxymethylfurfural to a cyclopentanone derivative by ring rearrangement over supported Au nanoparticles. Chem Commun. 2014;50(42):5633–6.
53. Tamura M, Shimizu K-i, Satsuma A. Comprehensive IR study on acid/base properties of metal oxides. Appl Catal A. 2012;433–434:135–45.
54. Shimizu K-i, Ohshima K, Satsuma A. Direct dehydrogenative amide synthesis from alcohols and amines catalyzed by γ-Alumina supported silver cluster. Chem Eur J. 2009;15(39):9977–80.
55. Nishimura S, Takagaki A, Ebitani K. Characterization, synthesis and catalysis of hydrotalcite-related materials for highly efficient materials transformations. Green Chem. 2013;15(8):2026–42.
56. Tanabe K, Misono M, Ono Y, Hattori H. New solid acids and bases, Studies in surface science and catalysis, vol 51. Tokyo/Amsterdam: Kodansha/Elsevier; 1989.
57. Zhang G-S, Zhu M-M, Zhang Q, Liu Y-M, He H-Y, Cao Y. Towards quantitative and scalable transformation of furfural to cyclopentanone with supported gold catalysts. Green Chem. 2016;18(7):2155–64.
58. Busca G. The surface acidity of solid oxides and its characterization by IR spectroscopic methods. An attempt at systematization. Phys Chem Chem Phys. 1999;1(5):723–36.

59. Liu YF, Mellmer MA, Alonso DM, Dumesic JA. Effects of water on the Copper-catalyzed conversion of Hydroxymethylfurfural in Tetrahydrofuran. ChemSusChem. 2015;8(23):3983–6.
60. Chatterjee M, Ishizaka T, Kawanami H. Hydrogenation of 5-hydroxymethylfurfural in supercritical carbon dioxide-water: a tunable approach to dimethylfuran selectivity. Green Chem. 2014;16(3):1543–51.
61. Goyal R, Sarkar B, Bag A, Siddiqui N, Dumbre D, Lucas N, Bhargava SK, Bordoloi A. Studies of synergy between metal-support interfaces and selective hydrogenation of HMF to DMF in water. J Catal. 2016;340:248–60.
62. Faria J, Pilar Ruiz M, Resasco DE. Carbon Nanotube/Zeolite hybrid catalysts for glucose conversion in water/oil emulsions. ACS Catal. 2015;5(8):4761–71.
63. Hu X, Kadarwati S, Song Y, Li C-Z. Simultaneous hydrogenation and acid-catalyzed conversion of the biomass-derived furans in solvents with distinct polarities. RSC Adv. 2016;6(6):4647–56.
64. Ren D, Song Z, Li L, Liu Y, Jin F, Huo Z. Production of 2,5-hexanedione and 3-methyl-2-cyclopenten-1-one from 5-hydroxymethylfurfural. Green Chem. 2016;18(10):3075–81.
65. Yao SX, Wang XC, Jiang YJ, Wu F, Chen XG, XD M. One-step conversion of biomass-derived 5-Hydroxymethylfurfural to 1,2,6-Hexanetriol over Ni-Co-Al mixed oxide catalysts under mild conditions. ACS Sustain Chem Eng. 2014;2(2):173–80.
66. Mizugaki T, Yamakawa T, Nagatsu Y, Maeno Z, Mitsudome T, Jitsukawa K, Kaneda K. Direct transformation of Furfural to 1,2-Pentanediol using a Hydrotalcite-supported platinum nanoparticle catalyst. ACS Sustain Chem Eng. 2014;2(10):2243–7.
67. Buntara T, Noel S, Phua PH, Melian-Cabrera I, de Vries JG, Heeres HJ. From 5-Hydroxymethylfurfural (HMF) to polymer precursors: catalyst screening studies on the conversion of 1,2,6-hexanetriol to 1,6-hexanediol. Top Catal. 2012;55(7–10):612–9.
68. Xiao B, Zheng M, Li X, Pang J, Sun R, Wang H, Pang X, Wang A, Wang X, Zhang T. Synthesis of 1,6-hexanediol from HMF over double-layered catalysts of Pd/SiO_2 + $Ir\text{-}ReOx/SiO_2$ in a fixed-bed reactor. Green Chem. 2016;18(7):2175–84.
69. Chia M, Pagán-Torres YJ, Hibbitts D, Tan Q, Pham HN, Datye AK, Neurock M, Davis RJ, Dumesic JA. Selective hydrogenolysis of polyols and cyclic ethers over bifunctional surface sites on Rhodium–Rhenium catalysts. J Am Chem Soc. 2011;133(32):12675–89.
70. Tuteja J, Choudhary H, Nishimura S, Ebitani K. Direct synthesis of 1,6-Hexanediol from HMF over a heterogeneous Pd/ZrP catalyst using Formic acid as hydrogen source. ChemSusChem. 2014;7(1):96–100.
71. Buntara T, Noel S, Phua PH, Melián-Cabrera I, de Vries JG, Heeres HJ. From 5-Hydroxymethylfurfural (HMF) to polymer precursors: catalyst screening studies on the conversion of 1,2,6-hexanetriol to 1,6-hexanediol. Top Catal. 2012;55(7):612–9.
72. Ohyama J, Ohira Y, Satsuma A. Hydrogenative ring-rearrangement of biomass derived 5-(hydroxymethyl)furfural to 3-(hydroxymethyl)cyclopentanol using combination catalyst systems of Pt/SiO and lanthanoid oxides. Catal Sci Technol. 2017;7(14):2947–53.

Chapter 6
Catalytic Cascade Transformations of Biomass into Polyols

Javier Fernández-Rodríguez, Xabier Erdocia, Pedro Luis de Hoyos, Ane Sequeiros, and Jalel Labidi

Abstract Among the many oxygen-rich chemicals that can be obtained from biomass, polyols, such as ethylene glycol and propylene glycol, are widely used in industry. Liquid polyols have been used in polyurethane foam preparation and as components of adhesives. Hydrolysis, coupled with hydrogenation and hydrogenolysis allows transformation of biomass or its constituents into polyols. Liquefaction is another approach which is efficient for converting biomass into liquid polyols that have high content of reactive hydroxyl groups.

The cascade transformation of biomass to polyols is widely used. In some cases, fractionation of biomass into its main components (cellulose, hemicelluloses and lignin) is accomplished to facilitate its further conversion. Production of polyols from cellulose can be conducted in a two-step process with suitable catalysts or in an one-step process with bifunctional catalysts. Polyol production depends on the material used as feedstock, catalyst employed and reaction conditions.

This chapter reviews key strategies used to convert biomass into polyols. Special emphasis is given to emerging processes aimed to valorize biomass through a cascade approach.

J. Fernández-Rodríguez • X. Erdocia • A. Sequeiros • J. Labidi (✉)
Chemical and Environmental Engineering Department, University of the Basque Country UPV/EHU, Donostia-San Sebastián, Spain
e-mail: jalel.labidi@ehu.eus

P.L. de Hoyos
Chemical and Environmental Engineering Department, University of the Basque Country UPV/EHU, Donostia-San Sebastián, Spain

University of Pau and Pays de l'Adour, IPREM-Physics and Chemistry of Polymers team (PCP), UMR CNRS 5254, IUT des Pays de l'Adour, 371 Rue de Ruisseau, 40004 Mont de Marsan, France

Z. Fang et al. (eds.), *Production of Biofuels and Chemicals with Bifunctional Catalysts*, Biofuels and Biorefineries 8, https://doi.org/10.1007/978-981-10-5137-1_6

6.1 Introduction

In the last decade, the conversion of lignocellulosic biomass to fuels, chemicals and materials has received significant attention. The manufacturing of products by environmentally-friendly processes is included within the concept of the modern biorefinery. In this context, biorefineries of today have to develop new competitive and economically feasible processes to compete against the top position of refineries in the manufacture of chemicals.

Among the oxygen-rich chemicals that can be obtained from biomass, polyols are one of the most important ones due to their wide use in polymer synthesis like polyurethanes (PU), food industry or in pharmaceuticals manufacturing [1]. Polyols are chemical compounds of the alcohol family, which have several hydroxyl groups available for organic chemical reactions [2]. Polyols that can be produced from biomass include sorbitol, mannitol, xylitol, ethylene glycol and propylene glycol. The production of these chemicals could be carried out using different routes, but in general, in the latest studies, single pot operations with internal successive processes are employed in the so-called cascade approach. This way, the isolation and purification of intermediates is eliminated, and the formation of unwanted by-products can be prevented [3].

Usually, the first step is to fractionate biomass into its main components (cellulose, hemicelluloses and lignin) and then, by depolymerization and/or fermentation to convert them into platform molecules that are subsequently employed as building blocks for the synthesis of intermediates and fine chemicals via heterogeneous and/or homogenous catalytic processes. However, the direct path of biomass transformation into polyols has also been studied employing different acid catalyst and organic solvents.

All these processes, especially the ones when biomass has previously been fractionated into its components, have been carried out using different simple catalysts. The first types of catalysts are used to break macromolecules into small molecules which are then converted into target polyols.

Nonetheless, bifunctional catalysts, which are able to perform these two steps in a single step, are gaining considerable attention [3]. New advances processes like microwave or ultrasound are being studied to improve either the fractionation step or the polyol production from the different fractions of biomass feedstock [4, 5].

In this chapter, diverse approaches for polyols production will be deeply studied. Firstly, the direct route of biomass transformation into polyols will be reviewed and then each main structural component of lignocellulosic biomass will be evaluated for forming the selected chemicals through cascade transformation with bifunctional catalysts. Finally, new advanced processes will be explored.

6.2 Biomass Direct Liquefaction

The direct conversion of biomass into a liquid fuel is not new and has been studied for several centuries [6]. There are different technologies to convert biomass into a liquid fuel which can be divided in two main groups: thermo-chemical and biochemical methods. The main objective of all the thermochemical methods is to produce energy from biomass and mainly include direct combustion, gasification, pyrolysis, and liquefaction [7]. However, only the last two methods are able to produce a liquid product directly from biomass.

6.2.1 *Pyrolysis*

Pyrolysis is defined as a thermal decomposition process which takes place in the absence of oxygen or air. In biomass pyrolysis, the three main products recovered are solid charcoal, liquid (bio-oil), and gases. Pyrolysis bio-oil usually contains high concentrations of phenolic compounds from lignin decomposition and furanic compounds derived from cellulose, along with organic acids and low molecular weight oxygenates.

6.2.2 *Liquefaction*

Liquefaction is defined as a low-temperature (250–400 °C) and high-pressure (5–20 MPa) thermochemical process where biomass, combined with water or another solvent, is converted into three products, i.e. a bio-oil fraction, a gas fraction and a solid residue fraction [8]. In this process, the employed solvents usually act as catalysts, rendering higher quality products as compared with those obtained from pyrolysis. The obtained bio-oil has a very low oxygen content and therefore it has a high heating-value.

Although many solvents can be employed in the liquefaction of biomass, methanol, ethanol and especially water are the most commonly used ones. To convert biomass into the desired products a highly reactive environment is required. At high pressures water remains as a liquid even at elevated temperatures with the consequent increase in the ion product ($Kw = [H^+][OH^-]$) and decrease in the dielectric constant [9]. The property changes help to form a good medium for biomass liquefaction process since the solubility of biomass in water is enhanced and acid- or base-catalyzed reactions are favored.

Biomass direct liquefaction is known since the 1980s when a liquefaction process known as "Hydrothermal Upgrading", or the HTU process, was developed by Shell [10]. Since then, many works have been carried out to study the influence of process parameters (reaction temperature, residence time, type of solvent, biomass composition or type of catalyst) in the liquefaction process of biomass.

Paying attention to the type of catalyst employed, the most common ones are strong bases like NaOH [11], $Ca(OH)_2$ and $Ba(OH)_2$ [12], strong acids like HCl and especially H_2SO_4 [13] or some other compounds like K_2CO_3 [14], Fe, Na_2CO_3 [15], FeS and $FeSO_4$ [16]. All of these are homogeneous catalysts that mainly help to reduce the formation of solid residue and to increase bio-oil yield. Among the alkali catalysts, Huang et al. [11] studied the liquefaction of rice husk in ethanol and concluded that NaOH was an ideal catalyst for the process enhancing the formation of long-chain alkanes and esters. Xu and Land [12] used Jack pine powder as biomass feedstock and were able to upgrade it into liquid organic compounds with $Ca(OH)_2$, $Ba(OH)_2$ and $FeSO_4$ as catalyst. They stated that $Ba(OH)_2$ and especially $Ca(OH)_2$ were very effective to improve the yield of heavy oils production from pine powder.

Acid catalysts have been widely used in biomass liquefaction. Zhang et al. [17] investigated the liquefaction of wood powder from Chinese fir (*Cunninghamia lanceolata*) and poplar (*Triploid Populus tomentosa Carr*) in the presence of phenol with phosphoric acid (85%), sulfuric acid (36%), hydrochloric acid (37%) and oxalic acid (99.5%) inorganic acids as catalysts. Phosphoric acid and sulfuric acid were better than the other catalysts leading to solid residues as low as 3.2 and 4.0 wt %, respectively. Besides alkali and acid catalyst, some other compounds like salts have also been used for biomass liquefaction. The production of oxygenated hydrocarbons by hydrothermal treatment of biomass at low reaction temperature with a short reaction time employing K_2CO_3 was studied by Karagöz et al. [14]. The study showed that the concentration of catalyst had an important effect on the inhibition of solid residue formation. Thus, the maximum conversions of biomass and oil products (ether and ethyl acetate soluble) were high with 1 M concentration of K_2CO_3 and solid residue remarkably decreased to 4%. Sun et al. [15] carried out an experimental study on direct liquefaction of paulownia in hot compressed water with and without catalyst at 280–360 °C. The addition of catalyst enhanced the liquefaction process of biomass. Thereafter, the maximum heavy oil yield was found to be 36% employing Fe as catalyst while little solid residue yield was obtained when Na_2CO_3 was used. The heavy oil mainly consisted of phenol derivatives, ketones, carboxylic acid/ester, benzene derivatives and long-chain alkanes, aldehydes and their derivatives. They found that different catalysts could affect the product distribution and their relative abundance.

Liquefaction of bagasse powder in supercritical ethanol was conducted in a batch reactor under a pressurized hydrogen atmosphere by Chumpoo and Prasassarakich [16]. The addition of iron based catalysts, Fe_2S_3/AC, and especially, $FeSO_4$, significantly enhanced both oil production and biomass conversion. Catalytic effects on oil formation became more apparent at high temperatures, and the highest oil yields and biomass conversions were obtained with 10% (w/w) $FeSO_4$. The obtained oil contained phenolics, esters, and furan derivatives as the dominant compounds and their composition could be altered by addition of appropriate catalysts. The use of ethanol-water solvent as opposed to pure ethanol, yields oils with higher heating values than those obtained from noncatalytic liquefaction.

6.2.3 Polyols Production from Biomass Liquefaction

Previously described processes are for focused on direct biomass liquefaction and use a simple catalyst to improve the oil production. However, there are other few works where the direct production of polyols such as xylitol or sorbitol from biomass has been reported. Sorbitol can be produced by hydrogenation of C6 sugars while xylitol, the C5-analogue of sorbitol, is produced by hydrogenation of xylose [18].

Usually the hydrolysis of biomass is the first step for the production of polyols or other types of chemicals. After this first step, the hydrodeoxygenation of the hydrolysis solution is carried out. In this context, Robinson et al. [19] were able to convert combined hemicellulose and cellulose obtained after the hydrolysis of different types of lignocellulosic biomass into xylitol and sorbitol, respectively. Conversion of the carbohydrate polymers, hemicellulose and cellulose, to polyols and anhydro polyols was more than 99% for loadings of up to 15% at 185–190 °C for 3–5 h with 0.8% H_3PO_4, stirring and hydrogenation with Ru/C catalyst. Li et al. [20] used maple wood for the production of xylitol. Firstly, hydrolysis with dilute sulfuric or oxalic acid was performed producing high yields of xylose monomers and minor amounts of glucose monomers. Afterwards, a low temperature hydrogenation of the product solutions was carried out with Ru/C as the catalyst. The catalyst selectively hydrogenated xylose into xylitol and glucose into sorbitol but did not selectively hydrogenate oligomers to xylitol and sorbitol.

Stalks of *Nicotiana tabacum* and castor oil-based polyol were synthesized via two-step process by Patel et al. [21]. Initially, *N. tabacum* stalks were liquefied at 150 °C and 180 min using *p*-toluenesulfonic acid (PTSA) catalyst to obtain glycol-glycoside. Afterwards, the glycol-glycoside produced was further reacted with castor oil in the presence of lithium hydroxide to yield a dark brown-colored polyol with hydroxyl value was running in between 200 and 400 mg/g-KOH.

The direct liquefaction of some types of biomass (very rich in cellulose and hemicelluloses) to obtain polyols is shown for selected works in Table 6.1.

Liquefaction of corn stover in organic solvents (90 wt% ethylene glycol and 10 wt% ethylene carbonate) with catalysts at moderate temperature under atmospheric pressure was carried out by Yu et al. [22]. Different catalysts, such as H_2SO_4, HCl, H_3PO_4, and $ZnCl_2$, were evaluated but the higher liquefaction yield was achieved in 2 wt% H_2SO_4, 1/4 (w/w) stover to liquefying reagent ratio, 160 °C temperature and 2 h of reaction time. The liquefied product had a very high polyols concentration and a hydroxyl value between 150 and 200 mg/g-KOH. In another study, the liquefied corn stover was directly used to prepare polyurethane foams [23]. Corn stover was liquefied by using ethylene carbonate (EC) as liquefying solvent and 97% sulfuric acid as catalyst at 170 °C for 90 min. The obtained liquefied product was rich in polyols and had a hydroxyl number of 137.34 mg/g-KOH.

Table 6.1 Direct conversion of biomass into polyols via liquefaction

Feedstock	Solvent	Catalyst	Hydroxyl content (mg/g-KOH)	References
Corn Stover	EG ethylene carbonate	H_2SO_4, HCl, H_3PO_4, $ZnCl_2$	150–200	[22]
Corn Stover	Ethylene carbonate	H_2SO_4	137.34	[23]
Pinus radiata Bark	Phenol	$C_6H_6O_3S$ (BSA)[a], CH_4O_3S (MSA)[b], $C_{10}H_8O_6S_2$ (NDSA)[c], $C_{10}H_7SO_3H$ (NSA)[d], $CH_3C_6H_4SO_3H$ (PTSA)[e] H_2SO_4	–	[24]
Bamboo	PEG400, glycerol	H_2SO_4	56.63	[25]
Poplar wood chips	PEG, EG, glycerin	H_2SO_4	155–235	[26]

[a]Benzenesulfonic acid
[b]Methanesulfonic acid
[c]1,5-naphthalenedisulfonic acid
[d]1-naphthalenesulfonic acid
[e]p-toluenesulfonic acid

Besides agricultural residues like corn stover which is very rich in carbohydrates, wood has also been liquefied to produce polyols. *Pinus radiata* bark has been treated using phenol as liquefying agent with five different organic sulfonic acids as catalyst [24]. The obtained results confirmed that the organic sulfonic acids play an important role in reducing the condensation reactions during the phenol liquefaction thus high amount of hydroxyl groups were free. Using polyethylene glycol (PEG400) and glycerol as liquefaction agents, the residue from bamboo processing has also been liquefied at the presence of sulfuric acid as catalyst [25]. The result showed that liquefaction yield could reach as high as 99.3% when liquefaction temperature was 160 °C, mass ratio of liquid to solid was 4:1 and liquefaction time was 90 min. Hydroxyl value of liquefaction products obtained was 56.63 mg/g-KOH. The liquefaction process of poplar wood chips has been studied using PEG, EG and glycerin as solvents [26]. At optimum conditions, wood liquefaction reached 93.8% yield and product hydroxyl values ranged from 155 to 235 mg/g-KOH.

Despite the studies mentioned above where lignocellulosic biomass is directly used to produce bio-oils or polyols for further application, the most common procedure is to treat each component of biomass separately. Thus, cellulose, hemicelluloses and lignin are previously fractionated from biomass and then used independently to facilitate the production of different polyols. In the following sections of the chapter, the treatment to produce polyols for each component of biomass will be described.

6.3 Cellulose Conversion to Polyols

Cellulose is most abundant natural polymer worldwide. It is the main component of lignocellulosic biomass together with hemicelluloses and lignin, which constitute the most important source of this biopolymer. Its constitution is based on a carbohydrate structure composed exclusively of cellobiose units, which are formed by two β-D-glucopyranose molecules linked between them by β-1,4-glycosidic bonds. The great density of hydroxyl groups lead to the generation of intra and intermolecular hydrogen bonds, forming the laminar chains which by means of their linkage are the origin of the microfibrillar structure of the plant cell wall. Its macromolecular structure is divided into two different regions, one having high crystallinity and the other being amorphous with the degree of polymerization (DP) being around 10,000. The high crystalline area makes cellulose insoluble in water as well as in most common of organic solvents and is the main cause of the difficulties to decompose it into low-molecular weight products [27]. As consequence, the main industrial application of cellulose is the pulp and paper products in which cellulose is commercialized with a low degree of degradation according to its high molecular size. However, the hydrolysis of the β-1,4-glycosidic bond in cellulose to form glucose, represents a pivotal process in biomass transformation industry, opening a tremendous opportunity to produce a large variety of low molecular mass chemicals that could substitute products obtained from fossil resources [28].

Conversion of cellulose to produce chemicals, as for instance, ethanol, ethylene glycol, glycerol, levulinic acid, hydroxymethyl furfural or lactic acid, has been widely studied in the last decade. The most industrially manufactured product from cellulose is bioethanol as second generation biofuels with a world production of 108 billion liters in 2014 [29]. Nevertheless, the current global reliance on fossil fuels for power generation remains dominant. The difficulty of cellulose fractionation is the main cause of its high production cost of this type of desired products.

Nowadays, great interest relies on cellulose transformation into polyol components, such as EG, PG, sorbitol and mannitol among others [30]. The interest in obtaining these chemicals from cellulose is hugely increasing based on three main points: green principle of manufacturing chemicals from natural resources reducing the environmental impact in the common synthesis process, the higher ratio of O/C that exists in biomass compared to hydrocarbons and recent improvement in catalyst design, which is able to boost reaction yields as well as selectivity to target products [31].

The transformation of cellulose to polyols in a cascade process (Fig. 6.1) occurs as the hydrolysis of cellulose to sugar alcohols is followed by hydrogenation and hydrogenolysis to different polyols, such as sorbitol, EG or PG [32]. The whole process has been traditionally divided into two distinct steps, whose reaction mechanisms and strategies will be developed below.

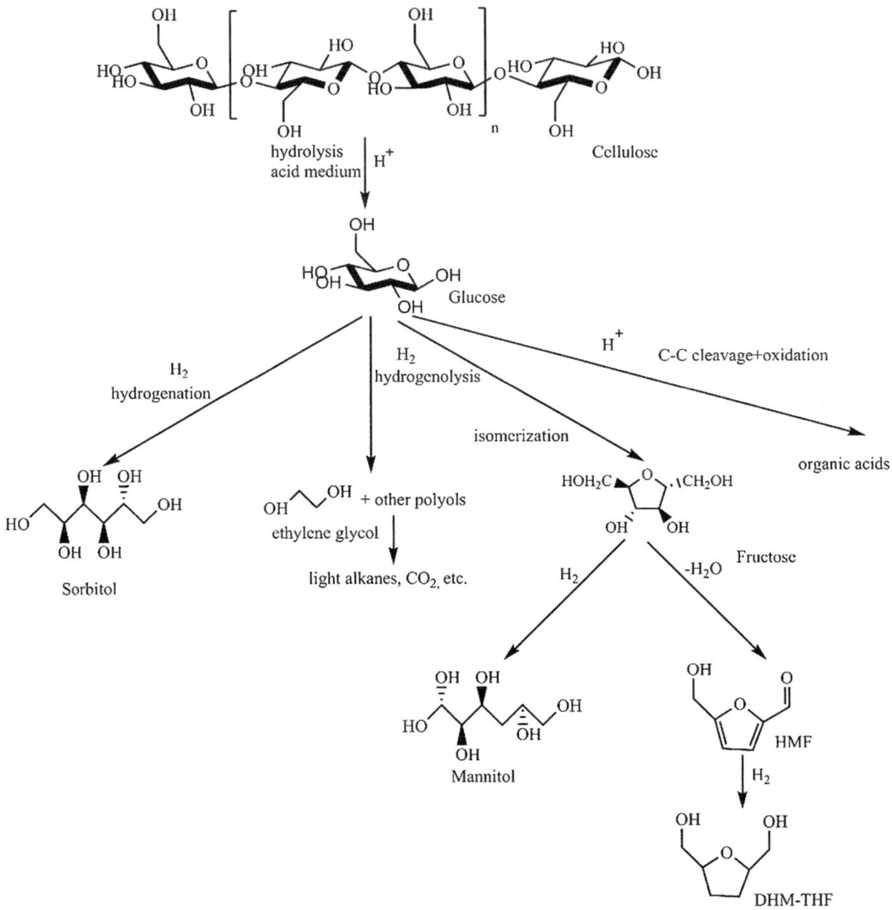

Fig. 6.1 Schematic representation of different pathways to cellulose conversion into polyols

6.3.1 *Cellulose Transformation into Sugar Alcohols*

The hydrolysis of cellulose into glucose is based on the cleavage of the β-1,4-glycosidic linkage. After the generation of glucose, the products can be converted into value-added chemicals. In the case of polyols generation from glucose hydrogenolysis/hydrogenation, the most critical stage is cellulose hydrolysis due to the difficulty in decomposing crystalline regions composed around 40–60% of the total cellulose depending on the source [33].

The most widely used methods for cellulose hydrolysis are enzymes, mineral acids, alkali or supercritical fluids. However, all of them present several drawbacks such as high cost, low reaction rates and enzymatic recovery in case of biological processes; waste water disposal, equipment corrosion and acid or alkali recovery in chemical processes and high energy consumption in supercritical reactions [30]. For all these reasons, so far the substitution of polyols from biomass cannot industrially compete with processes that use fossil resources as the feedstock.

The approach of using solid acid catalysts has risen in the last years as an alternative, since heterogeneous catalysts can overcome the main drawbacks of traditional processes described above [34]. Solid acids could involve a wide range of components, such as metal oxides, polymer-based acids, sulfonated carbon-based acids, heteropoly acids (HPAs), H-form zeolites as well as metal catalysts. The strong points of solid acid catalyzed reactions are based on the separation of catalysts after the reaction that can be accomplished very easily by filtration without generating a great waste stream which should be treated as it produced for case of liquid acids [35].

Cellulose kinetic hydrolysis models described by Saeman [36] involve two consecutive pseudo-first order reactions depending on H^+ concentration as it is shown in Eq. 6.1.

$$Cellulose \xrightarrow{k_1} Oligosaccharides \xrightarrow{k_2} Glucose \tag{6.1}$$

Many studies in this area identified kinetic parameters, however, large variability between the different works is found because of the different reaction conditions in which the studies were carried out in comparison with Saeman's work [37]. A more flexible model considers that cellulose is not directly decomposed into monomers of glucose but instead consists of two consecutive steps. At first cellulose is converted into soluble oligosaccharides, which are assumed to be intermediate products during the conversion of cellulose to glucose; and finally the conversion of them to glucose, is represented by Eq. 6.2 [38].

$$Cellulose \xrightarrow{k_1} Oligosaccharides \xrightarrow{k_2} Glucose \xrightarrow{k_3} Degradation\ products \tag{6.2}$$

In this model, values of kinetic constants (k_1, k_2) are correlated to the available acid sites and accessible in aqueous medium, the diffusion limitations of poly/oligosaccharides and the DP of the cellulose and intermediate products. The second reaction is faster than first one, being the first reaction rate the determining step whose kinetics still follows that of the Seaman's [39].

Numerous studies have demonstrated that with solid acids, the activation energy of cellulose hydrolysis can be reduced, involving lower energy requirements for carrying out the depolymerization reaction [40]. Although heterogeneous catalyzed reactions present mass transfer limitations between the insoluble substrate (cellulose) and solid catalyst, solid acids present higher specific surface areas, pore sizes/volume and strong acid sites which lead to achieve better performance in the hydrolysis reaction. The most important parameters are those that drive the number of Brønsted acid sites, affinity with the substrates and catalyst stability. As an example of the advantages described previously, carbon-based solid acids have strong affinity for cellulose chain locating acid sites closer to the β-1,4-glycosidic bonds which help the cleavage [41]. Magnetic-containing acids allow easier recuperation and reutilization since they can be recovered by magnetic filtration [42].

6.3.2 Glucose Conversion to Polyols

Among the different polyols that can be obtained from natural resources, sorbitol is mostly produced from glucose aqueous solutions by catalytic hydrogenation. The concentration of glucose can be as high as 65% and can be carried out both in either autoclave systems or continuous (trickle bed) reactors under high pressure [43].

The most common catalysts used in these reactors are based on nickel as the active metal due to its low price, easy support and high activity. The main drawback of this type of catalyst is the restriction of nickel allowed in sorbitol used in food, medical or cosmetic applications. Therefore, the elimination of nickel from sorbitol has to be addressed [44]. When the reaction is conducted at high temperatures, leaching of nickel occurs, polluting the final solution of sorbitol, which involves additional product cleaning stages for the entire process increasing the production cost of sorbitol. To overcome this problem as well as the deactivation process, many authors have included the utilization of other active metals such as cobalt, platinum, palladium, rhodium and ruthenium [45, 46] as well as to improve the catalyst support to achieve a better anchoring of the catalyst metal. Ruthenium is considered one of the best metals to develop high activity, besides not being soluble under reaction conditions, which involve not polluting the final product.

The other current challenge in the production of this product is to obtain high selectivity for sorbitol with a total conversion and a high stability of the catalyst during a long period of time. To meet the stability challenge, the catalyst should be resistant to metal sintering and poisoning, but the most crucial point is to avoid metal and support leaching in the acidic and chelating reaction medium [47].

The hydrogenation of glucose to sorbitol follows a Langmuir–Hinshelwood mechanism where the surface reaction is the rate-determining step. Three possible models have been used to describe the hydrogenation of glucose to polyols based on Langmuir–Hinshelwood–Hougen–Watson (LHHW) kinetics. Model 1: Noncompetitive adsorption of hydrogen and glucose. Model 2: competitive adsorption of molecular hydrogen and glucose. Model 3: competitive adsorption of dissociatively chemisorbed hydrogen and glucose [39]. The three models can describe reaction data satisfactorily with no inhibition by sorbitol or mannitol. Following these methods, activation energy of 55 kJ/mol was found by Crezee et al. [48] over Ru catalysts. Values obtained over Ni catalysts were slightly higher 63–83 kJ/mol [45, 49].

6.3.3 One-Step Cellulose Conversion to Polyols

As mentioned above, the interest in cellulose to be used as feedstock for polyol production could open a new avenue for renewable resources that is not in conflict with food and feed industries since biomass wastes can be used for this purpose. However, to carry out the process from cellulose is not economically feasible if two distinct stages have to be implemented. In this line, novel catalytic developments have opened a new approach for converting cellulose into polyols in one-step by environmentally-sustainable processes [50]. This modification is based on intensification process principles that lead to a reduction in cost of the entire process since less time and energy are required and hazards and waste stream are reduced.

Performance advantages can be obtained with the single-step method since during cellulose hydrolysis and glucose generation, the activation energy is similar to that of cellulose hydrolysis. As consequence, the rates of sugar formation and its degradation are on the same order of magnitude under harsh reaction conditions [51]. Therefore, the hydrogenation of glucose at the same time is formed and is favorable to avoid its degradation allowing high yields of polyols at high cellulose conversions.

Different effects are given by catalysts used in this one-step reaction system. At first, acid conditions have to be created to facilitate cellulose hydrolysis. On the other hand, hydrogenation/hydrogenolysis needs metal catalysts to transform glucose into polyols in the presence of hydrogen donor, being H_2 the most common agent used [52].

Many advances for converting cellulose into polyols in a single-step have been possible thanks to the development of catalysts that combine two effects in one single catalyst that allows also developing reactions that traditionally were carried out in two steps as one unique step, as in the case of cellulose hydrolytic hydrogenation. The bifunctional catalyst has both Lewis or Brønsted basic functionality and a hydrogen-bond donor group. In comparison with single functional group catalysts, the cooperative effect of the two complementary groups can boost the rate of reactions that were previously challenging or unprecedented [53]. In cellulose hydrolytic hydrogenation, the support is functionalized to not only fulfill the function of allocating the metallic active components, but also to afford acidic sites where cellulose is hydrolyzed. Solid acids are used as support for bifunctional catalytic entities.

The first study of cellulose conversion by solid catalysts to polyols in a single-step reaction was accomplished by Fukuoka and Dhepe [54] using noble metal catalyst supported over a solid acid. In particular, different bifunctional Pt/Al_2O_3, Pt/SiO_2-$Al2O_3$, Pt/HUSY, and Ru/HUSY catalysts were used at 463 K and 5 MPA H_2, during 24 h, achieving around 30% yields of hexitols, with sorbitol being the main product. Noble metal catalysts participate in the hydrogenation reaction and also play a solid acid role in cellulose hydrolysis. The acid sites are created from both the acidic supports as well as generated by hydrogen atoms spilling over metallic catalysts to the support surface [55].

Several catalytic reaction systems have been successfully established using acid catalyzed hydrolysis coupled with metal-catalyzed hydrogenation/hydrogenolysis [56]. Multiple reactions take place in cellulose one-step hydrolytic hydrogenation/hydrogenolysis, involving polysaccharides hydrolysis, sugar isomerization, hydrogenation, retro-aldol condensation, decarbonylation, polymerization, and some other side reactions. Products resulted from these reactions can vary from C6 to C2 polyols, including, sorbitol, mannitol, glycerol, 1,2-propylene glycol, and ethylene glycol. The distribution of the final polyol products strongly depends on the type of catalyst, reaction conditions and kinetics of the major reactions that occur in the process. Cellulose hydrolysis is the rate-determining step of the process. The product distribution for this case could be controlled by the balance between the rates of cellulose hydrolysis and glucose hydrogenation/hydrogenolysis [30]. Table 6.2 summarizes possible catalytic systems used, conditions and product diversity that can be performed in a single-step cellulose hydrolysis/hydrogenation reaction system to obtain polyols.

Mineral acids such as phosphoric acid or sulfuric acid can be introduced in a reaction system where hydrogenation is enabled by noble metals over a carbon support. At relative low temperatures (433 K) and short reaction times (5 h) 72% of cellulose conversion can be obtained at 5 MPa of H_2 [57]. A synergic effect has been observed between combinations of acids with metal supported catalysts, emphasizing the supporting effect of the metal catalyst in the hydrolysis of cellulose. Other acids have been proposed in the literature to promote hydrolysis of cellulose over metal catalyst, since mineral acids as HCl [58] or heteropolyacid (HPA) could reach 99% cellulose conversion. For instance, $H_3PW_{12}O_{40}$ and $H_4SiW_{12}O_{40}$ were selected by Geboers et al. [59], compounds that allow their recovery for further reuse in extraction and recrystallization.

In spite the advantage of liquid acids, their main drawback is the restriction of their reuse as consequence of the problems in the recovery of the acid. Subsequently, solid acids have been proposed to solve this problem, since these compounds are able to be reused after their easy separation by filtration once reaction is finished. Besides their easy separation from the reaction system, other advantage is that the specific solid acid present higher acid density than even H_2SO_4 [60] avoiding problems, such as equipment corrosion or neutralization of waste streams. Different components have been studied in the literature with the purpose, being to exploit the advantages of the higher acid density of solid acids (Table 6.2).

Subcritical water has been proposed as agent to promote acids formation *in situ,* due to the higher concentration of both protons and hydroxyl ions above 473 K, which are able to promote cellulose hydrolysis [61]. The concentration of protons and hydroxyl ions vary according to the ion product of water (Kw), being water neutral when temperature is reduced again to room temperature. Several studies have been developed following the principles of this approach, in which temperature is raised above 500 K, using different catalysts. The resulting sugar then undergoes hydrogenation over transition metal catalysts. This reaction principle was used by Tai et al. [62] highlighting the effect of temperature over cellulose conversion, which can achieve a 100% yield or conversion value at 518 K in only 30 min.

The reusability of these catalysts constitutes one of the main challenging issues to overcome. In spite of having a relatively good stability, long-term use of the catalyst undergoes activity decay. The Brønsted acid sites on the oxide support surface gradually leach into the hot aqueous solution. Additionally, hot water transforms oxide supports into hydroxide support, leading to block the pristine structure and to the loss acidic sites. As an alternative, acidic carbon materials have been proposed to address this issue. The higher strength of carbon allows them to bear harsh tem-

Table 6.2 Examples of cellulose conversion to polyols with bifunctional catalysts

Catalyst	Product (% Yield)	Temp (°C)	$P(H_2)$ (MPa)	Time (h)	References
Ru/H-USY	SUA (60)	190	5	13	[34]
	SOT (33)				
Ru/C	SOR (33.2)	160	5	5	[57]
	SOT (13.6)				
Ru/C-$H4SiW_{12}O_{40}$	SUA (85)	190	5	1	[59]
	SOT (15)				
Ru/C	SUA (46)	245	6	0.5	[55]
	SOT (13.4)				
WC_x/MC	EG (72.9)	210	5	0.5	[75]
	SOR (1.2)				
	MAN (1.4)				
Pt/Al_2O_3	SOR (21)	190	6	24	[54]
	MAN (6)				
Ni/CNF	SOR (29.8)	210	6	4	[63]
	MAN (5)				
	PG (4.3)				
	EG (4.6)				
Pt/BP2000	SOR (49)	190	5	24	[76]
	MAN (9)				
Ni-WO_3/SBA-15	EG (70.7)	235	6	6	[77]
	PG (5.9)				
M(8,9,10)-W/SBA-15	EG (76.1)	245	6	0.5	[67]
	PG (3.2)				
	SUA (8)				
Ru-Fe_3O_4-SiO_2	EG (19)	255	6	0.83	[74]
	PG (20)				

Notes: *SUA* sugar alcohols, *SOT* sorbitan, *SOR* sorbitol, *MAN* mannitol, *EG* ethylene glycol, *PG* 1,2-propylene glycol

peratures and make them more suitable for hydrothermal conditions. Furthermore, the surface of carbon materials can be modified to have many OH and carboxyl groups. These weak acidic sites promote the activation of cellulose cleavage by the activation of the glycosidic bonds [31, 63].

Cost is a clear constraint to scale-up many processes industrially. The amount of precious metals required for cellulose conversion is considered as 4–10 g per kg of cellulose. To transform large quantities of cellulose, cheap catalysts have to be proposed. In this line, carbides from Groups-6 metals show catalytic activities similar to those of Pt group metals when hydrogen is also involved in the reaction [54, 64]. For this reason, Zhang et al. [65] use carbide catalysts for biomass conversion at similar conditions than Liu et al. [66]. The approach lead to the production of EG as the main product of the reactions instead of hexitols which were obtained in previous works. Cellulose was totally converted over a catalyst from a combination of Ni with tungsten carbide over activated carbon reaching 61% of EG yield formation at

518 K for 30 min reaction time. This involves not only the hydrolysis of cellulose but also a C-C bond cleavage that leads to EG generation after the hydrogenolysis reaction. Afterwards, several research works have optimized the process for EG production, achieving yields around 75% [44, 67]. The reaction mechanism shows that oligosaccharides and glucose derived from cellulose hydrolysis undergo retro-aldol condensation catalyzed by dissolved tungsten species to form glycol aldehyde, which is further hydrogenated into EG over the bifunctional catalyst. The product distribution could be adjusted by monitoring the balance among the reactions of hydrolysis, retro-aldol condensation and hydrogenation [68].

As it was previously mentioned, the rate-determining step of cellulose conversion into polyols is the hydrolysis of cellulose. To reduce the influence of this step in the final yields, pretreatment of cellulose to decrease its crystallinity index has been studied by enhancing the contact between substrate and catalytic agents. Higher sugar alcohols yield was reached using ball-milled pretreated cellulose by Geboers et al. [59] in comparison with not pretreated cellulose, being hexitols yield 92%. Indeed, this pretreatment can be applied not only to cellulose substrate but also to catalyst. Ribeiro et al. [69] combined ball-milling pretreatment of cellulose together with the catalyst for 4 h. Sorbitol yield of almost 70% was achieved after 1 h reaction.

Among the noble metals Ruthenium-containing catalysts are considered the most active compounds for developing this reaction. Therefore, the recovery of the catalyst plays a role in demonstrating the economic feasibility of this process. One strategy is to dope the support with paramagnetic materials so that the catalysts can be retrieved easily after reaction [70]. In this line, precious metals can be conserved for using in further reaction increasing their lifetime. Ru-containing magnetically recoverable catalysts are used in olefin metathesis, azide – alkyne cycloaddition, hydrogenation, oxidation, and nitrile hydration reactions [71]. Magnetically recoverable catalysts have been reported for transformations of cellulose to glucose [72] and to sorbitol [73]. In most cases, magnetically recoverable catalysts are formed by functionalization of magnetic non-entities with ligands followed by the formation of catalytic complexes or nanoparticles. As an example, Manaenkov et al. [74] used Ru in magnetic mesoporous Fe_3O_4/SiO_2 to produce EG and PG in yield of 19% and 20% respectively at 528 K for 50 min reaction time and 60 bar hydrogen pressure. The activity, selectivity as well as the excellent stability of the $Ru/Fe_3O_4/SiO_2$ catalysts significantly exceed from those commercial Ru/C catalysts previously reported.

6.4 Conversion of Noncellulosic Part of Biomass to Polyols

Lignocellulosic biomass is considered to be the most abundant organic carbon source and has been broadly examined lately as a potential renewable feedstock for the production of valuable chemicals [78]. It is known that lignocellulosic biomass is mainly composed of three fractions, namely, cellulose, hemicellulose and lignin. In this section the valorization of the hemicellulose and lignin parts for obtaining polyols will be presented.

6.4.1 *Transformation of Hemicelluloses into Polyols*

The extraction of hemicellulose from lignocellulosic feedstock is mainly carried out by hot water extraction (autohydrolysis) [79] due to its simplicity and environmental friendliness. Other methods such as acid hydrolysis, alkaline solvent extraction [80]; aided in some cases by mechanical treatments e.g. extrusion [81] and biological processes involving enzyme utilization [82] can be found in the literature.

Once the oligomeric units of hemicelluloses are isolated, they can be broken down into the monomeric sugars, yielding several value-added chemicals afterwards. For this reason, hemicellulose transformations are attracting much attention lately. This section will be focused on the hemicelluloses conversion process into polyols. This process is composed of two reactions namely transformation of hemicelluloses oligomers into monomeric sugars and monosaccharide conversion into polyols. These two reactions can be carried out several different ways, as shown in the next sections.

6.4.1.1 Hemicellulose Oligomer Transformation into Monomeric Sugars

This step is normally carried out by acid hydrolysis of the previously extracted hemicelluloses. Throughout this reaction, the glycosidic bonds between the different units are cleaved. If the hydrolysis is complete, then the oligomeric units totally split to yield the respective monosaccharides [83]. The process proceeds analogously to that of the celluloses, but it is favored by the non-crystalline and branched structure of hemicellulose [84].

The approach employed for converting cellulose into glucose, has been applied to hemicelluloses conversion into monomeric sugars. Mineral acids such as sulphuric acid, hydrochloric acid or phosphoric acid can be used [85]. Enzymes can also be utilized for the reaction [86]. To overcome the issues associated with these methods, solid heterogeneous acid catalysts can be used. Some examples such as zeolites or clays [87] and metal oxides [88] can be found in the literature.

6.4.1.2 Monosaccharides Conversion into Polyols

Several different monosaccharides can be produced from the previous reaction depending on the source of hemicelluloses and conditions of the hydrolysis. However xylose, arabinose and galactose are the main ones in the literature. These sugar monomers are interesting since they are valuable platform chemicals for the production of fuels and specialty products [88]. The main path for their valorization is the catalytic hydrogenation of the monomeric sugars to stable polyols. These polyols, xylitol, arabinol and galactitol mainly, are well-integrated in the current chemical market and they serve as a potential feedstock for other chemicals as well [89]. The hydrogenation reaction is normally carried out in presence of a reductant

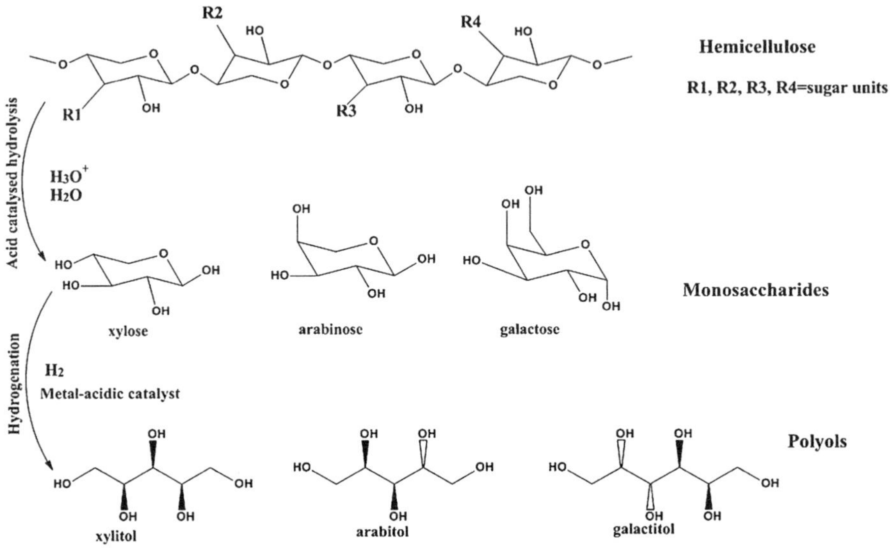

Fig. 6.2 Reaction scheme for single-step hydrolytic hydrogenation of hemicellulose into polyols

and a metallic catalyst promoting the formation of sugar alcohols. Commonly, hydrogen gas is used the reaction as reductant but recently, other hydrogen sources like alcohols have been proposed [90]. Regarding the catalyst, the most commonly used type is Raney® nickel [91] due to its low price, ease of use and good activity [92]. However, it deactivates quickly due to leaching and poisoning, requiring additional purification steps [93]. Alternatively, Raney nickel can be replaced by noble metals like Ru, Rh and Pd, which have good activity. From all these metals, Ru has been reported to be the most efficient [94] with little or no deactivation [95].

The previous two reactions are generally carried out in separate stages, resulting in laborious, cost- and energy-intensive and environmentally unfriendly steps [96]. In contrast single-step hydrolytic-hydrogenation processes have been proposed as shown in Fig. 6.2 In this figure the main monosaccharides and sugar alcohol produced from hemicellulose are presented.

This process is an attractive alternative since allows conversion of polysaccharides such as hemicellulose into monomeric sugar units and sugars alcohols, in the presence of a hydrogen source and using water as solvent [97]. The process is carried out in a single-step using a selective heterogeneous bifunctional catalyst which can be eliminated afterwards by liquid filtration. Such a catalyst has the advantage of avoiding furan formation which promote the sugar dehydration [98]. The catalysts should have large enough pores and a porous structure to reduce pore blockage and mass transfer limitations [99]. It is also important that the catalyst be bifunctional to achieve selective conversion into monosaccharides first, and then to sugar alcohols later. One bifunctional catalyst is referred to as dual acid-metal sites. On the one hand, the acid sites are said to favor the hydrolysis step, whereas the metal

Table 6.3 Selected works on the conversion of hemicellulose into sugars and sugar alcohol via one-pot hydrolytic/hydrogenation with bifunctional catalysts

Catalyst	Substrate[a]	H agent	Polyol product	Temp (°C)	Time (h)	Yield[b] (%)	References
Ru/H-USY	AX[c]	H_2	Xylitol Arabitol	160	5	90	[100]
Ru/C + H_2SO_4	X[d]	C_3H_8O	Xylitol	140	3	>80	[96]
Ru/C+ $H_3PW_{12}O_{40}$	AX	C_3H_8O	Xylitol Arabitol	140	3	82	[101]
H/β-11 + Ru/β-11	AG[e]	H_2	Arabitol Galactitol	185	2.5	<10	[83]
Ru/C	SBF[f] (A)[g]	H_2	Arabitol	155	24	83	[102]
Ru/MCM-48	AG	H_2	Arabitol Galactitol	185	24	30	[99]
Ru/C-H_2SO_4	AG	H_2	Arabitol Galactitol	185	2	<10	[98]
(Pt/BP2000)	SG[h] (X)	H_2	Xylitol	190	24	14	[103]

[a]Substrate of hemicellulose used as raw material in the process of conversion
[b]Maximum sugar alcohol yield obtained
[c]AX = arabinoxylan
[d]X = xylan
[e]AG = arabinogalactan
[f]SBF = sugar beet fiber
[g]A = arabinan
[h]SG = silver grass

sites promote the hydrogenation reaction [83, 99]. The necessity of a good balance between the two catalytic functions to obtain a proper overall productivity of the desired products should be taken into account [99]. In Table 6.3 several processes reported in the literature for the conversion of hemicellulose into polyols following this methodology are presented.

As it can be seen from the Table 6.3 there are several works in the recent years dealing with this conversion process. The majority of them use specific hemicellulose polysaccharides as substrates in comparison to the ones using real biomass. Nevertheless, the future trend is to develop an integrated biorefinery.

6.4.2 *Transformation of Lignin into Polyols*

Lignin also constitutes one of the main components of terrestrial biomass. However compared to hemicelluloses or cellulose, it has a much more complex nature and heterogeneous composition [104]. Owing to that, there is not a detailed knowledge of its native structure and linkages.

Regarding lignin isolation from biomass, much research has been carried out over the years and several methods of extraction have been tested. The main industrial isolation techniques currently are the sulphite method [105], the soda method

[106], the Kraft method [107] and the organosolv method [108], with the Kraft being the most widely implemented. Each process is based on different chemicals and conditions and hence the lignin extracted in each case has a different structure and nature [109, 110].

Since lignin structure is characterized by the presence of phenolic and aliphatic hydroxyl groups, it can be used for the production of polyols [111]. This is really interesting since it could be applied for example to the preparation of polyurethanes (PU). Thus, the petro-derived polyols normally used for this kind of polymer can be substituted with lignin-derived ones. The application of lignin in products like polyurethanes (PU) can be implemented by means of two different approaches i.e. direct use of lignin as a polyol substitute [112] and lignin depolymerization or modification into smaller polyols [111]. Although the first procedure is possible, it entails many difficulties e.g. low reactivity, low replacement ratios and fragile and rigid structures. For this reason, the latter method is normally desired as it has fewer technical issues. The later method increases reactivity and reduces molecular weight of the final polyols, which is convenient for the production of most lignin-based polymeric materials [113].

There are several techniques available to depolymerize lignin but in this section special attention is given to those that yield polyols as final products. These are usually characterized by the thermal treatment of lignin, with or without organic solvents, chemical additives or catalysts [110]. The main processes currently used for the production of lignin derived polyols are commented on below.

6.4.2.1 Lignin Liquefaction

This process is generally carried out at high temperatures, employing appropriate reagents and catalysts that have reactivity towards the lignin. In most of the literature works, the liquefaction reagents are alcohols such as phenol [114] or glycols, e.g. glycerol and polyethylene glycol (PEG) [65, 115]. The catalysts mainly used are strong acids like sulfuric acid [116]. In this process the lignin is first degraded into fragments of lower molecular weight. Then, the polyols are said to be formed by the condensation reactions between the hydroxyl groups of glycerol or polyethylene glycol (PEG) and those of lignin [117]. In last step fragments of residual lignin are produced due to self-polymerization reactions [118].

There is a trend towards the integration of microwave heating in the liquefaction processes [116, 119, 120]. Microwave heating presents several advantages compared with conventional heating, such as reduced reaction times, lower degradation of reagents and rapid heating [119]. Microwave heating will be comprehensively treated in Sect. 6.5.2.

In Table 6.4, different works related to the lignin liquefaction aiming to produce polyols are summarized.

Table 6.4 Selected works on the conversion of lignin into polyols via liquefaction

Lignin	Combined reagents	Catalyst	Temp (°C)	Time (min)	Yield (%)	References
EHL[a]	PE[i], glycerol	H_2SO_4	130–170	60–180	97	[115]
BL[b]	PEG[j], glycerol	H_2SO_4	160	60	97	[117]
HL[c]	Water, ethanol	–	250	60	70	[121]
KL[d]	PEG and glycerol	–	MW 135	3	–	[122]
OOL[e]	PEG and glycerol	H_2SO_4	MW 155	5	99.07	[116]
SL[f]	EJ300 and glycerol	–	MW 140	2	–	[119]
KL	PEG and glycerol	H_2SO_4	160	60	–	[118]
NSL[g]	PEG and glycerol	H_2SO_4	MW 140	5	97.47	[120]
SBOL[h]	Crude glycerol	H_2SO_4	160	90	–	[123]

[a]EHL: Enzymatic hydrolysis lignin
[b]BL: Birch lignin
[c]HL: Hydrolysis lignin
[d]KL: Kraft lignin
[e]OOL: Olive organosolv lignin
[f]SL: Soda lignin
[g]NSL: Non specified lignin
[h]SBOL: Sugarcane bagasse organosolv lignin
[i]PE: Polyethylene
[j]PEG: Polyethylene glycol

6.4.2.2 Hydrolytic Depolymerization

Lignin hydrolysis is an attractive method for producing various high value-added compounds including polyols. The solvent normally used in this process is water alone or in combination with other organic solvents. Moreover, it is carried out under pressurized liquid or supercritical state [124] by means of acid- or base-catalyzed reactions [125]. The study of the hydrolytic process under acidic conditions has been broadly treated [126, 127]. A variety of catalysts can be used, namely, mineral, Lewis and organic acids, zeolites or acidic ionic liquids [128] to cleave the α- and β-aryl ether linkages of lignin. However there is inevitable repolymerization of lignin intermediate products [129] resulting in high yields of solid residues and corrosion-derived problems. Consequently, the use of alternatives such as alkaline catalysis or organic solvents is desirable. The former is frequently preferred since the use of organic solvents increases the complexity and the cost of the process [130].

Yuan et al. [129] reported on the production of low molecular weight polyols (Mn ≈ 450 g/mol) through hydrolytic depolymerization of Kraft lignin, using a water-ethanol mixture, sodium hydroxide (NaOH) as base catalyst and phenol as a capping agent. The lignin degradation was almost complete and the solid residue was less than 1%. On the other hand, Mahmood et al. [130] achieved a high yield of polyols (92%) using only water as the solvent, sodium hydroxide (NaOH) as catalyst and no capping agent. Accordingly, in this case, the residue amount was higher (9.8%). An alternative to this approach has been presented recently by Xue et al. [124], who achieved a good depolymerization of corncob lignin using a mixture of water and isopropanol (hydrogen donor) and sodium hydroxide (NaOH) as catalyst. In this sense, the hydrogen donor appeared to be a good substitute for the capping agents to reduce the amount of solid residue.

6.4.2.3 Oxyalkylation

Unlike the other processes, oxyalkylation is based on the modification of the lignin structure by means of different alkylene oxides, especially propylene oxide (PO). Through this reaction, the hydroxyl groups of lignin, particularly the phenolic ones, are extended by the alkylene oxide moieties being less affected by steric and/or electronic constraints [131]. The result is the transformation of the solid slightly-reactive lignin into liquid reactive polyols with extended structure and exposed hydroxyl groups [132].

The oxyalkylation reactions are generally performed in presence of a catalyst, mostly a base catalyst such as potassium hydroxide (KOH). Acid catalysts can be also employed, although they yield a considerable amount of by-products [133].

The process of lignin modification can be accomplished following two different mechanisms i.e. direct oxyalkylation and a two-step reaction with maleic anhydride (MA) followed by oxyalkylation.

The former mechanism starts with the catalytic activation of the lignin hydroxyl groups yielding an alcoholate. Then, the reaction between the oxirane ring of the alkylene oxide and the alcoholate group proceeds, resulting in a chain-extended alcoholate compound. The chain extension continues until the depletion of the alkylene oxide. The scheme of these reactions is shown in Fig. 6.3.

Oxyalkylation reaction is generally accompanied by other side reactions that originate by the homopolymerization and isomerization of the alkylene oxide. Some research can be found in the literature regarding this direct method. On the one hand Nadji et al. [134] achieved the conversion of different lignins into polyols using propylene oxide (PO) as a chain extender and potassium hydroxide (KOH) as catalyst. These polyols were able to yield rigid polyurethane (RPU) afterwards. On the other hand, Kühnel et al. [131] proposed a direct oxyalkylation path using polycarbonate (PC) as chain extender instead of propylene oxide (PO). Thus, the risks derived from the use of the latter, such as flammability or toxicity, were avoided.

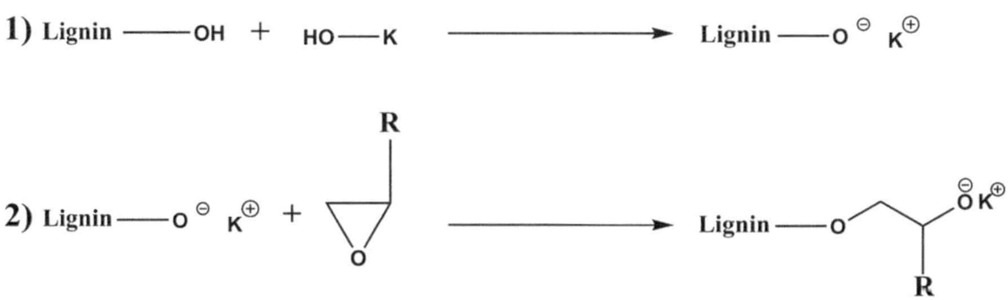

Fig. 6.3 Mechanism for direct oxypropylation of lignin: (1) catalytic activation of hydroxyl group. (2) open ring reaction between the alcoholate and the alkylene oxide

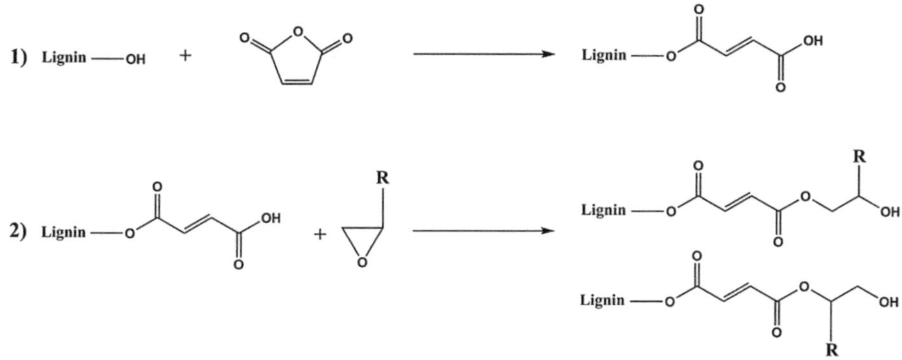

Fig. 6.4 Mechanism for two step oxypropylation of lignin: (1) lignin reaction with maleic anhydride (MA) yielding the carboxylic acid derivate and (2) open ring reaction between the carboxylic acid and the alkylene oxide

In the two-step method as show in Fig. 6.4, the maleic anhydride is used to transform the lignin phenolic hydroxyl groups into their respective aliphatic carboxylic acids [135]. In the second step these carboxylic acids are subjected to oxyalkylation in the same way that was presented before. Nevertheless, in this case the modification leads to a more branched structure with hydroxyl groups at the end.

The two step oxyalkylation mechanism though, is complex and the presence of carboxyl groups (COOH), slow down the oxypropylation reaction [134]. For this reason direct oxyalkylation mechanism is often preferred.

Concerning the two-step mechanism (Fig. 6.4) some work has been reported by Ahvazi et al. [135] who carried out a comparison between the direct oxyalkylation of straw soda lignin and the two-step process. The final results indicated that the direct oxyalkylation method was more convenient for the preparation of lignopolyols with a higher aliphatic hydroxyl content than the two step process.

For closure of this section the main applications of the polyols obtained from lignin are introduced.

In this section, the importance of the lignin derived polyols was shown to be mainly related to their incorporation in the PU preparation. The polyurethanes are one of the most versatile classes of polymers in industry and have multiple applications. Chemically, they are characterized by urethane units resulting from the condensation of isocyanates and polyols.

Lignin polyols are currently used in the polyurethanes (PU) production for some applications such as foams, elastomers, adhesives or coatings [136]. From all of these, they are predominantly used as foams, which represent one of the most important commercial applications. Foams are usually classified as rigid, semi-rigid and flexible depending on their mechanical and structural properties. The majority of the works from the literature have reported the lignin polyols use for rigid and semi-rigid polyurethane due to their structure and the crosslink between them and the isocyanate groups in the condensation reaction [124, 132, 134, 137].

6.5 Advanced Processes

All of the previously mentioned fractionation or transformations could be upgraded by advanced processes increasing the yield of product and reducing reaction time. There are two key factors that could enhance overall systems. Between the different processes, ionic liquids, microwave and ultrasound should be emphasized.

6.5.1 Ionic Liquids

Ionic liquids are classified as green solvents as well as water. They could deal with the water solubility issue that there is in cellulose and mass transfer difficulty of solid catalysts in heterogeneous hydrolytic hydrogenation. Even the crystallinity of cellulose could be addressed once cellulose is soluble in many ionic liquids, which would improve the contact between the solid catalysts and cellulose. Examples of the use of ionic liquids in hydrolytic hydrogenation/hydrogenolysis of cellulose are shown in Table 6.5.

Yan et al. [141] used cellulose as the starting material for studying its catalytic conversion over Ru nanocluster catalyst in an ionic liquid ([Bmim]Cl) to obtain C6-alcohols like sorbitol, glucose or 3-β-D-glucopyranosyl-D-glucitol. Conversion of these products was only of 15%. Nevertheless, the concept established a new opportunity for the transformation of cellulose into valuable chemicals, such bio-oils.

Cellulose could be completely converted into a mixture of C6 sugar and alcohols (cellulose conversion 57%, selectivity of the products: glucose 43%, sorbitol 29% and mannitol 5%) in the ionic liquid [Bmim]Cl as the solvent in the presence of the homogeneous catalyst of ruthenium [138].

Zhu et al. [139] reported the formation of a conjugated complex between an ionic liquid moiety and boric acid and a binding that could stabilize ruthenium

Table 6.5 Ionic liquids used in the hydrolytic hydrogenation/hydrolysis of cellulose

Ionic liquid	Catalyst	Product (% Yield)	Temp (°C)	$P(H_2)$ (MPa)	Time (h)	References
[Bmim]Cl	Ru/C	GLU (43), SOR (29), MAN (5)	150	3.5	24	[138]
[Bmim]Cl	HRuCl (CO) $(PPh_3)_3$	GLU (14), SOR (29)	150	3.5	24	[138]
[THTdP]DbS and [Bmim]Cl	Ru	SOR (94)	80	–	5	[139]
–	Ru/ $[Bmim]_3PW_{12}O_{40}$	SOR (70.3)	245	6	0.5	[140]

Notes: *GLU* glucose, *SOR* sorbitol, *MAN* mannitol, *[Bmim]Cl* 1-Butyl-3-methylimidazolium chloride

nanoparticles catalyst employed for example in the hydrogenation of cellulose into hexitols. They demonstrated that the yield of sorbitol, with sodium formed as the hydrogen source, was improved to 94% running the reaction for 5 h at 80 °C.

Xie et al. [140] successfully dispersed Ru in an insoluble hybrid material of ionic liquid [Bmim][PF_6]/heteropolyacid ($H_3W_{12}O_{40}$ nH_2O) hybrid support. On this catalyst, the Ru sites for hydrogenation combine both Lewis and Bronsted acidic sites for hydrolysis, which could be able to convert microcrystalline cellulose into sorbitol with a conversion of 63.7% and selectivity of 70.3% at 433 K with 5 MPa H_2 in 24 h.

6.5.2 *Microwave*

Microwaves (MW) can interact with polar molecules very efficiently to allow rapid heating of solutions so that reactions can be carried out more quickly and with better yields and selectivity. The dipole rotation and ion conductivity are the two main phenomena causing MW dielectric heating. The dipole rotation is an interaction in which polar molecules try to align themselves in the electric field. In conduction, as the dissolved charged particles oscillate back and forth under the influencing microwave field, they collide with neighbor molecules, causing movement that heat up materials.

The MW can penetrate lignocellulosic materials and the energy can be absorbed throughout the volume of the materials. Water contained in lignocellulosic materials absorbs the MW irradiation extremely well and this is the main reason why MW is an efficient way to study the lignocellulosic pretreatments respect to those over conventional heating methods [142].

Among thermochemical processes, pyrolysis has been extensively investigated during the last decades for biomass valorization ranging from organic residues to plastics and many others. However, this process still faces some technical challenges in terms of improving yield and quality of the products and increasing energy efficiency of the process. The microwave technology is a promising attempt to resolve these challenges, because of a rapid and efficient heating and a better control of reaction parameters, such as time, temperature and power enhancing biomass conversion [3].

Zhou et al. [82] proposed a method in which the hydroxymethyl groups of glucose units of the cellulose chain are converted to carboxyl groups by a preoxidation step using air as oxidant. The carbonyl sites act as catalytically active sites both for cellulose depolymerization during pretreatment and the subsequent hydrolysis of cellulose to glucose. To prove the efficiency of the peroxidation-hydrolysis method, the hydrolysis of microcrystalline cellulose (MCC) was investigated under different conditions. The results revealed a 65.5% of selectivity and 25.2% yield of glucose at 170 °C for 8 h for MCC oxidized my air. They also investigated alternative pathways for the cellulose depolymerization by for example MW. The direct MW assisted hydrothermal depolymerization of microcrystalline cellulose gave 21% yield with 36% selectivity of glucose in water at 250 °C for 1 h.

6.5.2.1 Microwave-Assisted Solvolysis Technique

Due to its fast heating rate and high heating efficiency, microwave heating technology has been attracting attention for lignin solvolysis.

Gu et al. [143] investigated the application of Lanthanum-containing SBA15 mesoporous molecular sieves into lignin microwave-assisted catalytic oxidation with hydrogen peroxide as oxidant. The maximum yield of vanillin 9.94% and for syringaldehyde 15.66% was accomplished after 10 min of reaction time at 200 W. The same conditions with non-catalyst gave rise to only 0.38% yield of vanillin and 0.52% yield for syringaldehyde.

In the work of Shen et al. [144] acidic solvolysis of lignin performed by formic acid under microwave heating was investigated with the addition of modified HUSY catalyst. HUSY zeolites were modified by oxalic acid with different concentrations. The acidic solvolysis of lignin at set conditions of microwave reactor (power: 600 W; reaction temperature: 130 °C and reaction time: 30 min), was notably promoted with the addition of HUSY catalyst, achieving the maximum values of 15.36 wt% for bio-oil-1 and 67.52 wt% for bio-oil-2; when modified HUSY catalyst by 0.2 mol/L of oxalic acid (HUSY-0.2 M) was used. They observed that the production of aromatic monomers in the bio-oil-1 was supported by addition of HUSY catalyst and among those catalysts HUSY-0.2 M with its appropriate pore structure (averagely 2.45 nm) and acidic site distribution (Si/Al ratio) was the preferable one. Likewise, the production of aromatic oligomers in bio-oil-2 was ameliorated with the addition of HUSY catalysts.

Toledano et al. [145] studied microwave-assisted lignin depolymerization in formic acid or tetralin catalyzed by different supported metal nanoparticles on mesoporous Al-SBA-15, including-nickel, palladium, platinum and ruthenium. MW conditions were set at 400 W with a reaction time of 30 min for all experiments and a reaction temperature of 140 °C. The highest bio-oil yield (30%) using formic acid as hydrogen donor was obtained when nickel based catalyst (Ni10%AlSB) was utilized in lignin hydrogenolysis. However, palladium-based catalyst exhibited the lowest oil yield (5%) in spite of its excellent hydrogenation properties. The main monomeric products in tetralin experiments were mesitol and syringaldehyde.

6.5.3 Ultrasound

Ultrasound is a mechanical acoustic wave that induces cavitation and secondary effects in which high energy is imparted to the reaction medium. In a standard dynamic process of cavitation bubbles, solvent vapors containing microbubbles are generated that grow and undergo radical motions as acoustic energy propagates through the liquid medium. These microbubbles, that can be stable or transient, grow to a maximum of 4–300 μm in diameter. With low acoustic intensity, the radius of microbubbles periodically and repetitively expands and shrinks within several acoustic cycles. When the resonant frequency of bubbles exceeds the one of ultrasound field, bubbles collapse within several nanoseconds, which create special physical and chemical effects, and enhances thermochemical/biochemical reactions or treatment [33].

The use of sonochemistry has been applied to many chemical, physical and biological processes. Irradiation of the medium with ultrasound, which generates pressure waves, could be regarded as providing of a particular type of energy into the system. Acoustic waves can break the cohesion of a liquid, creating micro cavities. The cavities are actually micro bubbles that following the sound wave grow until becoming unstable. At that point drastic conditions inside the medium are created caused by the collapse of bubbles. Under such harsh conditions, molecules can undergo different fragmentation mechanisms to form free radicals [146]. Even if the mechanisms are not fully understood it is accepted that low frequency reactions (20–80 kHz) would suffer from physical effects, while high frequency reactions (150–2000 kHz) would lead to radical production [147].

Behling et al. [147] performed the catalytic oxidation of vanillyl alcohol to vanillin under milder reaction conditions than those generally used in literature. Working under mild condition (hydrogen peroxide, low temperature, ambient pressure) while promoting the mass transfer in a heterogeneously catalyzed system, required the employment of specific activation methods, such as ultrasound irradiation. Ultrasound generated in aqueous medium indeed favored the production of hydroxyl radicals through in situ formation of H_2O_2, as well as the increase of mass transfer between the Co_3O_4 catalysts and the organic substrate. Vanillin alcohol conversion of 38% was reached after only 15 min of reaction time under ultrasound conditions with selectivity to vanillin of 50%. The ultrasound-assisted H_2O_2 catalytic oxidation of vanillyl alcohol to vanillin in water proved to be faster (4x), more selective (2.3%) and more efficient (2.7%) than the corresponding reaction under silent conditions.

6.6 Conclusions and Future Outlook

In this chapter, a review on catalytic cascade production methods for polyols starting either from biomass or its fractions has been provided. Initially a two-step process was used in the conversion of cellulose to polyols. However in the last decade some research works have been performed in a single-step approach. In this context, bifunctional catalysts have great importance as they can develop the hydrolysis and hydrogenation/hydrogenolysis stages simultaneously, directly converting cellulose to polyols throughout its monomeric sugar (glucose). Hemicelluloses followed the same pathway as cellulose using hydrolysis process for the formation of polyols. Nevertheless in this case, a wider range of polyols can be obtained as the monomeric sugars which come from hemicelluloses is more heterogeneous. For instance, C5 sugars could lead to xylitol production. In the case of lignin liquefaction, hydrolysis and oxyalkylation are the main used methods in the transformation of lignin into polyols. Lignin liquefaction is the degradation into lower molecular weight fractions followed by the condensation between the hydroxyl groups of glycerol or polyethylene glycol and the ones of lignin to form polyols. Lignin hydrolysis towards polyols has the disadvantage of repolymerization due to the reactive fragments that are formed after the depolymerization that are deal with the addition of capping agents to the reaction. Finally the oxyalkylation is the modification of the lignin structure by different alkyne oxides. The all process could be enhanced with the use of ionic liquids which help on the solubilization of the crystalline part of cellulose, microwave and ultrasound. As a future line of research, the improvement of bifunctional catalysts in terms of their lifetime and reusability as well as their activity should be studied. Furthermore, the replacement of precious metal by other cost effective catalyst systems, based on transition metal needs to be addressed to guarantee the economic feasibility of the process. Finally, intensification of catalytic cascade process should be studied to detect advantages and drawbacks for further scale-up to an industrial level.

Acknowledgements The authors thank the University of the Basque Country and the University of Pau and Pays de l'Adour (Predoctoral fellowship), the University of the Basque Country (Postdoctoral fellowship no. ESPDOC15/044) and the Spanish Ministry of Economy and Competitiveness (project CTQ2016-78689-R) for financially supporting this work.

References

1. Gallezot P. Conversion of biomass to selected chemical products. Chem Soc Rev. 2012;41:1538–58.
2. Vira RB. Valorización de residuos agroindustriales mediante conversión química. Doctoral dissertation, University of the Basque Country, Spain; 2012.
3. Li H, Fang Z, Smith RL, Yang S. Efficient valorization of biomass to biofuels with bifunctional solid catalytic materials. Prog Energy Combust Sci. 2016;55:98–194.
4. Kumar AK, Sharma S. Recent updates on different methods of pretreatment of lignocellulosic feedstocks: a review. Bioresour Bioprocess. 2017;4(1):7. https://doi.org/10.1186/s40643-017-0137-9.
5. Brodin M, Vallejos M, Opedal MT, María Area C, Chinga-Carrasco G. Lignocellulosics as sustainable resources for production of bioplastics- a review. J Clean Prod. 2017. https://doi.org/10.1016/j.jclepro.2017.05.209.
6. Dimitriadis A, Bezergianni S. Hydrothermal liquefaction of various biomass and waste feedstocks for biocrude production: a state of the art review. Renew Sust Eener Rev. 2017;68:113–25.
7. McKendry P. Energy production from biomass (part 2): conversion technologies. Bioresour Technol. 2002;83(1):47–54.
8. Huang HJ, Yuan XZ. Recent progress in the direct liquefaction of typical biomass. Prog Energy Combust Sci. 2015;49:59–80.
9. Savage PE. A perspective on catalysis in sub- and supercritical water. J Supercrit Fluids. 2009;47:407–14.
10. Peterson AA, Vogel F, Lachance RP, Fröling M, Antal MJ Jr, Tester JW. Thermochemical biofuel production in hydrothermal media: a review of sub- and supercritical water technologies. Energy Environ Sci. 2008;1:32–65.
11. Huang HJ, Yuan XZ, Zeng GM, Liu Y, Li H, Yin J, Wang XL. Thermochemical liquefaction of rice husk for bio-oil production with sub- and supercritical ethanol as solvent. J Anal Appl Pyrolysis. 2013;102:60–7.

12. Xu C, Lad N. Production of heavy oils with high caloric values by direct liquefaction of woody biomass in sub/near-critical water. Energy Fuel. 2008;22(1):635–42.
13. Cheng S, D'Cruz I, Wang M, Leitch M, Xu C. Highly efficient liquefaction of Woody biomass in hot-compressed alcohol-water co-solvents. Energy Fuel. 2010;24:4659–67.
14. Karagöz S, Bhaskar T, Muto A, Sakata Y. Hydrothermal upgrading of biomass: effect of K_2CO_3concentration and biomass/water ratio on products distribution. Bioresour Technol. 2006;97(1):90–8.
15. Sun P, Heng M, Sun S, Chen J. Direct liquefaction of paulownia in hot compressed water: influence of catalysts. Energy. 2010;35:5421–9.
16. Chumpoo J, Prasassarakich P. Bio-oil from hydro-liquefaction of Bagasse in supercritical ethanol. Energy Fuel. 2010;24:2071–7.
17. Zhang Q, Zhao G, Chen J. Effects of inorganic acid catalysts on liquefaction of wood in phenol. Front For China. 2006;2:214–8.
18. Luterbacher JS, Alonso DM, Dumesic JA. Targeted chemical upgrading of lignocellulosic biomass to platform molecules. Green Chem. 2014;16:4816–38.
19. Robinson JM, Burgess CE, Bently MA, Brasher CD, Horne BO, Lillard DM, Macias JM, Mandal HD, Mills SC, O'Hara KD, Pon JT, Raigoza AF, Sanchez EH, Villarreal JS. The use of catalytic hydrogenation to intercept carbohydrates in a dilute acid hydrolysis of biomass to effect a clean separation from lignin. Biomass Bioenergy. 2004;26(5):473–83.
20. Li N, Tompsett GA, Zhang T, Shi J, Wyman CE, Huber GW. Renewable gasoline from aqueous phase hydrodeoxygenation of aqueous sugar solutions prepared by hydrolysis of maple wood. Green Chem. 2011;13:91–101.
21. Patel CM, Barot AA, Sinha VK. Sequential liquefaction of *Nicotiana tabacum* stems biomass by crude polyhydric alcohols for the production of polyols and rigid polyurethane foams. J Appl Polym Sci. 2016;43(974):1–10.
22. Yu F, Liu Y, Pan X, Lin X, Liu C, Chen P, Ruan R. Liquefaction of corn stover and preparation of polyester from the liquefied polyol. Biochem Biotechnol. 2006;130(1):574–85.
23. Wang T, Zhang L, Li D, Yin J, Wu S, Mao Z. Mechanical properties of polyurethane foams prepared from liquefied corn stover with PAPI. Bioresour Technol. 2008;99(7):2265–8.
24. Mun SP, Gilmour IA, Jordan PJ. Effect of organic sulfonic acids as catalysts during phenol liquefaction of *Pinus Radiata* Bark. J Ind Eng Chem. 2006;12(5):720–6.
25. Zhang J, Du M, Hu L. Bamboo liquefaction with polyhydric alcohols and its application in flexible polyurethane foam. Adv Mater Res. 2012;524–527:2113–7.
26. Li X, Li X, Qi W, Zhang J, Shi J, Pang J. Research on liquefaction technology of poplar wood chips with multicomponent solvent. Taiyangneng Xuebao/Acta Energiae Solaris Sinica. 2015;36(4):971–5.
27. Heinze T, Petzold K. Cellulose chemistry: novel products and synthesis paths. In: Belgacem MN, Gandini A, editors. Monomers, polymers and composites from renewable resources. 1st ed. Amsterdam: Elsevier; 2008. p. 343–68.
28. Robles E, Salaberria AM, Herrera R, Fernandes SC, Labidi J. Self-bonded composite films based on cellulose nanofibers and chitin nanocrystals as antifungal materials. Carbohydr Polym. 2016;144:41–9.
29. Fagundes PM, Padula AD, Padilha ACM. Interdependent international relations and the expansion of ethanol production and consumption: the Brazilian perspective. J Clean Prod. 2016;133:616–30.
30. Liu X, Wang X, Yao S, Jiang Y, Guan J, Mu X. Recent advances in the production of polyols from lignocellulosic biomass and biomass-derived com-pounds. RSC Adv. 2014;4(90):49501–20.
31. Wang X, Meng L, Wu F, Jiang Y, Wang L, Mu X. Efficient conversion of microcrystalline cellulose to 1,2-alkanediols over supported Ni catalysts. Green Chem. 2012;14(3):758–65.
32. Yue H, Zhao Y, Ma X, Gong J. Ethylene glycol: properties, synthesis, and applications. Chem Soc Rev. 2012;41(11):4218–44.

33. Luo J, Fang Z, Smith RL. Ultrasound-enhanced conversion of biomass to biofuels. Prog Energy Combust Sci. 2014;41:56–93.
34. Geboers J, Van de Vyver S, Carpentier K, Jacobs P, Sels B. Efficient hydrolytic hydrogenation of cellulose in the presence of Ru-loaded zeolites and trace amounts of mineral acid. Chem Commun. 2011;47(19):5590–2.
35. Onda A, Ochi T, Yanagisawa K. Selective hydrolysis of cellulose into glucose over solid acid catalysts. Green Chem. 2008;10(10):1033–7.
36. Saeman JF. Kinetics of wood saccharification hydrolysis of cellulose and decomposition of sugars in dilute acid at high temperature. Ind Eng Chem Res. 1945;37(1):43–52.
37. Rodrıguez-Chong A, Ramırez JA, Garrote G, Vázquez M. Hydrolysis of sugar cane bagasse using nitric acid: a kinetic assessment. J Food Eng. 2004;61(2):143–52.
38. Vilcocq L, Castilho PC, Carvalheiro F, Duarte LC. Hydrolysis of oligo-saccharides over solid acid catalysts: a review. ChemSusChem. 2014;7(4):1010–9.
39. Zheng M, Wang A, Pang J, Li N, Zhang T. Mechanism and kinetic analysis of the hydrogenolysis of cellulose to polyols. In: Schlaf M, editor. Reaction pathways and mechanisms in thermocatalytic biomass conversion I. Singapore: Springer; 2016. p. 227–60.
40. Shuai L, Pan X. Hydrolysis of cellulose by cellulase-mimetic solid catalyst. Energy Environ Sci. 2012;5(5):6889–94.
41. Suganuma S, Nakajima K, Kitano M, Yamaguchi D, Kato H, Hayashi S. Hydrolysis of cellulose by amorphous carbon bearing SO3H, COOH, and OH groups. J Am Chem Soc. 2008;130(38):12787–93.
42. Zhang F, Deng X, Fang Z, Zeng HY, Tian XF, Kozinski JA. Hydrolysis of crystalline cellulose over Zn-Ca-Fe oxide catalyst. Petrochem Technol. 2011;40:43–8.
43. Gallezot P, Nicolaus N, Fleche G, Fuertes P, Perrard A. Glucose hydrogenation on ruthenium catalysts in a trickle-bed reactor. J Catal. 1998;180:51–5.
44. Zhao G, Zheng M, Zhang J, Wang A, Zhang T. Catalytic conversion of concentrated glucose to ethylene glycol with semicontinuous reaction system. Ind Eng Chem Res. 2013;52(28):9566–72.
45. Wisnlak J, Simon R. Hydrogenation of glucose, fructose, and their mix-tures. Ind Eng Chem Prod Res D. 1979;18(1):50–7.
46. Li H, Li H, Deng JF. Glucose hydrogenation over Ni–B/SiO_2 amorphous alloy catalyst and the promoting effect of metal dopants. Catal Today. 2002;74(1):53–63.
47. Marques C, Tarek R, Sara M, Brar SK. Sorbitol production from bio-mass and its global market. In: Brar SK, Sarma SJ, Pakshirajan K, editors. Platform chemical biorefinery. Amsterdam: Elsevier; 2016. p. 217–27.
48. Crezee E, Hoffer BW, Berger RJ, Makkee M, Kapteijn F, Moulijn JA. Three-phase hydrogenation of D-glucose over a carbon supported ruthenium catalyst mass transfer and kinetics. Appl Catal A Gen. 2003;251(1):1–17.
49. Dechamp N, Gamez A, Perrard A, Gallezot P. Kinetics of glucose hydrogenation in a trickle-bed reactor. Catal Today. 1995;24:29–34.
50. Jiang Y, Wang X, Cao Q, Dong L, Guan J, Mu X. Chemical conversion of biomass to green chemicals. In: Xian M, editor. Sustainable production of bulk chemicals integration of bio-, chemo- resources and processes. 1st ed. Dordrecht: Springer; 2016. p. 19–49.
51. Rinaldi R, Schuth F. Acid hydrolysis of cellulose as the entry point into biorefinery schemes. ChemSusChem. 2009;2(12):1096–107.
52. Climent MJ, Corma A, Iborra S. Heterogeneous catalysts for the one-pot synthesis of chemicals and fine chemicals. Chem Rev. 2010;111(2):1072–133.
53. Dixon DJ. Bifunctional catalysis. J Org Chem. 2016;12:1079–80.
54. Fukuoka A, Dhepe PL. Catalytic conversion of cellulose into sugar alcohols. Angew Chem Int Ed. 2006;45(31):5161–3.
55. Luo C, Wang S, Liu H. Cellulose conversion into polyols catalyzed by reversibly formed acids and supported ruthenium clusters in hot water. Angew Chem Int. 2007;46(40):7636–9.

56. Ruppert AM, Weinberg K, Palkovits R. Hydrogenolysis goes bio: from carbohydrates and sugar alcohols to platform chemicals. Angew Chem Int. 2012;51(11):2564–601.
57. Palkovits R, Tajvidi K, Procelewska J, Rinaldi R, Ruppert A. Hydrogenolysis of cellulose combining mineral acids and hydrogenation catalysts. Green Chem. 2010;12(6):972–8.
58. Liang G, Wu C, He L, Ming J, Cheng H, Zhuo L, Zhao F. Selective conversion of concentrated microcrystalline cellulose to isosorbide over Ru/C catalyst. Green Chem. 2011;13(4):839–42.
59. Geboers J, Van de Vyver S, Carpentier K, de Blochouse K, Jacobs P, Sels B. Efficient catalytic conversion of concentrated cellulose feeds to hexitols with heteropoly acids and Ru on carbon. Chem Commun. 2010;46(20):3577–9.
60. Okuhara T. Water-tolerant solid acid catalysts. Chem Rev. 2002;102(10):3641–66.
61. Chamblee TS, Weikel RR, Nolen SA, Liotta CL, Eckert CA. Reversible in situ acid formation for β-pinene hydrolysis using CO2 expanded liquid and hot water. Green Chem. 2004;6(8):382–6.
62. Tai Z, Zhang J, Wang A, Zheng M, Zhang T. Temperature-controlled phase-transfer catalysis for ethylene glycol production from cellulose. Chem Commun. 2012;48(56):7052–4.
63. Van de Vyver S, Geboers J, Dusselier M, Schepers H, Vosch T, Zhang L, Van Tendeloo G, Jacobs PA, Sels BF. Selective bifunctional catalytic conversion of cellulose over reshaped Ni particles at the tip of carbon nanofibers. ChemSusChem. 2010;3(6):698–701.
64. Dhepe PL, Fukuoka A. Cellulose conversion under heterogeneous catalysis. ChemSusChem. 2008;1:969–75.
65. Zhang T, Zhou Y, Liu D, Petrus L. Qualitative analysis of products formed during the acid catalyzed liquefaction of bagasse in ethylene glycol. Bioresour Technol. 2007;98(7):1454–9.
66. Liu Y, Luo C, Liu H. Tungsten trioxide promoted selective conversion of cellulose into propylene glycol and ethylene glycol on a ruthenium catalyst. Angew Chem. 2012;124(13):3303–7.
67. Zheng MY, Wang AQ, Ji N, Pang JF, Wang XD, Zhang T. Transition metal–tungsten bimetallic catalysts for the conversion of cellulose into ethylene glycol. ChemSusChem. 2010;3(1):63–6.
68. Zheng M, Pang J, Wang A, Zhang T. One-pot catalytic conversion of cellulose to ethylene glycol and other chemicals: from fundamental discovery to potential commercialization. Chin J Catal. 2014;35(5):602–13.
69. Ribeiro LS, Órfão JJ, Pereira MFR. Enhanced direct production of sorbitol by cellulose ball-milling. Green Chem. 2015;17(5):2973–80.
70. AH L, Schmidt W, Matoussevitch N, Bönnemann H, Spliethoff B, Tesche B, Eckhard B, Kiefer W, Schüth F. Nanoengineering of a magnetically sepa-rable hydrogenation catalyst. Angew Chem. 2004;116(33):4403–6.
71. Wang D, Astruc D. Magnetically recoverable ruthenium catalysts in organic synthesis. Molecules. 2014;19(4):4635–53.
72. Zhang C, Wang H, Liu F, Wang L, He H. Magnetic core–shell Fe3O4@ C-SO3H nanoparticle catalyst for hydrolysis of cellulose. Cellulose. 2013;20(1):127–31.
73. Zhang J, SB W, Liu Y. Direct conversion of cellulose into sorbitol over a magnetic catalyst in an extremely low concentration acid system. Energy Fuel. 2014;28(7):4242–6.
74. Manaenkov OV, Mann JJ, Kislitza OV, Losovyj Y, Stein BD, Morgan DG, Pink M, Lependina OL, Shifrina ZB, Matveeva VG, Sulman EM, Bronstein LM. Ru-containing magnetically recoverable catalysts: a sustainable path-way from cellulose to Ethylene and Propylene Glycols. ACS Appl Mater Interfaces. 2016;8(33):21285–93.
75. Zhang Y, Wang A, Zhang T. A new 3D mesoporous carbon replicated from commercial silica as a catalyst support for direct conversion of cellulose into ethylene glycol. Chem Commun. 2010;46(6):862–4.
76. Kobayashi H, Ito Y, Komanoya T, Hosaka Y, Dhepe PL, Kasai K, Hara K, Fukuoka A. Synthesis of sugar alcohols by hydrolytic hydrogenation of cellulose over supported metal catalysts. Green Chem. 2011;13(2):326–33.
77. Cao Y, Wang J, Kang M, Zhu Y. Efficient synthesis of ethylene glycol from cellulose over Ni–WO_3/SBA-15 catalysts. J Molec Catal A Chem. 2014;381:46–53.

78. Li H, Yang S, Riisager A, Pandey A, Sangwan RS, Saravanamurugan S, Luque R. Zeolite and zeotype-catalysed transformations of biofuranic compounds. Green Chem. 2016;18(21):5701–35.
79. Li Z, Qin M, Xu C, Chen X. Hot water extraction of hemicelluloses from aspen wood chips of different sizes. Bioresources. 2013;8(4):5690–700.
80. Fišerová M, Opálená E. Hemicelluloses extraction from beech wood with water and alkaline solutions. Wood Res. 2012;57(4):505–14.
81. Jacquemin L, Zeitoun R, Sablayrolles C, Pontalier PY, Rigal L. Evaluation of the technical and environmental performances of extraction and purification processes of arabinoxylans from wheat straw and bran. Process Biochem. 2012;47(3):373–80.
82. Zhou S, Liu X, Guo Y, Wang Q, Peng D, Cao L. Comparison of the immunological activities of arabinoxylans from wheat bran with alkali and xylanase-aided extraction. Carbohydr Polym. 2010;81(4):784–9.
83. Faba L, Kusema BT, Murzina EV, Tokarev A, Kumar N, Smeds A, Díaz E, Ordóñez S, Mäki-Arvela P, Willför S, Salmi T, Murzin DY. Hemicellulose hydrolysis and hydrolytic hydrogenation over proton-and metal modified beta zeolites. Micropor Mesopor Mat. 2014;189:189–99.
84. Mäki-Arvela P, Salmi T, Holmbom B, Willför S, Murzin DY. Synthesis of sugars by hydrolysis of hemicelluloses-a review. Chem Rev. 2011;111(9):5638–66.
85. González G, López-Santín J, Caminal G, Sola C. Dilute acid hydrolysis of wheat straw hemicellulose at moderate temperature: a simplified kinetic model. Biotechnol Bioeng. 1986;28(2):288–93.
86. Kambourova M, Mandeva R, Fiume I, Maurelli L, Rossi M, Morana A. Hydrolysis of xylan at high temperature by co-action of the xylanase from *Anoxybacillus flavithermus* BC and the β-xylosidase/α-arabinosidase from *Sulfolobus solfataricus* Oα. J Appl Microbiol. 2007;102(6):1586–93.
87. Dhepe PL, Sahu R. A solid-acid-based process for the conversion of hemicellulose. Green Chem. 2010;12(12):2153–6.
88. Herrera VAS, Saleem F, Kusema B, Eränen K, Salmi T. Hydrogenation of L-arabinose and D-galactose mixtures over a heterogeneous Ru/C catalyst. Top Catal. 2012;55(7–10):550–5.
89. Schiweck H, Bär A, Vogel R, Schwarz E, Kunz M, Dusautois C, Clement A, Lefranc C, Lüssem B, Moser M, Peters S. Sugar alcohols. In: Elvers B, editor. Ullmann's encyclopedia of industrial chemistry. Weinheim: Wiley-VCH Verlag GmbH & Co. KGaA; 2012. p. 1–37.
90. Kobayashi H, Matsuhashi H, Komanoya T, Hara K, Fukuoka A. Transfer hydrogenation of cellulose to sugar alcohols over supported ruthenium catalysts. Chem Commun. 2011;47(8):2366–8.
91. Mikkola JP, Salmi T. Three-phase catalytic hydrogenation of xylose to xylitol-prolonging the catalyst activity by means of on-line ultrasonic treatment. Catal Today. 2001;64(3):271–7.
92. Mishra DK, Dabbawala AA, Hwang JS. Ruthenium nanoparticles supported on zeolite Y as an efficient catalyst for selective hydrogenation of xylose to xylitol. J Mol Catal A Chem. 2013;376:63–70.
93. Hernandez-Mejia C, Gnanakumar ES, Olivos-Suarez A, Gascon J, Greer HF, Zhou W, Rothenberg G, Shiju NR. Ru/TiO2-catalysed hydrogenation of xylose: the role of the crystal structure of the support. Catal Sci Technol. 2016;6(2):577–82.
94. Wisniak J, Hershkowitz M, Stein S. Hydrogenation of xylose over platinum group catalysts. Ind Eng Chem Prod Res D. 1974;13(4):232–6.
95. Yadav M, Mishra DK, Hwang JS. Catalytic hydrogenation of xylose to xylitol using ruthenium catalyst on NiO modified TiO_2 support. Appl Catal A Gen. 2012;425:110–6.
96. Yi G, Zhang Y. One-pot selective conversion of Hemicellulose (Xylan) to Xylitol under mild conditions. ChemSusChem. 2012;5(8):1383–7.
97. Käldström M, Kumar N, Tenho M, Mokeev MV, Moskalenko YE, Murzin DY. Catalytic transformations of birch kraft pulp. ACS Catal. 2012;2(7):1381–93.

98. Murzin DY, Murzina EV, Tokarev A, Shcherban ND, Wärnå J, Salmi T. Arabinogalactan hydrolysis and hydrolytic hydrogenation using functionalized carbon materials. Catal Today. 2015;257:169–76.
99. Kusema BT, Faba L, Kumar N, Mäki-Arvela P, Díaz E, Ordóñez S, Salmi T, Murzin DY. Hydrolytic hydrogenation of hemicellulose over metal modified mesoporous catalyst. Catal Today. 2012;196(1):26–33.
100. Ennaert T, Feys S, Hendrikx D, Jacobs PA, Sels BF. Reductive splitting of hemicellulose with stable ruthenium-loaded USY zeolites. Green Chem. 2016;18(19):5295–304.
101. Dietrich K, Mejia CH, Verschuren P, Rothenberg G, Shiju NR. One-pot selective conversion of Hemicellulose to Xylitol. Org Process Res Dev. 2017;21(2):165–70.
102. Guha SK, Kobayashi H, Hara K, Kikuchi H, Aritsuka T, Fukuoka A. Hydrogenolysis of sugar beet fiber by supported metal catalyst. Catal Commun. 2011;12(11):980–3.
103. Kobayashi H, Yamakoshi Y, Hosaka Y, Yabushita M, Fukuoka A. Production of sugar alcohols from real biomass by supported platinum catalyst. Catal Today. 2014;226:204–9.
104. Abdelaziz OY, Brink DP, Prothmann J, Ravi K, Sun M, García-Hidalgo J, Sandahl M, Hulteberg CP, Turner C, Lidén G, Gorwa-Grauslund MF. Biological valorization of low molecular weight lignin. Biotechnol Adv. 2016;34(8):1318–46.
105. Andelin J, Niblock RW, Curlin JW. The pulp and paper making process, in Technologies for reducing dioxin in the manufacture of bleached wood pulp. In: Background paper (OTA-BP-O-54), US Congress, Office of technology assessment, Washington DC, USA; 1989. p. 74.
106. Mousavioun P, Doherty WO. Chemical and thermal properties of fractionated bagasse soda lignin. Ind Crop Prod. 2010;31(1):52–8.
107. Koljonen K, Österberg M, Kleen M, Fuhrmann A, Stenius P. Precipitation of lignin and extractives on kraft pulp: effect on surface chemistry, surface morphology and paper strength. Cellulose. 2004;11(2):209–24.
108. Erdocia X, Prado R, Corcuera MA, Labidi J. Effect of different organosolv treatments on the structure and properties of olive tree pruning lignin. J Ind Eng Chem. 2014;20(3):1103–8.
109. Fatehi P, Chen J. Extraction of technical Lignins from pulping spent liquors, challenges and opportunities. In: Fang Z, editor. Production of biofuels and chemicals from Lignin. Singapore: Springer; 2016. p. 35–54.
110. Pandey MP, Kim CS. Lignin depolymerization and conversion: a review of thermochemical methods. Chem Eng Technol. 2011;34(1):29–41.
111. Mahmood N, Yuan Z, Schmidt J, CC X. Depolymerization of lignins and their applications for the preparation of polyols and rigid polyurethane foams: a review. Renew Sust Energ Rev. 2016;60:317–29.
112. Cateto CA, Barreiro MF, Rodrigues AE, Brochier-Salon MC, Thielemans W, Belgacem MN. Lignins as macromonomers for polyurethane synthesis: a comparative study on hydroxyl group determination. J Appl Polym Sci. 2008b;109(5):3008–17.
113. Yang L, Wang X, Cui Y, Tian Y, Chen H, Wang Z. Modification of renewable resources-lignin-by three chemical methods and its applications to polyurethane foams. Polym Adv Technol. 2014;25(10):1089–98.
114. Maldas D, Shiraishi N. Liquefaction of biomass in the presence of phenol and H_2O using alkalies and salts as the catalyst. Biomass Bioenergy. 1997;12(4):273–9.
115. Jin Y, Ruan X, Cheng X, Lü Q. Liquefaction of lignin by polyethyleneglycol and glycerol. Bioresour Technol. 2011;102(3):3581–3.
116. Sequeiros A, Serrano L, Briones R, Labidi J. Lignin liquefaction under microwave heating. J Appl Polym Sci. 2013;130(5):3292–8.
117. Xue BL, Wen JL, Xu F, Sun RC. Polyols production by chemical modification of autocatalyzed ethanol-water lignin from *Betula alnoides*. J Appl Polym Sci. 2013;129(1):434–42.
118. Faris AH, Ibrahim MNM, Rahim AA, Hussin MH, Brosse N. Preparation and characterization of Lignin polyols from the residues of oil palm empty fruit Bunch. Bioresources. 2015;10(4):7339–52.

119. Bernardini J, Anguillesi I, Coltelli MB, Cinelli P, Lazzeri A. Optimizing the lignin based synthesis of flexible polyurethane foams employing reactive liquefying agents. Polym Int. 2015;64(9):1235–44.
120. Xue BL, Wen JL, Sun RC. Producing lignin-based polyols through microwave-assisted liquefaction for rigid polyurethane foam production. Mater. 2015;8(2):586–99.
121. Mahmood N, Yuan Z, Schmidt J, Xu C. Valorization of hydrolysis lignin for polyols and rigid polyurethane foam. J Sci Technol Forest Prod Process. 2013a;3(5):26–31.
122. Cinelli P, Anguillesi I, Lazzeri A. Green synthesis of flexible polyurethane foams from liquefied lignin. Eur Polym J. 2013;49(6):1174–84.
123. Muller LC, Marx S, Vosloo H. Polyol preparation by liquefaction of technical Lignins in crude glycerol. J Renew Mater. 2017;14:67–80.
124. Xue BL, Huang PL, Sun YC, Li XP, Sun RC. Hydrolytic depolymerization of corncob lignin in the view of a bio-based rigid polyurethane foam synthesis. RSC Adv. 2017;7(10):6123–30.
125. Fang Z, Sato T, Smith RL, Inomata H, Arai K, Kozinski JA. Reaction chemistry and phase behavior of lignin in high-temperature and supercritical water. Bioresour Technol. 2008;99(9):3424–30.
126. Matsushita Y, Yasuda S. Preparation and evaluation of lignosulfonates as a dispersant for gypsum paste from acid hydrolysis lignin. Bioresour Technol. 2005;96(4):465–70.
127. Wongsiriwan U, Noda Y, Song C, Prasassarakich P, Yeboah Y. Lignocellulosic biomass conversion by sequential combination of organic acid and base treatments. Energy Fuel. 2010;24(5):3232–8.
128. Li Y, Liao Y, Cao X, Wang T, Ma L, Long J, Liu Q, Xua Y. Advances in hexitol and ethylene glycol production by one-pot hydrolytic hydrogenation and hydrogenolysis of cellulose. Biomass Bioenergy. 2015;74:148–61.
129. Yuan Z, Cheng S, Leitch M, CC X. Hydrolytic degradation of alkaline lignin in hot-compressed water and ethanol. Bioresour Technol. 2010;101(23):9308–13.
130. Mahmood N, Yuan Z, Schmidt J, CC X. Production of polyols via direct hydrolysis of kraft lignin: effect of process parameters. Bioresour Technol. 2013b;139:13–20.
131. Kühnel I, Podschun J, Saake B, Lehnen R. Synthesis of lignin polyols via oxyalkylation with propylene carbonate. Holzforschung. 2015;69(5):531–8.
132. Cateto CA, Barreiro MF, Rodrigues AE, Belgacem MN. Optimization study of lignin oxypropylation in view of the preparation of polyurethane rigid foams. Ind Eng Chem Res. 2009;48(5):2583–9.
133. Cateto CAB. Lignin-based polyurethanes: characterisation, synthesis and applications. Doctoral dissertation, University of Porto, Portugal. 2008.
134. Nadji H, Bruzzese C, Belgacem MN, Benaboura A, Gandini A. Oxypropylation of lignins and preparation of rigid polyurethane foams from the ensuing polyols. Macromol Mater Eng. 2005;290(10):1009–16.
135. Ahvazi B, Wojciechowicz O, Ton-That TM, Hawari J. Preparation of lignopolyols from wheat straw soda lignin. J Agric Food Chem. 2011;59(19):10505–16.
136. Zhang Q, Zhang G, Xu J, Gao C, Wu Y. Recent advances on ligin-derived polyurethane polymers. Rev Adv Mater Sci. 2015;40(2):146–54.
137. Sakdaronnarong C, Srimarut N, Laosiripojana N. Polyurethane synthesis from Sugarcane Bagasse Organosolv and Kraft Lignin. Key Eng Mater. 2015;659:527–32.
138. Ignatyev IA, Van Doorslaer C, Mertens PG, Binnemans K, De Vos DE. Reductive splitting of cellulose in the ionic liquid 1-Butyl-3-Methylimidazolium Chloride. ChemSusChem. 2010;3(1):91–6.
139. Zhu Y, Kong ZN, Stubbs LP, Lin H, Shen S, Anslyn EV, Maguire JA. Conversion of cellulose to Hexitols catalyzed by ionic liquid-stabilized Ruthenium nanoparticles and a reversible binding agent. ChemSusChem. 2010;3(1):67–70.
140. Xie X, Han J, Wang H, Zhu X, Liu X, Niu Y, Song Z, Ge Q. Selective conversion of microcrystalline cellulose into hexitols over a Ru/[Bmim]3PW12O40 catalyst under mild conditions. Catal Today. 2014;233:70–6.

141. Yan N, Zhao C, Luo C, Dyson PJ, Liu H, Kou Y. One-step conversion of cellobiose to C6-alcohols using a ruthenium nanocluster catalyst. J Am Chem Soc. 2006;128(27):8714–5.
142. Merino-Pérez O, Martínez-Palou R, Labidi J, Luque R. Microwave-assisted pretreatment of lignocellulosic biomass to produce biofuels and value-added products. In: Fang Z, Smith RL, Qi X, editors. Production of biofuels and chemicals with microwave. Berlin: Springer; 2015. p. 197–224.
143. Gu X, Kanghua C, Ming H, Shi Y, Li Z. La-modified SBA-15/H_2O_2 systems for the microwave assisted oxidation of organosolv beech wood lignin. Maderas Cienc Tecnol. 2012;14(1):31–41.
144. Shen D, Liu N, Dong C, Xiao R, Gu S. Catalytic solvolysis of lignin with the modified HUSYs in formic acid assisted by microwave heating. Chem Eng J. 2015;270:641–7.
145. Toledano A, Serrano L, Pineda A, Romero AA, Luque R, Labidi J. Microwave-assisted depolymerisation of organosolv lignin via mild hydrogen-free hydrogenolysis: catalyst screening. Appl Catal B Environ. 2014;145:43–55.
146. Mateus MM, Acero NF, Bordado JC, dos Santos RG. Sonication as a foremost tool to improve cork liquefaction. Ind Crop Prod. 2015;74:9–13.
147. Behling R, Chatel G, Valange S. Sonochemical oxidation of vanillyl alcohol to vanillin in the presence of a cobalt oxide catalyst under mild conditions. Ultrason Sonochem. 2017;36:27–35.

Chapter 7
Production and Upgrading of γ-Valerolactone with Bifunctional Catalytic Processes

Laura Prati, Andrea Jouve, and Alberto Villa

Abstract The production of γ-valerolactone (GVL) is mainly based on the hydrogenation of levulinic acid (LA) which has been identified as one of the top abundant, renewable building-block biomass compounds by the US-DOE. LA can be prepared by dilute acid-catalyzed hydrolysis of cellulose. GVL can be used as solvent, as an electrolyte, but also as a starting material to produce chemicals and fuel additives in a similar capacity to ethanol. It can be converted to liquid fuels, namely valeric biofuels (mixture of pentanoic acid PA esters), liquid alkanes or to high-value precursor to bio-polymers such as 1,4 pentanediol, or alkyl pentenoates.

Heterogeneous catalytic processes able to transform LA into GVL have been studied for a long time in terms of metal (Ru, Cu....), support (activated carbons, CNT, oxides) and conditions (solvent, temperature, pressure). It has been also shown that by adding an acidic co-catalyst, such as niobium phosphate or oxide, to Ru/AC it is possible to speed up the reaction rate under mild reaction conditions (70 °C, 3 bar of H_2 pressure) keeping selectivity >98% toward GVL. Moreover, coupling acidity and Meerwein-Verley-Ponndorf reduction (MPV), GVL can be obtained avoiding the use of molecular hydrogen or precious metals. Under more severe conditions (200 °C and 40 bar of H_2 pressure) Ru supported on acidic supports (H-beta or ZSM-5) or Pd on Nb_2O_5 promotes ring opening and hydrogenation of GVL to PA. The major limitation of heterogeneous catalysts applied to this reaction is the low catalyst stability, that can be overcome with proper design.

The chapter reviews the impact of catalyst design on the activity and the selectivity to LA transformation with particular attention to the production of specific, important products such as GVL and valeric fuels.

L. Prati (✉) • A. Jouve • A. Villa
Dipartimento di Chimica, Università degli Studi di Milano, Milan, Italy
e-mail: laura.prati@unimi.it

Z. Fang et al. (eds.), *Production of Biofuels and Chemicals with Bifunctional Catalysts*, Biofuels and Biorefineries 8, https://doi.org/10.1007/978-981-10-5137-1_7

7.1 Importance of GVL and Its Derivatives

The progressive replacement of fossil resources with renewable ones, is seen as one of the topics of major urgency in the last decade. One of the most abundant renewable raw materials is lignocellulosic biomass which is therefore the most studied source for the production of fuels and chemicals [1–3]. However, the high oxygen content present in the biomass has to be removed to develop biomass-based processes [4–7].

One of the most appealing ways to lower oxygen content in lingo-cellulosic biomass has been shown to be the transformation of the raw material into an intermediate such as levulinic acid (LA) [8], which has a low oxygen content and can be catalytically transformed into fuels and chemicals (Fig. 7.1). In fact, LA has been recognized one of the top-12 building blocks by the US-DOE [9] for sustainable chemical production.

Cellulose can be hydrolyzed into the monomer (glucose) by acid catalysis (H_2SO_4) which under hydrothermal conditions is dehydrated to hydroxymethylfurfural (HMF). The hydrolysis of HMF then produces LA. In this latter reaction, formic acid is co-produced as a low-value product used for formaldehyde production or for plasticizers, but it can be also be used as a H_2 source.

Biofuel γ-valerolactone (GVL), 2-methyltetrahydrofuran (MTHF), pentanoic acid (PA) and high value chemical intermediates such as 1,4 pentanediol, or alkyl pentenoates which are precursors of bio-polymers, can be produced by hydrogenation of levulinic acid (LA) in a cascade of hydrogenation/reduction or hydrogenolysis and dehydration reactions (Fig. 7.2). The sequence and the occurrence of these reactions determine the selectivity of the process.

The sequence is affected by the nature of the catalyst employed and the reaction conditions. In particular, it can be seen that acid sites promote dehydration, Lewis acid sites in the presence of H-donors promote MPV reduction, and metallic sites promote hydrogenation. The complexity of the reaction pathways has therefore led to the introduction of multifunctional catalytic systems able to enhance the selectivity to the desired product.

GVL can be used as solvent, or as an electrolyte, or to produce chemicals and fuel additives in a similar capacity as that of ethanol. GVL holds important characteristics as an ideal sustainable liquid: it does not hydrolyze under neutral conditions and it is stable in air for long periods of time without formation of peroxide [10, 11]. Dumesic et al. showed that using GVL as solvent (90 wt% in water) hydro-

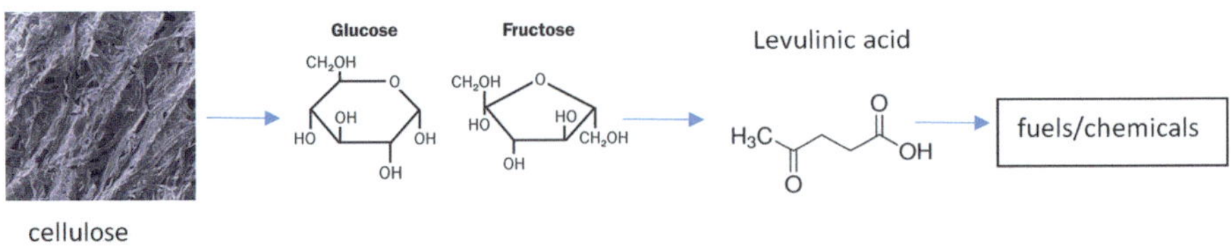

Fig. 7.1 Production of fuels and chemicals from cellulose

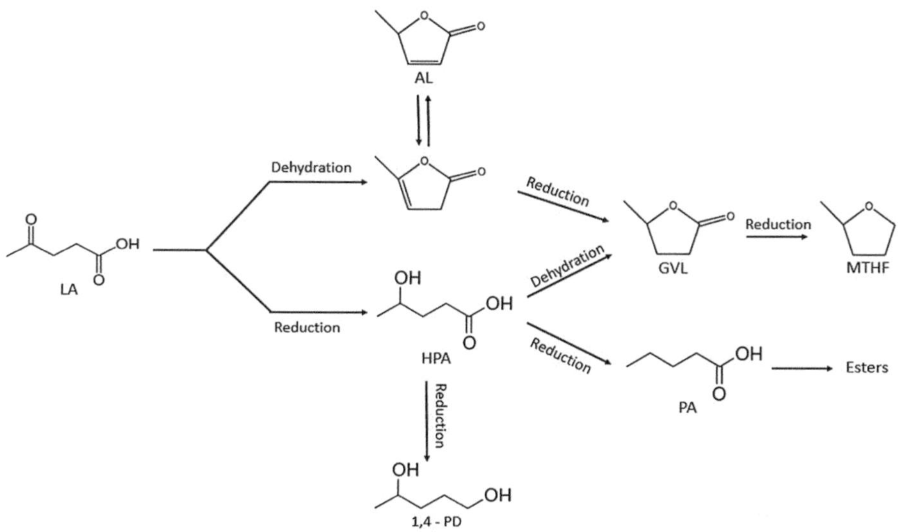

Fig. 7.2 Different hydrogenation/dehydration pathways for levulinic acid (LA)

lysis of cellulose in the presence of Amberlyst 70 can be speeded up and the selectivity to LA increases to 69% compared with 20% in water for the reducing formation of humins [12, 13]. Moreover, GVL does not form azeotropes with water, thus being more attractive and economically advantageous than ethanol as a fuel additive. A 10%v/v GVL or EtOH mixture with a 90%v/v gasoline has very similar fuel properties [14]. However, GVL has been shown to have blending limitations when used in conventional combustion engines thus making it more attractive to convert GVL into alkyl valerate esters (valeric biofuels) or liquid alkanes.

7.2 Production of γ-Valerolactone

GVL is mainly produced from LA even though there have been some attempts to obtain it directly from cellulose [12, 13]. In this section, attention is on the synthetic pathways from LA, as the direct use of cellulose is affected by technological issues such as the accessibility of the 1,4-beta glucosidic bond which should be hydrolyzed or the high production of humins, which are dark brown inhomogeneous, insoluble compounds derived from condensation pathways. These latter compounds are undesirable not only because they decrease the yield of useful products, but also because they deposit on the surface of catalysts thus interfere with the active sites. It should be noticed that the use of GVL as a reaction medium has been shown to reduce humins formation during the hydrolysis of cellulose [12, 13].

Some initial attempts have been made to use homogeneous reaction systems. The homogeneous catalytic conversion of LA to GVL has been shown to be very effi-

Table 7.1 Hydrogenation of levulinic acid to γ-valerolactone (GVL)

Entry	Catalyst	Solvent	Temp (°C)	$P(H_2)$ (bar)	Yield (%)	References
1	Ru/Al_2O_3	Water/supercritical CO_2	200	200	>99	[25]
2	Ru/AC +Nb_2O_5	Water	70	3	57	[26]
3	Ru/AC	Dioxane	150	34.5	72	[27]
4	Ru/AC	Water	265	N.D.	99	[28]
5	Ru/AC	MeOH	160	N.D.	91	[29]
6	Ru/hydroxyapatite	Water	70	5	99	[31]
7	Pd/hydroxyapatite	Water	70	5	23	[31]
8	Pt/hydroxyapatite	water	70	5	37	[31]
9	Ru/ H-β zeolite	Dioxane	200	5	40	[32]
10	Ru/OMC	Water	70	7	99	[33]
11	Raney Ni	Iso-propanol	100	15	83	[37]
12	Ru/SiO_2	Supercritical CO_2	200	48	94	[25]
13	$Pd/NiO\text{-}SiO_2$	Water	30	10	82	[39]
14	Ni-Fe/ montmorillonite	Iso-propanol	200	34.5	97	[40]
15	Cu-Fe oxide	Water	220	9	91	[41]
16	Cu-Cr oxide	Water	220	9	89	[42]
17	Cu-ZrO2	Water	200	34.5	100	[44]
18	Mo_2C	Water	180	30	85	[48]

cient and selective [15, 16] but the recycling of the catalyst presents several drawbacks. A possible solution was shown to be the use of biphasic systems [17, 18] typically constituted by water and an immiscible organic solvent and water soluble Ru-complexes bearing ligands such as sulphonated phosphine [19]. Even, though the results obtained are promising in terms of activity and selectivity, heterogeneous catalytic materials still represent the best choice from an industrial point of view due to their easy recovery and reusability (Table 7.1). Therefore, a lot of effort have been spent in studying heterogeneous catalysts and their performance to establish the influence of solvent, reaction conditions and more specifically the effect of the supporting materials. The main challenge, however, still is the design of the catalyst, which has to be selective for the desired product, stable in performance, recyclable and inexpensive.

The disclosure of the reaction mechanism represents a crucial role. Jesse and coworkers studied the hydrogenation of LA in the presence of Ru/C catalysts in detail [20, 21]. A cascade of reactions involving reduction and dehydration steps take place, under acidic conditions, with the order of these steps determining the overall reaction pathway (Fig. 7.3).

Levulinic acid (LA) can be reduced to 4-hydroxypentanoic acid (HPA) that readily forms the thermodynamically favored lactone GVL by acid catalyzed dehydration (pathway B) [22]. Alternatively, LA can be converted to angelica lactone (AL) via an endothermic dehydration followed by reduction to GVL (pathway A) [23]. This latter pathway becomes more and more important at high temperatures and in

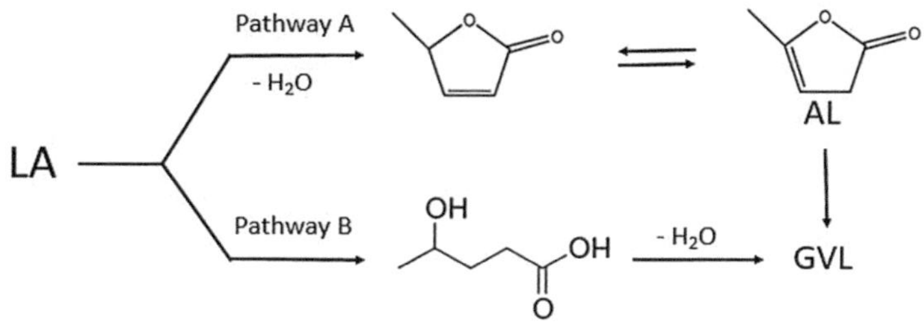

Fig. 7.3 Levulinic acid (LA) to γ-valerolactone (GVL) transformation: pathway A, dehydration of LA to angelica lactone (AL) followed by reduction to GVL; pathway B, reduction of LA to 4-hydroxypentanoic acid (HPA) and then dehydration to GVL

the presence of an acidic medium. Dehydration can be catalyzed by acid whereas reduction can be performed in the presence of H_2 (catalytic hydrogenation) or in presence of an H_2-source molecule (MPV reduction). In the case of hydrogenation, a transition metal is generally required whereas MPV is catalyzed by the presence of Lewis acid sites [24]. The advantages in using MPV route is therefore to avoid the presence of precious metal (such as Pd, Ru, and Pt) and the presence of large quantities of gaseous H_2 constituting an advantage in terms of cost and safety. Over Ru/C, it has been shown [25, 26] that at low temperatures the rate of GVL formation appears to be controlled by the intramolecular esterification of HPA, whereas at high temperatures, mass transfer limits the hydrogenation rate. Therefore, LA can be hydrogenated at near room temperature following pathway B (Fig. 7.3) with bifunctional catalysts that exhibit hydrogenation functionality and acidity as being applicable for LA transformation to GVL.

The Ru-based catalysts have been investigated in detail taking into account different reaction environment (neat, water, organic solvent, supercritical CO_2) and supports. Ru on Al_2O_3 was tested as catalyst for GVL production at 200 °C and 20 MPa in water/supercritical CO_2 and gave a conversion over 99% with full selectivity to GVL [27].

It has also been shown that by adding an acidic co-catalyst, such as niobium phosphate or oxide to Ru/AC, that it is possible to speed up the reaction rate under mild reaction conditions (70 °C, 3 bar of H_2), keeping selectivity >98% towards the GVL product even the yield appears lower (57%) [28] Ru on activated carbon in dioxane at 150 °C and 34.5 bar of hydrogen yields to 72% of GVL in 2 h [29] or at 265 °C to 99% in 50 h [30]. In MeOH at 160 °C, Ru/C is very active leading to 91% yields of GVL in only 1.5 h [31] Ru nanoparticles of around 3 nm supported on DOWEX 50WX2-100 gel-type resin show a very high activity for LA transformation to GVL [32]. Another supporting material for Ru particles is hydroxyapatite [33] which is very active even under mild conditions (70 °C, 5 bar of H_2 in water or water/toluene) producing up to 99% yields of GVL. On the same support Pd and Pt nanoparticles produce 23 and 37% yields of GVL respectively. Acidic supports such as H-β zeolite or H-ZSM5 (200 °C, 5 bar of H_2 in dioxane) on the contrary do not perform well as Ru supporting materials gave GVL yields of around 40% even under harsh conditions [34]. Acid functionalization of mesoporous carbon has been demonstrated to improve not only the activity, but also the stability of Ru catalyst allowing quantitative GVL production under mild conditions (70 °C, 7 bar of H_2) [35]. Moreover, the exposition of the active Ru species appears to be very important as demonstrated in the use of Ru supported on few-layer Graphene [36]. We have addressed the high activity and stability of the catalytic system to its greater metal exposition with respect to the one in Ru/C.

Pt and Pd have been investigated as catalytically active metals and Pd shows higher activity than Pd Pt. Pd nanoparticles are very efficient for LA to GVL transformation when supported on SiO_2 tested at 160 °C [37]. A comparison between Pd and Pt has been carried out using carbon as the support with the order of performance being found as Ru > Pd >Pt [30].

Iridium polymer complexes have been investigated [38]. N-heterocyclic carbenes (NHC)-iridium coordination polymers have been proposed as solid molecular catalysts with a peculiar stability due to the chemical bond between the active center (Ir) and the polymer. This solid complex works at 50 bar of H_2 and 100 °C in i-propanol and in the presence of a base (KOH).

The high cost of these precious metals and the relatively easy deactivation of the Ru-based catalysts greatly affect the effectiveness of conversion of LA into GVL reaction. Therefore, recent studies point out the possibility of using non-noble metal-based catalysts. Ni and Ni-based catalysts were proposed initially [39, 40] Raney Ni has the highest activity among Raney Co, Raney Cu, Raney Fe in terms of GVL yields (83%) [39]. Under relatively high temperatures (200 °C) and H_2 pressure (48 bar) 94% yields of GVL can be obtained [27]. Ni as an active metal has been investigated for the synthesis of bimetallic particles to decrease the Ru content while maintaining the high activity of the Ru metal [40].

$Ru_{0.9}Ni_{0.1}$ on ordered mesoporous carbon (OMC) show not only a high activity, but also good stability being recyclable up to 15 times without appreciable loss of activity or selectivity. Most probably, the confinement of bimetallic nanoparticles in the channels of the carbon reduce metal leaching or aggregation, thus favoring the stability of the entire catalyst.

Pd deposited on 9.9% NiO/SiO_2 in low amounts (0.2%) by wet impregnation results in a nanoscale intimacy between the two metals [41]. The catalyst works at almost ambient temperature (30 °C) at low H_2 partial pressure (0.3–1 MPa) in hydrogenating angelica-lactone (AL) to GVL with 82% conversion and 100% selectivity efficiently with reaction time being on the order of minutes.

Nickel and iron in a 1:1 weight ratio appear highly active when supported on montmorillonite (MMT) providing > 99% conversion with 98% selectivity to GVL [42]. Interestingly, the choice of the solvent is fundamental: in the presence of isopropanol the reaction proceeds smoothly whereas in the presence of water a consistent Fe leaching is observed. We addressed the hydrogenation step with the bimetallic sites while the strong acidic sites of MMT favor the subsequent cyclization to GVL. A plausible mechanism includes reductive cyclization process through the formation of levulinate ester that undergoes lactonization GVL [42].

Attempts to develop a Ru substitute have been made using copper [43, 44]. Cu-Fe and Cu-Cr systems under 9 bar of H_2 at 220 °C afford 90% yields of GVL at 98% conversion with limited metal leaching. This latter aspect was studied in detail for the Cu/ZrO_2 system, the stability of which being connected with a stable tetragonal phase of ZrO_2, is able to firmly bind with Cu [45]. There is the possibility to lower the metal content in Cu/ZrO2 which implies that the active Cu species are present in very low amounts with respect to the total amount [46]. The catalytic active site is constituted by reduced Cu particles with strong interaction with Cu incorporated into the ZrO_2 framework [47]. Copper as active metal was also tested on Al_2O_3 [48]. An interesting example of Cu catalysis is reported for gas phase reaction [49]. A complex composition of the catalytic bed (50–75% CuO, 20–25% SiO_2, 1–5% graphite, 0.1–1% $CuCO_3$/$Cu(OH)_2$) gave GVL in 93% yields for more than 90 h in continuous operation.

The β-Mo_2C is known to be able to replace noble metals in several processes. Indeed, also in the case of LA transformation to GVL β-Mo_2C is active at 180 °C under 30 bar of H_2, pressure giving a selectivity >85% for GVL [50]. In this case, hydrogenated products such as 1,4-pentandiol and methyltetrahydrofuran are observed even in minor amounts [50].

Formic acid can be used as a hydrogen source. This option appears highly desirable as HCOOH can be produced besides levulinic acid during HMF hydrolysis opening the way to the use of crude LA mixture. For example, Ru on carbon has been tested in a continuous reactor at 180 °C using propyl guaiacol as solvent and formic acid as the H_2 source yielding 93% GVL [51]. Ni supported on Al_2O_3, MgO, hydrotalcite has also been tested in continuous vapor phase hydrogenation without an external source of H_2 but using HCOOH [52]. The best performance was obtained with Ni on Al_2O_3 as the water produced during the reaction promoted the degradation/transformation of the support in the case of MgO or hydrotalcite thus leading to fast deactivation of the catalytic bed. Ni was used also to promote Cu/SiO_2 catalyst with a ratio 20:60 converting 99% of LA into 96% GVL with the rest constituted by angelica-lactone (AL) [53]. Ni was claimed to enhance the dispersion of Cu active phase whereas the nanocomposite nature of the catalyst contributes to the catalyst long-term stability.

A 1 mol% Au/ZrO_2 was shown to work in a batch mode at 150 °C using an equimolar amount of LA and HCOOH to give >99% yield of GVL in 6 h [54].

Zirconium dioxide, ZrO_2, has been studied as hydrogen transfer catalyst through Meerwein-Ponndorf-Verley reduction reaction (MPV), where the H_2 source is an alcohol. Using ZrO_2 and isobutanol at 160 °C after 16 h, 92% GVL is produced [55]. The system is enhanced by supporting ZrO_2 on SBA-15 [56].

Using bifunctional Zr- and Al- containing beta-zeolite catalysts, is possible to transform xylose directly into GVL for which at 190 °C and 48 h reaction time, a GVL yield of 35% is obtained [57].

7.3 Upgrading of γ-Valerolactone

Thanks to its favorable properties, GVL is suitable for the production of stable biofuels as well as for food additives [58, 59]. Catalytic transformation of GVL to valuable products has become an important research topic (Fig. 7.4).

For example, 2-methyltetrahydrofuran is one of the four main groups of fuels or chemicals derived from upgrading GVL. Hydrocarbons fuels constitute the second group followed by the valerate esters category (third group). The fourth group is made up of polymers. Each of these groups (Fig. 7.4) will be described in the following paragraphs.

7.3.1 *Group I: MTHF Biofuels*

When GVL is hydrogenated, both 2-methyltetrahydrofuran (MTHF) and 1,4-pentanediol can be produced [60]. MTHF can be used as a green solvent [61] as well as a biofuel compound [62]. Indeed, MTHF has been identified as a component

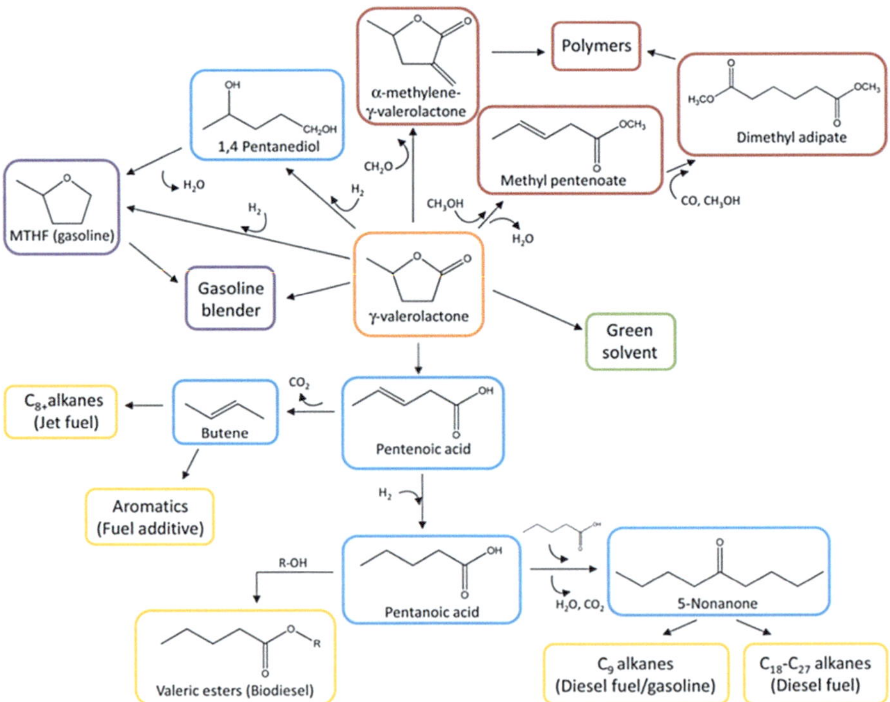

Fig. 7.4 Biofuels derived from γ-valerolactone (GVL) (Reproduced with permission from [22], Copyright © 2013 The Royal Society of Chemistry)

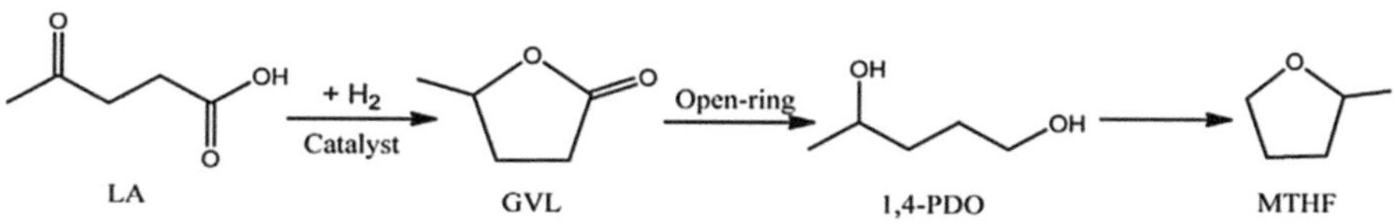

Fig. 7.5 Hydrogenation of levulinic acid (LA) to produce 2-methyltetrahydrofuran (MTHF)

for P-series fuels [63, 64]. By co-blending MTHF with ethanol in gasoline, the vapor pressure of ethanol is considerably reduced [65]. MTHF can be produced following the path [66] detailed in Fig. 7.5.

The path shown in Fig. 7.5 involves the acid catalyzed GVL ring-opening to produce 1,4 pentanediol (1,4-PDO) and the further dehydration of the diol to MTHF. The selectivity either to 1,4-PDO or MTHF depends on the catalyst. According to the chosen catalyst and reaction conditions, higher selectivity to one of the two products is favored. Bozell et al. [67], starting from levulinic acid, obtained up to 90% MTHF yield using a bi-functional catalyst (PdRe/C) at 200–250 °C and 100 H_2 partial pressure.

An 80 wt% copper loading on SiO_2 led to a 64% yield of MTHF (Fig. 7.5), yields that could be improved to 89% by addition of nickel [30, 68, 69]. Du et al. [70] modifying the activation parameters of the 30 wt% Cu/ZrO_2 catalysts, were able tune the selective hydrogenolysis of GVL either to MTHF or 1,4-PDO. Calcining the catalyst in air at 400 °C, they obtained a GVL conversion of 98% with 93% selectivity to MTHF within 6 h (6 bar H_2 partial pressure and reaction temperature of 240 °C). The catalyst was recycled over three runs and gave a MTHF yield of up to 85%. In comparison, activation of the Cu/ZrO_2 catalyst under a H_2-argon gas mixture at 300 °C resulted to the formation of 1,4-pentanediol with a 93% yield, with the catalyst being stable, for over two more runs [70].

Once MTHF is obtained in high yields, it can be further converted into C_4-C_9 alkanes. A bifunctional metal-acid catalyst allows one to reach this objective for conditions at high pressures (approx. 80 bar) and moderate temperatures (approx. 150 °C) [16].

Besides upgrading GVL to 1,4-PDO or MTHF and derivative compounds, GVL can also be transformed into hydrocarbon fuels.

7.3.2 Group II: Hydrocarbon Fuels

The upgrading of GVL to hydrocarbon fuels generally follows two paths (Fig. 7.6). The first route lies in the transformation of GVL to pentenoic acid which is consecutively decarboxylated to yield butene and CO_2 (Fig. 7.6). Through oligomerization of butene at high pressures (35 bar), C_{8+} alkanes (jet fuel, C-12) are formed [71, 72]. The overall mechanism takes place on acid catalysts such as silica alumina, Amberlyst 70 or ZSM-5. It is worth to mention that catalyst stability is enhanced

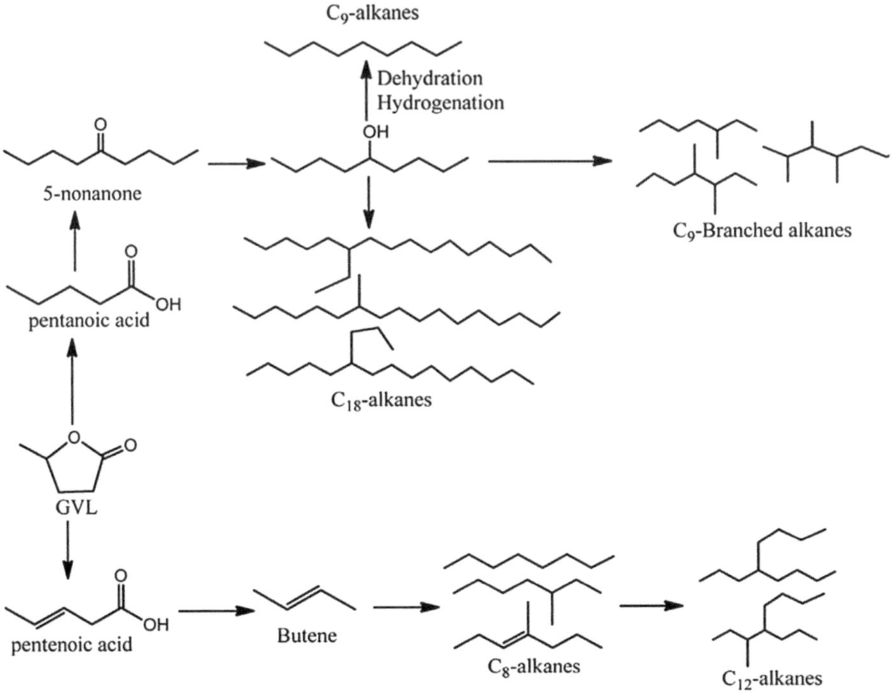

Fig. 7.6 Routes for the transformation of γ-valerolactone (GVL) into liquid alkanes

with the addition of water, which prevents coke formation during the reaction. Moreover, the any carbon deposits is removed by simple calcination in air [72].

The second route in Fig. 7.6 requires a bi-functional catalyst and can be described as the production of pentanoic acid through the ring-opening of GVL on acidic sites, and further hydrogenation on metal sites. Serrano-Ruiz et al. [73] used a dual-bed reactor to carry out their reaction. The first catalytic bed contained a 0.1 wt% Pd/Nb_2O_5 catalyst to perform the GVL ring-opening while the second catalytic bed was loaded with a $CeZrO_x$ catalyst. Pentanoic acid is ketonized to 5-nonanone and one equivalent each of CO_2 and water. That research group reported a 84% yield of 5-nonanone [73] which is very promising considering that they started from aqueous solution of GVL. In that study, the formation of ketones such as 2-hexanone and 3-heptanone, which are the results of the 5-nonanone scission was also reported [73].

Once 5-nonanone is obtained it can be hydrogenated to 5-nonanol, which is the precursor for the production of hydrocarbon fuels over, for example Ru/C [74]. 5-nonanol can be transformed into C_9-alkanes, the C_9-branched alkanes or the C_{18}-alkanes. C_9-alkanes are produced from the dehydration of 5-nonanol and can be used as fuel additive to diesel fuel. Consecutively, through isomerization and aromatization reactions of the C_9-alkanes over zeolites, branched C_9-alkanes are formed. The by-products, which are shorter ketones, can be converted into C_6-C_7

alkanes that can be easily removed by evaporation [65]. Alternatively, C_9-alkanes can overcome oligomerization over an acid catalyst to produce a mixture of C_{18}-C_{27} olefins, that can be used as diesel fuel. To tune the molecular weight range, reaction parameters can be modified. Linear alkenes are obtained with high reactivity of the terminal olefins. In contrast, lower reactivity leads to the production of wax and solid olefins through oligomerization reaction [74]. Upgrading GVL to target alkane fuels has been established, although additional efforts should be made to improve formation steps and yields.

7.3.3 Group III: Valerate Esters

Lange et al. [75] pioneered the synthesis of valerate esters also called valeric biofuels as fuel additives. The production of these valeric biofuels starts with the GVL ring-opening over a bifunctional metal-acid catalyst (Pt/ZSM5) to form pentanoic acid. Once produced, both the pentanoic acid and the alcohol present in the reaction media undergo esterification on an acidic ion-exchange resin to produce valeric esters. A time on stream of 1500 h, interrupted with some calcination in air, led to a pentanoic acid yield of over 90%.

Serrano-Ruiz et al. [23, 65, 76] converted GVL to pentanoic acid over Pd/Nb_2O_5 catalysts and reported a 92% yield. To minimize by-products, the authors optimized several parameters such as Pd loading, H_2 partial pressure and reaction temperature.

Using an inexpensive Cu based catalyst, Chan-Thaw et al. [77] succeeded in producing pentyl valerate (PV) from GVL. The supports used in that work [77] are weakly acid amorphous materials and were able to upgrade GVL to valeric fuels in a one-pot reaction. Such achievement was possible by using pentanol as solvent under H_2 and 8 wt% Cu/SiO_2 –ZrO_2 as catalyst. The reaction begins with nucleophilic addition of the alcohol to the carbonyl acid group producing the hydroxypentanoate. The latter is further dehydrated and hydrogenated to the saturated ester, which is known as pentyl valerate. They reported a 90% GVL conversion and up to 83% selectivity to PV. For comparison, they also used ethanol as solvent and obtained ethyl valerate. However, lower yield to ethyl valerate (EV) was reported (about 40%) mainly due to formation of 4-ethoxy pentanoate and pentenoic esters. For the same reaction, they also performed a stability test and the catalyst showed good recyclability over nine runs (about 20% loss in activity) and almost stable EV yield (from 41 to 35%). Higher EV yield (about 62%) was reported later by Sun et al. [78] over a Co/HZSM-5. Starting from levulinic acid, those authors also investigated the production of pentanoic acids with a batch reactor and EV with a fixed-bed reactor system. The series of reactions gave a pentanoic acid yield of 93% at 80 °C and a 97% EV yield at 240 °C.

Precious metal based catalysts as well as inexpensive metal based catalysts are effective in producing valerate esters. More efforts are still necessary to reduce the number of steps and costs as well as increasing the valerate esters yield.

7.3.4 Group IV: Polymers

The last group of materials that can be produced from GVL are not renewable fuels but monomers for use in polymer production. Although similar to products obtained from with petroleum, these polymers differ with their chemical properties. The compound α-methylene-γ-valerolactone, which is a biomass-derived monomer, can be obtained from GVL as reported by Manzer et al. [29, 79]. Similar properties of methyl methacrylate were found for α-methylene-γ-valerolactone. Additionally, thanks to the lactone structure, an improved thermal stability of the polymer was achieved.

New bio-based polymers precursors can be synthesized through a reaction between an amine and the GVL. When performed at mild temperatures (approx. 50 °C), the GVL ring-opening in presence of primary or secondary aliphatic amines allows formation of various γ-hydroxyl-amides. The reaction of GVL and 1,2-diaminoethane is catalyzed by $SnCl_2$ at 50 °C and gives γ-hydroxyl(amino) amide compounds [80] for which the nucleophilicity of the amine plays an important role in the addition-elimination reaction as claimed by Chalid et al. [80]. These monomers are the starting materials to produce polymers such as polyethers or polyurethanes.

7.4 Conclusions and Future Outlook

The decrease of fossil resources and related environmental issues in their use as fuel and energy source, prompt researchers to develop sustainable alternative solutions. Lignocellulosic biomass represents an appealing starting material according to its abundance and renewability. However, the high oxygen content represents a big limitation for the development of efficient processes that can replace fuels and chemicals from fossil materials. Cellulose and hemicellulose constitute up to 85% in mass of lignocellulose and therefore most studies focus on these materials. Among the possible products deriving from chemical or enzymatic hydrolysis, levulinic acid has been considered as one of the most promising platform chemicals from which it is possible to produce fuels, fuel-additives, fragrances, solvents, oil additives, pharmaceuticals and plasticizers.

In particular, the authors of this chapter have focused on the production of GVL and its upgrading routes. GVL has important use (fuel additives or as a solvent) but it also is a versatile raw material for many important classes of compounds (biofuels, plasticizers). The transformation of LA into GVL and subsequent transformation appears to be constituted by two elemental steps: reduction and dehydration. Therefore, it is evident that reliable materials for possible catalytic processes should be bifunctional materials with redox as well as acidic functionalities. First attempts have been made with the use of classical precious metals (Ru, Pt, Pd) on different support (carbons, oxides, HT, etc.) but mainly the cost of these metals and stability concerns have directed the most recent studies towards the use of non-precious, more accessible metals such as Cu or Ni even though they normally require higher reaction temperatures and H_2 pressures.

The use of molecular H_2 in some cases can be replaced by H_2-donor molecules such as alcohols or formic acid, which also contribute to the economy and the safeness of the processes. Catalytic materials need to be designed specifically for the type of the reaction.

Methods for upgrading GVL still require improvement and research to become cost-competitive with the existing petroleum-based processes, especially in terms of development of selective heterogeneous catalysts with high recyclability. Some steps have been made with careful design of the catalytic materials, but understanding of the processes at the molecular level of the active site and its deactivation mechanism should be studied to develop ideal bifunctional catalysts that are both selective and robust.

Moreover, only a few attempts up to now have been made for developing one-pot processes starting from sugars or (better) directly from cellulose with bifunctional catalyst. Issues in selectivity and also durability should be solved for making these processes feasible.

References

1. Geilen FMAMA, Engendahl B, Harwardt A, Marquardt W, Klankermayer J, Leitner W. Selective and flexible transformation of biomass-derived platform chemicals by a multifunctional catalytic system. Angew Chem Int Ed. 2010;122(32):5642–6.
2. Geilen FMA, vom Stein T, Engendahl B, Winterle S, Liauw MA, Klankermayer J, Leitner W. Highly selective decarbonylation of 5-(Hydroxymethyl)furfural in the presence of compressed carbon dioxide. Angew Chem Int Ed. 2011;50(30):6831–4.
3. vom Stein T, Grande P, Sibilla F, Commandeur U, Fischer R, Leitner W, Domínguez de María P. Salt-assisted organic-acid-catalyzed depolymerization of cellulose. Green Chem. 2010;12(10):1844–189.
4. Yan K, Wu G, Lafleur T, Jarvis C. Production, properties and catalytic hydrogenation of furfural to fuel additives and value-added chemicals. Renew Sust Energ Rev. 2014;38:663–76.
5. Huber GW, Dumesic JA. An overview of aqueous-phase catalytic processes for production of hydrogen and alkanes in a biorefinery. Catal Today. 2006;111(1):119–32.
6. Huber GW, Iborra S, Corma A. Synthesis of transportation fuels from biomass: chemistry, catalysts, and engineering. Chem Rev. 2006;106(9):4044–98.
7. Yan K, Wu X, An X, Xie X. Facile synthesis of reusable coal-hydrotalcite catalyst for dehydration of biomass-derived fructose into platform chemical 5-Hydroxymethylfurfural. Chem Eng Commun. 2014;201(4):456–65.
8. Pileidis FD, Titirici M-M. Levulinic acid biorefineries: new challenges for efficient utilization of biomass. ChemSusChem. 2016;9(6):562–82.
9. Werpy T, Petersen G, Aden A, Bozell J, Holladay J, White J, Manheim A, Eliot D, Lasure L, Jones S. Top value added chemicals from biomass. Volume 1 – Results of screening for potential candidates from sugars and synthesis gas. Washington, DC: Department of Energy, Office of Energy Efficiency and Renewable; 2004.
10. Fábos V, Koczó G, Mehdi H, Boda L, Horváth IT. Bio-oxygenates and the peroxide number: a safety issue alert. Energy Environ Sci. 2009;2(7):767–9.

11. Zaitseva A, Pokki J-P, Le HQ, Alopaeus V, Sixta H. Vapor–liquid equilibria, excess enthalpy, and density of aqueous γ-valerolactone solutions. J Chem Eng Data. 2016;61(2):881–90.
12. Alonso DM, Gallo JMR, Mellmer MA, Wettstein SG, Dumesic JA. Direct conversion of cellulose to levulinic acid and gamma-valerolactone using solid acid catalysts. Catal Sci Technol. 2013;3(4):927–31.
13. Rackemann DW, Doherty WO. The conversion of lignocellulosics to levulinic acid. Biofuels Bioprod Biorefin. 2011;5(2):198–214.
14. Horváth IT, Mehdi H, Fábos V, Boda L, Mika LT. γ-Valerolactone—a sustainable liquid for energy and carbon-based chemicals. Green Chem. 2008;10(2):238–42.
15. Braca G, Raspolli Galletti AM, Sbrana G. Anionic ruthenium iodorcarbonyl complexes as selective dehydroxylation catalysts in aqueous solution. J Organomet Chem. 1991;417(1–2):41–9.
16. Mehdi H, Fábos V, Tuba R, Bodor A, Mika LT, Horváth IT. Integration of homogeneous and heterogeneous catalytic processes for a multi-step conversion of biomass: from sucrose to Levulinic acid, γ-valerolactone, 1,4-Pentanediol, 2-Methyl-tetrahydrofuran, and Alkanes. Top Catal. 2008;48(1–4):49–54.
17. Fache E, Santini C, Senocq F, Basset JM. Homogeneous catalysis in water Part II. Synthesis and characterization of ruthenium water-soluble complexes. J Mol Catal. 1992;72(3):331–6.
18. Nuithitikul K, Winterbottom M. Crucial parameters in the selective biphasic hydrogenation of cinnamaldehyde by biphasic Ru-TPPTS and RhCl(TPPTS)3 catalysts. Catal Today. 2007;128(1):74–9.
19. Chalid M, Broekhuis AA, Heeres HJ. Experimental and kinetic modeling studies on the biphasic hydrogenation of levulinic acid to γ-valerolactone using a homogeneous water-soluble Ru–(TPPTS) catalyst. J Mol Catal A Chem. 2011;341(1–2):14–21.
20. Abdelrahman OA, Heyden A, Bond JQ. Analysis of kinetics and reaction pathways in the aqueous-phase hydrogenation of Levulinic acid to form γ-valerolactone over Ru/C. ACS Catal. 2014;4(4):1171–81.
21. Piskun AS, van de Bovenkamp HH, Rasrendra CB, Winkelman JGM, Heeres HJ. Kinetic modeling of levulinic acid hydrogenation to γ-valerolactone in water using a carbon supported Ru catalyst. Appl Catal A Gen. 2016;525:158–67.
22. Alonso DM, Wettstein SG, Dumesic JA. Gamma-valerolactone, a sustainable platform molecule derived from lignocellulosic biomass. Green Chem. 2013;15(3):584.
23. Serrano-Ruiz JC, West RM, Dumesic JA. Catalytic conversion of renewable biomass resources to fuels and chemicals. Annu Rev Chem Biomol Eng. 2010;1(1):79–100.
24. Komanoya T, Nakajima K, Kitano M, Hara M. Synergistic catalysis by Lewis acid and base sites on ZrO_2 for Meerwein–Ponndorf–Verley reduction. J Phys Chem C. 2015;119(47):26540–6.
25. Zhang J, Chen J. Selective transfer hydrogenation of biomass-based furfural and 5-hydroxymethylfurfural over hydrotalcite-derived copper catalysts using methanol as a hydrogen donor. ACS Sustain Chem Eng. 2017. https://doi.org/10.1021/acssuschemeng.7b00778.
26. Kuwahara Y, Kaburagi W, Osada Y, Fujitani T, Yamashita H. Catalytic transfer hydrogenation of biomass-derived levulinic acid and its esters to ?-valerolactone over ZrO2 catalyst supported on SBA-15 silica. Catal Today. 2017;281:418–28.
27. Bourne RA, Stevens JG, Ke J, Poliakoff M. Maximising opportunities in supercritical chemistry: the continuous conversion of levulinic acid to γ-valerolactone in CO2. Chem Commun. 2007;28(44):4632–4.
28. Galletti AMR, Antonetti C, De Luise V, Martinelli M. A sustainable process for the production of γ-valerolactone by hydrogenation of biomass-derived levulinic acid. Green Chem. 2012;14(3):688–94.
29. Manzer LE. Catalytic synthesis of α-methylene-γ-valerolactone: a biomass-derived acrylic monomer. Appl Catal A Gen. 2004;272(1):249–56.
30. Upare PP, Lee J-M, Hwang DW, Halligudi SB, Hwang YK, Chang J-S. Selective hydrogenation of levulinic acid to γ-valerolactone over carbon-supported noble metal catalysts. J Ind Eng Chem. 2011;17(2):287–92.
31. Gong Y, Lin L, Yan Z. Catalytic hydrogenation and oxidation of biomassderived levulinic acid. Bioresources. 2011;6(1):686–99.

32. Moreno-Marrodan C, Barbaro P. Energy efficient continuous production of γ-valerolactone by bifunctional metal/acid catalysis in one pot. Green Chem. 2014;16(7):3434–8.
33. Sudhakar M, Lakshmi Kantam M, Swarna Jaya V, Kishore R, Ramanujachary KV, Venugopal A. Hydroxyapatite as a novel support for Ru in the hydrogenation of levulinic acid to γ-valerolactone. Catal Commun. 2014. https://doi.org/10.1016/j.catcom.2014.03.005.
34. Luo W, Deka U, Beale AM, van Eck ERH, Bruijnincx PCA, Weckhuysen BM. Ruthenium-catalyzed hydrogenation of levulinic acid: Influence of the support and solvent on catalyst selectivity and stability. J Catal. 2013;301:175–86.
35. Villa A, Schiavoni M, Chan-Thaw CE, Fulvio PF, Mayes RT, Dai S, More KL, Veith GM, Prati L. Acid-functionalized mesoporous carbon: an efficient support for ruthenium-catalyzed γ-valerolactone production. ChemSusChem. 2015;8(15):2520–8.
36. Xiao C, Goh T-W, Qi Z, Goes S, Brashler K, Perez C, Huang W. Conversion of Levulinic acid to γ-valerolactone over few-layer graphene-supported ruthenium catalysts. ACS Catal. 2016;6(2):593–9.
37. Yan K, Lafleur T, Wu G, Liao J, Ceng C, Xie X. Highly selective production of value-added γ-valerolactone from biomass-derived levulinic acid using the robust Pd nanoparticles. Appl Catal A Gen. 2013;468:52–8.
38. Liu Y, Sun Z, Huang C, Tu T. Efficient hydrogenation of biomass oxoacids to lactones by using NHC-Iridium coordination polymers as solid molecular catalysts. Chem An Asian J. 2017;12(3):355–60.
39. Rong Z, Sun Z, Wang L, Lv J, Wang Y, Wang Y. Efficient conversion of Levulinic acid into γ-valerolactone over Raney Ni catalyst prepared from melt-quenching alloy. Catal Lett. 2014;144(10):1766–71.
40. Yang Y, Gao G, Zhang X, Li F. Facile fabrication of composition-tuned Ru-Ni bimetallics in ordered mesoporous carbon for levulinic acid hydrogenation. ACS Catal. 2014;4(5):1419–25.
41. Zhang P, Yuan Q, Chen L, Xue T, Guan Y, Wu P. Low temperature hydrogenation of α-angelica lactone on silica supported Pd–NiO catalysts with synergistic effect. RSC Adv. 2016;6(70):65377–82.
42. Kadu BS, Hengne AM, Biradar NS, Rode CV, Chikate RC. Reductive cyclization of Levulinic acid to γ-Valerolactone over non-noble bimetallic nanocomposite. Ind Eng Chem Res. 2016;55(51):13032–9.
43. Yan K, Chen A. Selective hydrogenation of furfural and levulinic acid to biofuels on the eco-friendly Cu–Fe catalyst. Fuel. 2014;115:101–8.
44. Yan K, Chen A. Efficient hydrogenation of biomass-derived furfural and levulinic acid on the facilely synthesized noble-metal-free Cu–Cr catalyst. Energy. 2013;58:357–63.
45. Hengne AM, Rode CV. Cu–ZrO2 nanocomposite catalyst for selective hydrogenation of levulinic acid and its ester to γ-valerolactone. Green Chem. 2012;14(4):1064–72.
46. Jones DR, Iqbal S, Ishikawa S, Reece C, Thomas LM, Miedziak PJ, Morgan DJ, Edwards JK, Bartley JK, Willock DJ, Hutchings GJ. The conversion of levulinic acid into γ-valerolactone using Cu–ZrO2 catalysts. Catal Sci Technol. 2016;6(15):6022–30.
47. Ishikawa S, Jones DR, Iqbal S, Reece C, Morgan DJ, Willock DJ, Miedziak PJ, Bartley JK, Edwards JK, Murayama T, Ueda W, Hutchings GJ. Identification of the catalytically active component of Cu–Zr–O catalyst for the hydrogenation of levulinic acid to γ-valerolactone. Green Chem. 2017;19(1):225–36.
48. Putrakumar B, Nagaraju N, Kumar VP, Chary KVR. Hydrogenation of levulinic acid to γ-valerolactone over copper catalysts supported on γ-Al2O3. Catal Today. 2015;250:209–17.
49. Bonrath W, Castelijns AMCF, de Vries JG, Guit RPM, Schütz J, Sereinig N, Vaessen HWLM. Gas phase hydrogenation of Levulinic acid to γ-valerolactone. Catal Lett. 2016;146(1):28–34.
50. Quiroz J, Mai EF, Teixeira da Silva V. Synthesis of nanostructured molybdenum carbide as catalyst for the hydrogenation of Levulinic acid to γ-valerolactone. Top Catal. 2016;59(2–4):148–58.

51. Azadi P, Carrasquillo-Flores R, Pagán-Torres YJ, Gürbüz EI, Farnood R, Dumesic JA. Catalytic conversion of biomass using solvents derived from lignin. Green Chem. 2012;14(6):1573–6.
52. Varkolu M, Velpula V, Burri DR, Kamaraju SRR. Gas phase hydrogenation of levulinic acid to γ-valerolactone over supported Ni catalysts with formic acid as hydrogen source. New J Chem. 2016;40(4):3261–7.
53. Upare PP, Jeong M-G, Hwang YK, Kim DH, Kim YD, Hwang DW, Lee U-H, Chang J-S. Nickel-promoted copper–silica nanocomposite catalysts for hydrogenation of levulinic acid to lactones using formic acid as a hydrogen feeder. Appl Catal A Gen. 2015;491:127–35.
54. X-L D, He L, Zhao S, Liu Y-M, Cao Y, He H-Y, Fan K-N. Hydrogen-independent reductive transformation of carbohydrate biomass into γ-valerolactone and pyrrolidone derivatives with supported gold catalysts. Angew Chem. 2011;123(34):7961–5.
55. Chia M, Dumesic JA. Liquid-phase catalytic transfer hydrogenation and cyclization of levulinic acid and its esters to γ-valerolactone over metal oxide catalysts. Chem Commun. 2011;47(44):12233–5.
56. Enumula SS, Gurram VRB, Kondeboina M, Burri DR, Kamaraju SRR. ZrO2 /SBA-15 as an efficient catalyst for the production of γ-valerolactone from biomass-derived levulinic acid in the vapour phase at atmospheric pressure. RSC Adv. 2016;6(24):20230–9.
57. Hernández B, Iglesias J, Morales G, Paniagua M, López-Aguado C, García Fierro JL, Wolf P, Hermans I, Melero JA. One-pot cascade transformation of xylose into γ-valerolactone (GVL) over bifunctional Brønsted–Lewis Zr–Al-beta zeolite. Green Chem. 2016;18(21):5777–81.
58. Jessop PG. Searching for green solvents. Green Chem. 2011;13(6):1391.
59. Mellmer MA, Sener C, Gallo JMR, Luterbacher JS, Alonso DM, Dumesic JA. Solvent effects in acid-catalyzed biomass conversion reactions. Angew Chem Int Ed. 2014;53(44):11872–5.
60. Mascal M, Nikitin EB. Direct, high-yield conversion of cellulose into biofuel. Angew Chem Int Ed. 2008;47(41):7924–6.
61. Pace V, Hoyos P, Castoldi L, Domínguez De María P, Alcántara AR. 2-Methyltetrahydrofuran (2-MeTHF): a biomass-derived solvent with broad application in organic chemistry. ChemSusChem. 2012;5(8):1369–79.
62. Climent MJ, Corma A, Iborra S. Conversion of biomass platform molecules into fuel additives and liquid hydrocarbon fuels. Green Chem. 2014;16(2):516–47.
63. Balat M. Current alternative engine fuels. Energy Sources. 2005;27(6):569–77.
64. Kar Y, Deveci H. Importance of p-series Fuels for flexible-fuel vehicles (FFVs) and alternative fuels. Energy Sources Part. 2006;28(10):909–21.
65. Serrano-Ruiz JC, Braden DJ, West RM, Dumesic JA. Conversion of cellulose to hydrocarbon fuels by progressive removal of oxygen. Appl Catal B Environ. 2010;100(1–2):184–9.
66. Yan K, Yang Y, Chai J, Lu Y. Catalytic reactions of gamma-valerolactone: a platform to fuels and value-added chemicals. Appl Catal B Environ. 2015;179:292–304.
67. Bozell JJ, Moens L, Elliott DC, Wang Y, Neuenscwander GG, Fitzpatrick SW, Bilski RJ, Jarnefeld JL. Production of levulinic acid and use as a platform chemical for derived products. Resour Conserv Recycl. 2000;28(3–4):227–39.
68. Upare PP, Lee JM, Hwang YK, Hwang DW, Lee JH, Halligudi SB, Hwang JS, Chang JS. Direct hydrocyclization of biomass-derived levulinic acid to 2-methyltetrahydrofuran over nanocomposite copper/silica catalysts. ChemSusChem. 2011;4(12):1749–52.
69. Upare PP, Yoon J-W, Kim MY, Kang H-Y, Hwang DW, Hwang YK, Kung HH, Chang J-S. Chemical conversion of biomass-derived hexose sugars to levulinic acid over sulfonic acid-functionalized graphene oxide catalysts. Green Chem. 2013;15(10):2935.
70. X-L D, Bi Q-Y, Liu Y-M, Cao Y, He H-Y, Fan K-N. Tunable copper-catalyzed chemoselective hydrogenolysis of biomass-derived γ-valerolactone into 1,4-pentanediol or 2-methyltetrahydrofuran. Green Chem. 2012;14(4):935.
71. Bond JQ, Martin Alonso D, West RM, Dumesic JA. γ-Valerolactone ring-opening and decarboxylation over SiO2/Al2O3 in the presence of water. Langmuir. 2010;26(21):16291–8.
72. Bond JQ, Wang D, Alonso DM, Dumesic JA. Interconversion between γ-valerolactone and pentenoic acid combined with decarboxylation to form butene over silica/alumina. J Catal. 2011;281(2):290–9.

73. Serrano-Ruiz JC, Wang D, Dumesic JA. Catalytic upgrading of levulinic acid to 5-nonanone. Green Chem. 2010;12(4):574.
74. Alonso DM, Bond JQ, Serrano-Ruiz JC, Dumesic JA. Production of liquid hydrocarbon transportation fuels by oligomerization of biomass-derived C9 alkenes. Green Chem. 2010;12(6):992–9.
75. Lange J-P, Price R, Ayoub PM, Louis J, Petrus L, Clarke L, Gosselink H. Valeric biofuels: a platform of cellulosic transportation fuels. Angew Chem Int Ed. 2010;49(26):4479–83.
76. Buitrago-Sierra R, Serrano-Ruiz JC, Rodríguez-Reinoso F, Sepúlveda-Escribano A, Dumesic JA. Ce promoted Pd–Nb catalysts for γ-valerolactone ring-opening and hydrogenation. Green Chem. 2012;14(12):3318.
77. Chan-Thaw CE, Marelli M, Psaro R, Ravasio N, Zaccheria F. New generation biofuels: γ-valerolactone into valeric esters in one pot. RSC Adv. 2013;3(5):1302–6.
78. Sun P, Gao G, Zhao Z, Xia C, Li F. Stabilization of cobalt catalysts by embedment for efficient production of valeric biofuel. ACS Catal. 2014;4(11):4136–42.
79. Heeres H, Handana R, Chunai D, Borromeus Rasrendra C, Girisuta B, Jan Heeres H. Combined dehydration/(transfer)-hydrogenation of C6-sugars (D-glucose and D-fructose) to γ-valerolactone using ruthenium catalysts. Green Chem. 2009;11(8):1247.
80. Chalid M, Heeres HJ, Broekhuis AA. Ring-opening of γ-valerolactone with amino compounds. J Appl Polym Sci. 2012;123(6):3556–64.

Chapter 8
Production of Furanic Biofuels with Zeolite and Metal Oxide Bifunctional Catalysts for Energy-and Product-Driven Biorefineries

Jesús Requies, Ion Agirre, and Aitziber Iriondo

Abstract In the last years, research on the transformation of biomass into different compounds has grown significantly with the motivation being to reduce the dependency of oil and to develop of sustainable and environmental friendly energy sources. In this context biomass appears to be as the only renewable source of carbon that is able to provide a substitute for fossil fuels. In the near future, bio-refineries, in which biomass is catalytically converted into pharmaceuticals, agricultural chemicals, plastics and transportation fuels will take the place of current petrochemical plants. Among the transportation fuels, furanic biofuels as 2,5-dimethylfuran (DMF) and 2-methylfuran (2-MF) have good performance as a fuel for direct injection spark ignition type engines without important modifications of the engine. The transformation of biomass into furanic biofuel compounds takes place via 5 hydroxymethylfurfural (HMF) in the case of the DMF and 2-MF; in the case of the 2-MF it can be also produced from furfural (FF) via furfuryl alcohol (FOL). In these reactions it is necessary to employ of bifunctional catalysts for the hydrogenolysis. Metals are required to fix the hydrogen reaction and sometimes for the C–O and C–C bonds cleavage and the acid-base supports of the dehydration, for the C–C and C–O bonds scissions. This chapter provides an overview of current methods for converting biomass to furanic biofuels with zeolite and metal oxide bifunctional catalysts. The chapter provides state-of-the-art overview on furanic biofuel production from biomass with a brief description of the DMF production process and the 2-MF production process. Use of different bifunctional catalysts for the DMF production process and the 2-MF production process is described. The influence of the support and that of different metals will be discussed along with properties of the bifunctional catalysts like metal dispersion, catalysts acidity and operating conditions. Finally, the use of different solvents to improve the yield of biofuels will be analyzed.

J. Requies (✉) • I. Agirre • A. Iriondo
Engineering Faculty of Bilbao, University of Basque Country (UPV/EHU), Bilbao, Spain
e-mail: jesus.requies@ehu.eus

Z. Fang et al. (eds.), *Production of Biofuels and Chemicals with Bifunctional Catalysts*, Biofuels and Biorefineries 8, https://doi.org/10.1007/978-981-10-5137-1_8

8.1 Introduction

In 2017, more than 80% of primary energy was generated from petroleum, coal and natural gas resources [1–3]. Oil production and the carbon proportion in the world electricity mix have increased each year [3–6]. As these energy sources are fossil fuels, there are serious concerns about their depletion in the following decades [2, 7–9]. Among the non-fossil sources, biofuels and waste maintained their share of world energy production in 2014 (10.2%), though their development (Fig. 8.1). Other renewable sources such as wind, solar thermal, solar PV, geothermal, are expanding at a fast pace, but still account for just over 1% of global energy production [6].

Petroleum not only plays an important role in the transportation sector, but also in the chemical industry. Many oil-derived raw materials are the base or platform chemicals for the production of a wide variety of products like plastics, carpets, cloths, paints, perfumes and hairsprays [10].

Apart from all these data, there is evidence that the planet is undergoing global warming, as a result of CO_2 emissions coming from the combustion of fossil fuels [7, 11]. As shown in Fig. 8.1 two sectors produced nearly two-thirds of global CO_2 emissions in 2013: electricity and heat generation, by far the largest, which accounted for 42%, while transport accounted for 23%.

Based on these environmental concerns, economical fluctuations and the geopolitical instability in the producer countries, the use of other raw materials is being explored and therefore, research and development in sustainable and environmentally friendly alternatives have become one of the technological goals of the near future.

To prepare for the future energy sources, one possible alternative raw material is biomass that can be used as a source of energy or as a source of platform products. A biorefinery is defined as "the sustainable processing of biomass into a spectrum of marketable products (food, feed, materials, chemicals) and energy (fuels, power, heat)" by IEA Bioenergy Task 42 or as "a facility that integrates biomass conversion processes and equipment to produce fuels, power, and chemicals from biomass" by the National Renewable Energy Laboratory of EE.UU. According to these definitions, the biorefinery can be a concept, a process, a plant or even a cluster of facilities, where the core is biomass conversion into several product streams and the integration of various technologies and processes in the most sustainable way. The conversion of biomass can be carried out following different physical, thermal, biochemical and chemical operations, depending on the type of biomass and the demand of the derived products [13–15]. The biorefinery concept is analogous to today's petroleum refineries, which produce multiple fuels and products from petroleum and can be a good alternative to cover the current needs of energy, fine chemicals and unconventional fuels [1, 16]. The main reasons why biorefineries must be developed are:

- To decrease the oil dependence of the chemical products.
- To have higher sustainability as a chemical process
- To decrease the net emission of greenhouse gases

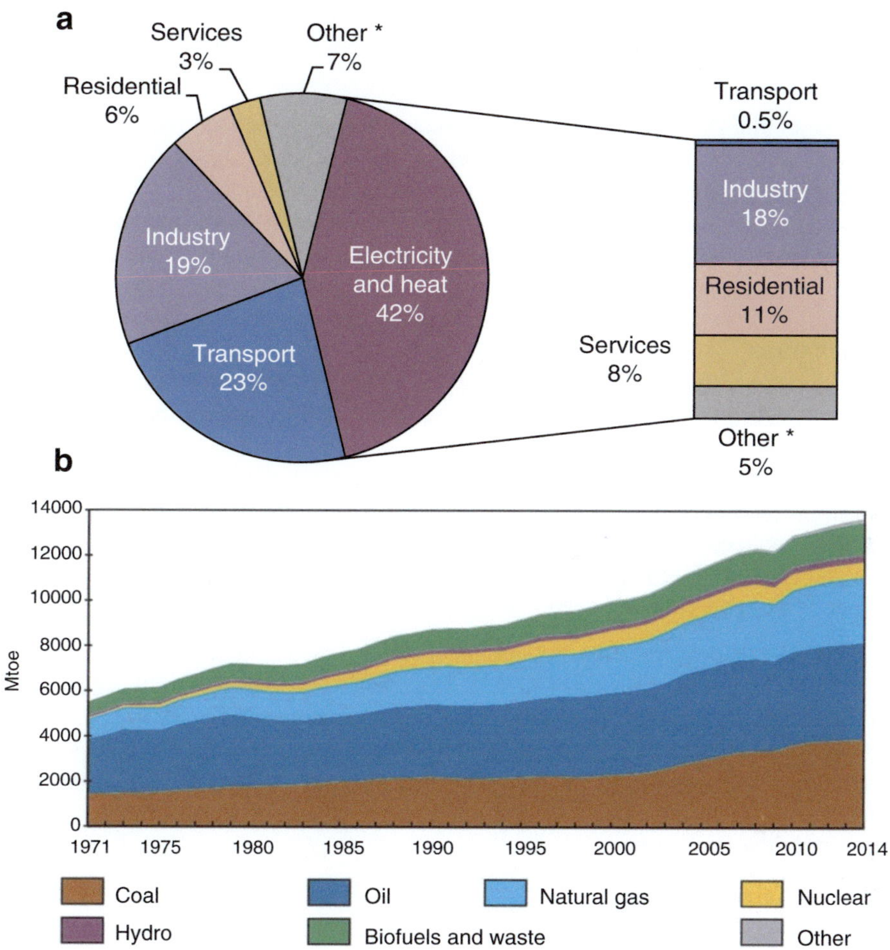

Fig. 8.1 (**a**) World CO_2 emissions by sector in 2013 [12]. (**b**) Word total primary energy supply from 1971 to 2014 by fuel (Mtoe) [3]. Based on IEA data from *Key world energy statistics* © OECD/IEA 2016, www.iea.org/statistics, Licence: www.iea.org/t&c; as modified by Universidad del País Vasco/Euskal Herriko Unibertsitatea

However, there are several obstacles in the development of biorefineries:

- Lack of integrated technology to convert biomass into industrial products; Lack of sustainability regarding biodiversity with nature competition with food.

It is in this context where second generation biofuels appeared. Biodiesel and bioethanol generated from first generation biomass (biomass rich on sugar in case of bioethanol; biomass rich on oil for biodiesel) greatly contributed to the transportation sector, but these proposals came into conflict with the food industry [17, 18]. Due to the drawbacks found in first generation biofuels efforts switched to development of second generation biofuels and bio-additives [17, 19], which can be produced from non-food residual lignocellulosic biomass (sugars) [9, 18, 19].

Biorefineries can be classified as, "energy-driven" or "product-driven". According to the implementation, one can speak about first, second or third generation biorefineries. Biorefineries can be named as:

- Lignocellulosic biorefineries: those which use lignocellulosic biomass as feedstock and then fractionate it into cellulose, hemicellulose and lignin.
- Whole crop biorefinery: those which use cereals as feedstock.
- Green biorefinery: those which use green biomass as feedstock and fractionate it by pressure to obtain a nutrient-rich juice organic solution and a fiber-rich lignocellulosic press cake.
- Two-platform concept biorefinery: those which includes sugars and syngas platforms.
- Conventional biorefinery: those based on existing industries, i.e., sugar or starch.
- Thermochemical biorefinery: those based on a mix of several thermochemical processes.
- Algal biorefinery: those which use aquatic biomass, algae, as feedstock.
- Liquid-phase catalytic processing biorefinery: those based on the production of functionalized hydrocarbons from biomass-derived intermediates
- Forest-based biorefinery: those which use forest biomass for simultaneous production of paper, fibers, chemicals and energy.

The IEA Bioenergy Task 42 has developed a new classification scheme to describe different biorefineries [20–22]. The classification of a biorefinery consists of the following features: platforms, feedstocks, products and processes. With the combination of these features, different biorefinery configurations can be described and named in a consistent manner. The most important feature is the platform. Platforms are intermediates which are able to connect to different biorefinery systems and their processes.

- Syngas platform
- Pyrolysis oil platform
- C6 and C5/C6 sugar platform
- Oil platform
- Biogas platform
- Organic solutions platform
- Lignin platform
- Hydrogen platform
- Electricity and heat platform

The two biorefinery product groups are:

- Energy products: e.g. bioethanol, biodiesel, and synthetic biofuels
- Material products: e.g. chemicals, materials, food and feed

Feedstocks can be grouped as:

- Energy crops from agriculture: e.g. starch crops, short rotation forestry
- Biomass residues from agriculture, forestry, trade and industry: e.g. straw, bark, used cooking oils, waste streams from biomass processing

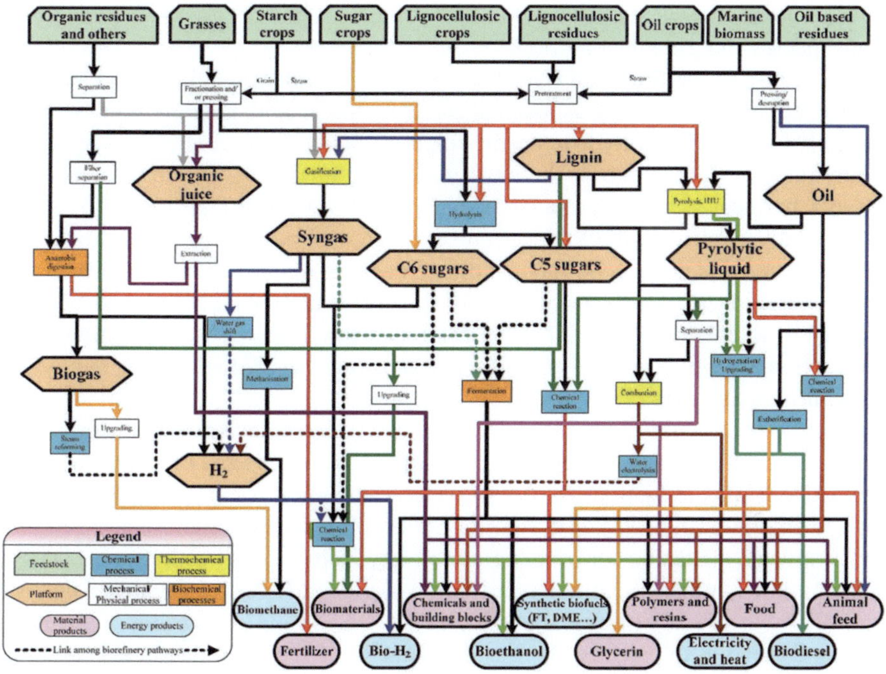

Fig. 8.2 Overview of the biorefinery classification system proposed by IEA-Biorefinery Task 42 group [21]. Based on IEA data from *Bioenergy Task 42* © OECD/IEA 2010, www.iea.org/statistics, Licence www.iea.org/t&c; as modified by Universidad del País Vasco/Euskal Herriko Unibertsitatea

Concerning conversion processes, the classification system identifies four main groups:

- Biochemical: e.g. fermentation, enzymatic conversion
- Thermochemical: e.g. gasification, pyrolysis
- Chemical: e.g. acid hydrolysis, synthesis, esterification
- Mechanical: e.g. fractionation, pressing, size reduction

An overview of current platforms, products, feedstocks and conversion processes is given in Fig. 8.2.

The present book chapter is focused on the production of furanic biofuels. Therefore, according to the classification developed by *IEA Bioenergy Task 42,* this topic fits the best in the "*C6 and C5/C6 sugar*" platform by means of "*chemical conversion*" to form "*energy products*" using "*lignocellulosic*" feedstock.

Lignocellulose is a natural composite which consists of three main constituents [14]:

- Cellulose (38–50%), a polysaccharide formed by glucose
- Hemicellulose (23–32%): sugar polymer consisting of hexoses (glucose, galactose and mannose) and pentoses (xylose and arabinose)
- Lignin (15–25%): a polymer formed by oligomerization of propenyl phenols.

The structures and compositions of these biopolymers vary greatly depending on plant species and growth conditions.

According to the literature [23], the top seven chemical building blocks can be produced from sugars using a chemical route: levulinic acid, glucaric acid, sorbitol, arabitol, furfural and 5-hydroxymethylfurfural, being furfural (FF) and hydroxymethylfurfural (HMF) two of the most interesting ones [24], due to their possible use as raw material to produce a large spectrum of chemicals and fuels [25].

Typically, triple dehydration of pentoses (xylose) results in the production of FF [26, 27], which is not a good fuel candidate because of its tendency to polymerize [28]. Therefore, FF has been proposed as a viable platform chemical to be integrated in biorefineries [29]. It can be used as a raw material for the synthesis of several nonpetroleum-derived chemicals such as furfuryl alcohol (FOL), methyltetrahydrofuran and furan [30].

HMF is an aromatic aldehyde and it is believed as an outstanding platform chemical. It is considered as a bridge between biomass resources and biochemical products [31]. HMF production has the great advantage that it can be produced from triple dehydration [26, 27] of different hexoses like glucose, fructose and sucrose, and afterwards, it can be converted into biofuels, such as 2,5-dimethylfuran (DMF), and some other fine chemical molecules like levulinic acid (LA), 2,5-furandicarboxylic acid (FDCA) or ethyl levulinate (EL) [1, 8, 16, 31–33]. Nowadays, the most extended reaction to produce HMF is dehydration of the aforementioned carbohydrates over acid, homogeneous or heterogeneous catalysts [1, 16, 32, 33].

Both, FF and HMF are suitable raw materials for the production of jet fuels as well as transportation fuels. Few research works report furanic compounds to be used as aviation fuels. Malinowki et al. [34] report some FF derived furanic ethers like 2.6-bis(alkoxymethyl)tetrahydrofuran and 2-(alkoxymethyl)tetrahydrofuran as a high potential compounds to be used as jet fuels. However, there are already some literature reported by Dumesic [35–38] and Huber [39, 40] about the synthesis of jet fuel range alkanes from HMF and FF where the direct catalytic treating of whole biomass creates a basic fuel platform chemicals including FF and HMF that can be upgraded to jet fuels. More recently, Yang et al. [41–43] reported several works on the synthesis of high density aviation fuel using cyclopentanol, which is a the aqueous-phase selective hydrogenation product of FF, using the combinations of magnesium-aluminum hydrotalcite (MgAl-HT) and Raney Ni (or Raney Co) catalytic system.

In the present book chapter, the production of furanic fuels derived from HMF and FF for transportation are studied, the production of DMF and 2-methylfuran (2-MF), in particular (Fig. 8.3). DMF and 2-MF are considered as a high-quality fuel, as ethanol, an ideal renewable and sustainable substitute or additive of conventional gasoline [2, 31, 44]. They present some interesting chemical properties as a high energy density, similar to that of gasoline and higher than ethanol [8, 19, 44], and high octane number, higher than gasoline [1]. Moreover, DMF and 2-MF show very low solubility in water and therefore it can be used as a blender in transportation fuels [19, 44, 45]. In this sense, some literature reports the good performance of the DMF and 2-MF as a fuel on direct injection spark ignition (DISI) type engines

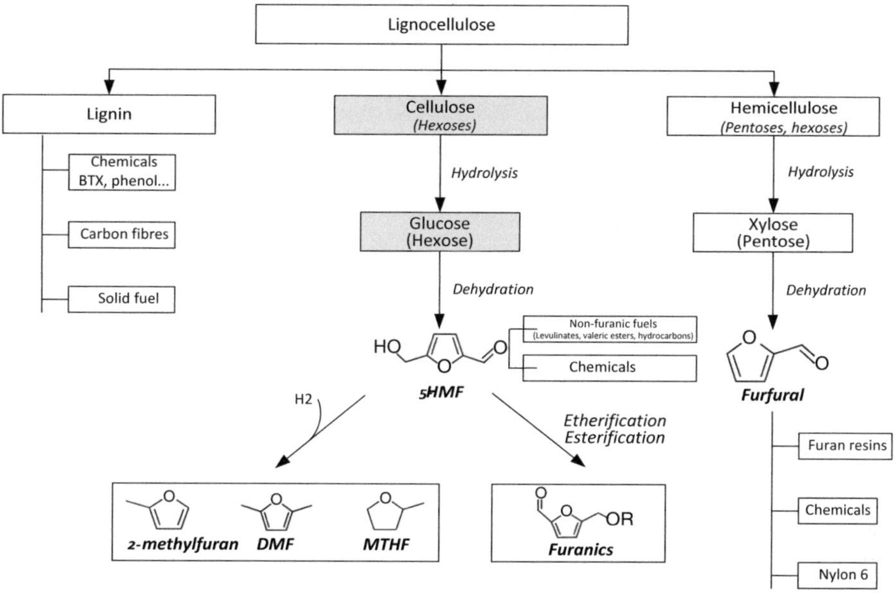

Fig. 8.3 Lignocellulosic feedstock biorefinery, focused on the production of furanic fuels

[8, 11, 46] without any important modifications of the engine and also in compression engines (Direct-injection compression engine (DICI)) [47]. Other authors, like Jezak et al. [48], conclude that DMF could be an additive to petroleum-based fuels but taking into consideration pressure and temperature conditions. Moreover, DMF and 2-MF reduce significantly the amount of large particles produced [46, 49], reducing the total mass of particles emitted, although it is not able to reduce significantly the number of particles smaller than 10 nm [49]. The 2-MF has better combustion characteristics and knock suppression ability than DMF and gasoline, nevertheless there is higher NO_x emissions for 2-MF than gasoline [46], which is also applicable to DMF, due to its high adiabatic flame temperature [45]. To address this emission issue some authors mix low amount of DMF and 2-MF with the gasoline or with diesel [44]. The aim of these mixtures is to improve the combustion and reduce the emissions. Xiao [44] for DICI engines and Wei [50] for DISI engines present good results with different mixtures of 2-MF and gasoline or 2-MF and diesel, respectively.

Dehydration processes to produce HMF or FF have been carried out using a great variety of systems:

- Catalysts: homogeneous vs. heterogeneous
- Solvent types: monophasic (aqueous or water with a miscible solvent or water with salts) vs. biphasic

A classification of the homogeneous and heterogeneous catalysts used to produce FF and HMF, respectively, have been published in reviews [27, 29, 51].

Homogeneous catalysts include organic acids (oxalic acid, levulinic acid, maleic acid) and mineral acids (sulfuric acid, phosphoric acid, hydrochloric acid). Heterogeneous catalysts include Lewis acids ($ZnCl_2$, $AlCl_3$, $CrCl_3$) and solid catalysts (ion exchange resins, zeolites, TiO_2, ZrO_2). Both catalytic systems have advantages and disadvantages.

Regarding the use of homogeneous acid catalysts, among their advantages are high activity and low sensitivity to poisons, to coking and to fouling [26]. However, these type of acid catalysts: (i) are uneconomical due to the downstream separation process, (ii) cause corrosion equipment, (iii) generate wastes that have safety and environmental issues and iv) have low selectivity and cause product decomposition because of its severe acidity [52–55].

Heterogeneous catalysts (i) allows the use of sequential type reactions in a single pot, resulting in few or no purification stages [56] -, and (ii) have favorable selectivity, recyclability and corrosion characteristics [57]. However heterogeneous catalysis shows also some disadvantages [26], such as; (i) variable activity due to possible deactivation by coke deposition, presence of poisons, fouling and metal leaching and (ii) diffusion limitations, although they are used in many catalytic processes [58].

These catalytic systems (homogeneous and heterogeneous) can be carried out in different reaction media: in monophasic or in biphasic systems in order to enhance the dissolution of the reactants or facilitate the extraction of the products. Water or water and some miscible solvents (salts forming ionic liquids or organic solvents like γ-valerolactone, methyl isobutyl ketone, dimethyl sulfoxide or sulfolane) form the usual monophasic and monophasic-binary systems, respectively. Water and immiscible organic solvents (toluene, p-xylene) form biphasic systems [51, 56, 59–63].

As the catalytic systems are assorted the objective of the present book chapter is to focus on zeolite [59] and metal oxide type heterogeneous catalysts since they show excellent thermal-chemical stability and pore selectivity properties. Moreover, the most effective applications of chemical catalysts in biomass conversion processes are similar to those that have already been demonstrated in conventional oil based processes [64, 65].

8.2 Furfural Synthesis

Furfural is produced typically from the dehydration of xylose containing in hemicellulose, for which the transformation takes place via a cyclodehydration reaction [51, 66–68]. However, another type of substrates (hemicellulose and glucose) can be used [69]; although there is not a consensus about reaction mechanism, the most accepted theory is that this cyclodehydration reaction seems to follow a sequential route of reactions through xylose isomerization and subsequent dehydration to produce furfural (Fig. 8.4), which are carried out with Lewis and Brønsted-acid catalyst respectively, but also has possible side reactions.

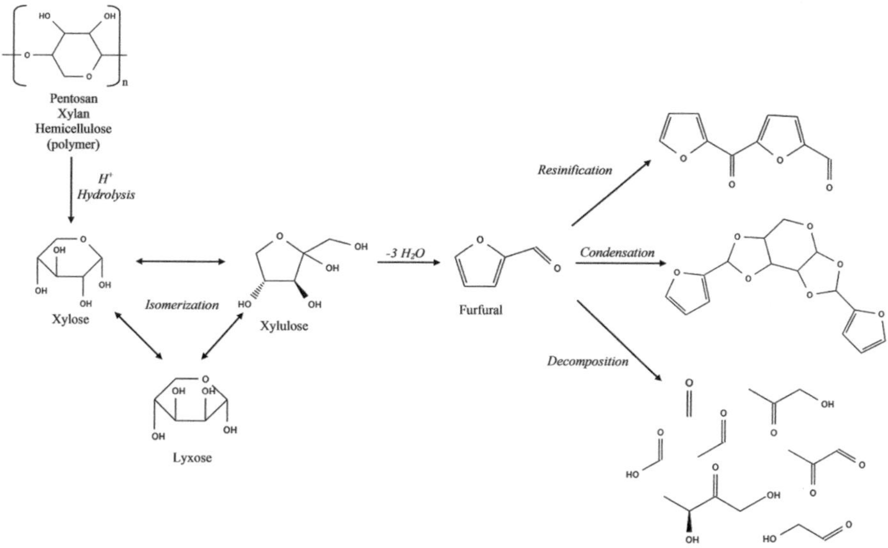

Fig. 8.4 Furfural production pathway from xylose and the possible side reactions adapted from literature [66–68, 71]

According to Choudhary et al. [68], xylulose and lyxose are the main products of xylose isomerization. Among these products, xylulose formation seems to be more favored thermodynamically and by the kinetics [70]. This is supported by the reported data [68] and suggests that xylulose and lyxose maximum yields of 27% and 11% can be obtained, respectively, at a xylose conversion of around 60% under 110 °C and 15 min using functionalized Lewis-acid Sn-Beta zeolite as catalyst. However, when Brønsted acid-catalyst, such as Amberlyst-15, is used in xylose transformation into FF the achieved conversion is negligible; while this type of catalyst is more active and selective in xylulose conversion into FF. Brønsted acid-catalysts seem to be more active in dehydration of xylulose than in xylose isomerization to xylulose, where this last reaction is activated with Lewis acid-catalysts. Therefore, it can be concluded that the sequential transformation of xylose into FF needs a catalyst with both Lewis-acid and Brønsted-acid active sites [51, 66–68].

Although zeolite type catalysts, also known as molecular sieves [56], and their amorphous counterparts, are useful in dehydration of xylose into FF, the morphology of this type of catalysts is typically microporous. This characteristic gives zeolites strong adsorption capacity for FF, resulting in undesirable reactions, such as polymerization, which leads the formation of humins [72] and carbon, that accumulate on the catalyst [73]. Moreover, the regeneration process of the catalysts requires the gasification of the deposited carbon that causes the loss of Brønsted-acid sites [51]. Many forms of zeolites are unstable under hydrothermal conditions [56]. Therefore, recent literature centers on the synthesis of metal catalysts supported on solid materials, such as metal oxides, that are active, selective and stable under hydrothermal conditions.

8.2.1 *Modified Zeolites Catalysts*

To avoid the disadvantages of zeolite type catalysts in xylose conversion into furfural, research has focused on the development of modified zeolite type catalysts (Table 8.1). Metkar et al. [71] have screened different types of ammonium forms of zeolite powders or extrudates (binder with γ-Al_2O_3), and metal oxides (γ-Al_2O_3, ZrO_2, SiO_2 and TiO_2), in a continuous reactive distillation reactor. Zeolites bound with γ-Al_2O_3 and TiO_2 metal oxides provide FF yields lower than 40% at 175 °C. However, FF yields higher than 45% are obtained with zeolites without any binder, reaching a FF yield of 75–80% for a xylose conversion about 98% with H-mordenite (Si/Al = 10) catalyst. The good performance of the zeolites without γ-Al_2O_3 binder could be because alumina adds acidity to the catalysts, which avoids the isomerization of xylose to xylulose. Regarding ZSM-5 catalyst [69], it has good activity towards xylose conversion (≈70–75%), but lower FF yields (≈30%) at 190 °C. However, when metal oxides (ZnO and Fe_2O_3), or chlorides (NaCl and $FeCl_3$) are added, the conversion increases and reaches almost 100%, while the FF yield remains almost constant, except for the case of NaCl (82.4%).

Table 8.1 Zeolites used in the conversion of carbohydrates substrates into FF with summary of reaction conditions and catalytic activity

Catalyst	Temp (°C)	System Monophasic	Biphasic	Raw material	Conversion (%)	FF yield (%)	References
H-mordenite[a]	170	W-sulfolane	–	Xylose	90	16	[71]
H-Beta[a]					90	21	
H-Y[a]					90	31	
H-mordenite					80	5	
H-Beta					80	70	
H-Y					80	75	
ZSM-5	190	W	–	Xylose	75	30	[69]
		W (+ NaCl)			98	82	
		W (+ $FeCl_3$)			98	30	
		–	W-t		92	66	
HMCM-22	170	W	–	Xylose	97	53	[74]
ITQ-22					97	54	
HMCM-22		–	W-t		92	70	
ITQ-22					99	66	
HMCM-22(38)					82	60	
Na-MCM-22					98	47	
Sn-Beta	170	W-GVL	–	Glucose	99	66	[75]
Fe-Beta					99	64	
H-USY	170	–	W-t	Xylan	72	56	[76]

[a] → binder to γ-Al_2O_3; Temp → temperature; W → water; t → toluene; GVL → γ-valerolactone

In a biphasic system at 170 °C, H-MCM-22 and its delaminated counterpart ITQ-2 show high xylose conversion (around 97%) and FF yields of around 54% are obtained in aqueous media and around 70% in water/toluene reaction media [74]. The increase of Si/Al ratio and incorporation of sodium in H-MCM-22 zeolite modifies the catalyst behavior. Namely, when Si amount increases in H-MCM-22 catalyst, amount of Lewis plus Brønsted acid sites and FF yield remains constant while the xylose conversion decreases (80%). However, when sodium is present, the catalyst shows a similar activity towards xylose conversion, but gives a lower FF yield. This behavior is due to the presence of sodium basic acid sites in this catalyst, which increases to undesirable products produced by side reactions. In conclusion, zeolite type catalysts needs to reach an optimal Lewis/Brønsted acid site amount and distribution to obtain good activity and selectivity in FF production using.

The activity of H-Beta zeolites modified with Sn, Zr and Fe metal oxides has been studied for FF production from glucose [75]. Although, hexose type monosaccharides are used as the raw material, they undergo isomerization reactions, just like for xylose. Among the modified zeolites, Sn- and Fe-beta catalysts show the highest FF yields, close to 50%, for complete conversion of glucose at 170 °C. The FF yield increases to 66% for Sn-beta catalysts when the glucose-used/catalyst mass ratio is 0.25. The Fe-beta catalysts show a similar FF yield under the same conditions of FF production. According to the literature Fe-beta zeolite has favorable characteristics for FF production from hemicellulose and cellulose, followed by Al-beta and Cr-beta zeolites [52]. The behavior of these metal modified zeolites is ascribed to the synergetic effect between textural-surface (metal-support interaction and low particle size, which confers to the zeolites-supported catalysts an homogeneous dispersion of the metals) properties and acid sites [75]. Regarding textural properties of Sn- and Fe-modified zeolites, they have a microporous structure associated with zeolites and greater width of pores provided by the added metal elements. The microporous structure facilitates FF formation, while the pore size promotes the diffusion of FF, avoiding degradation reactions. The acid sites number, distribution and ratio play also an important role. Concretely, Sn-Beta catalyst provides the best catalytic activity in terms of FF formation because of its optimum Brønsted/Lewis acid ratio, corroborating the conclusion established by Antunes et al. [74].

As indicated above [52, 74, 75], metal oxides or metal elements are widely employed to modify the catalytic characteristics of zeolite catalysts in FF synthesis. In fact, several studies reported in the literature investigate the performance of supported and non-supported metal oxides catalysts [76–81].

8.2.2 Supported and Non-supported Metal Oxide Catalyst

Among non-supported metal oxide catalysts, silicoaluminophosphate (SAPO), SiO_2, Al_2O_3, SiO_2-Al_2O_3, Nb_2O_5, ZrO_2 and sulfated ZrO_2 have been investigated.

SAPO type 34, 34C (commercial) and 56 catalysts (Table 8.2) show moderate activity when compared with amberlyst-70 and ZSM-5 catalysts, achieving 31%,

Table 8.2 Metal oxides used in the conversion of carbohydrates substrates into FF with summary of reaction conditions and catalytic activity

Catalyst Type	Catalyst	Temp (°C)	System: Monophasic	System: Biphasic	Raw material	Conversion (%)	FF yield (%)	References
Non-supported metal oxides	SAPO-34	170	W-GVL	–	Xylose	n.a.	31	[77]
	SAPO-34C					n.a.	40	
	SAPO-56					n.a.	38	
	SAPO-44	170	–	W-t	Xylan	76	63	[76]
	ALPO-44					51	41	
	SA-44					52	40	
	SiO_2					42	28	
	Al_2O_3					44	35	
	TiO_2	170	W-sulfolane	–	Xylose	80	18	[71]
	SiO_2-Al_2O_3	180	W	–	Xylose	65	52	[73]
	ZrO_2	170	–	W-t	Xylose	n.a.	25	[81]
					Xylan	50	20	
	Nb_2O_5	170	–	W-t	Xylose	90	53	[79]
Supported metal oxides	WO_3/SiO_2[a]	170	–	W-t	Xylan	85	55	[81]
	MoO_3/SiO_2[a]					71	45	
	Ga_2O_3/SiO_2[a]					80	50	
	WO_3/ZrO_2[a]					57	20	
	MoO_3/ZrO_2[a]					59	22	
	Ga_2O_3/ZrO_2[a]					81	41	
	Nb_2O_5/SiO_2[b]	160	W	–	Xylose	84	61	[78]
	Nb_2O_5/Al_2O_3[c]					75	30	
	Nb_2O_5/SiO_2[c]		–	W-t		n.a.	80	
	Nb_2O_5/Al_2O_3[c]					n.a.	60	

[a]→ prepared by sol-gel method
[b]→ prepared by incipient wetness impregnation method
[c]→ FF selectivity; Temp → temperature; W → water; t → toluene; GVL → γ-valerolactone; n.a. → not available

40% and 38% FF yields, respectively, at 170 °C and using xylose as raw material and γ-valerolactone/water as solvent [77]. The reason for these results is ascribed to the small pore structure of the SAPO catalysts, which hinders xylose diffusion in the pores. Moreover, these catalysts do not seem to suffer from acid-site leaching, while amberlyst-70 and ZSM-5 catalysts show leaching phenomenon. On the other hand, carefully designed SAPO-44 catalysts provide higher FF yield [76] than the aforementioned SAPO catalysts. Concretely, SAPO-44 catalyst, with silica ratio of 1, is able to convert xylan raw material into FF at 170 °C using water/toluene solvent, reaching a yield of 63%, which is a higher yield that those achieved with ALPO-44 (aluminophosphate) (41%), SA-44 (silica-alumina) (40%), H-USY type zeolite (56%), and SiO_2 (28%) and Al_2O_3 (35%) type metal oxides [76] catalysts. These activity results are closely related to the acidic and structural characteristics of the mentioned solid acid catalysts. SAPO-44 catalyst has a higher amount of acid sites, including weak and strong acid sites, when compared with the rest of the catalysts. In the case of aluminophosphate, silica-alumina and metal oxide catalysts, they only have weak acid sites which are responsible of the obtained FF yield. Moreover, the crystallinity of SAPO-44 catalyst is higher than aluminophosphate and silica-alumina catalysts, which decreases when SiO_2 ratio is lower or higher than 1.

You et al. [73] employed SiO_2-Al_2O_3 type catalyst (with 3.55 wt% of Al_2O_3) for xylose dehydration with different raw material concentrations, temperatures and space velocities. The best FF yield (around 52% and xylose conversion of 65.02%) obtains at 180 °C, 11 mL/g_{cat}·h and aqueous xylose solution of 1 wt%. It seems to be that the combination of SiO_2 and Al_2O_3 gives a catalyst with suitable acid and surface properties, catalytic activity and selectivity in FF synthesis from sugars.

The ZrO_2 catalyst [81] and sulfated and phosphate ZrO_2 [80] are also employed as catalysts in FF synthesis. However, these catalysts give FF yield below 25% when xylose or xylan is used as raw materials at 160–170 °C. Less FF formation is obtained by these catalysts due to the lower Lewis acid sites. Regarding Nb_2O_5 catalyst calcined at 450 °C [79], it shows good performance in terms of activity and stability, under dehydration of xylose to FF, reaching a conversion higher than 90% and a yield of 53% at 170 °C. The mesoporous structure and acidity gives this catalyst high activity, while its good stability can be ascribed to the absence of any leaching phenomena.

Non-supported metal oxides are active in xylan and/or xylose dehydration into FF [76, 77, 79–81]. However, the activity of these catalysts can be improved by modifying their structural and acidic characteristics. In this sense, ZrO_2, SiO_2 and Al_2O_3 have been used as supports of WO_3, MoO_3, Ga_2O_3 and Nb_2O_3 materials [78, 81]. SiO_2- and ZrO_2-supported WO_3, MoO_3 and Ga_2O_3 catalyst are prepared by wet impregnation and sol-gel methods. In general, catalysts prepared by sol-gel methods provide better FF yields at 170 °C than catalysts synthesized by impregnation method due to the sol-gel catalysts having a higher number of acid sites and less leaching phenomena. When xylan is used as raw material, the SiO_2 supported catalysts prepared by sol-gel, Ga_2O_3/SiO_2 and WO_3/SiO_2, give FF yields around 55% after 8 h reaction time, which is about 10 points above the values obtained with ZrO_2-supported counterparts. This difference can be associated with the well-dispersed metal oxides on the SiO_2 framework [81].

Regarding to SiO_2- and Al_2O_3-supported Nb_2O_3 catalysts [78], which are prepared by incipient wetness impregnation, they are able to convert more than 70 and 80% of xylose into FF, respectively, after 24 h of time on stream at 160 °C with water as solvent. However, only SiO_2 supported Nb_2O_5 catalyst provides FF selectivities higher than 55%, using water (61%) or water/toluene (80%) as reaction medium. The activity and selectivity showed by the last catalyst is ascribed to the larger pore size and optimum Lewis and Brønsted acid sites, with this last parameter being similar to the Nb_2O_3 supported on MCM and SBA type mesoporous materials.

Taking into account the data presented about FF catalytic synthesis, it is important to note that the conversion of the raw material (xylan, xylose and even glucose) and the yield or the selectivity towards FF are closely related to the textural, superficial and acidic characteristics of the catalyst. In this sense, it is important to recall that there is a suitable compromise among these characteristics, since the pore size and the Lewis acid sites promote raw material conversions, while the micropores and Brønsted acid sites promote FF formation. Even crystallinity and leaching stability of the catalysts seem to play an important role in FF synthesis.

8.3 Hydroxymethylfurfural Synthesis

Typically, the formation of HMF leads to sequential reactions of isomerization and dehydration, which need Lewis and Brønsted-acid catalyst, respectively. However, although the reaction mechanism has not fully elucidated, the most extended reaction mechanism is that HMF is produced from fructose dehydration and this hexose is obtained from the isomerization of glucose (Fig. 8.5) [82, 83]. Another type of substrates, such as cellulose, inulin and sugarcane bagasse [69], are used for HMF production.

As the production process of HMF is similar to FF one, the used catalysts are also similar. Concretely, literature on the HMF production from glucose primarily focuses on the use of heterogeneous catalytic systems based on metal oxides and their modifications. However, zeolite-type catalysts are also used in this process due to the good results achieved for FF production.

8.3.1 Zeolites

Although zeolite microporous materials are selective towards HMF and have good activity for cellulose or glucose raw materials [56], they are typically modified to improve their catalytic performance. In this sense, some authors use dealumination

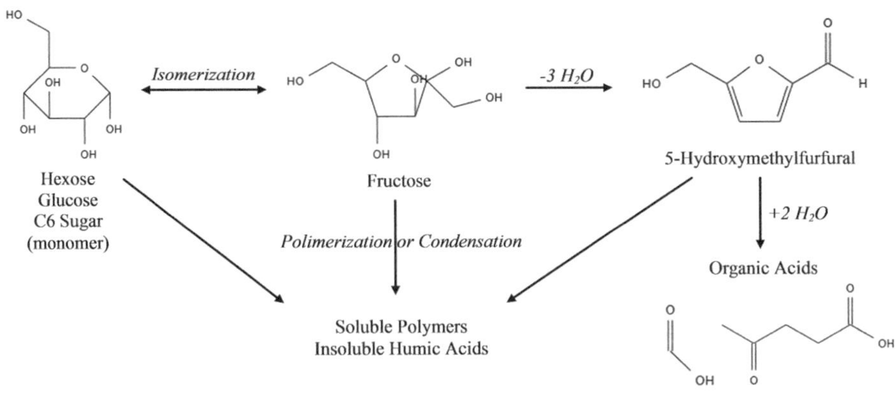

Fig. 8.5 HMF production pathway from glucose and the possible side reactions (Adapted with permission from [82, 83], Copyright © 2015, 2017 Elsevier)

Table 8.3 Zeolites used in the conversion of carbohydrates substrates into HMF with summary of reaction conditions and catalytic activity

Catalyst	Temp (°C)	System Monophasic	Biphasic	Raw material	Conversion (%)	FF yield (%)	References
H-Al-Beta[100]	160	W-DMSO	–	Glucose	n.a.	11.3	
Sn-Beta						12.3	[82]
[Sn,Al]-Beta[50]						31.4	
[Sn,Al-Beta][100]						37.1	
[Ti,Al]-Beta[50]						18.0	
[Ti,Al-Beta][100]						14.1	
Fe-Beta	170	W-DMSO	–	Glucose	n.a.	34.5	[75]
Sn-Beta				Fructose		36.4	
Zr-Beta				Sucrose		33.2	
Fe-Beta						11.8	
Sn-Beta						15.7	
Zr-Beta						8.6	
Fe-Beta						9.4	
Sn-Beta						12.1	
Zr-Beta						6.8	
H-ZSM5	195	W-MIBK	–	Glucose	60	5[a]	[84]
Fe-ZSM5					96	35[a]	
Cu-ZSM5					98	30[a]	
H-ZSM5		W-NaCl-MIBK			82	55[a]	
Fe-ZSM5					43	55[a]	
Cu-ZSM5					85	60[a]	

[50,100] → Sn/Al and Ti/Al ratio; [a] → HMF selectivity; Temp → temperature; W → water; DMSO → dimethyl sulfoxide; MIBK → methyl isobutyl ketone; n.a. → not available

process and/or incorporation of metal oxides to enhance the characteristics, and therefore the catalytic activity and stability of Beta, H-Beta and ZSM-5 zeolites [82–84] in glucose dehydration (Table 8.3). In the case of Beta-zeolite, after its dealumination, the obtained catalyst is modified with different amounts of Sn and Ti. The catalyst without Al or Sn element is less effective for glucose dehydration reaction, resulting in lower HMF yields. However, when the zeolite content of Sn and Al, concretely dealuminated [Sn,Al]-beta zeolites, the activity and selectivity increase. This improvement of the activity and selectivity is more pronounced in dealuminated zeolites with lower amounts of Sn, reaching a HMF selectivity of 62% (yield 37.1%) and glucose conversion around 60% for the [Sn,Al]-Beta(100) catalyst, when DMSO is employed as solvent and the temperature is fixed at 160 °C. Adequate amounts of Sn in this catalyst allows the presence of Brønsted and Lewis acid sites, while the incorporation of high amounts of Sn provides more Lewis acid sites than Brønsted acid sites. When Ti is added instead of Sn, the obtained HMF yield (≈16%) is slightly higher than that achieved with the commercial H-Al-Beta, which is Sn-free.

Following a similar strategy, Zhang et al. [75] use modified H-Beta zeolite with Fe, Sn and Zr metals in glucose, fructose and sucrose dehydration and they concluded that, when glucose is used as raw material, HMF is produced as an intermediate product. The results show that all modified zeolites provide a similar HMF yield, with Sn-Beta catalyst giving a yield of 36.4% at 170 °C using DMSO as a solvent. The presence of Sn leads to a higher pore width (similar to Fe-Beta), and higher number of Brønsted and Lewis sites.

Another type of zeolite, H-ZSM-5 and its Fe-ZSM-5 and Cu-ZSM-5 counterparts, have been investigated for glucose dehydration after determining the best operating conditions [84]. Among these zeolites, Cu-ZSM-5 shows the best glucose conversions and HMF yields, reaching values of around 85% and 65%, respectively, at 195 °C using a biphasic aqueous NaCl/MIBK system. It is important to mention that H-ZSM-5 shows similar activity but gives a slightly lower selectivity. In conclusion, the addition of metallic elements enhances the properties of zeolites in terms of acidity amount, allow a good compromise between Brønsted and Lewis acid sites, mesopore structure, micropore volume and dispersion of active species. This fact makes metal oxides important in isomerization processes, and the Al content of the zeolite catalysts is necessary to carry out the dehydration process. Moreover, modified zeolite catalysts show good stability, as it is confirmed by the reusability tests.

The improvement reported in activity/selectivity and textural, morphological and chemical properties when metal oxides are used as modifiers in the zeolites, means that the metal oxides are considered good candidates for the HMF-type furan compound production.

8.3.2 Metal Oxides

Although metal oxides have low and moderate selectivity towards HMF production [56], they can be modified to have high activity and selectivity. Zhang et al. [85] tested amorphous Cr_2O_3, SnO_2, SrO, Fe_2O_3, FeO, CdO and CoO metal oxides for glucose dehydration (Table 8.4). The main conclusions are that: (i) the increase of activation treatment of the catalyst has negative influence on the HMF yield (ii) the supported metal oxides on graphene oxide improve in some cases the HMF yield. In fact, amorphous Cr_2O_3, SnO_2 and SrO, activated at 60 °C, show HMF yields over 80% at 140 °C and for 4 h reaction time, while iron oxides have the lowest HMF yields, below 5%. However, the activity and selectivity of the iron oxides grows significantly when they are supported in graphene oxide, reaching for supported Fe_2O_3 an HMF yield of 85%. This improvement seems to be due to the dispersion of iron oxides on the support and the interactions formed between both oxides.

Table 8.4 Metal oxides used in the conversion of carbohydrates substrates into HMF with summary of reaction conditions and catalytic activity

Catalyst	Temp (°C)	System		Raw material	Conversion (%)	FF yield (%)	References
		Monophasic	Biphasic				
Fe_2O_3	140	W-[EMIM]Br	–	Glucose	n.a.	3	[85]
FeO						3	
SnO_2						81	
CdO						10	
Cr_2O_3						84	
SrO						90	
CoO						30	
S-ZrO_2	220	W	–	Cellulose	20.6	16.1[a]	[87]
Pt/S-ZrO_2					13.9	4.9 [a]	
CeO_2/S-ZrO_2					26.5	15.8[a]	
CeO_2 Nb_2O_3/S-ZrO_2					28.3	14.7[a]	
TS350	120	W	–	Glucose	19	7.2	[88]
TSA350					22	8.0	
S/SnO_2	190	W	–	Cellulose	n.a.	7.7	[86]
S/TiO_2						5.5	
S/ZrO_2						4.9	
S/Fe_2O_3						3.6	
S/Al_2O_3						6.9	
γ-Al_2O_3	140	W-DMSO	–	Glucose	95	18	[89]
Al_5O_5					95	41	
Al_6O_4					90	30	
Al_7O_3					88	30	
Al_4O_6					85	28	
Al_3O_7					90	25	
Nb_1W_7	120	W	–	Glucose	11.6	3.3	[90]
Nb_3W_5					32.2	10.3	
Nb_4W_4					36.1	18.8	
Nb_5W_3					29.1	10.7	
Nb_7W_1					30.4	12.4	

[a] → HMF selectivity; Temp → temperature; W → water; [EMIM]Br → 1-ethyl-3-methylimidazolium bromide; S-ZrO_2 → $ZrO_2(SO_4^{2-})$; S → SO_4^{2-}; n.a. → not available

Some of these metal oxides, such as ZrO_2, TiO_2 and Al_2O_3, were modified to transform them into superacid catalyst [86]. The reported HMF yields are not so high (around 7 and 11%) when cellulose is used as substrate at 190 °C and low reaction time (2–4 h). However, the acid strength of the catalysts seems to be an important parameter to avoid side reactions of cellulose. Shirai et al. [87] also used ZrO_2 superacid as catalyst and a metal (Pt and Ru) or a metal oxide (CeO_2 and Nb_2O_3-CeO_2) as support. The cellulose conversion (89.1%) and HMF selectivity (16.1%)

reported for superacid ZrO_2, measured at 220 °C and 5 min, are higher than previous reports [86]. However, a significant improvement on the activity and the selectivity is not observed when the ZrO_2 superacid catalyst is used as support. Based on these results, it would seem that similar textural properties of these catalysts lead to similar activities and selectivities. Calcined SnO_2 superacid catalyst, bare and doped with Al_2O_3, have been used for glucose dehydration, but HMF yields are very poor due to leaching of the sulfate groups [88].

Apart from the mentioned catalysts, nano-sized and ordered mesoporous metal oxide combinations seem to be a good catalytic system to carry out dehydration reaction of glucose. Among the catalytic systems, Al_2O_3-B_2O_3 [89] and Nb_2O_3-WO_3 should be mentioned [90]. The Al_2O_3-B_2O_3 catalysts are prepared with different Al/B molar ratios and they are tested using glucose as a raw material. The prepared combinations show higher glucose conversions and higher HMF yields than the pure metal oxides, among others, reaching for Al_2O_3-B_2O_3 catalyst with Al/B ratio of 5:5 a conversion close to 100% and HMF yield of 41.4%, at 140 °C, 4 h and with DMSO being used as solvent. The mentioned catalyst has favorable textural and morphological characteristics in terms of high total acidity, high acid strength and surface area. The catalyst is probably mesoporous and it contains Lewis acid sites. Moreover, it has an intermediate pore volume. Regarding Nb_2O_3-WO_3 mixed oxides, they have been prepared with different molar ratios. Under operation conditions of 120 °C, 2 h of reaction time in an aqueous medium, the Nb_4W_4 catalyst provides a glucose conversion of 36.1% and HMF yield of 18.8%. Moreover, this catalyst has high surface area and high amounts of Lewis and Brønsted acid sites, which are necessary to obtain good yields of fructose and HMF, respectively, from glucose substrate. Moreover, the mentioned catalyst shows good stability. When the performance of both mixed metal oxides are compared, it is observed that Al_2O_3-B_2O_3 catalysts have better activity and selectivity than Nb_2O_3-WO_3 due to the milder operating conditions that these last catalysts were tested.

Other investigations focus on the use of fructose as a raw material to produce HMF with a solid acid, such as silica-induced heteropolyacids, and aggregated nanosheets and layered metal oxides [91, 92]. The catalysts are able to convert almost all the fructose and achieve HMF yields above 80%, in the best of cases. It is important to mention that fructose is more reactive than glucose.

Taking into account the data presented on FF and HMF catalytic synthesis, it is important to remark that the conversion of raw materials (xylan, xylose and even glucose, in the case of FF, and glucose and fructose, in the case of HMF) and yield or selectivity toward FF and HMF are closely related to the textural, superficial and acid characteristics of the catalyst. Moreover, it can be concluded that there is a suitable compromise among these characteristics, since on the one hand, the size of pores and the Lewis acid sites allow substrate access, while, on the other hand, the microporous and Brønsted acid sites are responsible for FF and HMF formation.

8.4 DMF Production

DMF can be produced from hydrogenolysis, which is also called selective hydrogenation of HMF [19] using non-noble and noble metal catalysts [1, 19]. This reaction involves different pathways [1, 6, 19] in which dehydration and hydrogenation reactions are predominant. Figure 8.6 shows a simplified reaction scheme of DMF production from biomass. The main route of DMF production is by hydrogenolysis of 5-hydroxymethylfurfural (HMF) to 5-methylfurfuryl alcohol 5-(MFOL) via 5-methylfurfural or via 2,3-bis(hydroxymethyl)furan (BHF). After the 5-MFOL by different hydrogenolysis reaction it can be produced the DMF, but at this point it could be open the furan rings obtained different ring operation products or by over hydrogenation of the DMF it can be obtained the 2,5 dimethyltetrahydrofuran (2,5 DMTHF). Although at specific operating conditions could take place another route of 2-methylfuran (2-MF) via furfuryl alcohol (FOL), in this route it is not produce DMF.

Usually the production of HMF is carried out in batch reactors, and after its purification the obtained HMF is fed into another reactor (Fig. 8.7). DMF production from HMF is typically carried out by catalytic hydrogenolysis. In the HMF hydrogenolysis some carbon-oxygen bonds are cleaved and the hydrogen is introduced in the molecule by hydrogenation- dehydration mechanisms (Fig. 8.6) that occur in the 2,5-dihydroxymethylfuran synthesis process. In the hydrogenolysis processes from biomass, bifunctional routes have been suggested by various authors being the dehydration step, that is catalyzed by acid or basic sites [94–96], whereas the

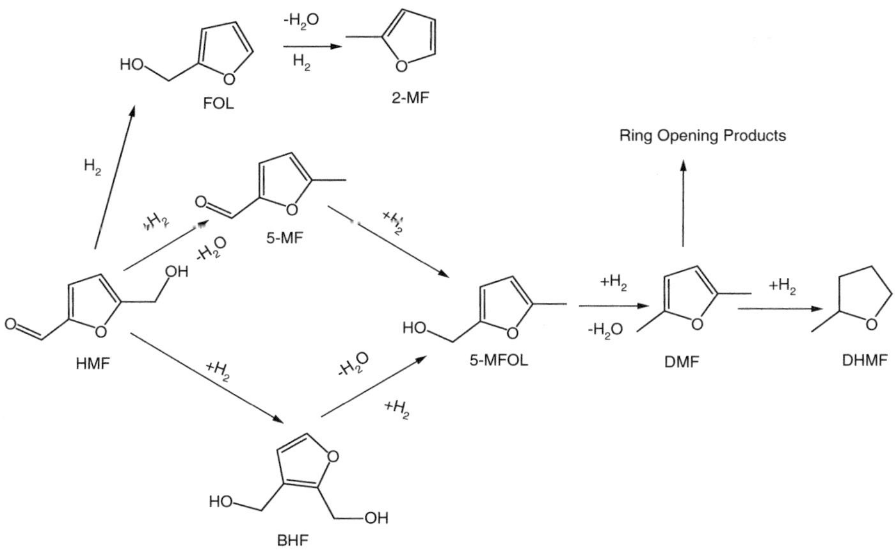

Fig. 8.6 Production route of 2,5 dimethylfuran and 2-methylfuran via 5-hydroxymethylfurfural from biomass (Reprinted with permission from [93], Copyright © 2017 Elsevier)

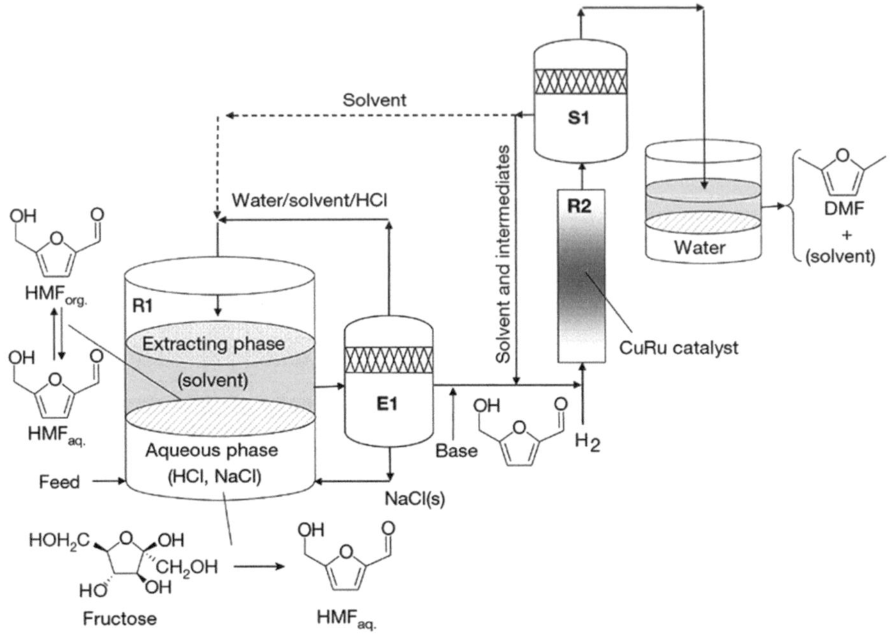

Fig. 8.7 Schematic of the process for fructose conversion to 2,5 dimethylfuran. (Reprinted from [32], Copyright © 2007 Nature)

hydrogenation process is catalyzed by the metal [6, 31, 97]. Some metals, such as Cu allow both steps to occur, namely dehydration and hydrogenation [98–101]. The presence of strong acid sites, Brønsted or Lewis, favor the C–C bond cleavage of the HMF molecule; therefore the presence of acid supports could cause the loss of one of the radicals present in the HMF molecule or degradation of the DMF, which is the reason why neutral-basic supported and weak acid supports are employed in HMF hydrogenolysis to DMF. This type of neutral-basic supports seems to avoid C–C bond cleavage of tertiary carbon of HMF raw material [17]. One of the most used supports in HMF hydrogenolysis is carbon [32, 102, 103], while some other authors report the use of zirconia [17] or metal oxides such as Co_3O_4 [2].

For the hydrogenation of the HMF it is necessary to use hydrogenating metal catalysts. There are two types of metal-based catalysts, those containing noble metals such as Pd, Pt, Ru, Rh and Au [1, 19, 45, 97, 104], and those containing transition metals, such as Cu, Co, Ni and Fe [97, 105, 106]. Some authors have studied the use of both type of metals [32] to obtain a high selectivity (over 70%) in the production of 2,5 DMF using Cu-Ru supported on carbon via a vapor-phase hydrogenolysis of HMF [32] using 1-butanol as solvent at 220 °C and 6.8 bar of H_2. Chidambaram et al. [85] reported a 32% of selectivity to DMF and a 47% of HMF conversion using the Pd/C catalyst with ionic liquids. In this last case, the operating conditions were 120 °C and a pressure of 62 bar of hydrogen; the high H_2 pressure was needed to maintain the liquid phase and to improve the H_2 diffusion across the

Table 8.5 DMF yield in a batch reactor

Catalyst	Solvent	Temp (°C)	$P(H_2)$ (bar)	DMF yield (%)	References
Cu-Ru/C	Butanol	220	6.8	70	[32]
Pd/C	Ionic liquid	120	62	32	[102]
Ru/Co_3O_4	THF	130	7	93.4	[2]
Ni/SBA-15	W	200	30	98.7	[105]
Pd/C	THF	70	H_2 donors	51	[107]

Temp → temperature; THF → tetrahydrofuran; W → water

liquid reaction media. Zu et al. [2] reported a very high yield to DMF (93.4%) using tetrahydrofuran (THF) as a solvent at 130 °C, and 7 bar of hydrogen partial pressure using HMF. In this case, the Ru was the responsible for the hydrogenation and the Co_3O_4 was the responsible for the HMF adsorption and breaking the C–O bonds. R. Goyal et al. [105] obtained total conversion of HMF and selectivity to DMF over 98.7% employing a Ni catalyst supported on mesoporous nitrogen rich carbon material with liquid water as solvent at 200 °C and 30 bar of H_2 pressure using HMF such as feed. The reason of the reported high selectivity to DMF is the presence of Ni nanoparticles (below 5 nm). This high Ni dispersion was due to the high interaction among the support and the Ni, because when the nickel particle size was higher than 5 nm the selectivity of DMF decreased considerably [105]. All these experiments were developed in a Batch Reactor using HMF such as feed.

As previously mentioned, when biomass is using such as feed, it is necessary two reactors to obtain DMF, in the first reactor HMF is produced and after its purification the HMF is feed in another reactor (Fig. 8.7). Nevertheless Thananatthanachon and Rauchfuss [107] provide a new solution using a one pot reactor in the process to obtain DMF from Biomass. They use fructose as raw material for HMF production, and in this first step, fructose and formic acid are fed into the pot which is stirred and heated at 150 °C for 2 h. After that the resulting products are cooled down to room temperature and then THF, H_2SO_4, and Pd/C catalyst are added. Then, the pot is heated up at 70 °C and allowed to react for 15 h. Finally DMF is extracted to obtain a DMF yield of 51%. However, the presence of formic acid and H_2SO_4 makes the process highly corrosive and the method is not environmentally-friendly. This is summarized in the Table 8.5.

Some solvents favor the no formation of secondary reactions, allow high H_2 diffusion, and facilitate separation of DMF from the reaction media in the different reaction systems in batch or in fix bed reactors. THF and butanol have good H_2 solubility and thus high H_2 pressures are unnecessary. On the contrary ionic liquids generally have poor H_2 solubility and the use of this kind of solvents require large H_2 pressures (≥50 bar) [102]. Moreover, the separation of DMF from the ionic liquids is not easy; therefore it seems that the ionic liquids are not very suitable for hydrogenolysis reaction. In case of 1-butanol, 2-butanol, and hexanol are favorable for DMF biomass-to-liquid production due to their alcoholic properties [108]. Goyal et al. [105] showed that the use of water can also be very helpful to reach high DMF yield (98.7%).

Hydrogenation in liquid reaction systems required of high H_2 pressures to improve the H_2 solubility and its availability on the catalytic surface. Although the presence of molecular hydrogen has many advantages like high availability and an easy activation of the metal surface, there are also some problems related to the safety, economy and sustainability that open a new route in hydrogenation research: the use of liquid-phase hydrogen donors [109]. Some compounds as formic acid or different alcohols (ethanol, formic acid and propanol) can be employed as hydrogen donor compounds. One example is the glycerol hydrogenolysis to obtain 1,2 propanediol [109]. Catalytic hydrogen transfer can take place using environmentally friendly and sustainable compounds. Formic can be produced from lignocellulosic biomass or from electrochemical CO_2 reduction; methanol can be produced by biomass pyrolysis. Zu et al. showed that formic acid as hydrogen donor could lead to high HMF conversions (99%) with high selectivities (90%) using a Ni-Co/C catalyst and THF as the solvent [2].

Almost all the works that present high selectivity of DMF are developed in batch reactors, and sometimes the contact time or the reaction time is quite high due to the low volume of the reactors [103, 105]. For a large production of DMF is necessary the use continuous systems but only few authors has reported about hydrogenolysis of HMF over continuous systems [17, 104]. In these cases the DMF selectivity was lower than in batch reactors, for example, in the case of the Cu/Zr catalysts it was reached a maximum selectivity of 60.6% after 2 h on stream [17]. Therefore, still plenty of research is required to improve and develop these continuous reactor systems. As previously presented, the DMF extraction from biomass occurs in two steps, in the first one HMF is obtained by Biomass hydrolysis, and after that, in a new reactor, this HMF is transformed into DMF.

Finally, it must be concluded that many reports show high DMF selectivities obtained using synthetic HMF rather than HMF resulting from biomass. The use of real raw biomass presents new challenges that affect to the DMF selectivity. For example, the presence of impurities in the biomass-derived HMF would low and deactivate many types of catalysts [17].

8.5 2-MF and MTHF Production

As shown in Figs. 8.6 and 8.9, there are two different routes to obtain 2-MF with the first one being from furfural and the second one being from HMF. In both cases hydrogenation of the FF or the HMF to furfuryl alcohol (FOL) is required followed by a hydrogenolysis step to obtained 2-MF. If the hydrogenation process continues after forming 2-MF, the furanic biofuel methyltetrahydrofuran (MTHF) can be obtained.

The most studied route of 2-MF production is the FF transformation via FOL into 2-MF. FF normally is obtained under batch or continuous processes. It can be simultaneously extracted by steam-stripping being its concentration, being around 3% in this stream. The extracted furfural can be separated using consecutive purification stages to achieve a high furfural pure stream. After obtaining FF from biomass, the second step involves 2-MF production (Fig. 8.8). The fresh FF is mixed

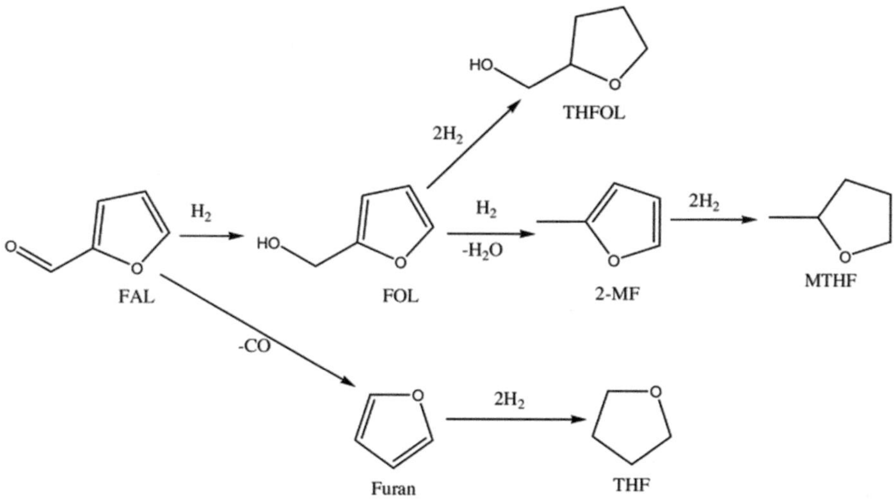

Fig. 8.8 Production route of 2-MF and MTHF via FF from biomass (Reprinted with permission from [93], Copyright © 2017 Elsevier)

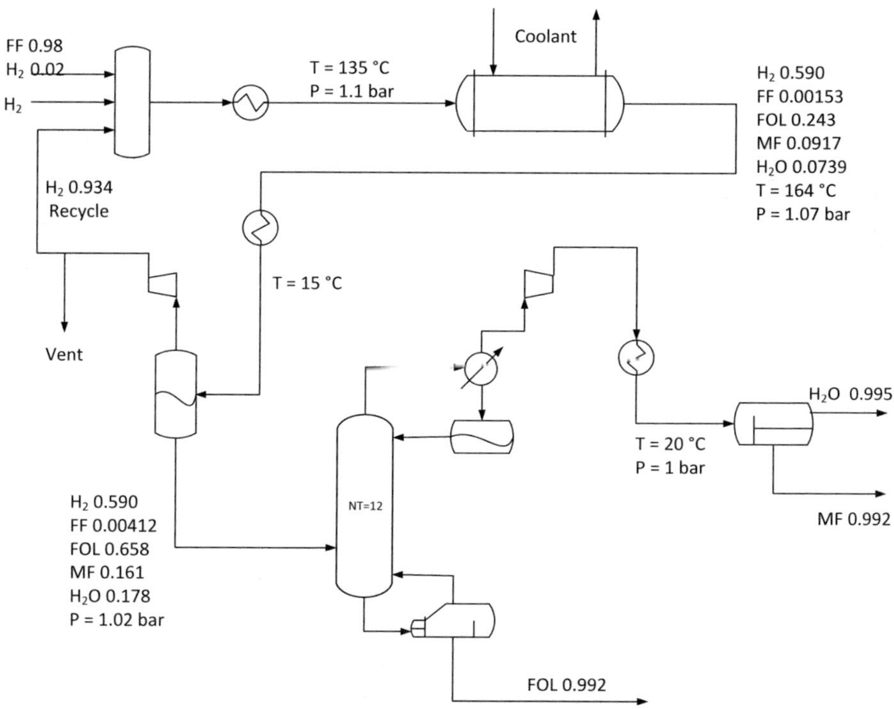

Fig. 8.9 Process flowsheet for obtaining 2-methylfuran (MF) from furfural (FF) via furfuryl alcohol (FOL). (Adapted with permission from [110], Copyright © 2016 Elsevier)

with a fresh and recycled H_2, before being heated up and being introduced to a tubular reactor. The flow rates are adjusted to have a reaction time of around 2.1 s. In this process, there is a clear relationship between reaction temperature and 2-methylfuran yield, with higher yields of 2-methylfuran being obtained when the reactor temperature is increased [110]. After reaction, a flash separator is in the system, to recover the no reacted hydrogen ant the rest of product. The no reacted hydrogen is mix with the fresh hydrogen and this mixture is introduced in the previous decrypted reactor. The rest of products are introduced in a distillation column, where the 2-MF is purified. A downstream separation processes by distillation leads to 2-MF purities higher than 99%.

As previously mentioned, it is necessary to cleave a C–O bond for 2-MF production. For this purpose metal catalysts seem to be promising candidates for the selective removal of oxygen from the carbonyl group of FF or HMF under relatively mild operating conditions [111]. In this process it is also necessary to avoid the C–C bond cleavage, therefore it is not usual to use strong acid supports that favored these scissions [111]. Metals such as Cu are able to promote C–O bond cleavage without C–C bond scission [99–101, 111, 112], while other typical hydrogenation metals as Pt, Ni or Pd have a high hydrogenation activity at relative low temperatures [the noble metals need lower temperatures] but they are not be able to cleave C–O bonds, with FOL being the major product [113–115]. In contrast, at high temperatures more than 200 °C, nickel favor the decarbonylation reaction to form furan as the main product [116]. Commonly used catalysts for 2-MF production from FF are the Cu catalysts; however, these catalysts present problems of stability [99, 111, 112]. The deactivation of Cu-based catalysts is associated with the sintering and leaching of copper species [93]. The addition of another metal such as Co, Ni, or, Fe improves the activity, selectivity and the stability of copper-based catalysts in liquid-phase catalytic reactions [117].

The hydrogenation of FF can be carried out in liquid phase or in gas phase. Usually when the process is developed in a batch reactor the reaction is carried out in the liquid phase while when the process is carried out in gas phase, tubular continuous reactors are used. In the case of the FF hydrogenation in the liquid phase, necessary hydrogen can be supplied by molecular H_2 or by hydrogen donors. Srivastava et al. [93] have achieved a high selectivity to 2-MF (81.6%) and a total FF conversion using 2-propanol as hydrogen donor in presence of Cu-Ni/Al_2O_3 catalysts. This system was also tested for HMF and the main product was DMF with the selectivity of 2-MF being lower than 10% [93]. Srivastava et al. tested different monometallic catalysts as Cu/Al_2O_3 and Ni/Al_2O_3. In case of the Cu monometallic catalyst requires higher temperatures than Ni or NiCu/Al_2O_3 catalysts to get a total FF conversion. For this case, the Ni favored the FF hydrogenation but at the same time it also promoted C–C bond scission. Thus, the 2-MF yield was very low with this catalyst while the major production was towards FOL and a little of THFOL, and furan. However, the use of bimetallic catalysts as Ni-Cu/Al_2O_3 improves the FF conversion and the 2-MF yield, due to the high hydrogenation capacity of the Ni and the C–O bond cleavage capacity of Cu. It seems that the presence of the Ni in the catalyst increases the Cu dispersion and reduces the Cu particle size which is an

Table 8.6 Selectivity of 2-methylfuran (2-MF) for different catalysts, operating conditions and reaction phases from furfural (FF)

Catalysts	Reaction phase	Temp (°C)	H_2 donor	2-MF Selectivity (%)	References
Cu-Ni/Al	Liquid phase	220	2-propanol	81.6	[117]
Ni-Co/C	Liquid phase	200	Formic acid	79.2	[122]
Cu-Zn/Al	Gas phase	350		93.5	[122]
Cu/SiO_2	Gas phase	220		89.5	[121]
Cu-Zn/Al-Ca-Na	Gas phase	250		87	[120]

Temp → temperature

important factor for C–O hydrogenolysis [93]. In the 2-MF production flowsheet (Fig. 8.8), higher temperatures favor 2-MF yield. At 150 °C, the 2-MF selectivity is lower than 10% while the FOL selectivity is over 90% and at 200 °C the 2-MF selectivity is close to 80% and the FOL selectivity becomes less than 10%. Xiong et al. confirmed Srivastava's results; they also used $NiCu/Al_2O_3$ and Cu/Al_2O_3 catalysts and observed that higher conversions of furfural and higher 2-MF yields achieved with the NiCu bimetallic surfaces could be explained due to the increased degree of interaction of the carbonyl group of furfural compared with the degree of interaction with monometallic Cu catalysts [28].

Yang et al. have employed hydrogen donors as formic acid to obtain a total FF conversion and a 2-MF selectivity of 79.2% with Ni-Co/C catalyst and using THF as solvent [31] at 200 °C to increase the DMF yield. Those authors tested Ni-Co/C catalyst employing H_2 molecular, but obtained lower 2-MF selectivity values.

As it was previously presented in the DMF production, when DMF production is carried out with liquid phase reactions, it is very common the use a co-solvent as THF to prevent side reactions like resinification of FF or FOL [118].

Many researchers have reported interesting results on the vapor phase furfural hydrogenation [119–121] (Table 8.6). Yang et al. showed a total FF conversion and 2-MF selectivity of 93.5% at 350 °C using a Cu-Zn/Al catalyst. For this case, calcination temperature effect was analyzed and they observed that at higher calcination temperatures than 350 °C it is formed a solid solution that decreased the selectivity to 2-MF. Dong et al. [121] obtained total FF conversion, and a selectivity of 2-MF of 89.5% using Cu/SiO_2 catalysts at 220 °C and atmospheric pressure. Those authors tested other Cu catalysts as Cu/Al_2O_3 and Cu/ZnO that have lower selectivities of 2-MF than the previous one. The performance of the Cu/SiO_2 is attributed to the interaction of SiO_2 with copper species that improve the uniformity copper particle dispersion with high loading of Cu^0 and Cu^+, which might be important for catalytic reaction of FF hydrogenation. This higher Cu^+/Cu^0 ratio might function as electrophilic or Lewis acid sites to polarize the saturated $–CH_2$-OH bond via the electron lone pair on oxygen [98]. Dong et al. [99] explain that probably the adequate weak acid sites on the Cu/SiO_2 catalyst could be helpful for a C–O bond cleavage, due to the strong oxyphilic nature for the saturated C–O band that might promote C–O bond hydrogenolysis. Zheng et al. show also significant results for conversion of FF to 2-MF with gas phase reaction and obtain a 2-MF yield of 87% with a Cu/Zn/Al/Ca/Na at 250 °C at atmospheric pressure.

8.6 Conclusions and Future Outlook

Furfural (FF) and 5-hydroxymethylfurfural (HMF) are interesting platform molecules because they are good candidates to produce a wide range of chemical products and biofuels. FF and HMF are synthesized from aqueous solutions of sugars, such as pentoses and hexoses, carrying out isomerization reactions followed by a triple dehydration process.

Sequential reactions can take place in unique single-step using heterogeneous bifunctional catalysis. Commonly studied for catalysts FF and HMF production are modified zeolites and metal oxides as they show Lewis and Bronsted acid sites and mesoporous structure. Regarding Lewis and Brønsted acid sites, the ratio between these types of acid sites, even density, should be suitable. Lewis acid sites are necessary to carry out isomerization reactions, while Brønsted acid sites promote the dehydration reaction. Mesoporous structures in the catalyst reduce raw material diffusional limitations. The catalyst designed to use in FF and HMF production must have a good compromise between type/density of acid sites and textural properties. Although reaction conditions have not been discussed in detail, it is important to mention that mainly temperature, solvents, reaction time and catalyst amount must be suitable chosen to reach high sugar conversion and yields toward the desired products.

In the hydrogenolysis of the HMF to obtain 2,5-dimethylfuran (DMF) or 2-methylfuran (2-MF), and in the FF hydrogenolysis, the employed supports are weak acid supports or neutral supports. The acidity of the supports is related with higher C–C bond scissions which decreases the DMF or 2-MF yield. In the case of the hydrogenation in both hydrogenolysis processes, different noble metals as Ru, and Pd are mainly employed, while in the case of non-noble metals Cu and Ni are mainly employed. The Cu metal promotes the C–O bond cleavage. However, the metal dispersion is also a very important parameter in the biofuels production; good dispersion of the metal in the catalyst is necessary for high biofuels yields. The HMF and FF hydrogenolysis can take place in batch liquid phase reactors or in tubular continuous reactors in gas phase. Usually, the liquid phase reaction media employs as co-solvent THF, butanol, or some other solvent to avoid the formation of furanic resins. In the liquid phase, the hydrogen source for the hydrogenolysis step can be either, molecular H_2 or hydrogen donors such as formic acid or 2-propanol. In the case of the hydrogenolysis in the gas phase it is not usual to employ a co-solvent. The use of HMF derived from actual biomass will be challenging in terms of yields for DMF or 2-MF, due to the presence of different impurities. Operating conditions also affect 2-MF and DMF yields. Temperatures higher than 200 °C are required to achieve high 2-MF yields.

The use of bifunctional catalysts is essential to obtain these biofuels from biomass due to the necessity of C–O bond cleavage and the hydrogenation capacity. Bifunctional catalysts are the keys to the future of practical biorefineries.

Acknowledgments This work was supported by funds from University of the Basque Country (UPV/EHU), Spanish Ministry of Economy and Innovation (Projects: CTQ2015-64226-C3-2-R), and Basque Country Government (Project: IT993-16).

References

1. Nishimura S, Ikeda N, Ebitani K. Selective hydrogenation of biomass-derived 5-hydroxymethylfurfural (HMF) to 2,5-dimethylfuran (DMF) under atmospheric hydrogen pressure over carbon supported PdAu bimetallic catalyst. Catal Today. 2014;232:89–98. https://doi.org/10.1016/j.cattod.2013.10.012.
2. Zu Y, Yang P, Wang J, et al. Efficient production of the liquid fuel 2,5-dimethylfuran from 5-hydroxymethylfurfural over Ru/Co3O4 catalyst. Appl Catal B Environ. 2014;146:244–8. https://doi.org/10.1016/j.apcatb.2013.04.026.
3. International Energy Agency (IEA). Key world energy statistics. International Energy Agency, Paris; 2016.
4. International Energy Agency (IEA) (2016) World Energy Outlook 2016. doi:http://www.iea.org/publications/freepublications/publication/WEB_WorldEnergyOutlook2015ExecutiveSummaryEnglishFinal.pdf
5. International Energy Agency (2016) Key Oil Trends.
6. International Energy Agency (IEA) (2016) Key world energy trends – excerpt from: world energy balances.
7. Zhang M, Tong X, Ma R, Li Y. Catalytic transformation of carbohydrates into 5-hydroxymethyl furfural over tin phosphate in a water-containing system. Catal Today. 2016;264:131–5. https://doi.org/10.1016/j.cattod.2015.06.031.
8. Yang P, Cui Q, Zu Y, et al. Catalytic production of 2,5-dimethylfuran from 5-hydroxymethylfurfural over Ni/Co3O4 catalyst. Catal Commun. 2015;66:55–9. https://doi.org/10.1016/j.catcom.2015.02.014.
9. De S, Saha B, Luque R. Hydrodeoxygenation processes: advances on catalytic transformations of biomass-derived platform chemicals into hydrocarbon fuels. Bioresour Technol. 2015;178:108–18. https://doi.org/10.1016/j.biortech.2014.09.065.
10. (2012) Green Chemistry. Fraunhofer magazine 2.12.
11. Qian Y, Zhu L, Wang Y, Lu X. Recent progress in the development of biofuel 2,5-dimethylfuran. Renew Sust Energ Rev. 2015;41:633–46. https://doi.org/10.1016/j.rser.2014.08.085.
12. International Energy Agency (IEA) (2015) CO_2 Emissions from fuel combustion. Highlights doi: https://doi.org/10.1787/co2-table-2011-1-en.
13. Centi G (Gabriele), Santen RA van (Rutger A., Wiley InterScience (Online service). Catalysis for renewables: from feedstock to energy production; 2008. doi:https://doi.org/10.1002/9783527621118
14. Kamm B, Gruber PR, Kamm M. Biorefineries – industrial processes and products: status quo and future directions. Weinheim: Wiley-VCH; 2010
15. Corma A, Iborra S, Velty A. Chemical routes for the transformation of biomass into chemicals. Chem Rev. 2007;107:2411–502. https://doi.org/10.1021/cr050989d.
16. Rosatella AA, Simeonov SP, Frade RFM, Afonso CAM. 5-Hydroxymethylfurfural (HMF) as a building block platform: Biological properties, synthesis and synthetic applications. Green Chem. 2011;13:754–93. https://doi.org/10.1039/C0GC00401D.

17. Iriondo A, Mendiguren A, Güemez MB, et al. 2,5-DMF production through hydrogenation of real and synthetic 5-HMF over transition metal catalysts supported on carriers with different nature. Catal Today. 2017;279:286–95. https://doi.org/10.1016/j.cattod.2016.02.019.
18. Menon V, Rao M. Trends in bioconversion of lignocellulose: biofuels, platform chemicals & biorefinery concept. Prog Energy Combust Sci. 2012;38:522–50. https://doi.org/10.1016/j.pecs.2012.02.002.
19. Climent MJ, Corma A, Iborra S. Conversion of biomass platform molecules into fuel additives and liquid hydrocarbon fuels. Green Chem. 2014;16:516–47. https://doi.org/10.1039/C3GC41492B.
20. Jungmeier G, van Ree R, Jorgensen H, et al. The biorefinery fact sheet; 2014.
21. IEA Bioenergy. IEA Bioenergy Task 42 Biorefinery; 2010.
22. Cherubini F, Jungmeier G, Wellisch M, et al. Toward a common classification approach for biorefinery systems. Biofuels Bioprod Biorefin. 2009;3:534–46. https://doi.org/10.1002/bbb.172.
23. Choi S, Song CW, Shin JH, Lee SY. Biorefineries for the production of top building block chemicals and their derivatives. Metab Eng. 2015;28:223–39. https://doi.org/10.1016/j.ymben.2014.12.007.
24. Kamm B, Kamm M. Principles of biorefineries. Appl Microbiol Biotechnol. 2004;64:137–45. https://doi.org/10.1007/s00253-003-1537-7.
25. Jiménez-Gómez CP, Cecilia JA, Márquez-Rodríguez I, et al. Gas-phase hydrogenation of furfural over Cu/CeO2 catalysts. Catal Today. 2017;279:327–38. https://doi.org/10.1016/j.cattod.2016.02.014.
26. Schlaf M. Homogeneous catalysts for the hydrodeoxygenation of biomass-derived carbohydrate feedstocks. In: Schlaf M, Zhang ZC, editors. Reaction pathways mechanisms thermocatalytic biomass conversion II homogeneously catalyzed transformations acrylics from biomass, theoretical aspectd lignin valorization pyrolysis pathways. Singapore: Springer; 2016. p. 13–38.
27. Alam MI. Chapter 4 – catalysis for the production of sustainable chemicals and fuels from biomass. In: Sustain Catal Process 2015. p 99–123.
28. Xiong K, Wan W, Chen JG. Reaction pathways of furfural, furfuryl alcohol and 2-methylfuran on Cu(111) and NiCu bimetallic surfaces. Surf Sci. 2016;652:91–7. https://doi.org/10.1016/j.susc.2016.02.011.
29. Aresta M, Angela D, Franck D. Biorefinery, From Biomass to Chemicals and Fuels. 2012. doi:https://doi.org/10.1515/9783110260281.
30. Agirrezabal-Telleria I, Larreategui A, Requies J, et al. Furfural production from xylose using sulfonic ion-exchange resins (Amberlyst) and simultaneous stripping with nitrogen. Bioresour Technol. 2011;102:7478–85. https://doi.org/10.1016/j.biortech.2011.05.015.
31. Zhang Y, Zhang J, Su D. 5-Hydroxymethylfurfural: Akey intermediate for efficient biomass conversion. J Energy Chem. 2015;24:548–51. https://doi.org/10.1016/j.jechem.2015.09.005.
32. Roman-Leshkov Y, Barrett CJ, Liu ZY, Dumesic JA. Production of dimethylfuran for liquid fuels from biomass-derived carbohydrates. Nature. 2007;447:982–5.
33. Zhao J, Zhou C, He C, et al. Efficient dehydration of fructose to 5-hydroxymethylfurfural over sulfonated carbon sphere solid acid catalysts. Catal Today. 2016;264:123–30. https://doi.org/10.1016/j.cattod.2015.07.005.
34. Malinowski A, Wardzińska D. Catalytic conversion of furfural towards fuel biocomponents. Chemik. 2012;66:987–90.
35. Bond JQ, Upadhye AA, Olcay H, et al. Production of renewable jet fuel range alkanes and commodity chemicals from integrated catalytic processing of biomass. Energy Environ Sci. 2014;7:1500–23. https://doi.org/10.1039/C3EE43846E.
36. Chheda JN, Roman-Leshkov Y, Dumesic JA. Production of 5-hydroxymethylfurfural and furfural by dehydration of biomass-derived mono- and poly-saccharides. Green Chem. 2007;9:342–50. https://doi.org/10.1039/B611568C.
37. Kunkes EL, Simonetti DA, West RM, et al. Catalytic conversion of biomass to monofunctional hydrocarbons and targeted liquid-fuel classes. Science. 2008;322(80):417 LP–421.

38. Bond JQ, Alonso DM, Wang D, et al. Integrated catalytic conversion of γ-valerolactone to liquid alkenes for transportation fuels. Science. 2010;327(80):1110 LP–1114.
39. Xing R, Subrahmanyam AV, Olcay H, et al. Production of jet and diesel fuel range alkanes from waste hemicellulose-derived aqueous solutions. Green Chem. 2010;12:1933–46. https://doi.org/10.1039/C0GC00263A.
40. Olcay H, Subrahmanyam AV, Xing R, et al. Production of renewable petroleum refinery diesel and jet fuel feedstocks from hemicellulose sugar streams. Energy Environ Sci. 2013;6:205–16. https://doi.org/10.1039/C2EE23316A.
41. Yang J, Li N, Li G, et al. Synthesis of renewable high-density fuels using cyclopentanone derived from lignocellulose. Chem Commun. 2014;50:2572–4. https://doi.org/10.1039/C3CC46588H.
42. Sheng X, Li N, Li G, et al. Synthesis of high density aviation fuel with cyclopentanol derived from lignocellulose. Sci Rep. 2015;5:9565.
43. Yang J, Li S, Li N, et al. Synthesis of Jet-Fuel Range Cycloalkanes from the Mixtures of Cyclopentanone and Butanal. Ind Eng Chem Res. 2015;54:11825–37. https://doi.org/10.1021/acs.iecr.5b03379.
44. Xiao H, Zeng P, Li Z, et al. Combustion performance and emissions of 2-methylfuran diesel blends in a diesel engine. Fuel. 2016;175:157–63. https://doi.org/10.1016/j.fuel.2016.02.006.
45. Shi J, Wang Y, Yu X, et al. Production of 2,5-dimethylfuran from 5-hydroxymethylfurfural over reduced graphene oxides supported Pt catalyst under mild conditions. Fuel. 2016;163:74–9. https://doi.org/10.1016/j.fuel.2015.09.047.
46. Wang C, Xu H, Daniel R, et al. Combustion characteristics and emissions of 2-methylfuran compared to 2,5-dimethylfuran, gasoline and ethanol in a DISI engine. Fuel. 2013;103:200–11. https://doi.org/10.1016/j.fuel.2012.05.043.
47. Xu N, Wu Y, Tang C, et al. Experimental study of 2,5-dimethylfuran and 2-methylfuran in a rapid compression machine: Comparison of the ignition delay times and reactivity at low to intermediate temperature. Combust Flame. 2016;168:216–27. https://doi.org/10.1016/j.combustflame.2016.03.016.
48. Jężak S, Dzida M, Zorębski M. High pressure physicochemical properties of 2-methylfuran and 2,5-dimethylfuran – second generation biofuels. Fuel. 2016;184:334–43. https://doi.org/10.1016/j.fuel.2016.07.025.
49. Conturso M, Sirignano M, Anna AD. Effect of 2,5-dimethylfuran doping on particle size distributions measured in premixed ethylene/air flames. Proc Combust Inst. 2016; In press:1–8. doi:https://doi.org/10.1016/j.proci.2016.06.048.
50. Wei H, Feng D, Shu G, et al. Experimental investigation on the combustion and emissions characteristics of 2-methylfuran gasoline blend fuel in spark-ignition engine. Appl Energy. 2014;132:317–24. https://doi.org/10.1016/j.apenergy.2014.07.009.
51. Agirrezabal-Telleria I, Gandarias I, Arias PL. Heterogeneous acid-catalysts for the production of furan-derived compounds (furfural and hydroxymethylfurfural) from renewable carbohydrates: a review. Catal Today. 2014;234:42–58. https://doi.org/10.1016/j.cattod.2013.11.027.
52. Zhang L, Xi G, Yu K, et al. Furfural production from biomass–derived carbohydrates and lignocellulosic residues via heterogeneous acid catalysts. Ind Crop Prod. 2017;98:68–75. https://doi.org/10.1016/j.indcrop.2017.01.014.
53. Yong TL-K, Mohamad N, Yusof NNM. Furfural production from oil palm biomass using a biomass-derived supercritical ethanol solvent and formic acid catalyst. Procedia Eng. 2016;148:392–400. https://doi.org/10.1016/j.proeng.2016.06.495.
54. Zhang L, Yu H, Wang P, et al. Conversion of xylan, d-xylose and lignocellulosic biomass into furfural using AlCl3 as catalyst in ionic liquid. Bioresour Technol. 2013;130:110–6. https://doi.org/10.1016/j.biortech.2012.12.018.
55. Zhang L, Xi G, Zhang J, et al. Efficient catalytic system for the direct transformation of lignocellulosic biomass to furfural and 5-hydroxymethylfurfural. Bioresour Technol. 2017;224:656–61. https://doi.org/10.1016/j.biortech.2016.11.097.

56. Li H, Fang Z, Smith RL, Yang S. Efficient valorization of biomass to biofuels with bifunctional solid catalytic materials. Prog Energy Combust Sci. 2016;55:98–194. https://doi.org/10.1016/j.pecs.2016.04.004.
57. Chen D, Liang F, Feng D, et al. Sustainable utilization of lignocellulose: preparation of furan derivatives from carbohydrate biomass by bifunctional lignosulfonate-based catalysts. Catal Commun. 2016. https://doi.org/10.1016/j.catcom.2016.06.012.
58. Shimanskaya M, Lukevits É. Catalytic reactions of furan compounds (review). Chem Heterocycl Compd. 1993;29:1000–11. https://doi.org/10.1007/BF00534382.
59. Filiciotto L, Balu AM, Van der Waal JC, Luque R. Catalytic insights into the production of biomass-derived side products methyl levulinate, furfural and humins. Catal Today. 2017. https://doi.org/10.1016/j.cattod.2017.03.008.
60. IKM Y, Tsang DCW. Conversion of biomass to hydroxymethylfurfural: a review of catalytic systems and underlying mechanisms. Bioresour Technol. 2017;238:716–32. https://doi.org/10.1016/j.biortech.2017.04.026.
61. Peleteiro S, Rivas S, Alonso JL, et al. Furfural production using ionic liquids: a review. Bioresour Technol. 2016;202:181–91. https://doi.org/10.1016/j.biortech.2015.12.017.
62. Yoo CG, Pu Y, Ragauskas AJ. Ionic liquids: promising green solvents for lignocellulosic biomass utilization. Curr Opin Green Sustain Chem. 2017;5:5–11. https://doi.org/10.1016/j.cogsc.2017.03.003.
63. Mukherjee A, Dumont M-J, Raghavan V. Review: sustainable production of hydroxymethylfurfural and levulinic acid: challenges and opportunities. Biomass Bioenergy. 2015;72:143–83. https://doi.org/10.1016/j.biombioe.2014.11.007.
64. Emam E, Centi G, Perathoner S, Vaccari A. Clays as catalysts in petroleum refining industry. ARPN. J Sci Technol. 2013;3:356–75. https://doi.org/10.1016/S0169-1317(98)00058-1.
65. Wheeldon I, Christopher P, Blanch H. Integration of heterogeneous and biochemical catalysis for production of fuels and chemicals from biomass. Curr Opin Biotechnol. 2017;45:127–35. https://doi.org/10.1016/j.copbio.2017.02.019.
66. Moller M, Schroder U. Hydrothermal production of furfural from xylose and xylan as model compounds for hemicelluloses. RSC Adv. 2013;3:22253–60. https://doi.org/10.1039/C3RA43108H.
67. Cortés W, Piñeiros-Castro Y, Campos Rosario AM. Conversion of d-xylose into furfural with aluminum and hafnium pillared clays as catalyst. DYNA. 2013;80:105–12.
68. Choudhary V, Pinar AB, Sandler SI, et al. Xylose isomerization to xylulose and its dehydration to furfural in aqueous media. ACS Catal. 2011;1:1724–8. https://doi.org/10.1021/cs200461t.
69. Gao H, Liu H, Pang B, et al. Production of furfural from waste aqueous hemicellulose solution of hardwood over ZSM-5 zeolite. Bioresour Technol. 2014. https://doi.org/10.1016/j.biortech.2014.09.026.
70. Choudhary V, Caratzoulas S, Vlachos DG. Insights into the isomerization of xylose to xylulose and lyxose by a Lewis acid catalyst. Carbohydr Res. 2013;368:89–95. https://doi.org/10.1016/j.carres.2012.12.019.
71. Metkar PS, Till EJ, Corbin DR, et al. Reactive distillation process for the production of furfural using solid acid catalysts. Green Chem. 2015;17:1453–66. https://doi.org/10.1039/C4GC01912A.
72. Weingarten R, Tompsett GA, Conner WC, Huber GW. Design of solid acid catalysts for aqueous-phase dehydration of carbohydrates: the role of Lewis and Brønsted acid sites. J Catal. 2011;279:174–82. https://doi.org/10.1016/j.jcat.2011.01.013.
73. You SJ, Park N, Park ED, Park M-J. Partial least squares modeling and analysis of furfural production from biomass-derived xylose over solid acid catalysts. J Ind Eng Chem. 2015;21:350–5. https://doi.org/10.1016/j.jiec.2014.02.044.
74. Antunes MM, Lima S, Fernandes A, et al. Aqueous-phase dehydration of xylose to furfural in the presence of MCM-22 and ITQ-2 solid acid catalysts. Appl Catal A Gen. 2012;417:243–52. https://doi.org/10.1016/j.apcata.2011.12.046.

75. Zhang L, Xi G, Chen Z, et al. Highly selective conversion of glucose into furfural over modified zeolites. Chem Eng J. 2017;307:868–76. https://doi.org/10.1016/j.cej.2016.09.001.
76. Bhaumik P, Dhepe PL. Effects of careful designing of SAPO-44 catalysts on the efficient synthesis of furfural. Catal Today. 2015;251:66–72. https://doi.org/10.1016/j.cattod.2014.10.042.
77. Bruce SM, Zong Z, Chatzidimitriou A, et al. Small pore zeolite catalysts for furfural synthesis from xylose and switchgrass in a γ-valerolactone/water solvent. J Mol Catal A Chem. 2016;422:18–22. https://doi.org/10.1016/j.molcata.2016.02.025.
78. García-Sancho C, Agirrezabal-Telleria I, Güemez MB, Maireles-Torres P. Dehydration of d-xylose to furfural using different supported niobia catalysts. Appl Catal B Environ. 2014:152:1–10. https://doi.org/10.1016/j.apcatb.2014.01.013.
79. García-Sancho C, Rubio-Caballero JM, Mérida-Robles JM, et al. Mesoporous Nb2O5 as solid acid catalyst for dehydration of d-xylose into furfural. Catal Today. 2014;234:119–24. https://doi.org/10.1016/j.cattod.2014.02.012.
80. Pholjaroen B, Li N, Wang Z, et al. Dehydration of xylose to furfural over niobium phosphate catalyst in biphasic solvent system. J Energy Chem. 2013;22:826–32. https://doi.org/10.1016/S2095-4956(14)60260-6.
81. Bhaumik P, Kane T, Dhepe PL. Silica and zirconia supported tungsten{,} molybdenum and gallium oxide catalysts for the synthesis of furfural. Cat Sci Technol. 2014;4:2904–7. https://doi.org/10.1039/C4CY00530A.
82. Li L, Ding J, Jiang J-G, et al. One-pot synthesis of 5-hydroxymethylfurfural from glucose using bifunctional [Sn,Al]-Beta catalysts. Chinese J Catal. 2015;36:820–8. https://doi.org/10.1016/S1872-2067(14)60287-4.
83. Zhang L, Xi G, Chen Z, et al. Enhanced formation of 5-HMF from glucose using a highly selective and stable SAPO-34 catalyst. Chem Eng J. 2017;307:877–83. https://doi.org/10.1016/j.cej.2016.09.003.
84. Moreno-Recio M, Santamaría-González J, Maireles-Torres P. Brönsted and Lewis acid ZSM-5 zeolites for the catalytic dehydration of glucose into 5-hydroxymethylfurfural. Chem Eng J. 2016;303:22–30. https://doi.org/10.1016/j.cej.2016.05.120.
85. Zhang M, Su K, Song H, et al. The excellent performance of amorphous Cr2O3, SnO2, SrO and graphene oxide–ferric oxide in glucose conversion into 5-HMF. Catal Commun. 2015;69:76–80. https://doi.org/10.1016/j.catcom.2015.05.024.
86. Yang F, Li Y, Zhang Q, et al. Selective conversion of cotton cellulose to glucose and 5-hydroxymethyl furfural with SO42−/MxOy solid superacid catalyst. Carbohydr Polym. 2015;131:9–14. https://doi.org/10.1016/j.carbpol.2015.05.036.
87. Shirai H, Ikeda S, Qian EW. One-pot production of 5-hydroxymethylfurfural from cellulose using solid acid catalysts. Fuel Process Technol. 2017;159:280–6. https://doi.org/10.1016/j.fuproc.2016.10.005.
88. Lopes M, Dussan K, Leahy JJ, da Silva VT. Conversion of d-glucose to 5-hydroxymethylfurfural using Al2O3-promoted sulphated tin oxide as catalyst. Catal Today. 2017;279:233–43. https://doi.org/10.1016/j.cattod.2016.05.030.
89. Liu J, Li H, Liu Y-C, et al. Catalytic conversion of glucose to 5-hydroxymethylfurfural over nano-sized mesoporous Al2O3–B2O3 solid acids. Catal Commun. 2015. https://doi.org/10.1016/j.catcom.2015.01.008.
90. Guo J, Zhu S, Cen Y, et al. Ordered mesoporous Nb–W oxides for the conversion of glucose to fructose, mannose and 5-hydroxymethylfurfural. Appl Catal B Environ. 2017;200:611–9. https://doi.org/10.1016/j.apcatb.2016.07.051.
91. Zhong J, Guo Y, Chen J. Protonated and layered transition metal oxides as solid acids for dehydration of biomass-based fructose into 5-hydroxymethylfurfural. J Energy Chem. 2017;26:147–54. https://doi.org/10.1016/j.jechem.2016.09.010.
92. Lv G, Deng L, Lu B, et al. Efficient dehydration of fructose into 5-hydroxymethylfurfural in aqueous medium over silica-included heteropolyacids. J Clean Prod. 2017;142:2244–51. https://doi.org/10.1016/j.jclepro.2016.11.053.

93. Srivastava S, Jadeja GC, Parikh J. Synergism studies on alumina-supported copper-nickel catalysts towards furfural and 5-hydroxymethylfurfural hydrogenation. J Mol Catal A Chem. 2017;426:244–56. https://doi.org/10.1016/j.molcata.2016.11.023.
94. Chen X, Wang X, Yao S, Mu X. Hydrogenolysis of biomass-derived sorbitol to glycols and glycerol over Ni-MgO catalysts. Catal Commun. 2013. https://doi.org/10.1016/j.catcom.2013.05.012.
95. Balaraju M, Rekha V, Prasad PSS, et al. Influence of solid acids as co-catalysts on glycerol hydrogenolysis to propylene glycol over Ru/C catalysts. Appl Catal A Gen. 2009;354:82–7. https://doi.org/10.1016/j.apcata.2008.11.010.
96. Gandarias I, Arias PL, Requies J, et al. Hydrogenolysis of glycerol to propanediols over a Pt/ASA catalyst: the role of acid and metal sites on product selectivity and the reaction mechanism. Appl Catal B Environ. 2010;97:248–56. https://doi.org/10.1016/j.apcatb.2010.04.008.
97. Wang G-H, Hilgert J, Richter FH, et al. Platinum–cobalt bimetallic nanoparticles in hollow carbon nanospheres for hydrogenolysis of 5-hydroxymethylfurfural. Nat Mater. 2014;13:293–300.
98. Kusunoki Y, Miyazawa T, Kunimori K, Tomishige K. Highly active metal-acid bifunctional catalyst system for hydrogenolysis of glycerol under mild reaction conditions. Catal Commun. 2005;6:645–9. https://doi.org/10.1016/j.catcom.2005.06.006.
99. Sato S, Akiyama M, Takahashi R, et al. Vapor-phase reaction of polyols over copper catalysts. Appl Catal A Gen. 2008;347:186–91. https://doi.org/10.1016/j.apcata.2008.06.013.
100. Vasiliadou ES, Lemonidou AA. Kinetic study of liquid-phase glycerol hydrogenolysis over Cu/SiO2 catalyst. Chem Eng J. 2013;231:103–12. https://doi.org/10.1016/j.cej.2013.06.096.
101. Rajkhowa T, Marin GB, Thybaut JW. A comprehensive kinetic model for Cu catalyzed liquid phase glycerol hydrogenolysis. Appl Catal B Environ. 2017;205:469–80. https://doi.org/10.1016/j.apcatb.2016.12.042.
102. Chidambaram M, Bell AT. A two-step approach for the catalytic conversion of glucose to 2,5-dimethylfuran in ionic liquids. Green Chem. 2010;12:1253–62. https://doi.org/10.1039/C004343E.
103. Saha B, Bohn CM, Abu-Omar MM. Zinc-assisted hydrodeoxygenation of biomass-derived 5-hydroxymethylfurfural to 2,5-dimethylfuran. ChemSusChem. 2014;7:3095–101. https://doi.org/10.1002/cssc.201402530.
104. Luo J, Lee JD, Yun H, et al. Base metal-Pt alloys: a general route to high selectivity and stability in the production of biofuels from HMF. Appl Catal B Environ. 2016;199:439–46. https://doi.org/10.1016/j.apcatb.2016.06.051.
105. Goyal R, Sarkar B, Bag A, et al. Studies of synergy between metal–support interfaces and selective hydrogenation of HMF to DMF in water. J Catal. 2016;340:248–60. https://doi.org/10.1016/j.jcat.2016.05.012.
106. Chen B, Li F, Huang Z, Yuan G. Carbon-coated Cu-Co bimetallic nanoparticles as selective and recyclable catalysts for production of biofuel 2,5-dimethylfuran. Appl Catal B Environ. 2017;200:192–9. https://doi.org/10.1016/j.apcatb.2016.07.004.
107. Thananatthanachon T, Rauchfuss TB. Efficient production of the liquid fuel 2,5-dimethylfuran from fructose using formic acid as a reagent. Angew Chem Int Ed Engl. 2010;49:6616–8. https://doi.org/10.1002/anie.201002267.
108. da Silva JL, Aznar M. Thermophysical properties of 2,5-dimethylfuran and liquid–liquid equilibria of ternary systems water+2,5-dimethylfuran+alcohols (1-butanol or 2-butanol or 1-hexanol). Fuel. 2014;136:316–25. https://doi.org/10.1016/j.fuel.2014.07.039.
109. Gandarias I, Arias PL, Fernández SG, et al. Hydrogenolysis through catalytic transfer hydrogenation: glycerol conversion to 1,2-propanediol. Catal Today. 2012;195:22–31. https://doi.org/10.1016/j.cattod.2012.03.067.
110. Tseng Y-T, Ward JD, Lee H-Y. Design and control of a continuous multi-product process with product distribution switching: sustainable manufacture of furfuryl alcohol and 2-methylfuran. Chem Eng Process Process Intensif. 2016;105:10–20. https://doi.org/10.1016/j.cep.2016.04.003.

111. Sitthisa S, An W, Resasco DE. Selective conversion of furfural to methylfuran over silica-supported NiFe bimetallic catalysts. J Catal. 2011;284:90–101. https://doi.org/10.1016/j.jcat.2011.09.005.
112. Miyazawa T, Koso S, Kunimori K, Tomishige K. Development of a Ru/C catalyst for glycerol hydrogenolysis in combination with an ion-exchange resin. Appl Catal A Gen. 2007;318:244–51. https://doi.org/10.1016/j.apcata.2006.11.006.
113. Li H, Luo H, Zhuang L, et al. Liquid phase hydrogenation of furfural to furfuryl alcohol over the Fe-promoted Ni-B amorphous alloy catalysts. J Mol Catal A Chem. 2003;203:267–75. https://doi.org/10.1016/S1381-1169(03)00368-6.
114. Liaw B-J, Chiang S-J, Chen S-W, Chen Y-Z. Preparation and catalysis of amorphous CoNiB and polymer-stabilized CoNiB catalysts for hydrogenation of unsaturated aldehydes. Appl Catal A Gen. 2008;346:179–88. https://doi.org/10.1016/j.apcata.2008.05.025.
115. Kijeński J, Winiarek P, Paryjczak T, et al. Platinum deposited on monolayer supports in selective hydrogenation of furfural to furfuryl alcohol. Appl Catal A Gen. 2002;233:171–82. https://doi.org/10.1016/S0926-860X(02)00140-0.
116. Sitthisa S, Resasco DE. Hydrodeoxygenation of furfural over supported metal catalysts: a comparative study of Cu, Pd and Ni. Catal Letters. 2011;141:784–91. https://doi.org/10.1007/s10562-011-0581-7.
117. Srivastava S, Jadeja GC, Parikh J. A versatile bi-metallic copper-cobalt catalyst for liquid phase hydrogenation of furfural to 2-methylfuran. RSC Adv. 2016;6:1649–58. https://doi.org/10.1039/C5RA15048E.
118. Mamman AS, Lee J-M, Kim Y-C, et al. Furfural: hemicellulose/xylosederived biochemical. Biofuels Bioprod Biorefin. 2008;2:438–54. https://doi.org/10.1002/bbb.95.
119. Yang J, Zheng H-Y, Zhu Y-L, et al. Effects of calcination temperature on performance of Cu–Zn–Al catalyst for synthesizing γ-butyrolactone and 2-methylfuran through the coupling of dehydrogenation and hydrogenation. Catal Commun. 2004. https://doi.org/10.1016/j.catcom.2004.06.005.
120. Zheng H-Y, Zhu Y-L, Teng B-T, et al. Towards understanding the reaction pathway in vapour phase hydrogenation of furfural to 2-methylfuran. J Mol Catal A Chem. 2006;246:18–23. https://doi.org/10.1016/j.molcata.2005.10.003.
121. Dong F, Zhu Y, Zheng H, et al. Cr-free Cu-catalysts for the selective hydrogenation of biomass-derived furfural to 2-methylfuran: the synergistic effect of metal and acid sites. J Mol Catal A Chem. 2015;398:140–8. https://doi.org/10.1016/j.molcata.2014.12.001.
122. Li D, Zhen H, Xingcai L, et al. Physico-chemical properties of ethanol–diesel blend fuel and its effect on performance and emissions of diesel engines. Renew Energy. 2005;30:967–76. https://doi.org/10.1016/j.renene.2004.07.010.

Chapter 9
Upgrading of Biomass-Derived Furans into Value-Added Chemicals

Song Song, Guangjun Wu, Naijia Guan, and Landong Li

Abstract Biomass-derived furans, e.g. furfural and 5-hydroxymethylfurfural, can be produced from the hydrolysis of biomass-derived chemicals. Upgrading of biomass-derived furans into value-added chemicals is a promising strategy for biomass valorization. Many chemical or enzymatic methods have been developed for the transformation of biomass-derived furans, while catalytic conversions are more attractive since they are efficient. Simple catalysts with monofunctional sites tend to exhibit low selectivity for multi-step transformations. Therefore, design of bifunctional or even multi-functional catalysts is needed for efficient catalytic conversion of biomass-derived furans into value-added chemicals via multiple-step reactions. In this chapter, an overview will be provided on the recent achievements made with bifunctional catalysts for the upgrading of biomass-derived furans via selective hydrogenation, selective oxidation, Diels-Alder reaction and others. Particularly, reaction pathways and plausible reaction mechanisms for biomass-derived furans upgrading with bifunctional catalysts will be discussed.

9.1 Introduction to Biomass-Derived Furans

9.1.1 Furans from Biomass

Furfural (FFR) and 5-hydroxymethylfurfural (HMF) are two main biomass-derived furans [1, 2], which are in the U.S. Department of Energy list of bio-based chemicals as "Top 10 + 4" according to the report by Petersen [3]. FFR, with the chemical formula of $C_5H_4O_2$, is an oxygen heterocyclic ring compound with an aldehyde functional group. It appears as a colorless liquid with an almond odor and its color becomes darker when exposed to air probably due to partial degradation. FFR is mainly produced through the acid-catalyzed transformation of xylose, which comes

S. Song • G. Wu • N. Guan • L. Li (✉)
School of Materials Science and Engineering & National Institute for Advanced Materials, Nankai University, Tianjin, People's Republic of China
e-mail: lild@nankai.edu.cn

Z. Fang et al. (eds.), *Production of Biofuels and Chemicals with Bifunctional Catalysts*, Biofuels and Biorefineries 8, https://doi.org/10.1007/978-981-10-5137-1_9

from renewable nonedible agricultural sources, e.g. food crop residues and wood wastes (Fig. 9.1) [4, 5]. HMF, with the chemical formula of $C_6H_6O_3$, consists of a furan ring, which contains both alcohol and aldehyde functional groups. Typically, HMF can be produced from the dehydration of glucose (Fig. 9.1) [6–8]. The biomass-derived furan, 3-acetamido-5-acetylfuran (3A5AF), has been reported as another compound of interest as it can be synthesized from waste crustacean shells by combining chitin hydrolysis and N-acetyl-D-glucosamine dehydration. The 3A5AF is proposed as a nitrogen-containing building block for sustainable biomass-derived chemicals (Fig. 9.1) [9–11]. Research on 3A5AF is just in its initial stages [12] although its importance is highlighted in a recent review on shell biorefineries [13] which discusses the potential of waste shell resources. In this chapter, the focus will be on the use of FFR and HMF over bifunctional catalysts, since research progress on these chemicals is substantially developed.

Some major physical properties of FFR and HMF are summarized in Table 9.1 for a direct view.

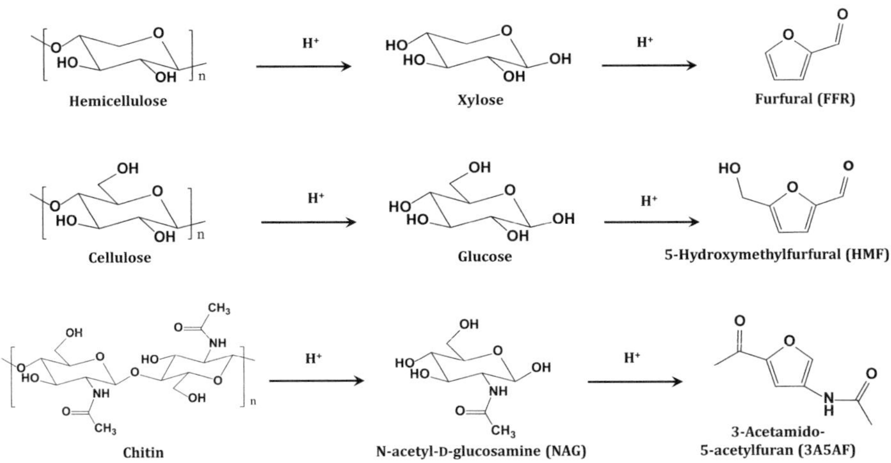

Fig. 9.1 Simplified reaction scheme for the production of furfural, 5-hydroxymethylfurfural and 3-acetamido-5-acetylfuran from hemicellulose, cellulose and chitin, respectively

Table 9.1 Physical properties of furfural (FFR) and 5-hydroxymethylfurfural (HMF)

Name	Chemical formula	Molar mass g/mol	Physical appearance at 25 °C	Odor	Liquid density g/mL	Melting point/K	Boiling point/K
FFR	$C_5H_4O_2$	96.09	Colorless oil	Almond	1.16[a]	236	435
HMF	$C_6H_6O_3$	126.11	Yellow powder	Chamomile	1.24[b]	301 ~ 307	387 ~ 389

[a]From GESTIS Substance Database (http://gestis-en.itrust.de)
[b]From Wikipedia

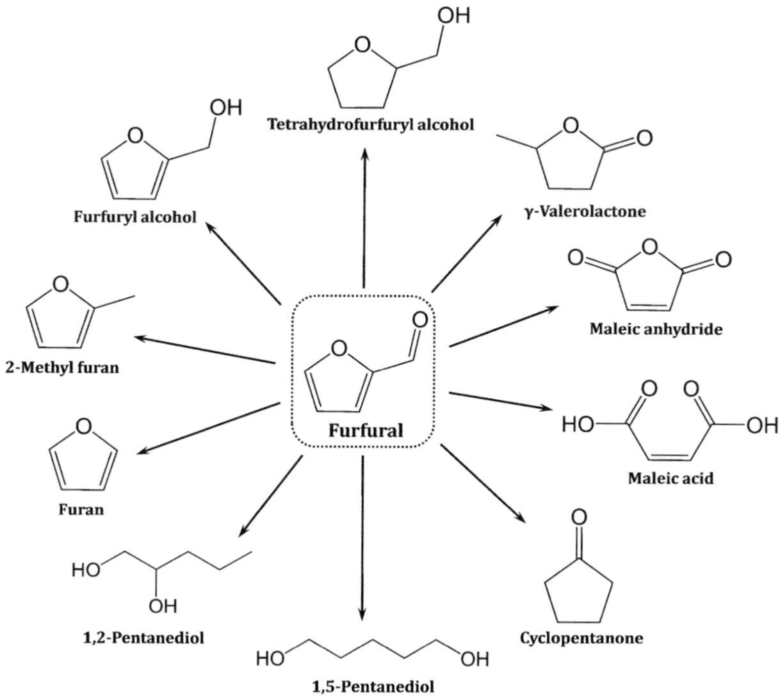

Fig. 9.2 Useful chemicals that can be derived from furfural

9.1.2 Chemical Properties of Furans

FFR can undergo typical reactions of the aldehyde group, such as reduction to an alcohol or the oxidation to a carboxylic acid. Hydrogenation, hydrogenolysis and Diels-Alder reactions can be conducted with the furan ring of FFR. The versatile properties of the aldehyde group and the furan ring widen the applications for upgrading FFR. As summarized in Fig. 9.2, a number of valuable chemicals can be produced from FFR [14–16].

HMF has similar functional groups like FFR but with additional alcohol group in the molecule. Starting from HMF, a lot of high-value chemicals can also be produced, as summarized in Fig. 9.3 [17, 18].

This chapter is briefly divided into three sections related to catalytic system with bifunctional catalysts: (1) hydrogenation or oxidation of FFR; (2) the catalytic hydrogenation or oxidation of HMF and (3) the Diels-Alder and dehydration reactions of biomass-derived furans with dienophiles.

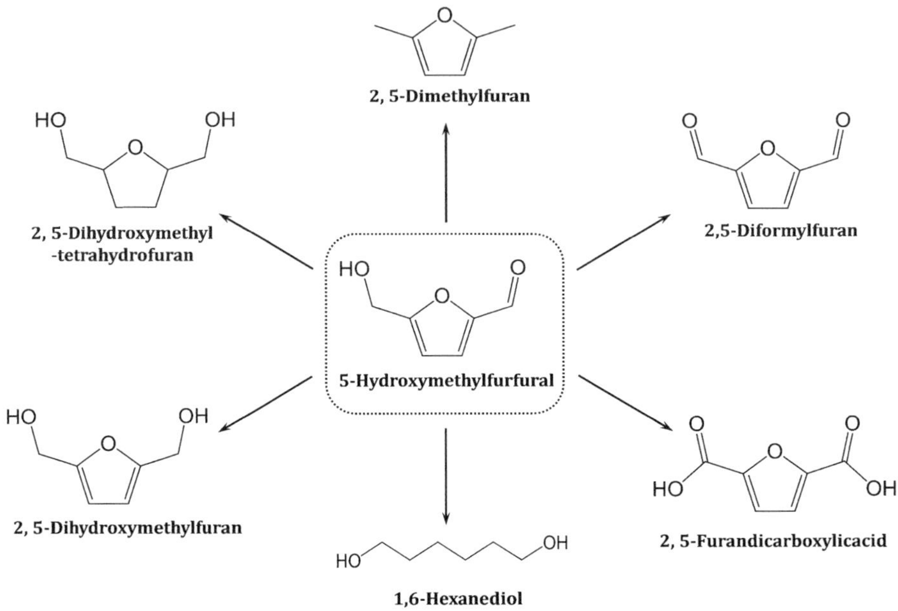

Fig. 9.3 Useful chemicals that can be derived from 5-hydroxymethylfurfural

9.2 Furfural Hydrogenation

9.2.1 Furfural to Furfuryl Alcohol

Furfuryl alcohol (FAL), which is also called 2-furylmethanol or 2-furancarbinol, is an organic compound containing a furan substituted with a hydroxymethyl group. FAL is the monomer for the synthesis of furan resins, which are widely used in polymer composites, cements, adhesives and coatings. Industrially, FAL is produced from FFR through a gas-phase or liquid-phase hydrogenation process in the presence of catalysts, e.g. catalytic hydrogenation. The catalytic hydrogenation of FFR to FAL can be achieved with Cu-based catalysts, and Cu-Cr catalysts have already been commercialized for FAL production [19–22]. However, the toxicity of Cr and the relatively low efficiency of Cu-Cr catalysts restrict their application to some extent. In this context, it is necessary to develop more environmental-benign and efficient catalysts for transformation of FFR to FAL.

9.2.1.1 Metal-Metal Synergistic Bifunctional Catalysts

Without the use of Cr as an additive, Cu-containing catalysts such as Cu-Pd, Cu-Ca, Cu-Ni and Cu-Fe have been explored and show high activity toward the hydrogenation of FFR to FAL (Table 9.2, Entries 1–5) [23–27]. Precious metal Pd was

Table 9.2 Hydrogenation of furfural (FFR) with bifunctional catalysts to obtain products furfuryl alcohol (FAL), tetrahydrofurfuryl alcohol (THFA) or 2-methylfuran (MF)

Entry	Catalyst	Product	T/K	P/MPa	t/h or HSV/h^{-1}	Conv/%	Select/%	Refs.
1	CuPd-Y	FAL	573	0.1	WHSV = 7.7	58	99	[23]
2	CuCa/SiO_2	FAL	403	0.1	GHSV = 0.3	100	99	[24]
3	CuNi/MgAlO	FAL	473	1.0	2	93	89	[25]
4	$CuFe_2O_4$	FAL	433	9.0	5	91	99	[26]
5	CuPd-MgO	FAL	383	0.6	1.5	100	98	[27]
6	Co-Mo-B	FAL	373	1.0	3	100	100	[30]
7	Ni-Fe-B	FAL	373	1.0	4	100	100	[31]
8	Ni-Ce-B	FAL	353	1.0	3	97	100	[32]
9	Co-B	FAL	353	1.0	–	100	100	[33]
10	Pt-TiO_2/SiO_2	FAL	423	0.1	LHSV = 2.0	68	94	[34]
11	Pt/TiO_2	FAL	393	0.1	3	100	90	[37]
12	Ir-ReOx/SiO_2	FAL	303	0.8	6	99	99	[39]
13	Ni/SiO_2	THFA	413	0.1	1	100	94	[40]
14	Ni-Sn/AlOH	THFA	403	3.0	1.5	99	100	[41]
15	Pd/MFI	THFA	493	3.5	5	100	95	[42]
16	Cu-Fe/SiO_2	MF	525	0.1	GHSV = 48	99	99	[48]
17	Cu-Fe	MF	493	9.0	14	99	51	[49]
18	Ni-Fe/SiO_2	MF	523	0.1	WHSV = 10	96	39	[50]
19	Ru/Co_3O_4	MF	44	1.0	24	99	93	[51]
20	Cu-Mn-Si	MF	552	0.1	LHSV = 0.5	99	94	[55]
21	Ru/RuOx/SiO_2	MF	453	2-Butanol	10	100	76	[56]

**HSV* hourly space velocity, *WHSV* weight basis, *LHSV* liquid basis, *GHSV* gas basis

demonstrated to be an effective additive and the gas-phase hydrogenation of FFR to FAL over bifunctional PdCu/zeolite Y catalyst has been reported [23]. Typically, 100% selectivity toward FAL with 58% FFR conversion was obtained under relatively mild conditions. Cu(II) ions existed as the active sites for FFR hydrogenation while Pd sites for hydrogen activation. Similarly, Pd-Cu/MgO was reported to be very active in the liquid-phase hydrogenation of FFR to FAL, with 100% FFR conversion and 98% selectivity toward FAL at 383 K. The surface composition of Pd-Cu/MgO was determined to Pd^0-Cu_2O with close contact between Pd^0 and Cu_2O. Cu^+ ions enabled the hydrogenation of C = O group with hydrogen transfer from the adjacent Pd-H sites. Alkaline support, e.g. MgO or $Mg(OH)_2$, could also promote the catalytic activity and selectivity to some extent [27]. Bifunctional Cu-based catalysts with non-precious metal additives such as Ca, Ni and Fe also exhibited high activity in FFR hydrogenation [24–26, 28–30]. However, relatively higher reaction temperature (>403 K) was required when compared with that employed in Pd-Cu catalyst. In addition to Cu-based catalysts, Co- and Ni-based bifunctional alloy catalysts have been applied in catalyzing the hydrogenation of FFR to FAL with considerable activity [31–35].

9.2.1.2 Metal-Support Synergistic Bifunctional Catalysts

The nature of support materials can influence the catalytic activity and selectivity in FFR hydrogenation. For example, Pt-TiO_2/SiO_2 exhibits a higher selectivity toward FAL (93.8%) than Pt/SiO_2 (33.8%) in the reaction. The role of the strong metal-support interaction and the strong oxide–oxide interaction have been confirmed for the case of Pt-TiO_2/SiO_2 [36]. It was also reported that Pt/TiO_2 exhibited higher selectivity to FAL than Pt/SiO_2 in FFR hydrogenation. That is to say, the oxide supports played a vital role for controlling the product selectivity and close contact between Pt nanoparticles and TiO_2 support was beneficial for selective hydrogenation of FFR to FAL. The Pt-TiO_2 also promoted the reaction rate due to the strong metal-support interactions [37], similar to the recent report on Cu-TiO_2-CuO interface [38]. Bifunctional Ir-ReOx/SiO_2 was reported as an efficient catalyst for FFR hydrogenation to FAL and the high activity was mainly ascribed to the synergistic effects between metal Ir and ReOx [39].

9.2.2 Furfural to Tetrahydrofurfuryl Alcohol

9.2.2.1 Metal-Support Synergistic Bifunctional Catalysts

The total hydrogenation of FFR to tetrahydrofurfuryl alcohol (THFA) over Ni/SiO_2 was reported, which underwent two successive steps, i.e. FFR to FAL and FAL to THFA [40]. The conversion of FAL to THFA was inhibited by the presence of FFR because FFR was more strongly adsorbed onto the catalyst surface than FAL. The conversion of FFR to FAL intermediate was less structure-sensitive. In contrast, the conversion of FAL to THFA was strongly structure-sensitive and higher turnover frequency values were obtained over Ni/SiO_2 catalysts with smaller Ni particle size (<4 nm). Here, the support SiO_2 appeared to be a media to adjust the size of Ni nanoparticles. Aluminum hydroxide supported Ni-Sn catalyst could also catalyze the total hydrogenation of FFR to THFA [41]. The basic property of the support and the hydrogenation property of Ni-Sn were both beneficial for the high catalytic activity obtained. In addition to Ni-based catalysts, Pd supported on zeolite, i.e. Pd/MFI, could catalyze the total hydrogenation of FFR to THFA [42]. Both the metal function and the support were critical in directing the selectivity to the ring hydrogenated product THFA, and for making Pd/MFI an efficient bifunctional catalyst.

9.2.3 Furfural to Pentanediol

9.2.3.1 Metal-Support Synergistic Bifunctional Catalysts

Pentanediol is a high-value compound that can be used as monomer for biomass-based polymers and also as an intermediate for synthesis of fungicides, printing inks, and cosmetics. The direct conversion of furfural to 1, 2-pentanediol, with a high yield of 73%, was achieved using Pt supported hydrotalcite (Pt/HT) as a bifunctional catalyst [43]. The remarkable activity of Pt/HT catalyst was attributed to the cooperative effect between metallic Pt nanoparticles and the basic support HT. The interface between Pt nanoparticles and HT was beneficial for the selective scission of C^5-O^1 bond of furan ring, leading to the formation of 1, 2-pentanediol. The one-pot conversion of FFR to 1, 5-pentanediol over the bifunctional Pt/Co_2AlO_4 catalysts was also achieved, although the selectivity towards 5-pentanediol was relatively low (~30%) [44]. CoOx sites, especially Co(III) ions, were claimed to be critical for the adsorption of C = C bond and the opening of the furan ring. Then, metallic Pt catalyzed the hydrogenation of ring-opening products to produce 1, 5-pentanediol. Pd modified Ir-ReOx/SiO_2 could efficiently catalyze the transformation of FFR and a high 1,5-pentanediol yield of 71.4% was achieved with a one-pot reaction for which two-step were used in temperature [45]. The concept of bifunctional catalyst is well illustrated in this catalytic system: Pd-ReOx active sites catalyzing the hydrogenation of FFR to THFA and Ir-ReOx active sites catalyzing the hydrogenolysis of THFA to 1,5-pentanediol. Further study demonstrated that Rh modified Ir-ReOx/SiO_2 could efficiently catalyze the transformation of FFR to 1, 5-pentanediol with an improved yield of 78.2% under identical reaction conditions [46].

9.2.4 *Furfural to 2-Methylfuran*

The product 2-methylfuran (MF) is a chemical intermediate for the synthesis of flavoring substances, antimalarial drugs and heterocycles. MF can be directly used as a motor biofuel or it can be further transformed into high-alkane liquid fuels through condensation. The main route for synthesis of MF is catalytic hydrogenation-hydrogenolysis, also called hydrodeoxygenation of FFR.

9.2.4.1 Metal-Metal Synergistic Bifunctional Catalysts

Gas-phase hydrogenation of FFR to MF is primarily conducted with Cu-based catalysts, similar to the hydrogenation of FFR to FAL but at much higher reaction temperature (ca. >523 K) [47]. The bifunctional supported Cu-Cr catalysts appear to be very active in the reaction, however, the toxic nature of chromium greatly restricts its practical application. Non-toxic Cu-Fe/SiO_2 exhibits remarkable performance in the conversion of FFR to MF in toluene solvent at 525 K, with 99% FFR conversion and 99% MF selectivity. The synergistic effect between Cu and Fe was crucial for the high conversion and selectivity [48]. In water solvent, the conversion of FFR to MF could be performed at lower temperature of 493 K [49]. SiO_2-supported Ni-Fe bimetallic catalysts exhibited high activity in the conversion of FFR to MF [50]. The main products were furan and FAL when using SiO_2-supported monometallic Ni as the catalyst. The addition of Fe to Ni/SiO2 could suppress the decarbonylation activity of Ni while promoting the C = O hydrogenation at low temperatures and the

C-O hydrogenolysis at high temperatures, which resulted in an increase in MF selectivity.

9.2.4.2 Metal-Support Synergistic Bifunctional Catalysts

Ru/Co_3O_4 was explored as a bifunctional catalyst for hydrodeoxygenation of FFR to MF [51]. The surface composition of the pre-reduced Ru/Co_3O_4 was determined to be metallic Ru and CoOx. Ru^0 was responsible for the hydrogenation of FFR to FAL. Then, the CoOx enabled the adsorption of FAL and the hydrogenolysis of C-O bond to give the final product MF.

Transfer hydrogenation represents an alternative hydrogenation route for the conversion of FFR without the using high pressure hydrogen [52–54]. The transfer hydrogenation of FFR to MF was achieved with cyclohexanol as hydrogen donor and using bifunctional Cu-Mn-Si catalyst [55]. Metal-silica and metal-metal interaction were thought to be essential for transfer hydrogenation in this system. Partially oxidized Ru/C could catalyze the liquid phase transfer hydrogenation of FFR to MF with 2-propanol as hydrogen donor [56]. Further study on the effect of hydrogen donors demonstrated that 2-butanol is the best hydrogen donor for transfer hydrogenation reaction [57]. The partial oxidation of Ru/C, i.e. Ru/RuOx/C, created a robust bifunctional catalyst containing Ru metal sites and RuOx Lewis acid sites [58]. In the reaction, RuOx Lewis acid sites catalyze the transfer hydrogenation of FFR to FAL, which could be further converted to MF via hydrogenolysis with the presence of both metallic Ru and RuOx Lewis acid sites.

We summarize the hydrogenation of FFR to valuable chemicals over bifunctional catalysts, as shown in Table 9.2 for a direct view.

9.3 Furfural Oxidation

9.3.1 Furfural to Maleic Anhydride

Maleic anhydride (MA) is an important chemical intermediate, which is mainly used for the synthesis of unsaturated polyester resins. Traditionally, MA was produced through the gas-phase oxidation petroleum-based n-butane and benzene. Nowadays, with the development of biomass feedstock, the oxidation of biomass-derived FFR and HMF represents an alternative renewable route for MA production [59]. For example, the combination of $H_5PV_2Mo_{10}O_{40}$ and Cu $(CF_3SO_3)_2$ could catalyze aerobic oxidation of FFR with 54% mole yield of MA [60]. Although the $H_5PV_2Mo_{10}O_{40}$ catalyst alone could catalyze the oxidation reaction, a significant increase of MA yield was obtained with the additive of Lewis acid catalyst $Cu(CF_3SO_3)_2$. This can be attributed to promotion of the electron transfer from radical intermediates to the $H_5PV_2Mo_{10}O_{40}$ by the presence of Lewis acid. It was also reported that the bifunctional Mo-V metal oxides could efficiently catalyze the aerobic oxidation of FFR to MA with yield up to 65% [61]. Undoubtedly, the Mo-V binary metal oxides exhibited synergism effect and have much higher activity than either V_2O_5 or MoO_3 alone.

9.3.2 Furfural to Maleic Acid

Maleic acid is an industrial raw material for the production of glyoxylic acid by ozonolysis and it can also be used for the production of unsaturated polyester resin, tartaric acid and other related chemicals. Industrially, maleic acid comes from the hydrolysis of MA, which is produced by the oxidation of benzene or butane (vide supra). Recently, a new route for maleic acid production from biomass-derived furans has been proposed. The combination of copper nitrate with phosphomolybdic acid could effectively catalyze the oxidation of FFR to maleic acid with oxygen [62]. The bifunctional combination system exhibits higher catalytic activity (49.2% yield) than copper nitrate (27.7% yield) or phosphomolybdic acid (38.4% yield) alone under identical reaction conditions. TS-1 zeolite was also reported to be a robust catalyst for the oxidation of FFR to maleic acid with aqueous H_2O_2 [63]. Tetrahedral Ti(IV) cations in the TS-1 could activate hydrogen peroxide and then oxidize FFR to maleic acid. The microporosity of TS-1 further improved the selectivity to maleic acid. Both tetrahedral Ti(IV) cations and the microporosity of TS-1 are beneficial for high yields of maleic acid through simple bifunctional catalytic system [64].

9.4 Miscellaneous Routes for Furfural Conversion

9.4.1 Furfural to Furan

Furan, a simple type of aromatic heterocyclic organic compound, is a common solvent and it is also used as a starting point for specialty chemicals. Presently, furan is manufactured industrially via the liquid-phase or gas-phase decarbonylation of FFR, typically catalyzed by Pd-based materials [65]. To obtain considerable furan yields, solid base additives, e.g. K_2CO_3, are necessary in the liquid-phase decarbonylation process. However, the solid base will inevitably poison Pd-based catalysts. In this context, it is desirable to design efficient bifunctional catalysts for the decarbonylation of FFR to furan in the absence of solid base.

9.4.1.1 Metal-Support Synergistic Bifunctional Catalysts

Bifunctional Pd/SBA-15 catalyst was reported to exhibit remarkable activity toward the decarbonylation of furfural under relatively mild conditions [66]. The high catalytic activity was ascribed to the fine Pd nanoparticles of 4–6 nm and the nature of the support, e.g. acidic or basic sites and large surface area. Tokunaga et al. report that Pd/ZrO_2 bifunctional catalysts display better performance than other transition metal oxide supported catalysts for decarbonylation of furfural to furan [67]. In contrast to the liquid-phase decarbonylation, the gas-phase decarbonylation of FFR to furan has its own advantages, such as easy separation and recycling of the catalyst [68]. To prolong the lifetime of catalysts, the gas-phase decarbonylation of FFR is usually conducted with hydrogen in the reaction system. On the other hand, the presence of hydrogen will decrease the selectivity to furan due to its further hydrogenation. The design of efficient bifunctional catalysts for decarbonylation of FFR to furan is challenging. The vapor-phase decarbonylation of FFR to furan was reported using K-doped Pd/Al_2O_3 catalyst in a fixed-bed reactor with H_2-rich stream [69]. K-doped Al_2O_3 is a more basic in nature than the parent Al_2O_3, which resulted in an electron transfer to Pd and finally promoted the decarbonylation of FFR over Pd. Alkali metal doping could also suppressed the hydrogenation of FFR to FAL and, accordingly, increase the selectivity to furan. The decarbonylation of FFR could also be catalyzed by Pt catalysts, which appears to be structure-sensitive. Pt nanoparticles with small particle sizes (<2 nm) and cubic in shape were very active in the decarbonylation of FFR to furan, rather than hydrogenation of FFR to FAL [70]. Recently, Pd nanoparticles confined in microporous Si-MFI zeolite, labeled as core-shell Pd@Si-MFI, was developed as a robust catalyst for the reaction [71]. The bifunctional Pd@Si-MFI catalyst combined the advantage of both highly dispersed Pd nanoparticles and zeolite microporous property, which exhibited a ultra-high selectivity toward furan (98.7%), in contrast to commonly-used impregnated Pd/Si-MFI (5.6%).

9.4.2 Furfural to γ-Valerolactone

9.4.2.1 Brønsted-Lewis Acid Synergistic Bifunctional Catalysts

The chemical γ-valerolactone (GVL), which is a potential green solvent and fuel, has attracted significant attention in recent years. GVL can be produced from the hydrogenation of levulinic acid via the ring opening product of FFR. An economical and environmentally-benign pathway for production of GVL from FFR is highly desirable. The one-pot synthesis route to GVL from biomass-derived FFR over bifunctional zeolite catalyst has been reported [72]. The reaction consisted of sequential catalytic transfer hydrogenation and hydrolysis reaction over bifunctional zeolite catalysts (a mixture of Zr-BEA and Al-MFI) containing both Brønsted and Lewis acid sites (Fig. 9.4). In the first step, FFR was converted into FAL and its ether catalyzed by Lewis acid sites in the presence of hydrogen donor. Then, the Brønsted acid sites catalyzed the hydrolysis of FAL and its ether to levulinic acid and its ester. Finally, Lewis acid sites catalyzed the transfer hydrogenation of levulinic acid and its ester to produce the corresponding 4-hydroxypentanoates, followed by lactonization to the final product GVL. This represents an attractive route for conversion biomass-derived FFR to high-valued chemicals. However, the

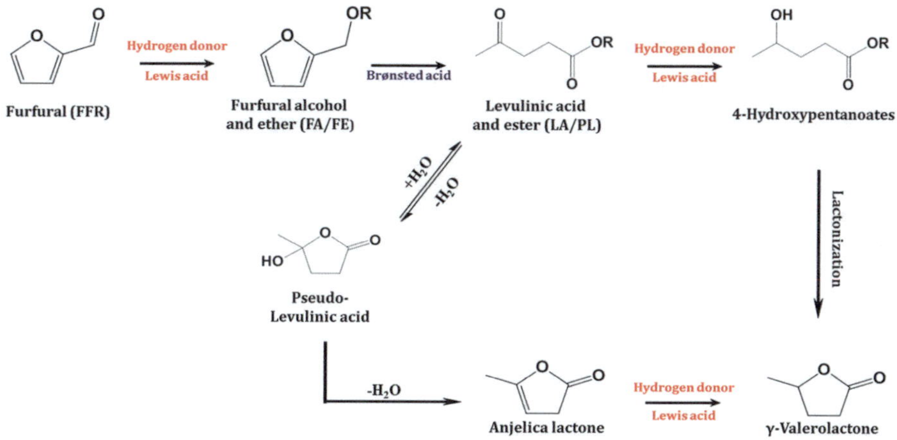

Fig. 9.4 Reaction networks for the one-pot conversion of furfural to γ-valerolactone.

outstanding performance of the catalytic system can be ascribed to the strong interface interaction of Au with ZrO_2 species, large amounts of medium-strength acid sites over ZSM-5, and synergy between active metal and acid sites.

9.4.3 Furfural to Cyclopentanone

9.4.3.1 Metal-Metal Synergistic Bifunctional Catalyst

Cyclopentanone (CPO) is a common precursor to fragrances and also a versatile chemical intermediate for producing pesticides and pharmaceuticals. CPO is produced industrially from the treatment of adipic acid with strong base at elevated temperatures while the direct conversion of FFR to CPO provides an alternative route [79]. Cu-based catalysts have been reported to be active in the reaction and bifunctional catalysts are required to achieve satisfactory CPO yields [79–83]. Typically, bifunctional NiCu/SBA-15 catalysts exhibited remarkable catalytic activity in FFR conversion and a CPO yield of 62% is achieved in aqueous phase at 433 K [80]. CuZnAl catalysts exhibit high activity in the reaction with a CPO yield of 62% under optimized conditions (423 K, 6 MPa H_2, 6 h) [81]. Metallic Cu acted as the hydrogenation sites and ZnO acts to disperse Cu species. Supported bimetallic Cu-Co catalyst, with strong interaction between Cu and Co, was reported to be very active in the conversion of FFR to CPO. Cu, Cu_2O and Co were claimed to be the active hydrogenation species, while Cu_2O also acted as Lewis acid to polarize the C = O bond [82]. Carbon supported Pd-Cu catalysts have been developed to catalyze the aqueous hydrogenation of FFR to CPO [83]. The high catalytic activity of this catalytic system was attributed to an appropriate distribution of Pd^0 and Cu^+.

9.4.4 Furfural to Furfurylamine

9.4.4.1 Metal-Support Bifunctional Catalyst

Furfurylamine, which is a very important derivative used in the manufacture of pharmaceuticals and pesticides, can be synthesized through the amination of furfural. Currently, furfurylamine is synthesized with a hydrogenation catalyst, along with additives and gaseous ammonia for which catalyst deactivation is the main drawback of the process. Reductive amination of furfural with aniline and electrochemical reduction of furfuryl oxime has been explored for the production of furfurylamine, however, the yields are relatively low [84]. In this context, the transformation of furfural to furfurylamine through reductive amination with ammonium hydroxide and hydrogen over bifunctional Rh/Al_2O_3 catalyst is of high interest [84]. An extraordinary high selectivity towards furfurylamine (92%) has been achieved at 353 K, which is the major advantage of the bifunctional catalytic process [84]. The synthesis of primary amines through reductive amination of biomass-derived aldehydes and ketones over bifunctional Ru/ZrO_2 catalysts has been reported. Comprehensive characterization results demonstrate that the RuO_2 species act as acidic promoters to accelerate the activation of carbonyl groups and the metallic Ru species act as hydrogenation active sites for the subsequent imine hydrogenation [85].

9.5 5-Hydroxymethylfurfural Hydrogenation

9.5.1 5-Hydroxymethylfurfural to 2, 5-Dihydroxymethylfuran

The chemical intermediate 2,5-dihydroxymethyl furan (DHMF) is useful for synthesis of polymers and chemicals, such as 1, 6-hexanediol. The conversion of HMF to DHMF proceeds through simple selective hydrogenation of $C = O$ bond of HMF.

9.5.1.1 Metal-Support Bifunctional Catalyst

Supported sub-nano Au clusters catalyze the selective hydrogenation of HMF to DHMF [86]. The selectivity toward DHMF exhibited a close relationship with the acid-base property of the support. Amphoteric oxide Al_2O_3 was the ideal choice of support for HMF hydrogenation because the opening of furan ring was favored over acidic supports. The catalytic activity increased with decreasing Au nanoparticle size, however, single gold atoms were inactive. Therefore, sub-nano Au/Al_2O_3 acted as an efficient bifunctional catalyst for conversion of HMF with DHMF yield of 96%. Pt/MCM-41 has also been demonstrated as an effective catalyst for HMF conversion to DHMF with almost perfect yield of ~99% being obtained [87]. Considering that the use of precious metals as catalysts will inevitably increase the cost of production, however, it is desirable to develop highly-efficient non-precious metal catalysts [88, 89]. Cu-based catalysts derived from mineral, e.g. Cu-ZnO, were able to effectively catalyze the conversion of HMF to DHMF [90]. The catalytic activity and product selectivity depended on surface Cu concentrations, acidity and microstructure.

9.5.2 5-Hydroxymethylfurfural to 2, 5-Dihydroxymethyl-Tetrahydrofuran

The compound 2,5-dihydroxymethyl-tetrahydrofuran (DHMTHF) is a good solvent and a precursor to many products. DHMTHF can be produced from HMF via the total hydrogenation of C = O and C = C bonds in HMF.

9.5.2.1 Metal-Support Synergistic Bifunctional Catalyst

Ru nanoparticles supported on various support catalyze the total hydrogenation of HMF to DHMTHF [91]. Support materials with a high isoelectric point, e.g. CeO_2, have been identified for Ru nanoparticles to achieve a high selectivity toward DHMTHF. Pd immobilized on amine-functionalized metal-organic framework, Pd/MIL-101, acted as a robust catalyst for the selective hydrogenation of HMF to DHMTHF [92]. Uniform and well-dispersed Pd nanoparticles were obtained with the help from free amine moieties in the metal-organic framework. On the other hand, free amine moieties favored the adsorption of the intermediate DHMF and, thereby, improved the selectivity toward the final product DHMTHF. The synergistic effect between Pd nanoparticles and the free amine moieties in the metal-organic framework made bifunctional amine-functionalized Pd/MIL-101 as an active and selective catalyst for HMF to DHMTHF conversion. Non-precious metal Ni based catalysts were also reported to be active for the hydrogenation of HMF to DHMTHF. For example, Ni-Al_2O_3 catalysts derived from hydrotalcite exhibited controllable selectivity toward different products DMF, DMTHF or DHMTHF by varying the catalyst calcination temperatures and reaction conditions [93]. The modulation of suitable metal-acid bifunctional sites was thought to be essential for target product. The highest DHMTHF yield of 96.2% was obtained over Ni-Al_2O_3 catalyst calcined at 723 K, which can be explained by the relatively low acid sites and the high amount of available Ni sites.

9.5.2.2 Metal-Metal Synergistic Bifunctional Catalyst

A series of supported bimetallic alloy catalysts, e.g. Pd-Ni and Pd-Ir alloy, have been investigated for the total hydrogenation of HMF to DHMTHF [94, 95]. The synergistic effect of Pd-Ir alloy was beneficial for the high activity. Addition of metal Ir promoted the activity of Pd-Ir/SiO_2 catalyst for C = O bond hydrogenation through increasing of the adsorption property of C = O bond on the catalyst surface [75].

9.5.3 5-Hydroxymethylfurfural to 1, 6-Hexanediol

The diol 1, 6-hexanediol (HDO) is widely used in industrial polyester and polyurethane production. The traditional route for HDO production is the hydrogenation of adipic acid and its esters. An alternative approach for production of HDO started from biomass-derived platform molecule HMF has been developed [96]. Hydrogenolysis of HMF to HDO proceeded via two key steps, i.e. the deoxygenation of the furan ring catalyzed by acid sites and the hydrogenation of unsaturated C = O and C = C bonds catalyzed by metal sites. That is, bifunctional catalysts containing both acid sites and metal hydrogenation sites promoted the reaction. Pd/ZrP catalyzed the conversion of HMF to HDO with formic acid as the hydrogen source [96]. A high HDO yield (43%) was achieved at 413 K. ZrP was established as the ideal Brønsted acidic support compared to other Brønsted or Lewis acidic supports, such as HY, Nb_2O_5, ZSM-5 and Al_2O_3. Metallic Pd promoted the dissociation of formic acid and, thereby, uses the released hydrogen for the hydrogenation of unsaturated C = O and C = C bonds. The bifunctional nature of Pd/ZrP catalyst made it as a robust catalyst for the hydrogenolysis of HMF to HDO. The combination of Pd/SiO_2 and Ir-ReOx/SiO_2 catalyzed the conversion of HMF to HDO in a fixed-bed reactor [97]. A very high HDO yield of 57.8% was obtained at low temperatures (373 K) in a solvent mixture of water and tetrahydrofuran. Brønsted acid sites were generated on Ir-ReOx/SiO_2 upon exposure to water vapor. A different reaction pathway was proposed with the combination catalyst system. In the first step, the hydrogenation of HMF to DHMTHF was accomplished over Pd/SiO_2 catalyst. Then, the ring opening of DHMTHF to 1, 2, 6-hexanetriol occured over the Brønsted acid sites derived from Ir-ReOx/SiO_2 catalyst. Finally, Ir-ReOx/SiO_2 catalyzed the hydrogenolysis of 1, 2, 6-hexanetriol to obtain the final product HDO.

9.5.4 5-Hydroxymethylfurfural to 2, 5-Dimethylfuran

The chemical 2,5-dimethylfuran (DMF) has attracted attention as a biofuel because it has an energy density comparable to gasoline. The hydrodeoxygenation of HMF from fructose is the main route for DMF production.

9.5.4.1 Metal-Metal Synergistic Bifunctional Catalysts

Bimetallic Pd-Au supported on carbon catalyzed the conversion of HMF to DMF in the presence of hydrochloric acid under atmosphere pressure [98]. Undoubtedly, bimetallic Pd-Au/C catalysts exhibited much higher activity than monometallic Pd/C and Au/C catalysts. Bimetallic Pd-Zn/C also appeared to be effective in catalyzing the hydrodeoxygenation of HMF to DMF in high yields (85%) [99]. Pd/C alone exhibited low activity in the conversion of HMF to DMF with 27% yield being obtained while Lewis acidic Zn(II) alone does not catalyze conversion of HMF to DMF, revealing the synergistic effects between Pd and Zn(II) in Pd-Zn/C responsible for the remarkable activity of bimetallic catalyst. Pt-Co bimetallic nanoparticles confined in hollow carbon nanospheres behaved as a robust catalyst for conversion of HMF to DMF. In contrast to Pt or Co alone, PtCo bimetallic catalyst exhibited remarkable activity in the reaction, which was the best result at that time [100]. Cu-Co coated with carbon layer has been proposed as an efficient catalyst for conversion of HMF to DMF with 99.4% yield [101]. The synergistic effect between two metal components Cu and Co was beneficial for high catalytic activity. The carbon layer also protected the catalyst from oxidation and further improved its stability.

9.5.4.2 Metal-Support Synergistic Bifunctional Catalyst

Cobalt and copper bimetallic catalyst on N-graphene-modified Al_2O_3 effectively catalyzed the conversion of HMF to DMF with perfect yield. The content of pyrrolic-N increased from 12.8 to 73.7% for CuCo/NGr/Al_2O_3 catalysts after hydrogen reduction, corresponding to an increase of DMF yield from 63.8 to 99%. The remarkable increase of DMF yield was attributed to the synergetic effect of bimetallic Cu-Co and pyrrolic-N [102]. The simple catalyst, Ru/Co_3O_4, was applied in catalyzing the conversion of HMF to DMF with 93.4% yield at mild conditions [103]. The bifunctional nature of Ru/Co_3O_4 contributed to its high activity: metallic Ru species was responsible for the hydrogenation of C = O bond while the CoOx species was responsible for the hydrogenolysis of the hydroxyl groups. Similarly, other supported precious metal catalysts, e.g. Ru and Pt, also exhibited considerable activity for the conversion of HMF to DMF [104–106]. Non-precious Ni based catalysts have been investigated for the conversion of HMF to DMF. Ni-W_2C/AC catalyzed the hydrodeoxygenation of HMF to DMF with excellent yields (96%) [107]. In catalyst, metallic Ni existed as the hydrogenation sites while W_2C acts as the deoxygenation sites. The synergistic effect between Ni and W_2C and the bifunctional nature of Ni-W_2C/AC catalysts were responsible for the high activity observed. Similar to the case of Ru/Co_3O_4, Ni/Co_3O_4 exhibited considerable activity in the catalytic hydrodeoxygenation of HMF to DMF with 76% yields being reported under mild conditions [108]. The selective hydrodeoxygenation of HMF to DMF over the bifunctional Ru-MoOx/C catalysts has been reported [109]. The synergetic effect between metallic Ru and acidic MoOx species played a crucial role in achieving the high yield of DMF [109]. Bifunctional Ru/RuO_2/C has been proposed as an active catalyst for the transfer hydrogenation of HMF with isopropanol being the hydrogen donor [110]. RuO_2 species acted as Lewis acid sites catalyzing the transfer hydrogenation of HMF to 2, 5-bis (hydroxymethyl) furan and then the Ru/C catalyzed the hydrogenolysis of 2, 5-bis (hydroxymethyl) furan to DMF [111].

A brief summary of HMF conversion to valuable furan chemicals over bifunctional catalysts is shown in Table 9.3.

Table 9.3 Hydrogenation of 5-hydroxymethylfurfural (HMF) substrate with bifunctional catalysts to yield products, 2,5-dimethylfuran (DMF) 2,5-dihydroxymethyl furan (DHMF) or 2,5-dihydroxymethyl tetrahydrofuran (DHMTHF)

Entry	Catalyst	Product	T/K	P/MPa	t/h	Conv/%	Select/%	Refs.
1	Ir-ReOx/SiO_2	DHMF	303	0.8	6	99	99	[39]
2	Au/Al_2O_3	DHMF	393	6.5	2	100	96	[86]
3	Pt/MCM-41	DHMF	308	0.8	2	100	99	[87]
4	Cu/ZnO	DHMF	373	1.5	2	100	99	[90]
5	Ru/CeOx	DHMTHF	403	3.0	12	100	91	[91]
6	Pd/MIL-101	DHMTHF	303	1.0	12	100	96	[92]
7	Ni-Al_2O_3	DHMTHF	333	6.0	6	100	96	[93]
8	NiPd/SiO_2	DHMTHF	313	8.0	2	99	97	[94]
9	IrPd/SiO_2	DHMTHF	275	8.0	4	99	95	[95]
10	PdAu/C + HCl	DMF	333	0.1	6	99	97	[98]
11	Pd/C + $ZnCl_2$	DMF	423	0.8	8	99	86	[99]
12	PtCo/C	DMF	453	1.0	2	100	98	[100]
13	CuCo/NGr/Al_2O_3	DMF	453	2.0	16	99	99	[101]
14	CuCo/C	DMF	453	5.0	8	100	99	[102]
15	Ru/Co_3O_4	DMF	403	0.7	24	99	95	[103]
16	Ru/NaY	DMF	493	1.5	1	100	78	[104]
17	Ru/HT	DMF	493	1.0	4	100	58	[105]
18	Pt/rGO	DMF	393	3.0	2	100	73	[106]
19	Ni-W_2C/AC	DMF	453	4.0	3	99	93	[50]
20	Ni/Co_3O_4	DMF	403	1.0	24	99	94	[51]
21	Ru/C	DMF	463	2-propanol	6	100	76	[56]

9.6 5-Hydroxymethylfurfural Oxidation

9.6.1 5-Hydroxymethylfurfural to 2, 5-Diformylfuran

The chemical 2, 5-diformylfuran (DFF), which is one of the most important derivatives of HMF, can be employed in the synthesis of pharmaceuticals, poly-Schiff bases and many other functional materials. DFF can be prepared from the selective oxidation of HMF [112].

9.6.1.1 Metal-Metal Synergistic Bifunctional Catalysts

Hollow Fe-Co bimetal nanoparticles derived from metal organic framework catalyzed the selective oxidation of HMF to DFF with perfect yield under mild conditions, comparable with the performance of precious metal catalysts [113]. The unique structure of hollow Fe-Co nanoparticles favored the adsorption of HMF and the quick desorption of DFF, which accordingly leads to considerable DFF yield (ca. 99%).

9.6.1.2 Metal-Support Synergistic Bifunctional Catalysts

Supported Ru catalysts have been reported to be efficient catalysts in the selective oxidation of HMF to DFF [112, 114, 115]. For example, hydrotalcite (HT) supported Ru catalyst exhibited remarkable activity in the oxidation of HMF to DFF by molecular oxygen, much higher than Ru catalysts on other supports, e.g. carbon, Al_2O_3 and $Mg(OH)_2$ [112]. The basicity of hydrotalcite and good dispersion of Ru(IV) species on the two-dimensional hydrotalcite were beneficial for the high activity. Another example was Ru nanoparticles supported on N-containing mesoporous polymers (Ru@mPMF), which exhibited high activity in conversion of HMF to DFF [115]. Basic N-containing mesoporous polymers have been demonstrated to be the crucial support for the remarkable activity and also responsible for the stabilization of Ru nanoparticles.

9.6.1.3 Zeolite Type Bifunctional Catalysts

A cryptomelane-type manganese oxide octahedral molecular sieve (OMS-2) with one-dimensional pores was reported to be a good catalyst for the oxidation of HMF to DMF by oxygen [116–118]. OMS-2 exhibited higher activity than other manganese oxides, such as Mn_2O_3, MnO or $KMnO_4$ [117]. The unique structure of Mn in the framework of OMS-2 made it easier to be reduced and oxidized. The good redox properties and the unique structure were responsible for the remarkable activity obtained for OMS-2 catalysts. Nano-fibrous Ag-OMS-2 has been confirmed to be a good catalyst for the oxidation of HMF to DFF with perfect yield. The high activity of Ag-OMS-2 catalyst was attributed to the synergistic effects of Ag and Mn [118]. In the first step, the catalyst was initiated with Ag^{2+} reduction to Ag^+ through oxygen release. Then, Mn^{3+} in the OMS-2 was oxidized to Mn^{4+} by the released oxygen. After that, HMF was oxidized to DFF accompanied with the reduction of Mn^{4+} to Mn^{3+}. Finally, Ag^+ and Mn^{3+} can be regenerated by exposure to oxygen. V-based zeolite catalysts have been reported to be active for the selective oxidation of HMF to DFF. Typically, vanadia grafted on SBA-15 mesoporous silica support catalyzed the conversion of HMF to DFF [119], and the bifunctional nature of immobilized catalyst made it even better than the corresponding homogeneous catalyst [98]. V_2O_5 supported on H-beta acted as an effective catalyst for the aerobic oxidation of HMF to DFF [120]. High surface area of H-beta zeolite promoted the dispersion of vanadia on the support and, accordingly, increased the catalytic activity. However, leaching tests of V_2O_5/H-beta catalysts revealed that the leaching of vanadium was inevitable in the reaction system.

9.6.1.4 Metal-Free Bifunctional Catalysts

The selective oxidation of HMF to DFF has been investigated using metal-free graphene oxide (GO) as a catalyst [121]. GO could not only catalyze the selective oxidation of HMF to DFF, but also was able to catalyze the one-pot production of

DFF from raw material fructose. The synergistic effects between sulfonic groups, carboxylic acid groups and the unpaired electrons at the GO edge defects were proposed to be responsible for the remarkable activity of the GO catalyst.

9.6.2 5-Hydroxymethylfurfural to 2, 5-Furandicarboxylic Acid

The chemical 2, 5-furandicarboxylic acid (FDCA) is an important renewable building block of biomass-based polymers and it is identified by the US Department of Energy as one of the 12 priority chemicals for establishing the "green" chemistry industry of the future. FDCA can be prepared via the catalytic or biological conversion of various furan derivatives. Especially, supported precious metal catalysts, e.g. Ru, Pd, Pt or Au, have been extensively investigated in the oxidation of HMF to FDCA [122–132]. The use of large amounts of base to obtain FDCA yield is generally necessary, however, the production of FDCA via these methods generate considerable waste. Therefore, it is an urgent need to design efficient catalysts that can promote the oxidation of HMF to FDCA under mild conditions without use of base additive.

9.6.2.1 Metal-Metal Synergistic Bifunctional Effect

Functionalized carbon nanotube-supported Au-Pd alloy nanoparticles could effectively catalyze base-free aerobic oxidation of HMF to FDCA. Functionalized carbon nanotubes with more carbonyl/quinone groups and less carboxyl groups promoted the adsorption of reactant and reactant intermediates. Au-Pd bimetallic alloy accelerated the rate of hydroxyl group oxidation. Both of these issues made Au-Pd/CNT a robust bifunctional catalyst for the aerobic oxidation of HMF to FDCA [127].

9.6.2.2 Metal-Support Synergistic Bifunctional Effect

To avoid the use of solid base additives, basic supports are general choices for selective oxidation catalysts. In an early work, the effect of support materials on Ru catalysts were investigated for HMF oxidation to FDCA in water [133]. Strong basic supports MgO, MgO/La_2O_3 and HT appeared to be better supports than acidic or neutral oxide supports, e.g. TiO_2 and CeO_2, and FDCA yields of above 95% could be obtained under optimized conditions. Subsequently, a series of experiments were made [134, 135]. For example, Au/HT effectively catalyzed the oxidation of HMF to FDCA under atmospheric pressure oxygen in water [134]. The basicity of HT promoted the first step oxidation to hemiacetal intermediate. The rate-determining step was the oxidation of 5-hydroxymethyl-2-furancarboxylic acid to FFCA, which was accelerated by the formation of Au alcoholate species with the help of HT. The base-free oxidation of HMF to FDCA has also been reported using Pt/C-O-Mg as catalyst [135]. The strong and stable basic sites were created due to the formation of C-O-Mg bonds from the MgO particles coated with carbon sphere. The strong basic sites combined with Pt made Pt/C-O-Mg a robust bifunctional catalyst for the oxidation of HMF to FDCA, and there was little loss of activity was observed even after ten recycles.

9.6.2.3 Bifunctional Mixed Oxide Catalysts

Recently, MnO_x-CeO_2 mixed oxides have been proposed as robust catalysts for the selective oxidation of HMF to FDCA [136]. Although CeO_2 alone could not effectively catalyze the conversion of HMF to FDCA (3.3% yield), the addition of CeO_2 to MnOx significantly increased the yield of FDCA from 54.5 to 91.0%. Further detailed characterizations demonstrated that higher concentrations of Mn^{4+} and Ce^{3+} on the catalyst surface were beneficial for the high activity in the selective oxidation of HMF to FDCA.

9.6.2.4 Metal Nanoparticle-Ionic Liquid Bifunctional Catalysts

It has been reported by Yan and Dyson *et al.* [137] that the combination of ionic liquid and metal nanoparticles could catalyze the transformation of HMF to FDCA under mild conditions in the absence of any additives. According to their results, ionic liquid could buffer the electron density of the Pt surface, then facilitate the reaction rate of the oxidation of HMF to FDCA [137].

9.7 Diels-Alder Reaction of Biomass-Derived Furans with Dienophiles

9.7.1 Diels-Alder and Dehydration of 2, 5-Dimethylfuran with Ethylene

The aromatic hydrocarbon *p*-xylene is a very important chemical feedstock with current annual demand being greater than 40 million tons per year. The *p*-xylene is raw material for large scale synthesis of many polymers. Especially, it can be oxidized to terephthalic acid (PTA) [138], which is used for the synthesis of important polymer polyethylene terephthalate. *p*-xylene is currently industrially manufactured through the separation of aromatic mixtures from naptha fraction of petroleum or the ring alkylation of toluene with methanol over ZSM-5 zeolite catalysts [139, 140]. Interestingly, a sequential process including Diels-Alder reaction, oxidation, dehydration and decarboxylation, has been proposed for synthesis of renewable *p*-xylene with DMF and acrolein [141]. The Diels-Alder and dehydration of DMF with ethylene to *p*-xylene represents an interesting alternative route to renewable *p*-xylene, which triggers significant research interest [142–152].

A catalytic route to p-xylene from biomass-derived DMF with ethylene has been explored [142]. Typically, Diels-Alder and dehydration of DMF with ethylene was achieved over H-Y zeolite with 75% *p*-xylene selectivity at 573 K. The reaction pathway was proposed to proceed via the Diels-Alder cycloaddition of DMF with ethylene to form cycloadduct, followed by the dehydration of cycloadduct to *p*-xylene [143]. Diels-Alder reaction of DMF with ethylene was promoted within the microporous structure of H-Y zeolite while dehydration of Diels-Alder cycloadduct to the final product p-xylene was catalyzed by the Brønsted acid sites in H-Y zeolite. The detailed mechanism of gas-phase Diels-Alder and dehydration of DMF with ethylene over H-Y zeolite was investigated by theoretical calculations [145]. It was concluded that Lewis acid sites promoted the Diels-Alder cycloaddition reaction through lower the HOMO–LUMO gap of the addend, while Brønsted acid sites had no benefit for this step. Strong Lewis acid sites catalyze the dehydration step, however, they are not as effective as Brønsted acid sites. For this case, Diels-Alder and dehydration of DMF with ethylene to *p*-xylene was kinetically limited by the Diels–Alder reaction with Brønsted acids as catalysts and by the dehydration with Lewis acids as catalysts. As a result, bifunctional catalysts with both Brønsted acid sites and Lewis acid sites were able to accelerate the Diels-Alder and dehydration overall reaction rate [146]. It was later reported that H-beta was as an ultra-selective catalyst for *p*-xylene production from Diels-Alder and dehydration of DMF with ethylene at relatively low temperature of 523 K [147]. H-beta exhibited higher catalytic activity than other solid acids, such as H-Y, H-ZSM-5, niobic acid and γ-Al_2O_3. There is no detailed discussion on the active sites in H-beta zeolite catalyst in the literature, however, the simultaneous presence of Brønsted and Lewis acid sites in H-beta zeolite, in contrast to other zeolites, might be a key point for the high activity and selectivity. Solvent effects on Diels-Alder and dehydration reaction of DMF with ethylene have also been considered [148, 149]. Nonpolar alkane solvents were better solvents for zeolite system [148]. However, when quasi-homogeneous Brønsted acids or Lewis acids were used as the catalyst, polar solvents, e.g., dioxane and γ-valerolactone, were the ideal choice [149].

Diels-Alder and dehydration reaction of DMF with ethylene have been conducted on other type catalysts [144, 151, 152]. However, zeolite catalysts appear to be more promising according to published results. A new pathway for synthesis of renewable aromatics through conversion of biomass derived furans with bio-ethanol over zeolite catalysts has been proposed [153]. Dienophile ethylene was reduced in situ from ethanol dehydration, and the use of ethanol instead of ethylene could not only avoid the use of high pressure ethylene but also could increase the reaction rate as well as the selectivity toward aromatics. According to the results, the bifunctional nature of the pore structure and acid strength was beneficial to the high activity of the catalyst. H-USY with a Si/Al ratio of 12 exhibited the highest carbon yields compared with other Si/Al ratios. H-USY (Si/Al = 12) also exhibits better activity than H-ZSM-5(Si/Al = 19 or 25) with similar acid strength, implying the preferred pore structure of H-USY for aromatic production.

9.7.2 Diels-Alder and Dehydration of Oxygenated Furans with Ethylene

Although significant achievements have been made on the transformation of DMF to *p*-xylene, the reduction of HMF to DMF is energy demanding and sometimes polluting, which greatly hinders the industrial application [154]. Therefore, it is desirable to develop new pathways for the synthesis of PTA from HMF without the expensive hydrogenation step. It has been reported that Lewis acid zeolites catalyze the Diels-Alder and dehydration of oxygenated furans with ethylene to produce intermediates of PTA [154]. For example, 5-(hydroxymethyl)furoic acid (HMFA), which was a partially oxidized product of HMF, reacted with ethylene over Sn-beta catalysts to produce 4-(hydroxymethyl)benzoic acid (HMA), with 31% selectivity at 61% HMFA conversion at 463 K. HMA was further oxidized to the target product PTA. Lewis acidic centers and a microporous structure of the catalyst were beneficial for the catalytic performance of Sn-beta. Detailed discussion on the reaction network and kinetic analysis were given in the literature [155]. Objectively speaking, the reaction rate and the selectivity toward the target product were relatively low over Lewis acidic zeolites and there is still a long way to go to achieve the sustainable production of PTA via this route.

9.7.3 Diels-Alder and Dehydration of Furan with Methyl Acrylate

Benzoic acid, which is a key chemical intermediate, can be used for synthesis of phenol, cyclohexanol and caprolactam. Benzoic acid is commercially produced by oxidation of petroleum based toluene with oxygen [156]. A multiple-step renewable route for synthesis of benzoic acid through Diels-Alder and dehydration reactions of furan with acrylic acid has been reported [157]. In the first step, Diels-Alder cycloaddition of furan with methyl acrylate was conducted on Lewis acidic zeolites, e.g. Hf-beta. Then, the cycloadduct was dehydrated over a mixture of methane sulfonic acid and acetic anhydride to form methyl benzoate. Finally, benzoic acid was obtained through dehydration of methyl benzoate. With Lewis acidic zeolite catalysts and mixed anhydrides, a considerable benzoic acid yield (96%) is obtained through Diels-Alder and dehydration reaction protocol. Diels-Alder and dehydration reactions of biomass-derived furans with dienophiles over different catalysts are summarized in Table 9.4.

Table 9.4 Diels-Alder reaction of biomass-derived furans with dienophiles over various catalysts

Entry	Catalyst	Diene	Dienophile	Product	T/K	t/h	Conv/%	Select/%	Refs.
1	H-Y	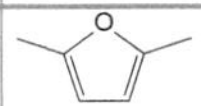	Ethylene	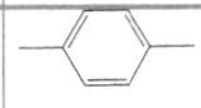	573	24	95	75	[142]
2	WOx-ZrO_2	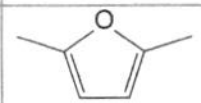	Ethanol	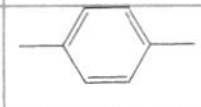	523	6	60	77	[144]
3	H-beta		Ethylene		523	24	99	90	[147]
4	$CF_2ClCOOH$		Ethylene	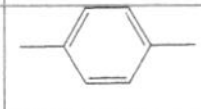	473	24	72	69	[149]
5	$Sc(OTf)_3$	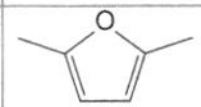	Ethylene	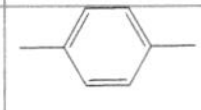	473	24	77	70	[149]
6	SiO_2-Al_2O_3	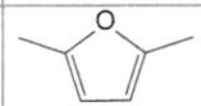	Ethylene	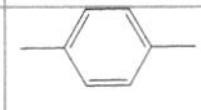	523	6	90	66	[151]
7	Zr-beta	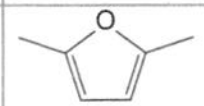	Ethylene	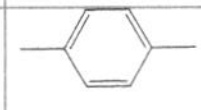	523	24	80	80	[152]
8	HUSY-12	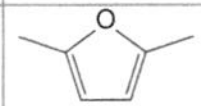	Ethanol	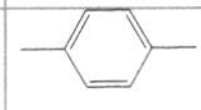	573	12	98	67	[153]
9	Sn-beta	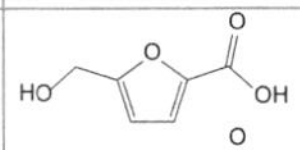	Ethylene	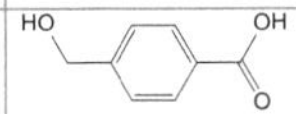	463	6	61	31	[154]
10	Sn-beta	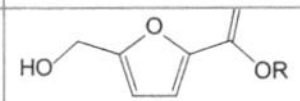	Ethylene	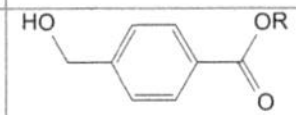	463	6	12	42	[154]
11	Sn-beta	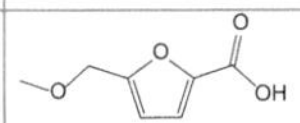	Ethylene	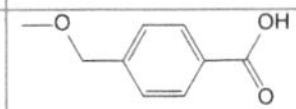	463	6	52	17	[154]
12	Sn-beta	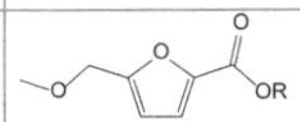	Ethylene	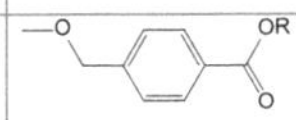	463	6	50	48	[154]
13	Hf-beta		Acrylic acid	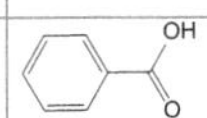	298	24	99	52	[156]

9.8 Conclusions and Future Outlook

Biomass-derived furans are readily available from the hydrolysis of abundant biomass resources and they are now recognized as most important platform molecules that can be obtained from biomass. The catalytic upgrading of biomass-derived furans into value-added chemicals presents a promising strategy for biomass valorization and it has stimulated significant attention since the concept was proposed. Many routes have been explored and their feasibility is primarily decided by economic factors. As for a certain catalytic conversion routes, the economical level is closely related to the efficiency of the catalytic materials. Considering that multiple reaction steps are usually involved in the upgrading of biomass-derived furans, the design of bifunctional or even multifunctional catalysts is highly desirable. The synergistic effects, e.g. metal-metal and metal-support synergistic effects, are the key reason for using bifunctional catalysts instead of combined catalyst systems. Thanks to the significant achievements made so far, the catalytic upgrading of biomass-derived furans is becoming closer and closer to industrial applications. However, there is still some way to go. For a given reaction of biomass upgrading, simultaneous achieving high productivity and high selectivity to target product under mild reaction conditions is challenging. On the other hand, new efficient catalytic conversion routes are still being explored. Nevertheless, with further cooperation between both academy and industry, commercial applications of value-added chemicals from biomass-derived furans can be expected in the near future. For example, polyethylene 2,5-furandicarboxylate is now considered as an interesting alternative to polyethylene terephthalate for packaging applications, which is under intensive research by several companies, e.g. Dupont, Corbion and Avantium.

References

1. Chheda JN, Román-Leshkov Y, Dumesic JA. Production of 5-hydroxymethylfurfural and furfural by dehydration of biomass-derived mono-and poly-saccharides. Green Chem. 2007;9(4):342–50.
2. Delidovich I, Hausoul PJ, Deng L, Pfützenreuter R, Rose M, Palkovits R. Alternative monomers based on lignocellulose and their use for polymer production. Chem Rev. 2015;116(3):1540–99.
3. Bozell JJ, Petersen GR. Technology development for the production of biobased products from biorefinery carbohydrates—the US Department of Energy's "Top 10" revisited. Green Chem. 2010;12(4):539–54.
4. Lima S, Antunes MM, Fernandes A, Pillinger M, Ribeiro MF, Valente AA. Catalytic cyclodehydration of xylose to furfural in the presence of zeolite H-Beta and a micro/mesoporous Beta/TUD-1 composite material. Appl Catal A-Gen. 2010;388(1):141–8.
5. Antunes MM, Lima S, Fernandes A, Pillinger M, Ribeiro MF, Valente AA. Aqueous-phase dehydration of xylose to furfural in the presence of MCM-22 and ITQ-2 solid acid catalysts. Appl Catal A-Gen. 2012;417:243–52.
6. Zhao H, Holladay JE, Brown H, Zhang ZC. Metal chlorides in ionic liquid solvents convert sugars to 5-hydroxymethylfurfural. Science. 2007;316(5831):1597–600.

7. Yong G, Zhang Y, Ying JY. Efficient catalytic system for the selective production of 5-hydroxymethylfurfural from glucose and fructose. Angew Chem Int Ed. 2008;120(48):9485–8.
8. Nikolla E, Román-Leshkov Y, Moliner M, Davis ME. "One-pot" synthesis of 5-(hydroxymethyl) furfural from carbohydrates using tin-beta zeolite. ACS Catal. 2011;1(4):408–10.
9. Chen X, Chew SL, Kerton FM, Yan N. Direct conversion of chitin into a N-containing furan derivative. Green Chem. 2014;16(4):2204–12.
10. Chen X, Gao Y, Wang L, Chen H, Yan N. Effect of treatment methods on chitin structure and its transformation into nitrogen-containing chemicals. ChemPlusChem. 2015;80(10):1565–72.
11. Liu Y, Rowley CN, Kerton FM. Combined experimental and computational studies on the physical and chemical properties of the renewable amide, 3-acetamido-5-acetylfuran. ChemPhysChem. 2014;15(18):4087–94.
12. Yan N, Chen X. Don't waste seafood waste: turning cast-off shells into nitrogen-rich chemicals would benefit economies and the environment. Nature. 2015;524(7564):155–8.
13. Chen X, Yang H, Yan N. Shell biorefinery: dream or reality? Chem Eur J. 2016;22:1–21.
14. Alonso DM, Bond JQ, Dumesic JA. Catalytic conversion of biomass to biofuels. Green Chem. 2010;12(9):1493–513.
15. Lange JP, van der Heide E, van Buijtenen J, Price R. Furfural-a promising platform for lignocellulosic biofuels. ChemSusChem. 2012;5(1):150–66.
16. Mariscal R, Maireles-Torres P, Ojeda M, Sadaba I, Granados ML. Furfural: a renewable and versatile platform molecule for the synthesis of chemicals and fuels. Energy Environ Sci. 2016;9(4):1144–89.
17. Rosatella AA, Simeonov SP, Frade RF, Afonso CA. 5-Hydroxymethylfurfural (HMF) as a building block platform: biological properties, synthesis and synthetic applications. Green Chem. 2011;13(4):754–93.
18. van Putten RJ, van der Waal JC, De Jong ED, Rasrendra CB, Heeres HJ, de Vries JG. Hydroxymethylfurfural, a versatile platform chemical made from renewable resources. Chem Rev. 2013;113(3):1499–597.
19. Rao R, Dandekar A, Baker RTK, Vannice MA. Properties of copper chromite catalysts in hydrogenation reactions. J Catal. 1997;171:406–19.
20. Nagaraja BM, Padmasri AH, Seetharamulu P, Hari Prasad Reddy K, David Raju B, Rama Rao KS. A highly active Cu-MgO-Cr_2O_3 catalyst for simultaneous synthesis of furfuryl alcohol and cyclohexanone by a novel coupling route-combination of furfural hydrogenation and cyclohexanol dehydrogenation. J Mol Catal A Chem. 2007;278:29–37.
21. Chilukuri SVV, Nagpure AS, Lucas NS. U.S. Patent Application No. 15/112,933; 2015.
22. Thirumalaiswamy R, Marimuthu M, Ashok KV. Ni containing anionic clay catalyst useful for selective hydrogenation of furfural to furfuryl alcohol and its preparation thereof. Council of Scientific & Industrial Research; 2016. PCT/IN2015/050050, WO2015198351 A3.
23. Seo G, Chon H. Hydrogenation of furfural over copper-containing catalysts. J Catal. 1981;67(2):424–9.
24. Wu J, Shen Y, Liu C, Wang H, Geng C, Zhang Z. Vapor phase hydrogenation of furfural to furfuryl alcohol over environmentally friendly Cu–Ca/SiO_2 catalyst. Catal Commun. 2005;6:633–7.
25. Xu C, Zheng L, Liu J, Huang Z. Furfural hydrogenation on nickel-promoted Cu-containing catalysts prepared from hydrotalcite-like precursors. Chin J Chem. 2011;29(4):691–7.
26. Yan K, Liao J, Wu X, Xie X. A noble-metal free Cu-catalyst derived from hydrotalcite for highly efficient hydrogenation of biomass-derived furfural and levulinic acid. RSC Adv. 2013;3(12):3853–6.
27. Fulajtárova K, Soták T, Hronec M, Vávra I, Dobročka E, Omastová M. Aqueous phase hydrogenation of furfural to furfuryl alcohol over Pd–Cu catalysts. Appl Catal A-Gen. 2015;502:78–85.
28. Wu J, Gao G, Li J, Sun P, Long X, Li F. Efficient and versatile CuNi alloy nanocatalysts for the highly selective hydrogenation of furfural. Appl Catal B-Environ. 2017;203:227–36.

29. Khromova SA, Bykova MV, Bulavchenko OA, Ermakov DY, Saraev AA, Kaichev VV, Venderbosch RH, Yakovlev VA. Furfural hydrogenation to furfuryl alcohol over bimetallic Ni-Cu sol-gel catalyst: a model reaction for conversion of oxygenates in pyrolysis liquids. Top Catal. 2016;59(15–16):1413–23.
30. Jiménez-Gómez CP, Cecilia JA, Márquez-Rodríguez I, Moreno-Tost R, Santamaría-González J, Mérida-Robles J, Maireles-Torres P. Gas-phase hydrogenation of furfural over Cu/CeO_2 catalysts. Catal Today. 2017;279:327–38.
31. Chen X, Li H, Luo H, Qiao M. Liquid phase hydrogenation of furfural to furfuryl alcohol over Mo-doped Co-B amorphous alloy catalysts. Appl Catal A-Gen. 2002;233:13–20.
32. Li H, Luo H, Zhuang L, Dai W, Qiao M. Liquid phase hydrogenation of furfural to furfuryl alcohol over the Fe-promoted Ni-B amorphous alloy catalysts. J Mol Catal A Chem. 2003;203:267–75.
33. Li H, Zhang S, Luo H. A Ce-promoted Ni-B amorphous alloy catalyst (Ni–Ce–B) for liquid-phase furfural hydrogenation to furfural alcohol. Mater Lett. 2004;58:2741–6.
34. Li H, Chai W-M, Luo H-S, Li H-X. Hydrogenation of furfural to furfuryl alcohol over Co-B amorphous catalysts prepared by chemical reduction in variable media. Chin J Chem. 2006;24:1704–8.
35. Lee J, Burt SP, Carrero CA, Alba-Rubio AC, Ro I, O'Neill BJ, Kim HJ, Jackson DHK, Kuech TF, Hermans I, Dumesic JA, Huber GW. Stabilizing cobalt catalysts for aqueous-phase reactions by strong metal-support interaction. J Catal. 2015;330:19–27.
36. Kijeński J, Winiarek P, Paryjczak T, Lewicki A, Mikołajska A. Platinum deposited on monolayer supports in selective hydrogenation of furfural to furfuryl alcohol. Appl Catal A-Gen. 2002;233(1):171–82.
37. Baker LR, Kennedy G, Van Spronsen M, Hervier A, Cai X, Chen S, Wang LW, Somorjai GA. Furfuraldehyde hydrogenation on titanium oxide-supported platinum nanoparticles studied by sum frequency generation vibrational spectroscopy: acid–base catalysis explains the molecular origin of strong metal–support interactions. J Am Chem Soc. 2012;134(34):14208–16.
38. Li F, Cao B, Ma R, Liang J, Song H, Song H. Performance of Cu/TiO_2-SiO_2 catalysts in hydrogenation of furfural to furfuryl alcohol. Can J Chem Eng. 2016;94(7):1368–74.
39. Tamura M, Tokonami K, Nakagawa Y, Tomishige K. Rapid synthesis of unsaturated alcohols under mild conditions by highly selective hydrogenation. Chem Commun. 2013;49(63):7034–6.
40. Nakagawa Y, Nakazawa H, Watanabe H, Tomishige K. Total hydrogenation of furfural over a silica-supported nickel catalyst prepared by the reduction of a nickel nitrate precursor. ChemCatChem. 2012;4:1791–7.
41. Khairi S, Hara T, Ichikunia N, Shimazu S. Highly efficient and selective hydrogenation of unsaturated carbonyl compounds using Ni-Sn alloy catalysts. Cat Sci Technol. 2012;2:2139–45.
42. Biradar NS, Hengne AM, Birajdar SN, Niphadkar PS, Joshi PN, Rode CV. Single-pot formation of THFAL via catalytic hydrogenation of FFR over Pd/MFI catalyst. ACS Sustain Chem Eng. 2013;2(2):272–81.
43. Mizugaki T, Yamakawa T, Nagatsu Y, Maeno Z, Mitsudome T, Jitsukawa K, Kaneda K. Direct transformation of furfural to 1, 2-pentanediol using a hydrotalcite-supported platinum nanoparticle catalyst. ACS Sustain Chem Eng. 2014;2(10):2243–7.
44. Xu W, Wang H, Liu X, Ren J, Wang Y, Lu G. Direct catalytic conversion of furfural to 1, 5-pentanediol by hydrogenolysis of the furan ring under mild conditions over Pt/Co_2AlO_4 catalyst. Chem Commun. 2011;47(13):3924–6.
45. Liu S, Amada Y, Tamura M, Nakagawa Y, Tomishige K. One-pot selective conversion of furfural into 1, 5-pentanediol over a Pd-added $Ir–ReO_x/SiO_2$ bifunctional catalyst. Green Chem. 2014;16(2):617–26.
46. Liu S, Amada Y, Tamura M, Nakagawa Y, Tomishige K. Performance and characterization of rhenium-modified Rh–Ir alloy catalyst for one-pot conversion of furfural into 1, 5-pentanediol. Cat Sci Technol. 2014;4(8):2535–49.

47. Holdren RF. U.S. Patent No. 2,445,714. Washington, DC: U.S. Patent and Trademark Office; 1948.
48. Lessard J, Morin JF, Wehrung JF, Magnin D, Chornet E. High yield conversion of residual pentoses into furfural via zeolite catalysis and catalytic hydrogenation of furfural to 2-methylfuran. Top Catal. 2010;53(15–18):1231–4.
49. Yan K, Chen A. Selective hydrogenation of furfural and levulinic acid to biofuels on the ecofriendly Cu–Fe catalyst. Fuel. 2014;115:101–8.
50. Sitthisa S, An W, Resasco DE. Selective conversion of furfural to methylfuran over silica-supported Ni Fe bimetallic catalysts. J Catal. 2011;284(1):90–101.
51. Wang J, Liu X, Hu B, Lu G, Wang Y. Efficient catalytic conversion of lignocellulosic biomass into renewable liquid biofuels via furan derivatives. RSC Adv. 2014;4(59):31101–7.
52. Grazia L, Lolli A, Folco F, Zhang Y, Albonetti S, Cavani F. Gas-phase cascade upgrading of furfural to 2-methylfuran using methanol as a H-transfer reactant and MgO based catalysts. Cat Sci Technol. 2016;6(12):4418–27.
53. Scholz D, Aellig C, Hermans I. Catalytic transfer hydrogenation/hydrogenolysis for reductive upgrading of Furfural and 5-(Hydroxymethyl) furfural. ChemSusChem. 2014;7(1): 268–75.
54. Villaverde MM, Garetto TF, Marchi AJ. Liquid-phase transfer hydrogenation of furfural to furfuryl alcohol on Cu-Mg-Al catalysts. Catal Commun. 2015;58:6–10.
55. Zhen HY, Zhu YL, Huang L, Zeng ZY, Wan HJ, Li YW. Study on Cu–Mn–Si catalysts for synthesis of cyclohexanone and 2-methylfuran through the coupling process. Catal Commun. 2008;9(3):342–8.
56. Panagiotopoulou P, Vlachos DG. Liquid phase catalytic transfer hydrogenation of furfural over a Ru/C catalyst. Appl Catal A-Gen. 2014;480:17–24.
57. Panagiotopoulou P, Martin N, Vlachos DG. Effect of hydrogen donor on liquid phase catalytic transfer hydrogenation of furfural over a Ru/RuO_2/C catalyst. J Mol Catal A Chem. 2014;392:223–8.
58. Gilkey MJ, Panagiotopoulou P, Mironenko AV, Jenness GR, Vlachos DG, Xu B. Mechanistic insights into metal lewis acid-mediated catalytic transfer hydrogenation of furfural to 2-methylfuran. ACS Catal. 2015;5(7):3988–94.
59. Yan GC, Lan JC, Chen ZQ. Method for preparing maleic anhydride through selected oxidizing furfural. Chinese Patent CN201410103972; 2014.
60. Lan J, Chen Z, Lin J, Yin G. Catalytic aerobic oxidation of renewable furfural to maleic anhydride and furanone derivatives with their mechanistic studies. Green Chem. 2014;16(9):4351–8.
61. Li X, Ho B, Zhang Y. Selective aerobic oxidation of furfural to maleic anhydride with heterogeneous Mo–V–O catalysts. Green Chem. 2016;18(10):2976–80.
62. Shi S, Guo H, Yin G. Synthesis of maleic acid from renewable resources: catalytic oxidation of furfural in liquid media with dioxygen. Catal Commun. 2011;12(8):731–3.
63. Alonso-Fagúndez N, Agirrezabal-Telleria I, Arias PL, Fierro JLG, Mariscal R, Granados ML. Aqueous-phase catalytic oxidation of furfural with H_2O_2: high yield of maleic acid by using titanium silicalite-1. RSC Adv. 2014;4(98):54960–72.
64. Alba-Rubio AC, Fierro JLG, León-Reina L, Mariscal R, Dumesic JA, Granados ML. Oxidation of furfural in aqueous H_2O_2 catalysed by titanium silicalite: deactivation processes and role of extraframework Ti oxides. App Catal B-Environ. 2017;202:269–80.
65. Lejemble P, Gaset A, Kalck P. From biomass to furan through decarbonylation of furfural under mild conditions. Biomass. 1984;4(4):263–74.
66. Huang YB, Yang Z, Chen MY, Dai JJ, Guo QX, Fu Y. Heterogeneous palladium catalysts for decarbonylation of biomass-derived molecules under mild conditions. ChemSusChem. 2013;6(8):1348–51.
67. Ishida T, Kume K, Kinjo K, Honma T, Nakada K, Ohashi H, Yokoyama T, Hamasaki A, Murayama H, Izawa Y, Utsunomiya M, Tokunaga M. Efficient decarbonylation of furfural to furan catalyzed by zirconia-supported palladium clusters with low atomicity. ChemSusChem. 2016;9(24):3441–7.

68. Jung KJ, Gaset A, Molinier J. Furfural decarbonylation catalyzed by charcoal supported palladium: part II-A continuous process. Biomass. 1988;16(2):89–96.
69. Zhang W, Zhu Y, Niu S, Li Y. A study of furfural decarbonylation on K-doped Pd/Al_2O_3 catalysts. J Mol Catal A Chem. 2011;335(1):71–81.
70. Pushkarev VV, Musselwhite N, An K, Alayoglu S, Somorjai GA. High structure sensitivity of vapor-phase furfural decarbonylation/hydrogenation reaction network as a function of size and shape of Pt nanoparticles. Nano Lett. 2012;12(10):5196–201.
71. Wang C, Wang L, Zhang J, Wang H, Lewis JP, Xiao FS. Product selectivity controlled by zeolite crystals in biomass hydrogenation over a palladium catalyst. J Am Chem Soc. 2016;138(25):7880–3.
72. Bui L, Luo H, Gunther WR, Román-Leshkov Y. Domino reaction catalyzed by zeolites with Brønsted and Lewis acid sites for the production of γ-valerolactone from furfural. Angew Chem Int Ed. 2013;52(31):8022–5.
73. Antunes MM, Lima S, Neves P, Magalhães AL, Fazio E, Fernandes A, Neri F, Silva CM, Riberio MF, Pillinger M, Urakawa A, Valente AA. One-pot conversion of furfural to useful bio-products in the presence of a Sn, Al-containing zeolite beta catalyst prepared via post-synthesis routes. J Catal. 2015;329:522–37.
74. Antunes MM, Neves P, Fernandes A, Lima S, Silva AF, Ribeiro MF, Silva CM, Pillinger M, Valente AA. Bulk and composite catalysts combining BEA topology and mesoporosity for the valorisation of furfural. Cat Sci Technol. 2016;6(21):7812–29.
75. Antunes MM, Lima S, Neves P, Magalhães AL, Fazio E, Fernandes A, Neri F, Silva CM, Riberio MF, Pillinger M, Urakawa A, Valente AA. Integrated reduction and acid-catalysed conversion of furfural in alcohol medium using Zr, Al-containing ordered micro/mesoporous silicates. Appl Catal B-Environ. 2016;182:485–503.
76. Winoto HP, Ahn BS, Jae J. Production of γ-valerolactone from furfural by a single-step process using Sn-Al-Beta zeolites: optimizing the catalyst acid properties and process conditions. J Ind Eng Chem. 2016;40:62–71.
77. Hernández B, Iglesias J, Morales G, Paniagua M, López-Aguado C, Fierro JLG, Wolf P, Hermans I, Melero JA. One-pot cascade transformation of xylose into γ-valerolactone (GVL) over bifunctional Brønsted-Lewis Zr-Al-beta zeolite. Green Chem. 2016;18(21):5777–81.
78. Song S, Di L, Wu G, Dai W, Guan N, Li L. Meso-Zr-Al-beta zeolite as a robust catalyst for cascade reactions in biomass valorization. Appl Catal B-Environ. 2017;205:393–403.
79. Hronec M, Fulajtarová K. Selective transformation of furfural to cyclopentanone. Catal Commun. 2012;24:100–4.
80. Yang Y, Du Z, Huang Y, Lu F, Wang F, Gao J, Xu J. Conversion of furfural into cyclopentanone over Ni–Cu bimetallic catalysts. Green Chem. 2013;15(7):1932–40.
81. Guo J, Xu G, Han Z, Zhang Y, Fu Y, Guo Q. Selective conversion of furfural to cyclopentanone with CuZnAl catalysts. ACS Sustain Chem Eng. 2014;2(10):2259–26.
82. XL L, Deng J, Shi J, Pan T, CG Y, HJ X, Fu Y. Selective conversion of furfural to cyclopentanone or cyclopentanol using different preparation methods of Cu-Co catalysts. Green Chem. 2015;17(2):1038–46.
83. Hronec M, Fulajtárová K, Vávra I, Soták T, Dobročka E, Mičušík M. Carbon supported Pd–Cu catalysts for highly selective rearrangement of furfural to cyclopentanone. Appl Catal B-Environ. 2016;181:210–9.
84. Chatterjee M, Ishizaka T, Kawanami H. Reductive amination of furfural to furfurylamine using aqueous ammonia solution and molecular hydrogen: an environmentally friendly approach. Green Chem. 2016;18(2):487–96.
85. Liang G, Wang A, Li L, Xu G, Yan N, Zhang T. Production of primary amines by reductive amination of biomass-derived aldehydes/ketones. Angew Chem Int Ed. 2017;129(11):3096–100.
86. Ohyama J, Esaki A, Yamamoto Y, Arai S, Satsuma A. Selective hydrogenation of 2-hydroxymethyl-5-furfural to 2, 5-bis (hydroxymethyl) furan over gold sub-nano clusters. RSC Adv. 2013;3(4):1033–6.

87. Chatterjee M, Ishizaka T, Kawanami H. Selective hydrogenation of 5-hydroxymethylfurfural to 2, 5-bis-(hydroxymethyl) furan using Pt/MCM-41 in an aqueous medium: a simple approach. Green Chem. 2014;16:4734–9.
88. Srivastava S, Jadeja GC, Parikh J. Synergism studies on alumina-supported copper-nickel catalysts towards furfural and 5-hydroxymethylfurfural hydrogenation. J Mol Catal A Chem. 2017;426:244–56.
89. Kumalaputri AJ, Bottari G, Erne PM, Heeres HJ, Barta K. Tunable and selective conversion of 5-HMF to 2, 5-furandimethanol and 2, 5-dimethylfuran over copper-doped porous metal oxides. ChemSusChem. 2014;7(8):2266–75.
90. Zhu Y, Kong X, Zheng H, Ding G, Zhu Y, Li YW. Efficient synthesis of 2, 5-dihydroxymethylfuran and 2, 5-dimethylfuran from 5-hydroxymethylfurfural using mineral-derived Cu catalysts as versatile catalysts. Cat Sci Technol. 2015;5(8):4208–17.
91. Alamillo R, Tucker M, Chia M, Pagán-Torres Y, Dumesic J. The selective hydrogenation of biomass-derived 5-hydroxymethylfurfural using heterogeneous catalysts. Green Chem. 2012;14(5):1413–9.
92. Chen J, Liu R, Guo Y, Chen L, Gao H. Selective hydrogenation of biomass-based 5-hydroxymethylfurfural over catalyst of palladium immobilized on amine-functionalized metal-organic frameworks. ACS Catal. 2014;5(2):722–33.
93. Kong X, Zheng R, Zhu Y, Ding G, Zhu Y, Li YW. Rational design of Ni-based catalysts derived from hydrotalcite for selective hydrogenation of 5-hydroxymethylfurfural. Green Chem. 2015;17(4):2504–14.
94. Nakagawa Y, Tomishige K. Total hydrogenation of furan derivatives over silica-supported Ni–Pd alloy catalyst. Catal Commun. 2010;12(3):154–6.
95. Nakagawa Y, Takada K, Tamura M, Tomishige K. Total hydrogenation of furfural and 5-hydroxymethylfurfural over supported Pd–Ir alloy catalyst. ACS Catal. 2014;4(8):2718–26.
96. Tuteja J, Choudhary H, Nishimura S, Ebitani K. Direct synthesis of 1, 6-hexanediol from HMF over a heterogeneous Pd/ZrP catalyst using formic acid as hydrogen source. ChemSusChem. 2014;7(1):96–100.
97. Xiao B, Zheng M, Li X, Pang J, Sun R, Wang H, Pang X, Wang A, Wang X, Zhang T. Synthesis of 1, 6-hexanediol from HMF over double-layered catalysts of Pd/SiO_2+Ir-ReO_x/SiO_2 in a fixed-bed reactor. Green Chem. 2016;18(7):2175–84.
98. Nishimura S, Ikeda N, Ebitani K. Selective hydrogenation of biomass-derived 5-hydroxymethylfurfural (HMF) to 2, 5-dimethylfuran (DMF) under atmospheric hydrogen pressure over carbon supported PdAu bimetallic catalyst. Catal Today. 2014;232:89–98.
99. Saha B, Bohn CM, Abu-Omar MM. Zinc-assisted hydrodeoxygenation of biomass-derived 5-hydroxymethylfurfural to 2, 5-dimethylfuran. ChemSusChem. 2014;7:3095–101.
100. Wang GH, Hilgert J, Richter FH, Wang F, Bongard HJ, Spliethoff B, Weidenthaler C, Schüth F. Platinum–cobalt bimetallic nanoparticles in hollow carbon nanospheres for hydrogenolysis of 5-hydroxymethylfurfural. Nat Mater. 2014;13(3):293–300.
101. Chen B, Li F, Huang Z, Yuan G. Carbon-coated Cu-Co bimetallic nanoparticles as selective and recyclable catalysts for production of biofuel 2, 5-dimethylfuran. Appl Catal B-Environ. 2017;200:192–9.
102. Guo W, Liu H, Zhang S, Han H, Liu H, Jiang T, Han B, Wu T. Efficient hydrogenolysis of 5-hydroxymethylfurfural to 2, 5-dimethylfuran over a cobalt and copper bimetallic catalyst on N-graphene-modified Al_2O_3. Green Chem. 2016;18(23):6222–8.
103. Zu Y, Yang P, Wang J, Liu X, Ren J, Lu G, Wang Y. Efficient production of the liquid fuel 2, 5-dimethylfuran from 5-hydroxymethylfurfural over Ru/Co_3O_4 catalyst. Appl Catal B-Environ. 2014;146:244–24.
104. Nagpure AS, Lucas N, Chilukuri SV. Efficient preparation of liquid fuel 2, 5-dimethylfuran from biomass-derived 5-hydroxymethylfurfural over Ru–NaY catalyst. ACS Sustain Chem Eng. 2015;3:2909–16.
105. Nagpure AS, Venugopal AK, Lucas N, Manikandan M, Thirumalaiswamy R, Chilukuri S. Renewable fuels from biomass-derived compounds: Ru-containing hydrotalcites as catalysts for conversion of HMF to 2, 5-dimethylfuran. Cat Sci Technol. 2015;5:1463–72.

106. Shi J, Wang Y, Yu X, Du W, Hou Z. Production of 2, 5-dimethylfuran from 5-hydroxymethylfurfural over reduced graphene oxides supported Pt catalyst under mild conditions. Fuel. 2016;163:74–9.
107. Huang YB, Chen MY, Yan L, Guo QX, Fu Y. Nickel-tungsten carbide catalysts for the production of 2, 5-dimethylfuran from biomass-derived molecules. ChemSusChem. 2014;7(4):1068–72.
108. Yang P, Cui Q, Zu Y, Liu X, Lu G, Wang Y. Catalytic production of 2, 5-dimethylfuran from 5-hydroxymethylfurfural over Ni/Co_3O_4 catalyst. Catal Commun. 2015;66:55–9.
109. Yang Y, Liu Q, Li D, Tan J, Zhang Q, Wang C, Ma L. Selective hydrodeoxygenation of 5-hydroxymethylfurfural to 2, 5-dimethylfuran on Ru-MoO_x/C catalysts. RSC Adv. 2017;7(27):16311–8.
110. Jae J, Zheng W, Lobo RF, Vlachos DG. Production of dimethylfuran from hydroxymethylfurfural through catalytic transfer hydrogenation with ruthenium supported on carbon. ChemSusChem. 2013;6(7):1158–62.
111. Jae J, Zheng W, Karim AM, Guo W, Lobo RF, Vlachos DG. The role of Ru and RuO_2 in the catalytic transfer hydrogenation of 5-hydroxymethylfurfural for the production of 2, 5-dimethylfuran. ChemCatChem. 2014;6(3):848–56.
112. Takagaki A, Takahashi M, Nishimura S, Ebitani K. One-pot synthesis of 2, 5-diformylfuran from carbohydrate derivatives by sulfonated resin and hydrotalcite-supported ruthenium catalysts. ACS Catal. 2011;1(11):1562–5.
113. Fang R, Luque R, Li Y. Selective aerobic oxidation of biomass-derived HMF to 2, 5-diformylfuran using a MOF-derived magnetic hollow Fe–Co nanocatalyst. Green Chem. 2016;18(10):3152–7.
114. Antonyraj CA, Jeong J, Kim B, Shin S, Kim S, Lee KY, Cho JK. Selective oxidation of HMF to DFF using Ru/γ-alumina catalyst in moderate boiling solvents toward industrial production. J Ind Eng Chem. 2013;19(3):1056–9.
115. Ghosh K, Molla RA, Iqubal MA, Islam SS, Islam SM. Ruthenium nanoparticles supported on N-containing mesoporous polymer catalyzed aerobic oxidation of biomass-derived 5-hydroxymethylfurfural (HMF) to 2, 5-diformylfuran (DFF). Appl Catal A-Gen. 2016;520:44–52.
116. Yang ZZ, Deng J, Pan T, Guo QX, Fu Y. A one-pot approach for conversion of fructose to 2, 5-diformylfuran by combination of Fe_3O_4-SBA-SO_3H and K-OMS-2. Green Chem. 2012;14(11):2986–9.
117. Nie J, Liu H. Efficient aerobic oxidation of 5-hydroxymethylfurfural to 2, 5-diformylfuran on manganese oxide catalysts. J Catal. 2014;316:57–66.
118. Yadav GD, Sharma RV. Biomass derived chemicals: environmentally benign process for oxidation of 5-hydroxymethylfurfural to 2, 5-diformylfuran by using nano-fibrous Ag-OMS-2-catalyst. Appl Catal B-Environ. 2014;147:293–301.
119. Navarro OC, Chornet SI. Chemicals from biomass: aerobic oxidation of 5-hydroxymethyl-2-furaldehyde into diformylfurane catalyzed by immobilized vanadyl-pyridine complexes on polymeric and organofunctionalized mesoporous supports. Top Catal. 2009;52(3):304–14.
120. Sádaba I, Gorbanev YY, Kegnæs S, Putluru SSR, Berg RW, Riisager A. Catalytic performance of zeolite-supported vanadia in the aerobic oxidation of 5-hydroxymethylfurfural to 2, 5-diformylfuran. ChemCatChem. 2013;5(1):284–93.
121. Lv G, Wang H, Yang Y, Deng T, Chen C, Zhu Y, Hou X. Direct synthesis of 2, 5-diformylfuran from fructose with graphene oxide as a bifunctional and metal-free catalyst. Green Chem. 2016;18(8):2302–7.
122. Gorbanev YY, Klitgaard SK, Woodley JM, Christensen CH, Riisager A. Gold-catalyzed aerobic oxidation of 5-hydroxymethylfurfural in water at ambient temperature. ChemSusChem. 2009;2(7):672–5.
123. Casanova O, Iborra S, Corma A. Biomass into chemicals: aerobic oxidation of 5-hydroxymethyl-2-furfural into 2, 5-furandicarboxylic acid with gold nanoparticle catalysts. ChemSusChem. 2009;2(12):1138–44.

124. Lilga MA, Hallen RT, Gray M. Production of oxidized derivatives of 5-hydroxymethylfurfural (HMF). Top Catal. 2010;53(15–18):1264–9.
125. Pasini T, Piccinini M, Blosi M, Bonelli R, Albonetti S, Dimitratos N, Lopez-Sanchez JA, Sankar M, He Q, Kiely CJ, Hutchings GJ, Cavani F. Selective oxidation of 5-hydroxymethyl-2-furfural using supported gold–copper nanoparticles. Green Chem. 2011;13(8):2091–9.
126. Davis SE, Houk LR, Tamargo EC, Datye AK, Davis RJ. Oxidation of 5-hydroxymethylfurfural over supported Pt, Pd and Au catalysts. Catal Today. 2011;160(1):55–60.
127. Wan X, Zhou C, Chen J, Deng W, Zhang Q, Yang Y, Wang Y. Base-free aerobic oxidation of 5-hydroxymethyl-furfural to 2, 5-furandicarboxylic acid in water catalyzed by functionalized carbon nanotube-supported Au-Pd alloy nanoparticles. ACS Catal. 2014;4(7):2175–85.
128. Siyo B, Schneider M, Pohl MM, Langer P, Steinfeldt N. Synthesis, characterization, and application of PVP-Pd NP in the aerobic oxidation of 5-hydroxymethylfurfural (HMF). Catal Lett. 2014;144(3):498–506.
129. Albonetti S, Lolli A, Morandi V, Migliori A, Lucarelli C, Cavani F. Conversion of 5-hydroxymethylfurfural to 2, 5-furandicarboxylic acid over Au-based catalysts: optimization of active phase and metal–support interaction. Appl Catal B-Environ. 2015;163:520–30.
130. Ait RH, Essayem N, Besson M. Selective aerobic oxidation of 5-HMF into 2, 5-furandicarboxylic acid with Pt catalysts supported on TiO_2-and ZrO_2-based supports. ChemSusChem. 2015;8(7):1206–17.
131. Liu B, Ren Y, Zhang Z. Aerobic oxidation of 5-hydroxymethylfurfural into 2, 5-furandicarboxylic acid in water under mild conditions. Green Chem. 2015;17(3):1610–7.
132. Zhang Z, Zhen J, Liu B, Lv K, Deng K. Selective aerobic oxidation of the biomass-derived precursor 5-hydroxymethylfurfural to 2, 5-furandicarboxylic acid under mild conditions over a magnetic palladium nanocatalyst. Green Chem. 2015;17(2):1308–17.
133. Gorbanev YY, Kegnæs S, Riisager A. Effect of support in heterogeneous ruthenium catalysts used for the selective aerobic oxidation of HMF in water. Top Catal. 2011;54(16–18):1318–24.
134. Gupta NK, Nishimura S, Takagaki A, Ebitani K. Hydrotalcite-supported gold-nanoparticle-catalyzed highly efficient base-free aqueous oxidation of 5-hydroxymethylfurfural into 2, 5-furandicarboxylic acid under atmospheric oxygen pressure. Green Chem. 2011;13(4):824–7.
135. Han X, Geng L, Guo Y, Jia R, Liu X, Zhang Y, Wang Y. Base-free aerobic oxidation of 5-hydroxymethylfurfural to 2, 5-furandicarboxylic acid over a Pt/C-O-Mg catalyst. Green Chem. 2016;18(6):1597–604.
136. Han X, Li C, Liu X, Xia Q, Wang Y. Selective oxidation of 5-hydroxymethylfurfural to 2, 5-furandicarboxylic acid over MnO_x-CeO_2 composite catalysts. Green Chem. 2017;2017:996. https://doi.org/10.1039/C6GC03304K.
137. Siankevich S, Savoglidis G, Fei Z, Laurenczy G, Alexander DT, Yan N, Dyson PJ. A novel platinum nanocatalyst for the oxidation of 5-hydroxymethylfurfural into 2, 5-furandicarboxylic acid under mild conditions. J Catal. 2014;315:67–74.
138. Partenheimer W. Methodology and scope of metal/bromide autoxidation of hydrocarbons. Catal Today. 1995;23(2):69–158.
139. Tsai TC, Liu SB, Wang I. Disproportionation and transalkylation of alkylbenzenes over zeolite catalysts. Appl Catal A-Gen. 1999;181(2):355–98.
140. Fong YY, Abdullah AZ, Ahmad AL, Bhatia S. Development of functionalized zeolite membrane and its potential role as reactor combined separator for para-xylene production from xylene isomers. Chem Eng J. 2008;139(1):172–93.
141. Shiramizu M, Toste FD. On the Diels-Alder approach to solely biomass-derived polyethylene terephthalate (PET): conversion of 2, 5-dimethylfuran and acrolein into p-xylene. Chem-A Eur J. 2011;17(44):12452–7.
142. Williams CL, Chang CC, Do P, Nikbin N, Caratzoulas S, Vlachos DG, Lobo RF, Fan W, Dauenhauer PJ. Cycloaddition of biomass-derived furans for catalytic production of renewable p-xylene. ACS Catal. 2012;2(6):935–9.
143. Do PT, McAtee JR, Watson DA, Lobo RF. Elucidation of Diels–Alder reaction network of 2, 5-dimethylfuran and ethylene on HY zeolite catalyst. ACS Catal. 2012;3(1):41–6.

144. Wang D, Osmundsen CM, Taarning E, Dumesic JA. Selective production of aromatics from alkylfurans over solid acid catalysts. ChemCatChem. 2013;5(7):2044–50.
145. Nikbin N, Feng S, Caratzoulas S, Vlachos DG. p-Xylene formation by dehydrative aromatization of a Diels–Alder product in Lewis and Brønsted acidic zeolites. J Phys Chem C. 2014;118(42):24415–24.
146. Nikbin N, Do PT, Caratzoulas S, Lobo RF, Dauenhauer PJ, Vlachos DG. A DFT study of the acid-catalyzed conversion of 2, 5-dimethylfuran and ethylene to p-xylene. J Catal. 2013;297:35–43.
147. Chang CC, Green SK, Williams CL, Dauenhauer PJ, Fan W. Ultra-selective cycloaddition of dimethylfuran for renewable p-xylene with H-BEA. Green Chem. 2014;16(2):585–8.
148. Xiong R, Sandler SI, Vlachos DG, Dauenhauer PJ. Solvent-tuned hydrophobicity for faujasite-catalyzed cycloaddition of biomass-derived dimethylfuran for renewable p-xylene. Green Chem. 2014;16(9):4086–91.
149. Song S, Wu G, Dai W, Guan N, Li L. Diels-Alder and dehydration reactions of furan derivatives with ethylene catalyzed by liquid Brønsted acids and Lewis acids. J Mol Catal A Chem. 2016;420:134–41.
150. Patet RE, Nikbin N, Williams CL, Green SK, Chang CC, Fan W, Caratzoulas S, Dauenhauer PJ, Vlachos DG. Kinetic regime change in the tandem dehydrative aromatization of furan Diels–Alder products. ACS Catal. 2015;5(4):2367–75.
151. Wijaya YP, Suh DJ, Jae J. Production of renewable p-xylene from 2, 5-dimethylfuran via Diels–Alder cycloaddition and dehydrative aromatization reactions over silica-alumina aerogel catalysts. Catal Commun. 2015;70:12–6.
152. Chang CC, Cho HJ, Yu J, Gorte RJ, Gulbinski J, Dauenhauer P, Fan W. Lewis acid zeolites for tandem Diels-Alder cycloaddition and dehydration of biomass-derived dimethylfuran and ethylene to renewable p-xylene. Green Chem. 2016;18(5):1368–76.
153. Teixeira IF, Lo BT, Kostetskyy P, Stamatakis M, Ye L, Tang CC, Mpourmpakis G, Tsang SCE. From biomass-derived furans to aromatics with ethanol over zeolite. Angew Chem Int Ed. 2016;55(42):13061–6.
154. Pacheco JJ, Davis ME. Synthesis of terephthalic acid via Diels-Alder reactions with ethylene and oxidized variants of 5-hydroxymethylfurfural. P Natl Acad Sci. 2014;111(23):8363–7.
155. Pacheco JJ, Labinger JA, Sessions AL, Davis ME. Route to renewable PET: reaction pathways and energetics of Diels-Alder and dehydrative aromatization reactions between ethylene and biomass-derived furans catalyzed by lewis acid molecular sieves. ACS Catal. 2015;5(10):5904–13.
156. Miki J, Osada Y, Konoshi T, Tachibana Y, Shikada T. Selective oxidation of toluene to benzoic acid catalyzed by modified vanadium oxides. Appl Catal A-Gen. 1996;137(1):93–104.
157. Mahmoud E, Yu J, Gorte RJ, Lobo RF. Diels–Alder and dehydration reactions of biomass-derived furan and acrylic acid for the synthesis of benzoic acid. ACS Catal. 2015;5(11):6946–55.

Part III
Production of Biodiesel

Chapter 10
Production of Biodiesel via Simultaneous Esterification and Transesterification

Hu Pan, Heng Zhang, and Song Yang

Abstract Biodiesel is considered as a substitute for fossil-based diesel fuels because of its renewability and environmental friendliness. Biodiesel is generally produced by transesterification of oils or esterification of free fatty acids with methanol. In these production procedures, basic and acidic catalysts enhance biodiesel yields under mild conditions. Particularly, heterogeneous acids promote simultaneous (trans)esterification in a single pot for biodiesel production (free of soap formation) from cheap raw materials containing high contents of free fatty acids, which cannot only avoid complex separation processes involved in pre-esterification, but also simplify catalyst separation and reuse, thus greatly reducing the cost. This chapter reviews the synthesis of biodiesel via simultaneous (trans)esterification catalyzed by mixed metal oxides, acidic ionic liquids, carbon-based solid acids, magnetic solid acids and hybrid solid acids. Advantages of heterogeneous catalytic reaction systems are discussed. Guidance is given on improving the performance of heterogeneous acid catalysts for production of biodiesel.

10.1 Introduction

Energy is the foundation of human survival. Currently, fossil fuel reserves are rapidly depleting due to continuous growth of population and high energy demands. Concern about environmental issues such as global warming related to fossil fuel emissions is motivating the search for renewable energy sources alternatives [1]. Biodiesel, which has many advantages such as reproducibility, biodegradability, low emissions except for NO_x, non-toxicity, high flash point and good lubricity, has a great potential to replace fossil diesel fuels [2–4]. Furthermore, the use of biodiesel as a fuel does not require engine modification on any large-scale [5]. Biodiesel is defined as long chain fatty acid methyl esters (FAMEs) that are synthesized from

H. Pan • H. Zhang • S. Yang (✉)
State-Local Joint Engineering Lab for Comprehensive Utilization of Biomass, State Key Laboratory Breeding Base of Green Pesticide and Agricultural Bioengineering (Ministry of Education), Center for R&D of Fine Chemicals, Guizhou University, Guiyang, China
e-mail: jhzx.msm@gmail.com

Z. Fang et al. (eds.), *Production of Biofuels and Chemicals with Bifunctional Catalysts*, Biofuels and Biorefineries 8, https://doi.org/10.1007/978-981-10-5137-1_10

oils through either esterification of free fatty acids (FFAs) or transesterification of triglycerides (TGs) with methanol, along with by-product (glycerol).

At present, about 95% of the total amount of biodiesel produced made from edible oils (e.g. rapeseed oil, soybean oil, sunflower oil and palm oil) for which the raw materials account for over 70–80% of the total cost [6–9]. Moreover, employing edible feedstocks causes concern since their use competes with food. Given these issues, non-edible oils are very attractive feedstocks for biodiesel production, however non-edible oils, typically contain high concentrations of FFAs and are easy to undergo saponification with base catalysts that increases purification costs and reduces biodiesel yields [10, 11]. Therefore, to avoid saponification issues, a two-step procedure that pre-esterifies FFAs into FAMEs using acid catalysts and trans-esterifies TGs into FAMEs with base catalysts is required to convert non-edible oils into biodiesel [12]. The two-step process is complex and inevitably increases production costs [13, 14].

Simultaneous esterification and transesterification transformation of non-edible oils into biodiesel by acid catalysts is possible option that avoids complex separation process of intermediates [15, 16]. Certain acid catalysts can promote transformation of non-edible oils into biodiesel through simultaneous esterification and transesterification. Employing proper acid catalysts for production of biodiesel can accelerate reaction rate and enhance biodiesel yield under mild reaction conditions.

Homogeneous acid catalysts (i.e., H_2SO_4, HCl) are relatively inexpensive and available, but they have corrosion and reuse issues that cause environmental problems and increase production cost [17]. Solid acid catalysts are promising substitutes for homogeneous acid catalysts because of their advantages associated with reuse, low corrosion, water resistance and low environmental impact [18]. Types of solid acid catalysts include mixed metal oxides, acidic ionic liquids, carbon-based solid acids, magnetic solid acids and hybrid solid acids. However, the production of biodiesel from non-edible oils using solid acids in one pot reaction is limited. This chapter reviews solid acidic catalysts for production of biodiesel as applied to low cost non-edible oils for simultaneous esterification and transesterification.

10.2 Catalysts

The activity of catalysts has an important effect on reaction conditions and biodiesel yield. They also play an important role in costs of biodiesel production. Different types of solid acid catalysts including mixed metal oxides, acidic ionic liquids, carbon-based solid acids, magnetic solid acids and hybrid solid acids were used to prepare biodiesel.

10.2.1 Homogeneous Acid Catalysts

Homogeneous acid catalysts have significant advantages over homogeneous alkali catalysts, since they are the insensitive to FFAs and water, capable of simultaneous catalyzing (trans)esterification reactions and they do not undergo saponification. Furthermore, acid catalyzed procedures for biodiesel production through simultaneous (trans)esterification are more economical than base-catalyzed approaches that require an additional step to convert FFAs to FAMEs [19, 20]. Homogeneous acid catalysts such as H_2SO_4, HCl and H_3PO_4 are used for non-edible oils having high FFA content [21, 22]. Wang et al. [23] used H_2SO_4 (4 wt.% relative to oil) as catalyst for transformation of waste cooking oil and obtained 90% of oil conversion in 10 h for a ratio of methanol to oil of 20:1. Trifluoroacetic acid has higher pK_a value than H_2SO_4, HCl and HNO_3, and exhibited 98.4% biodiesel yields for 2.0 M catalyst and, a methanol to oil ratio of 20:1 at 120 °C for 5 h reaction time [24]. Lewis acid can be used instead of the usual Brønsted acid for transformation of non-edible oils into biodiesel. Guo et al. [25] employed metal chlorides in a relatively common ionic liquid, 1-butyl-3-methylimidazolium tosylate, for conversion of *Jatropha* oil to biodiesel (13.8 mg KOH/g). The biodiesel yield data had the following activity order: $ZnCl_2 > MnCl_2 > FeCl_2 > CoCl_2 > CuCl_2$.

However, homogeneous acid-catalyzed systems are unpopular for biodiesel production because a large amount of energy is required to purify the biodiesel and to removal the catalyst and the catalysts are not readily reusable [26]. Moreover, homogeneous acid catalysts for biodiesel production inevitably lead to equipment corrosion and large amounts of waste water [27].

10.2.2 Heterogeneous Acid Catalysts

Heterogeneous acid catalysts are usually in a solid form and in a different phase from the reaction substrates. Heterogeneous acid catalysts have the following merits: (1) they are insensitive to the FFAs content which allows the use of low-cost feedstocks, (2) they promote esterification and transesterification occur simultaneously, (3) they catalyze the reaction without soap formation, (4) they are easy to separate and recycle and (5) they lower corrosion characteristics. Furthermore, the heterogeneous acid catalysts can be applied to continuous flow reactors to minimize biodiesel separation and purification costs. The following section outlines solid acid catalysts used to produce biodiesel from non-edible oils.

10.2.2.1 Metal Oxides

Mixed metal oxide catalysts typically possess Lewis (anionic) and Brønsted acid sites (cations) that provide the necessary catalytic sites for biodiesel production [28]. Many metal oxides are transition metal groups. Transition metals such as zirconium, titanium, tin, iron, tungsten and zinc are now receiving considerable attention in academic research and in industry.

Zirconium oxide (ZrO_2) has good thermal stability under oxidizing and reducing atmospheres, which can provide acid and base active sites. To increase the activity of zirconium oxide, it is usually applied with sulfuric acid or heteropoly acid. Sulfated zirconia (SO_4^{2-}/ZrO_2) synthesized by impregnating ZrO_2 with sulfuric acid solution, SO_4^{2-}/ZrO_2, was used to transform palm kernel oil and crude coconut oil into biodiesel with 90.3% and 86.3% yields, respectively, under the conditions of 3 wt.% catalyst dosage, 6:1 molar ratio of methanol to oil, 200 °C and 4 h [29]. However, ZrO_2 exhibited only 64.5% (palm kernel oil) and 49.3% (coconut oil) of biodiesel yield, respectively, using methanol to oil molar ratio of 6:1, and 3 wt.% catalyst amount at 200 °C after 4 h reaction time. Thus the surface acidity of the metal oxide has a direct impact on obtaining high yields of biodiesel.

Although SO_4^{2-}/ZrO_2 exhibits effective catalytic activity, significant inactivation was observed, which may be due to the sulfate leaching [30]. A more robust sulfated zirconia ($HClSO_3$-ZrO_2) can be prepared by sulfonating with chlorosulfonic acid instead of sulfuric acid [31]. The catalytic activities of $HClSO_3$-ZrO_2 in conversion crude rice bran oil into biodiesel via the simultaneous esterification and transesterification have been compared with SO_4^{2-}/ZrO_2 [18]. $HClSO_3$-ZrO_2 exhibits higher catalytic activity with 100% FAMEs yield, whereas SO_4^{2-}/ZrO_2 shows only 50% FAMEs yield, which may be because $HClSO_3$-ZrO_2 possesses stronger acid sites than SO_4^{2-}/ZrO_2. $HClSO_3$-ZrO_2 maintains high activity even when either 40 wt.% FFA or 3 wt.% water is present. The doping of ferric-manganese into sulfated zirconia nanoparticle for the sake of enhancing the specific surface area of SO_4^{2-}/ZrO_2 can be synthesized by impregnation followed by calcination at 600 °C for 3 h. The nanoparticle solid acid can be prepared by doping ferric-manganese into sulfated zirconia nanoparticle. A 96.5% yield is obtained with optimized conditions of reaction time of 4 h, 20:1 molar ratio of methanol to oil catalyst concentration of 3 wt.% and reaction temperature of 180 °C [32].

Tungstated zirconia (WO_x/ZrO_2), which can be synthetized by contact of ammonium metatungstate with $Zr(OH)_4$, is more catalytically active than sulfated zirconia. 97% oil conversion using 15 wt.% WO_3/ZrO_2 calcined at 750 °C is obtained under optimum conditions of methanol to oil molar ratio of 20:1, 5 h reaction time, and 200 °C reaction temperature [33]. Furthermore, WO_3/ ZrO_2 is water tolerant and has high acidity [33].

Tin oxide (SnO_2) is an acidic material and has been studied extensively as a catalyst. The SnO_2/SiO_2 catalyst can be prepared by dispersing amorphous SnO_2 species on a SiO_2 surface. Using SiO_2 as support allows high specific surface area and gives the material good thermal stability. An 8% SnO_2/SiO_2 mixture heated at 873 K showed excellent catalytic activity (81.7% oil conversion) using 5 wt % catalyst loading, 24:1 molar ratio of methanol:oil and at 453 K for 5 h reaction time [34]. The SO_4^{2-}/SnO_2 catalyst is a solid super acid that has potential in biodiesel production. The acid strength of SO_4^{2-}/SnO_2 is stronger than that of SO_4^{2-}/ZrO_2 as characterized by temperature programmed desorption of ammonia. The SO_4^{2-}/SnO_2-SiO_2 was synthesized by dipping amorphous SnO_2 and SiO_2 into 2.0 (M) H_2SO_4 solution and the precipitated solid was heated at 300 °C for 2 h. Yields of up to 84% of biodiesel were obtained using 3 wt.% catalyst, 19.5:1 methanol to oil molar ratio and 150 °C for 2.5 h reaction time [35].

Titanium dioxide (TiO_2) has gained interest for biodiesel production because they have several interesting attributes. TiO_2-ZnO and ZnO nanocatalysts can be prepared by urea-glycerol combustion method [36]. The TiO_2-ZnO exhibit better catalytic activity than ZnO for conversion of palm oil to biodiesel, which may be due to the defects created by the replacement of Ti ions in the zinc lattice. A 92% yield of biodiesel was obtained for a 6:1 methanol to oil molar ratio, at a TiO_2-ZnO catalyst dosage of 0.9 wt.% and at 60 °C for 5 h reaction time [36].

Moreover, loading of -SO_3H groups on the TiO_2 strengthen the acidity strength of TiO_2 [37]. The SO_4^{2-}/TiO_2-ZrO_2 with BET specific surface area of 161.9 m^2/g can be prepared by coprecipitation and dipping methods, for which the biodiesel yield reaches 96.7% under optimum reaction conditions: 25:1 mole ratio of methanol to oil, 5 wt.% catalyst dosage relative to oil at 120 °C for 8 h reaction time [38].

10.2.2.2 Acid Ionic Liquids

Ionic liquids (ILs) are organic salts that are typically composed of an anion and cation. Figure 10.1 shows that some of the cations and anions can be used for synthesis. Acidic ionic liquids, which can be considered environmentally benign catalysts, are a new generation of catalysts for producing biodiesel because of their particular characteristics, such as good thermal stability, strong acidity and low vapor pressure [39]. Another significant characteristic of many ILs is they have poor

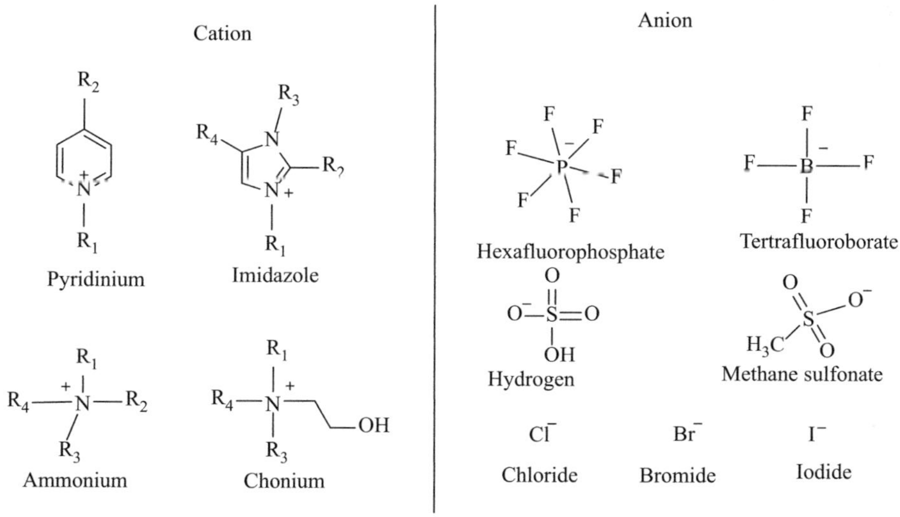

Fig. 10.1 Commonly used cations and anions in ionic liquids

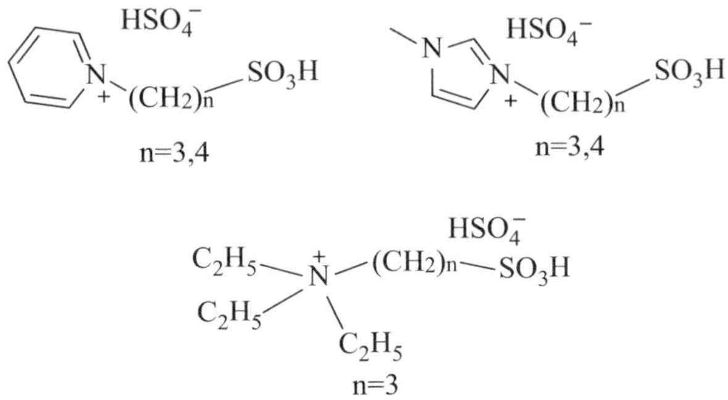

Fig. 10.2 Structures of the five SO_3H-functional Brønsted acidic ionic liquids

solubility in the biodiesel phase, which facilitates the separation of ionic liquids from biodiesel phase [40].

Several Brønsted acidic ionic liquids have been synthesized for the conversion of cottonseed oil with methanol to biodiesel. The structures of the ionic liquids are shown in Fig. 10.2. Wu et al. found that 1-(4-sulfonic acid) butyl pyridinium hydrogen sulfate showed outstanding catalytic activity, and obtained 92% yield of biodiesel from cotton seed oil for a methanol to oil to IL molar ratio of 12:1:0.057 at 170 °C for 5 h reaction time, which is almost the same activity as concentrated H_2SO_4. The excellent catalytic performance was attributed to the strong Brønsted acidity of the IL that is related to the nitrogen groups and the length of the carbon chain in the cations [41]. Liang et al. [42] reported an ionic liquid containing four sulfonic acid groups exhibited above 98% yields for the transformation of rapeseed oil to biodiesel at 70 °C for 7 h. It was found that the presence of Lewis acid played a positive role in the reaction when $[Et_3NH]Cl$ was chosen as the organic species. The biodiesel yield data had the following order in activity: $AlCl_3 > FeCl_3 > SnCl_4 > ZnCl_2 > MgCl_2$, which seemed to be related to the Lewis acidity. $[Et_3NH]Cl\text{-}AlCl_x$ ($x = 0.7$) was found to be the most efficient catalyst with the yield of 98.5% for the synthesis of biodiesel [43].

A Brønsted acid promotes esterification of FFAs while a Lewis acid facilitates transesterification of TGs [44, 45]. Bifunctional acid catalysts that have Brønsted acid and Lewis acid sites can effectively improve the catalytic activity for biodiesel production due to synergistic effect. Liu et al. [46] prepared Brønsted-Lewis acidic ionic liquids $[HO_3S\text{-}(CH_2)_3\text{-}NEt_3]Cl\text{-}xFeCl_3$ (molar fraction of $FeCl_3$ $x = 0.67$) in a "one-pot" reaction and found them to be efficient catalyst for biodiesel production from waste oil that contained high levels of FFAs. Even though the acidic value of the oil was 34.6 mg KOH/g, the yield was still over 95% that is attributed to synergetic effects of the Brønsted acid and Lewis acid catalytic sites. Li et al. [47] prepared various Brønsted-Lewis IL catalysts using Brønsted acidic IL $[BSO_3HMIM]HSO_4$ with a series of metal sulfates, that were applied to biodiesel production from *Camptotheca acuminata* seed oil. The results demonstrated that the metal salts, namely $Fe_2(SO_4)_3$, $CuSO_4$, $ZnSO_4$, $MgSO_4$ and $CaSO_4$ helped to improve the activity of $[BSO_3HMIM]HSO_4$, but decreased in order corresponding to the ability of the metal to accept electrons as follows: $Fe^{3+} > Cu^{2+} > Zn^{2+} > Mg^{2+} > Ca^{2+}$. As well-

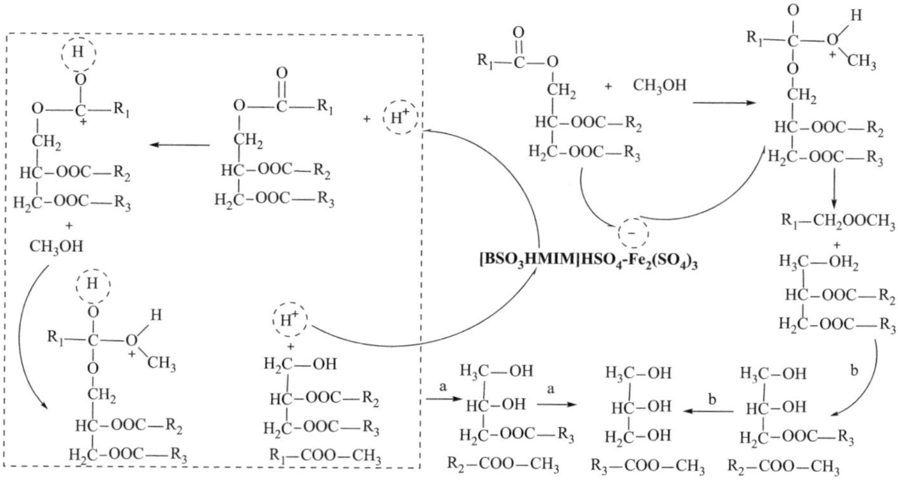

Scheme 10.1 Mechanism of transesterification reaction catalyzed by Brønsted-Lewis acidic ionic liquids (Reprinted with permission from Ref. [47], Copyright © 2014 Elsevier)

known, the Lewis acid activity of a metal is directly proportional to its ability to accept the electrons. Thus, [BSO_3HMIM]HSO_4-$Fe_2(SO_4)_3$ showed superior catalytic activity over other metal sulfates with 95.7% yields of biodiesel being obtained. The catalytic mechanism of transesterification reaction using the prepared Brønsted-Lewis acidic IL can be illustrated in Scheme 10.1.

Despite their excellent catalytic performance, acidic ILs have drawbacks, such as high viscosity, high cost, and solubility in some polar solvents, which limits their application [48]. On the other hand, acidic ILs immobilized onto supports possess the advantages of acidic ILs and heterogeneous catalysts [49, 50]. The emergence of organic polymers brings a great opportunity for supporting acidic ILs for exploiting their unique characteristics such as adjustable wettability and excellent physicochemical stabilities [51]. A series of novel acidic ionic liquid polymers (PILs) were prepared by Liang et al. via the copolymerization of acidic ionic liquid monomers and divinylbenzene, which were employed for biodiesel production from waste oils [52–54]. The catalysts showed over 99% yields of biodiesel from waste oil having high FFA content in a one-pot, reaction that can be ascribed to the unique character of the PILs such as high acidity, hydrophobicity and stability. Considering the large size of TGs and FFAs, mesoporous catalysts can be beneficial for the promotion of mass transport of reactants and products during the catalytic process [55] Acidic ionic liquid-functionalized mesoporous resins (OMR-[C_4HMTA][SO_4H]) were prepared by using a templating method, resulting in high surface area, as illustrated in Scheme 10.2. These IL resins exhibited superior catalytic performance compared with commercial heterogeneous catalysts for biodiesel production from brown grease [56]. The acidic ILs functionalized on superhydrophobic mesoporous polymers (PDVB-[C_1vim][SO_3CF_3]) have a higher activity in transesterification of tripalmitin with methanol than the homogeneous catalyst [C_1vim][SO_3CF_3] with the

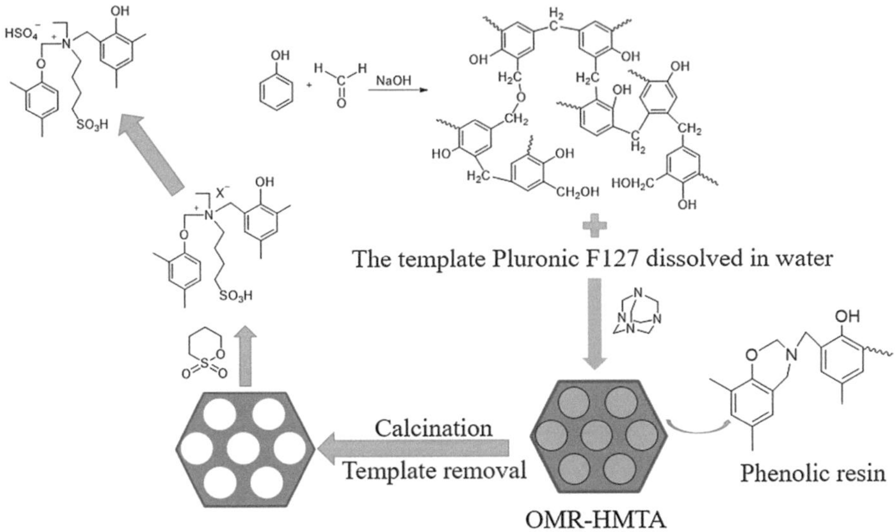

Scheme 10.2 Synthesis of OMR-[C_4HMTA][SO_4H] (Reprinted with permission from Ref. [56], Copyright © 2014 Elsevier)

same number of active sites [57]. The catalytic activity of these materials is associated with their hydrophobicity and mesoporous structure. The ionic liquids and sulfonic acid bifunctionalized mesoporous polymers have been prepared and investigated for biodiesel production from crude *Jatropha curcas* oil and they show 94% of biodiesel yield, which is higher than commercial resins [58].

10.2.2.3 Carbon-Based Solid Acids

Carbon-based solid acids are promising materials as catalysts because of their chemical and thermal stability [59]. Carbon-based solid acids are prepared through incomplete carbonation and then sulfonation. Carbon materials can be sulfonated and readily separated from the concentrated sulfuric acid used in the sulfonation procedure. The characteristic sulfonated carbon material has three acid sites, namely -COOH, -OH and -SO_3H [60]. Despite -SO_3H being the dominant active site for biodiesel production, the amount of -COOH and -OH can influence the hydrophilicity of the catalyst surface, which will affect the catalytic activity of the catalysts [61].

Early research regarding carbon-based materials for catalysts was investigated using refined carbohydrates as feedstock. Carbohydrates are biomolecules consisting mainly of C, H and O elements. Many kinds of refined carbohydrates have been investigated such as glucose, starch and or cellulose [62–64]. Carbon-based solid acids can be synthesized by mixing glucose-starch with ratio of 1:1, calcining the material at 400 °C for 75 min that is followed by sulfonation with H_2SO_4 (98%) at 150 °C for 5 h [65]. The catalyst gradually deactivates upon use, but it can be regenerated by simple H_2SO_4 treatment. High yields (90%) of biodiesel from waste cotton seed oil containing FFAs (55.2 wt.%) was achieved under reaction conditions of 5 wt.% of catalyst loading, 20 molar ratio of methanol to oil and 80 °C for 12 h reaction time. The carbon-based solid acid catalyst can be synthesized by incomplete calcination of glucose or *Calophyllum inophyllum* seed cake under inert atmosphere and then functionalized with concentrated H_2SO_4 (98 wt.%) or *p*-toluenesulfonic acid (99 wt.%) respectively, which can be employed to catalyze the *Calophyllum inophyllum* with FFAs of 15% into biodiesel [66], giving 99% of conversion under optimized conditions.

Wang et al. [67] introduced a method for preparing carbon-based solid acid in one step by carbonizing and sulfonating at the same time with glucose and H_2SO_4 (98 wt.%) as starting materials. A good linearity was found between the yield and the acid density of catalysts for which 97.2% yields for biodiesel were obtained from tributyrin. A SO_3H and NH_2^+ functional carbon-based solid acid was synthesized by carbonizing mixtures of D-glucose, diphenylammonium tosylate ($[Ph_2NH_2]^+[OTs]$) and p-toluenesulfonic acid in a one-step process [68]. 94% yield of biodiesel was obtained from rice oil. Sulfonic acid functionalized carbon with high BET surface area (2000 $m^2\ g^{-1}$) was investigated as a catalyst for biodiesel production from a mixture of triolein and oleic acid as substrate that resembles as a crude vegetable oil [69]. Over 94% yield of biodiesel was obtained because of the catalyst hydrophobicity and high BET surface area. The use of refined carbohydrates as stocks for carbon based catalysts has good potential, but pre-treatment is needed to prepare refined carbohydrate from biomass, which will increase their cost.

Biomass waste causes pollution and requires appropriate treatment to reduce its impact on the environment. Therefore, the research with regard to the use of waste biomass to prepare solid acids has raised some concern. Carbon-based solid acid prepared by sulfonating of carbon material derived from vegetable oil asphalt have been used to produce biodiesel from waste vegetable oil with high content of FFAs in a one-step process [70]. The conversion of TG and FFA was 80.5% and 94.8% respectively, for optimum reaction conditions of 220 °C, methanol oil mole ratio is 16.8:1, catalyst dosage 0.2 wt.% of oil and reaction time 4.5 h. Carbon-based solid acid can be produced from waste Kraft lignin [71]. A carbon-based solid acid derived from waste Kraft lignin was employed for biodiesel production using non-pretreated *Jatropha* oil (12.7 mg KOH/g) as substrate in one-step process with 96.3% yields. To produce carbon-based solid acid with high acid density from lignin, carbonization of lignin in ethanol at supercritical and subcritical temperatures was used that was followed by sulfonation [72]. The catalyst possessed high acid density (5.35 mmol $[H^+]$/g), and gave biodiesel yields of 90.9% and 93.2% were obtained from *Jatropha* and blended soybean oils with acid value of 17.2 and 22.1 mg KOH/g, respectively.

Carbon nanotubes have good prospects as a catalyst or a catalytic support because of their chemical stability, high surface area and non-toxicity. The sulfonic acid functionalized multi-walled carbon nanotubes (S-MWCNTs) catalyst can be employed to catalyze trilaurin with ethanol for biodiesel production. 97.8% yield of biodiesel was obtained under reaction condition of reaction time 1 h at 150 °C,

3.7 wt.% of catalyst loading and a 1:20 mass ratio of ethanol to trilaurin [73]. A series of polymer-nanotubes materials (CNT-P-SO_3H) were synthesized by immobilizing sulfonic acid-functionalized polymers onto multi-walled carbon nanotubes (CNT) by a covalent bond [74]. The CNT-P-SO_3H showed excellent catalytic activity in transesterification of triglycerides with methanol and esterification of oleic acid with methanol.

Acidic ionic liquid functionalized porous nitrogen and carbon materials (NPC-[C_xN][X]) have been prepared by quaternary amination of nitrogen sites in materials with quaternary ammoniation reagents (1,3-propane sultone, and 1,4-butane sultone) and subsequent acid (HSO_3CF_3, H_2SO_4, $H_3PW_{12}O_{40}$, and HBF_4) treatment [75]. NPC-[C_xN][X] showed better activity in producing biodiesel via esterification and transesterification over acid resin and zeolite catalysts. In another study, conversion of waste cow manure into N rich nanoporous carbons without templates was reported [76]. The N-rich nanoporous carbons were treated with 1,4-butane sultone and ion exchanged with HSO_3CF_3 or H_2SO_4 to obtain a carbon based solid acid catalyst for which a about 95% biodiesel yield was achieved from brown grease oil in a one step process.

10.2.2.4 Paramagnetic Solid Acids

Heterogeneous catalysts must be recovered after use through filtration or centrifugation and, there can be the unavoidable loss of catalysts during the separation procedures [77–82]. On the other hand, paramagnetic nanocatalysts can avoid these problems as they can be separated from reaction mixtures by an external magnetic field [83]. Furthermore, as a result of their many advantages, such as large surface-to-volume ratio, tailoring size, shape and composition, they are considered as ideal supports [79, 80, 82–86]. In addition, due to the small size of the particles, paramagnetic nanocatalysts are highly dispersible in solvents, making the active site readily accessible to the substrate. These types of supported catalysts have been recently named as a bridge between homogeneous and heterogeneous catalysis [87]. Owing to the low cost and easy preparation of Fe_3O_4, magnetite (Fe_3O_4) is considered as a ideal and most widely used support [88, 89].

Paramagnetic nanoparticles being coated with silica is beneficial to introduce functional molecules onto the surface of magnetic nanoparticles and reduce nanoparticles agglomeration. Sulfamic acid and sulfonic acid functionalized silica coated Fe/Fe_3O_4 core/shell magnetic nanoparticles can catalyze esterification of oleic acid and the transesterification of glyceryl trioleate [90]. The sulfamic acid-functionalized magnetic nanoparticles showed better reactivity (>95% conversion) in the reaction of transesterification of glyceryl trioleate than sulfonic acid-functionalized MNPs, presumably due to higher acid loading and less aggregating in the reaction solution. More importantly, the acid-functionalized paramagnetic nanoparticles could be easily separated from the mixtures by magneto-precipitation. The Brønsted-Lewis acidic ionic liquid (IL) immobilized on a paramagnetic support Fe_3O_4@SiO_2 (FS-B-L-IL) was synthesized for production biodiesel from non-edible oil (*Koelreuteria integrifoliola*) [91]. The FS-B-L-IL catalyst with uniform core-shell structure and with strong magnetic responsiveness showed high activity with 93.7% biodiesel yield, and could be separated from the mixtures by external magnetic field.

Carbon can also be used as a coating material for paramagnetic nanoparticles owing to its chemical and thermal stability. Magnetically separable SO_3H-functionalized solid acid nanospheres ($Fe_3O_4@SiO_2@PCS$) have been prepared by coating Fe_3O_4 with SiO_2 as well as carbon shell and sulfonation treatment [92]. The synthetic procedure for $Fe_3O_4@SiO_2@PCS$ showed 78% biodiesel yield form soybean oil under reaction conditions reaction temperature 150 °C, 3 wt.% of catalyst dosage, methanol oil mole ratio is 20:1 and 6 h reaction time owing to its high surface area (1118.3 m^2/g) and acidity (1.98 mmol H^+ /g). In a similar manner, a paramagnetic carbon based acid catalyst was prepared from glucose and $FeCl_3$ [93]. The catalyst AC-600-SO_3H@Fe/C possessed 2.79 mmol/g acid density and strong paramagnetism. The catalyst was reusable for biodiesel production and gave 90.5% yields from crude *Jatropha* oil with 17.2 mg KOH/g acid value.

Paramagnetic polymer particles have drawn great research interest because polymers usually possess easy surface modification, high density of functional groups, and high hydrothermal and chemical stabilities [94–97]. Ngu et al. [98] reported on phosphotungstic acid-functionalized magnetic nanoparticles (HPW-PGMA-MNPs) that had paramagnetic nanoparticles as the core, poly(glycidyl methacrylate) as the shell, and phosphotungstic acid functional groups. The catalyst possessed high acid density (1.13 mmol [H^+]/g), and was in nano-size (90 nm). High biodiesel yields of 98% were achieved from grease (21.3 wt.% FFAs) in a one-pot reaction, which was attributed to catalyst size and high acid density and the HPW-PGMA-MNPs could be readily separated with a magnetic field and were reusable. Cellulose is an abundant, and low cost renewable polymer [99]. A paramagnetic catalyst (MCM-HPW) was prepared by supporting $H_3PW_{12}O_{40}$ on paramagnetic cellulose microspheres (MCM) modified by triethylenetetramine through ionic bonds [100]. The MCM-HPW was employed for conversion of *Pistacia chinensis* seed oil into biodiesel and exhibited 93.1% biodiesel yields under mild conditions. The catalyst could be simply separated by a magnetic field, and 80.7% yields were obtained for the fourth cycle.

10.2.2.5 Hybrid Materials

Organic polymers and inorganic materials each have their own merits in application. Hybrid materials use organic moieties that are introduced into a porous inorganic metal network, followed by the introduction of more hydrophobicity onto the surface that is deemed as being highly beneficial to their catalytic activity [101]. A hybrid catalyst was synthetized by immobilizing complex [$Zn(OAc)_2$(N,N′-chelate)] onto silica (specific area, 163 m^2/g) employing *N,N′*-chelate as a suitably functionalized di-imine ligand and showed high activity in the simultaneous trans(esterification) of vegetable oils containing different contents of FFAs, and gave high biodiesel yields at 160 °C for 2 h [102]. As a catalyst support, SBA-15 has attracted wide attention due to its given features such as abundant mesoporous pores, pendent silanol groups, high surface area and good thermal stability [50]. A series of acidic hybrid catalysts were prepared by immobilizing sulfonic acid groups (propyl-SO_3H, arene-SO_3H, perfluoro-SO_3H) on the surface of SBA-15 *via* post-modification approach, which were investigated in the microwave assisted biodiesel production from soybean oil with up to 20 wt.% oleic acid [103]. Interestingly, the acid strength rather than the number of acid sites was found to be the most important factor regarding its catalytic performance. Propyl-SO_3H and arene-SO_3H functionalized SBA-15 catalysts exhibited excellent activity and stability for biodiesel production. Nonetheless, on account of the complete leaching of perfluoro-SO_3H groups during the reaction, perfluoro-SO_3H functionalized SBA-15 catalyst could not be reused.

Bridging organosilanes, which are usually regarded as silica precursor to introduce hydrophobicity and flexibility in a catalyst can be used to design a wide range of periodic mesoporous organic-inorganic hybrids. Specifically, ionic and non-ionic surfactants are generally needed during the process. With regards to this, a series of arenesulfonic or propylsulfonic acid functionalized ethane- or benzene-bridged organosilica nanotubes hybrid catalysts, Si(Ph/Et)Si-Ar/Pr-SO_3H NTs, were manufactured by one step nonionic surfactant-templated sol-gel co-condensation method [104]. Compared with the post-modification approach, this one-step strategy can bring many advantages such as tunable loading of sulfonic acid groups, textural properties and homogeneously distributed active phases throughout the structure of the materials. Due to its high Brønsted acid strength, acid-site density, enhanced surface hydrophobicity, unique porosity properties and perfect tube-shaped morphology, Si(Ph)Si-Ar-SO_3H-10 exhibited the best catalytic activity in the production of biodiesel under mild conditions. Similarly, for the purpose of increasing total acidity, introducing hydrophobicity (terminally bonding methyl groups) and Lewis acid sites (titanium species) onto the surface of organosulfonic acid functionalized silica materials *via* a direct sol-gel co-condensation method under low-temperature hydrothermal treatment has been proposed [105]. The unsaturated surface Ti^{4+} species of the hybrid material TiSBA-15-Me-PrSO_3H afforded Lewis acid sites, while the surface Ti-OH groups furnished extra Brönsted acids sites apart from the protons derived from propylsulfonic acid groups. On the basis of the combination of its strong Brönsted and Lewis acidity, high porosity properties and hydrophobic surface, TiSBA-15-Me-PrSO_3H showed comparatively high performance among heterogeneous acids. It is worth noting that its catalytic activity even outperformed commercially Amberlyst-15 in esterification of palmitic acid and transesterification of yellow horn oil with methanol.

$H_3PW_{12}O_{40}$ was regarded as one of the strongest Brønsted acids, but there are some drawbacks such as high solubility in polar solvents, low surface area, low thermal stability, and difficulty of separation in reaction mixture [106–108]. It is a feasible method to immobilize $H_3PW_{12}O_{40}$ on hybrid material. With regard to this kind of hybrid solid acid, ZrO_2 has both Brønsted and Lewis acidity attracting interest to heterogenize $H_3PW_{12}O_{40}$ [109, 110]. Because of the strong W-O-Zr covalent bindings between $H_3PW_{12}O_{40}$ clusters and ZrO_2 support, the enhanced release of the protons can further increase the Brønsted acidity of the Keggin units [111, 112]. The leakage of $H_3PW_{12}O_{40}$ that generally occurs in most of the supported heteropolyacids can be effectively inhibited by strong interaction between $H_3PW_{12}O_{40}$ and ZrO_2. Hence, a series of ZrO_2-based hybrid catalysts bearing mesostructure *via* one-step sol-gel-hydrothermal method in the presence of a triblock copolymer surfactant (F127), $H_3PW_{12}O_{40}$-ZrO_2-Si(Et)Si were designed [113]. Compared with ethane-free $H_3PW_{12}O_{40}$-ZrO_2, the as-prepared $H_3PW_{12}O_{40}$-ZrO_2-Si(Et)Si hybrids were demonstrated to show higher heterogeneous acid catalytic activity towards biodiesel synthesis, which should be attributed to the combination of strong Brønsted acidity, 3D interconnected mesostructure, and enhanced hydrophobicity. It was found that among these three factors, the inherent acidic properties of $H_3PW_{12}O_{40}$-ZrO_2 play a dominant role in catalytic activity. To prepare solid acid hybrid catalysts with different morphologies, tuning the molar ratios of Si/Zr in the initial gel mixture and dual-template (triblock copolymer surfactant F127 and 1,3,5-trimethylbenzene) is necessary [114]. To be specific, transforming the morphology of the catalysts ($H_3PW_{12}O_{40}$/ZrO_2-Et-HNS) from the 3D interconnected mesostructure to the hollow spherical nanostructure has been implemented and found to have catalytic activity for esterification and transesterification was obtained due to the combination of strong Brønsted and Lewis acidity, unique hollow nanospherical morphology and hydrophobic surface. Other types of organic-inorganic hybrid solid acids such as porous sulfonated phosphonate material [115], metal organic framework (MOF) encapsulated Keggin heteropolyacid [116] and encapsulating ILs into the frameworks of MIL-100(Fe) [117] have been synthesized. However, they have been mainly studied for in the production of biodiesel *via* esterification of FFAs with methanol, which may be due to low activity for preparation of biodiesel from non-edible oil.

10.3 Conclusions and Future Outlook

Non-food feedstocks, such as *Jatropha curcas*, micro algae and waste cooking oils, are considered to be highly promising for sustainable production of biodiesel. Although these feedstocks contain high concentrations of FFAs (>1 wt.%) and impurities such as water in their raw form, they can readily be converted to biodiesel in a one-step process with heterogeneous solid acid catalysts.

Mixed metal oxides solid acid catalysts have high thermal stability, but typically exhibit low catalytic activity for producing biodiesel from non-edible feedstocks, probably due to low acid strength. Introduction of sulfuric group or heteropoly acids on the surface of mixed metal oxides will improve their catalytic activity because of the increase of the acid strength and acid density. SO_3H-functionalized ILs are an important acidic catalysts because of their particular properties such as high thermal stability and strong acidity. Immobilized SO_3H-functionalized ILs on mesoporous materials or nanomaterials are advantageous in reducing the loss of ILs and in improving separation efficiency. Sulfonated carbon-based solid acids are promising for biodiesel preparation because of their merits such as high mechanical, chemical and thermal stability and their economy of source being synthesized from biomass. Paramagnetic nanocatalysts can simplify a chemical conversion process and can be efficiently separated by an external magnetic field. However, magnetic nanoparticles show low acidity and tend to aggregate, which restricts their catalytic applications. Modification of paramagnetic nanoparticles to improve acid density and reduce aggregation has attracted increasing attention. Acidic organic-inorganic hybrid catalysts combine the merits of organic materials and inorganic materials, such as high hydrophobicity and chemical stability compared with their inorganic counterparts and better thermal stability than their organic counterparts.

The aforementioned solid acid catalysts are often subject to deactivation, acid site leaching, poisoning and high cost. Thereby, it is still deemed as a challenge to develop efficient, robust and catalytically-stable solid acids to produce biodiesel from high FFAs content feedstocks *via* a one-pot reaction. To address these issues, the following matters should be taken into consideration. Firstly, high acid density along with acid strength is usually helpful to improve the activity of the catalyst. Bifunctional acid catalysts combining Brønsted acid and Lewis acid sites can effectively improve the catalytic activity to produce biodiesel due to synergistic effects. Secondly, high porosity including uniform mesoporous structure, high surface area and large pore volume can not only increase the number of active sites, but it can also reduce the mass-transport limitations and help to improve the contact between active sites and substrates. Thirdly, hydrophobic surfaces can cause oils to preferentially adsorb onto the catalyst surface as well as avoiding possible deactivation of active sites. Lastly, development of a facile, cost-effective approach to the synthesis of solid catalysts with high catalytic activity will allow production of biodiesel on a large-scale in practical applications and provide new green and economic processes.

References

1. Sani YM, Daud WMAW, Aziz AA. Activity of solid acid catalysts for biodiesel production: a critical review. Appl Catal A-Gen. 2014;470:140–61.
2. Guo WL, Li HL, Ji GL, Zhang GY. Ultrasound-assisted production of biodiesel from soybean oil using Brønsted acidic ionic liquid as catalyst. Bioresour Technol. 2012;125:332–4.
3. Meher LC, Vidya Sagar D, Naik SN. Technical aspects of biodiesel production by transesterification-a review. Renew Sust Energ Rev. 2006;10:248–68.
4. Alcantara R, Canoira J, Amores LT, Fidalgo E, Franco MJ, Navarro A. Catalytic production of biodiesel from soy-bean oil, used frying oil and tallow. Biomass Bioenergy. 2000;18:515–27.
5. Sarin R, Sharma M, Sinharay S, Malhotra RK. Jatropha-palm biodiesel blends: an optimum mix for Asia. Fuel. 2007;86:1365–71.
6. Zhang H, Zhou Q, Chang F, Pan H, Liu XF, Li H, DY H, Yang S. Production and fuel properties of biodiesel from *Firmiana platanifolia* L.f. as a potential non-food oil source. Ind Crop Prod. 2015;76:768–71.
7. Hajjari M, Tabatabaei M, Aghbashlo M, Ghanavati H. A review on the prospects of sustainable biodiesel production: a global scenario with an emphasis on waste-oil biodiesel utilization. Renew Sust Energ Rev. 2017;72:445–64.

8. Sajjadi B, Abdul Raman AA, Arandiyan H. A comprehensive review on properties of edible and non-edible vegetable oil-based biodiesel: composition, specifications and prediction models. Renew Sust Energ Rev. 2016;63:62–92.
9. Lee AF, Bennett JA, Manayil JC, Wilson K. Heterogeneous catalysis for sustainable biodiesel production via esterification and transesterification. Chem Soc Rev. 2014;43:7887–916.
10. Yin P, Chen L, Wang ZD, RJ Q, Liu XG, Ren SH. Production of biodiesel by esterification of oleic acid with ethanol over organophosphonic acid-functionalized silica. Bioresour Technol. 2012;110:258–63.
11. He LQ, Qin SJ, Chang T, Sun YZ, Gao XR. Biodiesel synthesis from the esterification of free fatty acids and alcohol catalyzed by long-chain Brønsted acid ionic liquid. Cat Sci Technol. 2013;3:1102–7.
12. Berchmans HJ, Hirata S. Biodiesel production from crude Jatropha curcas L. Seed oil with a high content of free fatty acids. Bioresour Technol. 2008;99:1716–21.
13. HF L, Liu YY, Zhou H, Yang Y, Chen MY, Liang B. Production of biodiesel from Jatropha curcas L. oil. Comput Chem Eng. 2009;33:1091–6.
14. de Oliveira JS, Leite PM, de Souza LB, Mello VM, Silva EC, Rubim JC, Meneghetti SM, Suarez PA. Characteristics and composition of *Jatropha gossypiifolia* and *Jatropha curcas L.* oils and application for biodiesel production. Biomass Bioenergy. 2009;33:449–53.
15. West AH, Posarac D, Ellis N. Assessment of four biodiesel production processes using HYSYS. Plant Bioresour Technol. 2008;99:6587–601.
16. Marchetti JM, Miguel VU, Errazu AF. Techno-economic study of different alternatives for biodiesel production. Fuel Process Technol. 2008;89:740–8.
17. Kolvari E, Koukabi N, Hosseini MM, Vahidian M, Ghobadi E. Nano-ZrO_2 sulfuric acid: a heterogeneous solid acid nano catalyst for Biginelli reaction under solvent free conditions. RSC Adv. 2016;6:7419–25.
18. Zhang Y, Wong WT, Yung KF. One-step production of biodiesel from rice bran oil catalyzed by chlorosulfonic acid modified zirconia via simultaneous esterification and transesterification. Bioresour Technol. 2013;147:59–64.
19. Zhang Y, Dube MA, McLean DD, Kates M. Biodiesel production from waste cooking oil:1. Process design and technological assessment. Bioresour Technol. 2003;89:1–16.
20. Zhang Y, Dubeì MA, McLean DD, Kates M. Biodiesel production from waste cooking oil: 2. Economic assessment and sensitivity analysis. Bioresour Technol. 2003;90:229–40.
21. Zabeti M, Daud WAW, Aroua MK. Activity of solid catalysts for biodiesel production: a review. Fuel Process Technol. 2009;90:770–7.
22. Konwar LJ, Boro J, Deka D. Review on latest developments in biodiesel production using carbon-based catalysts. Renew Sust Energ Rev. 2014;29:546–64.
23. Wang Y, Ou S, Liu P, Xue F, Tang S. Comparison of two different processes to synthesize biodiesel by waste cooking oil. J Mol Catal A-Chem. 2006;252:107–12.
24. Miao X, Li R, Yao H. Effective acid-catalyzed transesterification for biodiesel production. Energ Convers and Manage. 2009;50:2680–4.
25. Guo F, Fang Z, Tian XF, Long YD, Jiang LQ. One-step production of biodiesel from *Jatropha* oil with high-acid value in ionic liquids. Bioresour Technol. 2013;140:447–50.
26. Long Y, Fang Z, Su T, Yang Q. Co-production of biodiesel and hydrogen from rapeseed and Jatropha oils with sodium silicate and Ni catalysts. Appl Energy. 2014;113:1819–25.
27. Jyoti L, Boro J, Deka D. Review on latest developments in biodiesel production using carbon-based catalysts. Renew Sust Energ Rev. 2014;29:546–64.
28. Yan S, Salley SO, Ng KYS. Simultaneous transesterification and esterification of unrefined or waste oils over ZnO-La_2O_3 catalysts. Appl Catal A Gen. 2009;353:203–12.
29. Jitputti J, Kitiyanan B, Rangsunvigit P, Bunyakiat K, Attanatho L, Jenvanitpanjakul P. Transesterification of crude palm kernel oil and crude coconut oil by different solid catalysts. Chem Eng J. 2006;116:61–6.
30. Omota F, Dimian AC, Bliek A. Fatty acid esterification by reactive distillation: part 2-kinetics-based design for sulphated zirconia catalysts. Chem Eng Sci. 2003;58:3175–85.

31. Grecea ML, Dimian AC, Tanase S, Subbiah V, Rothenberg G. Sulfated zirconia as a robust superacid catalyst for multiproduct fatty acid esterification. Cat Sci Technol. 2012;2:1500–6.
32. Alhassan FH, Rashid U, Taufiq-Yap YH. Synthesis of waste cooking oil-based biodiesel via effectual recyclable bi-functional Fe_2O_3-MnO-SO_4^{2-}/ZrO_2 nanoparticle solid catalyst. Fuel. 2015;142:38–45.
33. Xie W, Wang T. Biodiesel production from soybean oil transesterification using tin oxide-supported WO_3 catalysts. Fuel Process Technol. 2013;109:150–5.
34. Xie W, Wang H, Li H. Silica-supported tin oxides as heterogeneous acid catalysts for transesterification of soybean oil with methanol. Ind Eng Chem Res. 2011;51:225–31.
35. Kafuku G, Lam MK, Kansedo J, Lee KT, Mbarawa M. Heterogeneous catalyzed biodiesel production from Moringa oleifera oil. Fuel Process Technol. 2010;91:1525–9.
36. Madhuvilakku R, Piraman S. Biodiesel synthesis by TiO_2-ZnO mixed oxide nanocatalyst catalyzed palm oil transesterification process. Bioresour Technol. 2013;150:55–9.
37. de Almeida RM, Noda LK, Gonçalves NS, Meneghetti SMP, Meneghetti MR. Transesterification reaction of vegetable oils, using superacid sulfated TiO_2-base catalysts. Appl Catal A-Gen. 2008;347:100–5.
38. Zhou HB, Cao Y, Li J. Research of preparation of SO_4^{2-}/TiO_2-ZrO_2 and its application on synthesis of biodiesel from waste cooking oil. Appl Mech Mater Trans Tech Publ. 2013;316:906–10.
39. Zhang Q, Luo J, Wei Y. A silica gel supported dual acidic ionic liquid: an efficient and recyclable heterogeneous catalyst for the one-pot synthesis of amidoalkyl naphthols. Green Chem. 2010;12:2246–54.
40. Fang D, Yang J, Jiao C. Dicationic ionic liquids as environmentally benign catalysts for biodiesel synthesis. ACS Catal. 2010;1:42–7.
41. Wu Q, Chen H, Han M, Wang D, Wang J. Transesterification of cottonseed oil catalyzed by Brønsted acidic ionic liquids. Ind Eng Chem Res. 2007;46:7955–60.
42. Liang X, Yang J. Synthesis of a novel multi-SO_3H functionalized ionic liquid and its catalytic activities for biodiesel synthesis. Green Chem. 2010;12:201–4.
43. Liang X, Gong G, Wu H, Yang J. Highly efficient procedure for the synthesis of biodiesel from soybean oil using chloroaluminate ionic liquid as catalyst. Fuel. 2009;88:613–6.
44. Lopez DE, Suwannakarn K, Bruce DA, Goodwin JG. Esterification and transesterification on tungstated zirconia: effect of calcination temperature. J Catal. 2007;247:43–50.
45. Di Serio M, Tesser R, Dimiccoli M, Cammarota F, Nastasi M, Santacesaria E. Synthesis of biodiesel via homogeneous Lewis acid catalyst. J Mol Catal A-Chem. 2005;239:111–5.
46. Liu S, Wang Z, Li K, Li L, Yu S, Liu F, Song Z. Brønsted-Lewis acidic ionic liquid for the "one-pot" synthesis of biodiesel from waste oil. J Renew Sustain Energ. 2013;5:023111.
47. Li J, Peng X, Luo M, Zhao CJ, CB G, YG Z, YJ F. Biodiesel production from Camptotheca acuminata seed oil catalyzed by novel Brönsted-Lewis acidic ionic liquid. Appl Energy. 2014;115:438–44.
48. Yin S, Sun J, Liu B, Zhang Z. Magnetic material grafted cross-linked imidazolium based polyionic liquids: an efficient acid catalyst for the synthesis of promising liquid fuel 5-ethoxymethylfurfural from carbohydrates. J Mater Chem A. 2015;3:4992–9.
49. Xu Z, Wan H, Miao J, Han M, Yang C, Guan GF. Reusable and efficient polystyrene-supported acidic ionic liquid catalyst for esterifications. J Mol Catal A-Chem. 2010;332:152–7.
50. Zhang L, Cui Y, Zhang C, Wang L, Wan H, Guan GF. Biodiesel production by esterification of oleic acid over brønsted acidic ionic liquid supported onto Fe-incorporated SBA-15. Ind Eng Chem Res. 2012;51:16590–6.
51. Pan H, Liu XF, Zhang H, Yang KL, Huang S, Yang S. Multi-SO_3H functionalized mesoporous polymeric acid catalyst for biodiesel production and fructose-to-biodiesel additive conversion. Renew Energy. 2017;107:245–52.
52. Liang XZ. Novel acidic ionic liquid polymer for biodiesel synthesis from waste oils. Appl Catal A-Gen. 2013;455:206–10.

53. Liang XZ, Xiao H, Qi C. Efficient procedure for biodiesel synthesis from waste oils using novel solid acidic ionic liquid polymer as catalysts. Fuel Process Technol. 2013;110:109–13.
54. Liang XZ. Novel efficient procedure for biodiesel synthesis from waste oils using solid acidic ionic liquid polymer as the catalyst. Ind Eng Chem Res. 2013;52:6894–900.
55. Liu F, Kong W, Qi C, Zhu L, Xiao FS. Design and synthesis of mesoporous polymer-based solid acid catalysts with excellent hydrophobicity and extraordinary catalytic activity. ACS Catal. 2012;2:565–72.
56. Noshadi I, Kanjilal B, Du S, Bollas GM, Suib SL, Provatas A, Liu F, Parnas RS. Catalyzed production of biodiesel and bio-chemicals from brown grease using Ionic Liquid functionalized ordered mesoporous polymer. Appl Energy. 2014;129:112–22.
57. Liu F, Wang L, Sun Q, Zhu L, Meng X, Xiao FS. Transesterification catalyzed by ionic liquids on superhydrophobic mesoporous polymers: heterogeneous catalysts that are faster than homogeneous catalysts. J Am Chem Soc. 2012;134:16948–50.
58. Pan H, Li H, Liu XF, Zhang H, Yang KL, Huang S, Yang S. Mesoporous polymeric solid acid as efficient catalyst for (trans) esterification of crude *Jatropha curcas* oil. Fuel Process Technol. 2016;150:50–7.
59. Chang B, Li Y, Guo Y, Yang B. Simple fabrication of magnetically separable mesoporous carbon sphere with excellent catalytic performance for biodiesel production. J Taiwan Inst Chem E. 2016;60:241–6.
60. Kastner JR, Miller J, Geller DP, Locklin J, Keith LH, Johnson T. Catalytic esterification of fatty acids using solid acid catalysts generated from biochar and activated carbon. Catal Today. 2012;190:122–32.
61. Kang S, Ye J, Chang J. Recent advances in carbon-based sulfonated catalyst: preparation and application. Int Rev Chem Eng. 2013;5:133–44.
62. Takagaki A, Toda M, Okamura M, Kondo JN, Hayashi S, Domen K, Hara M. Esterification of higher fatty acids by a novel strong solid acid. Catal Today. 2006;116:157–61.
63. Lou WY, Zong MH, Duan ZQ. Efficient production of biodiesel from high free fatty acid-containing waste oils using various carbohydrate-derived solid acid catalysts. Bioresour Technol. 2008;99:8752–8.
64. Ayodele OO, Dawodu FA. Production of biodiesel from *Calophyllum inophyllum* oil using a cellulose-derived catalyst. Biomass Bioenergy. 2014;70:239–48.
65. Chen G, Fang B. Preparation of solid acid catalyst from glucose-starch mixture for biodiesel production. Bioresour Technol. 2011;102:2635–40.
66. Dawodu FA, Ayodele O, Xin J, Zhang S, Yan D. Effective conversion of non-edible oil with high free fatty acid into biodiesel by sulphonated carbon catalyst. Appl Energy. 2014;114:819–26.
67. Wang C, Yuan F, Liu L, Niu X, Zhu Y. Transesterification of tributyrin and dehydration of fructose over a carbon-based solid acid prepared by carbonization and sulfonation of glucose. ChemPlusChem. 2015;80:1657–65.
68. Cai L, Meng D, Zhan S, Yang X, Liu T, Pu H, Tao X. SO_3H and NH^{2+} functional carbon-based solid acid catalyzed transesterification and biodiesel production. RSC Adv. 2015;5:72146–9.
69. Fukuhara K, Nakajima K, Kitano M, Hayashi S, Hara M. Transesterification of Triolein over hydrophobic microporous carbon with SO_3H groups. ChemCatChem. 2015;7(23):3945–50.
70. Shu Q, Gao J, Nawaz Z, Liao Y, Wang D, Wang J. Synthesis of biodiesel from waste vegetable oil with large amounts of free fatty acids using a carbon-based solid acid catalyst. Appl Energy. 2010;87:2589–96.
71. Pua FL, Fang Z, Zakaria S, Guo F, Chia CH. Direct production of biodiesel from high-acid value Jatropha oil with solid acid catalyst derived from lignin. Biotechnol Biofuels. 2011;4:56.
72. Huang M, Luo J, Fang Z, Li H. Biodiesel production catalyzed by highly acidic carbonaceous catalysts synthesized via carbonizing lignin in sub-and super-critical ethanol. Appl Catal B-Environ. 2016;190:103–14.

73. Guan Q, Li Y, Chen Y, Shi Y, Gu J, Li B, Miao R, Chen Q, Ning P. Sulfonated multi-walled carbon nanotubes for biodiesel production through triglycerides transesterification. RSC Adv. 2017;7:7250–8.
74. Liu H, Chen J, Chen L, Xu Y, Guo X, Fang D. Carbon nanotube-based solid Sulfonic acids as catalysts for production of Fatty Acid Methyl Ester via transesterification and esterification. ACS Sustain Chem Eng. 2016;4:3140–50.
75. Liu F, Li B, Liu C, Kong W, Yi X, Zheng A, Qi C. Template-free synthesis of porous carbonaceous solid acids with controllable acid sites and their excellent activity for catalyzing the synthesis of biofuels and fine chemicals. Cat Sci Technol. 2016;6:2995–3007.
76. Noshadi I, Kanjilal B, Liu F. Porous carbonaceous solid acids derived from farm animal waste and their use in catalyzing biomass transformation. Appl Catal A-Gen. 2016;513:19–29.
77. Fihri A, Bouhrara M, Nekoueishahraki B, Basset JM, Polshettiwar V. Nanocatalysts for Suzuki cross-coupling reactions. Chem Soc Rev. 2011;40:5181–203.
78. Roduner E. Size matters: why nanomaterials are different. Chem Soc Rev. 2006;35:583–92.
79. Somorjai GA, Frei H, Park JY. Advancing the frontiers in nanocatalysis, biointerfaces, and renewable energy conversion by innovations of surface techniques. J Am Chem Soc. 2009;131:16589–605.
80. Goesmann H, Feldmann C. Nanoparticulate functional materials. Angew Chem Int Ed. 2010;49:1362–95.
81. Molnár A. Efficient, selective, and recyclable palladium catalysts in carbon-carbon coupling reactions. Chem Rev. 2011;111:2251–320.
82. Chng LL, Erathodiyil N, Ying JY. Nanostructured catalysts for organic transformations. Acc Chem Res. 2013;46:1825–37.
83. Verma S, Verma D, Sinha AK, Jain SL. Palladium complex immobilized on graphene oxide-magnetic nanoparticle composites for ester synthesis by aerobic oxidative esterification of alcohols. Appl Catal A-Gen. 2015;489:17–23.
84. Cubides-Roman DC, Pérez VH, de Castro HF, Orrego CE, Giraldo OH, Silveira EG, David GF. Ethyl esters (biodiesel) production by Pseudomonas fluorescens lipase immobilized on chitosan with magnetic properties in a bioreactor assisted by electromagnetic field. Fuel. 2017;196:481–7.
85. Mitsudome T, Kaneda K. Advanced core–shell nanoparticle catalysts for efficient organic transformations. ChemCatChem. 2013;5:1681–91.
86. Bai C, Liu M. Implantation of nanomaterials and nanostructures on surface and their applications. Nano Today. 2012;7:258–81.
87. Baig RBN, Varma RS. Magnetically retrievable catalysts for organic synthesis. Chem Commun. 2013;49:752–70.
88. Gawande MB, Branco PS, Varma RS. Nano-magnetite (Fe_3O_4) as a support for recyclable catalysts in the development of sustainable methodlogies. Chem Soc Rev. 2013;42:3371–93.
89. Roy S, Pericas MA. Functionalized nanoparticles as catalysts for enantioselective processes. Org Biomol Chem. 2009;7:2669–77.
90. Wang H, Covarrubias J, Prock H, Wu X, Wang D, Bossmann SH. Acid-functionalized magnetic nanoparticle as heterogeneous catalysts for biodiesel synthesis. J Phys Chem C. 2015;119:26020–8.
91. Zhang H, Li H, Pan H, Liu X, Yang K, Huang S, Yang S. Efficient production of biodiesel with promising fuel properties from *Koelreuteria integrifoliola* oil using a magnetically recyclable acidic ionic liquid. Energ Conv Manage. 2017;138:45–53.
92. Chang B, TianY SW, Liu J, Xi F, Dong X. Magnetically separable porous carbon nanospheres as solid acid catalysts. RSC Adv. 2013;3:20999–1006.
93. Zhang F, Fang Z, Wang YT. Biodiesel production direct from high acid value oil with a novel magnetic carbonaceous acid. Appl Energy. 2015;155:637–47.
94. Chen H, Dong M, Li Y, Kong J, Chai Z, Fu G. Facile synthesis of high-magnetization Fe_3O_4@ polydivinylbenzene core-shell submicrospheres. React Funct Polym. 2013;73:18–22.

95. Rodriguez AFR, Coaquira JAH, Morales MA, Faria FSEDV, Cunha RM, Santos JG, Azevedo RB. Synthesis, characterization and magnetic properties of polymer-Fe_3O_4 nanocomposite. Spectrochim Acta A. 2013;100:101–3.
96. Zhang Y, Ying JY. Main-chain organic frameworks with advanced catalytic functionalities. ACS Catal. 2015;5:2681–91.
97. Sun Q, Dai Z, Meng X, Xiao FS. Porous polymer catalysts with hierarchical structures. Chem Soc Rev. 2015;44:6018–34.
98. Ngu TA, Li Z. Phosphotungstic acid-functionalized magnetic nanoparticles as an efficient and recyclable catalyst for the one-pot production of biodiesel from grease via esterification and transesterification. Green Chem. 2014;16:1202–10.
99. Klemm D, Heublein B, Fink HP, Bohn A. Cellulose: fascinating biopolymer and sustainable raw material. Angew Chem Int Ed. 2005;44:3358–93.
100. Han YZ, Hong L, Wang XQ, Liu JZ, Jiao J, Luo M, YJ F. Biodiesel production from Pistacia chinensis seed oil via transesterification using recyclable magnetic cellulose-based catalyst. Ind Crop Prod. 2016;89:332–8.
101. Kalbasi RJ, Kolahdoozan M, Vanani SM. Preparation, characterization and catalyst application of ternary interpenetrating networks of poly 4-methyl vinyl pyridinium hydroxide-SiO_2-Al_2O_3. J Solid State Chem. 2011;184:2009–16.
102. Benessere V, Cucciolito ME, Esposito R, Lega M, Turco R, Ruffo F, Di Serio M. A novel and robust homogeneous supported catalyst for biodiesel production. Fuel. 2016;171:1–4.
103. Zuo D, Lane J, Culy D, Schultz M, Pullar A, Waxman M. Sulfonic acid functionalized mesoporous SBA-15 catalysts for biodiesel production. Appl Catal B Environ. 2013;129:342–50.
104. Zhang X, Su F, Song D, An S, Lu B, Guo Y. Preparation of efficient and recoverable organosulfonic acid functionalized alkyl-bridged organosilica nanotubes for esterification and transesterification. Appl Catal B-Environ. 2015;163:50–62.
105. Léon CIS, Song D, Su F, An S, Liu H, Gao J, Leng J. Propylsulfonic acid and methyl bifunctionalized TiSBA-15 silica as an efficient heterogeneous acid catalyst for esterification and transesterification. Micropor Mesoporous Mater. 2015;204:218–25.
106. Narkhede N, Singh S, Patel A. Recent progress on supported polyoxometalates for biodiesel synthesis via esterification and transesterification. Green Chem. 2015;17:89–107.
107. Li B, Ma D, Li Y, Zhang Y, Li G, Shi Z, Ma S. Dual functionalized cages in metal-organic frameworks via stepwise postsynthetic modification. Chem Mater. 2016;28:4781–6.
108. J Da Silva M, A Liberto N. Soluble and solid supported keggin heteropolyacids as catalysts in reactions for biodiesel production: challenges and recent advances. Curr Org Chem. 2016;20:1263–83.
109. Alsalme AM, Wiper PV, Khimyak YZ, Kozhevnikova EF, Kozhevnikov IV. Solid acid catalysts based on $H_3PW_{12}O_{40}$ heteropoly acid: acid and catalytic properties at a gas-solid interface. J Catal. 2010;276:181–9.
110. Samaranch B, Ramírez de la Piscina P, Clet G, Houalla M, Gélin P, Homs N. Synthesis and characterization of Ta_2O_5-ZrO_2 systems: structure, surface acidity, and catalytic properties. Chem Mater. 2007;19:1445–51.
111. Kulkarni MG, Gopinath R, Meher LC, Dalai AK. Solid acid catalyzed biodiesel production by simultaneous esterification and transesterification. Green Chem. 2006;8:1056–62.
112. Devassy BM, Halligudi SB. Zirconia-supported heteropoly acids: characterization and catalytic behavior in liquid-phase veratrole benzoylation. J Catal. 2005;236:313–23.
113. Su F, Ma L, Guo Y, Li W. Preparation of ethane-bridged organosilica group and keggin type heteropoly acid co-functionalized ZrO_2 hybrid catalyst for biodiesel synthesis from eruca sativa gars oil. Cat Sci Technol. 2012;2:2367–74.
114. Su F, An S, Song D, Zhang X, Lu B, Guo Y. Heteropoly acid and ZrO_2 bifunctionalized organosilica hollow nanospheres for esterification and transesterification. J Mater Chem A. 2014;2:14127–38.

115. Pramanik M, Nandi M, Uyama H, Bhaumik A. Organic-inorganic hybrid porous sulfonated zinc phosphonate material: efficient catalyst for biodiesel synthesis at room temperature. Green Chem. 2012;14:2273–81.
116. Wan H, Chen C, Wu Z, Que Y, Feng Y, Wang W, Liu X. Encapsulation of heteropolyanion-based ionic liquid within the metal-organic framework MIL-100 (Fe) for biodiesel production. ChemCatChem. 2015;7:441–9.
117. Han M, Gu Z, Chen C, Wu Z, Que Y, Wang Q, Guan G. Efficient confinement of ionic liquids in MIL-100 (Fe) frameworks by the "impregnation-reaction-encapsulation" strategy for biodiesel production. RSC Adv. 2016;6:37110–7.
118 DD Bala, Misra M, Chidambaram D. Solid-acid catalyzed biodiesel production, part I: biodiesel synthesis from low quality feedstock. J Clean Prod 2017;142(Part 4):4169–77.

Chapter 11
Biodiesel Production from Waste Oil in Multiphase Reactors with Bifunctional Catalyst for Sustainable Energy

M.E. Borges, J.C. Ruiz-Morales, P. Martín-Zarza, and P. Esparza

Abstract Multiphase reactors combined with bifunctional catalysts have great potential for producing biodiesel from waste oil. In this chapter, an example is described for continuous biodiesel fuel production from sunflower oil, non-edible oil and waste oil with a fixed-bed multiphase reactor packed with a bifunctional heterogeneous catalyst. The performance of the catalyst is demonstrated for biodiesel production reaction using the multiphase reactor for several oils as feedstock at low reaction temperatures (ca. 55 °C). The catalytic material, 3D–microstructured bifunctional catalyst showed good conversions for low-quality feedstocks that have high free fatty acid and water content. Multiphase reactors combined with microstructured bifunctional catalyst are a promising new technology for biodiesel production.

11.1 Introduction

The use of renewable energy is expected to improve the energy availability [1–3] and it is expected to lead to sustainable development [4, 5]. Presently, biodiesel is considered to be a practical source of energy [6] and after years of development and commercial use worldwide, biodiesel has proven its value as an alternative fuel to supplement or replace fossil fuel. Biodiesel is receiving great attention among researchers and policy makers for its numerous advantages such as renewability, biodegradability and lower gaseous emission profile. Biofuels can be obtained from renewable sources and it has been shown to reduce carbon dioxide emissions compared to conventional diesel fuel through life cycle assessment (LCA) studies [7, 8].

Conventionally, commercial biodiesel is produced using homogeneous catalysts in batch reactors [9]. These processes consume liquid bases as homogeneous cata-

M.E. Borges (✉)
Chemical Engineering Department, University of La Laguna, Tenerife, Spain
e-mail: eborges@ull.es

J.C. Ruiz-Morales • P. Martín-Zarza • P. Esparza
Inorganic Chemistry Department, University of La Laguna, Tenerife, Spain

Z. Fang et al. (eds.), *Production of Biofuels and Chemicals with Bifunctional Catalysts*, Biofuels and Biorefineries 8, https://doi.org/10.1007/978-981-10-5137-1_11

lysts, such as NaOH, which form waste products. Presents methods have issues related to the environment, separation of the biodiesel product and when non edible oils are used as feedstock, side reactions forming sodium soaps generally occur decreasing reaction yields. Although homogeneous catalysts used in biodiesel production, can promote the desired reaction under mild reaction conditions, the purification of the reaction products (methyl esters and glycerol) to separate the homogeneous catalyst by water washing at the end of the reaction results in large amounts of wastewater, polluting the environment [10, 11]. Furthermore, it is difficult to carry out continuous production of biodiesel using homogeneous catalysts as additional purification steps are necessary to achieve commercial specifications required for biodiesel [12].

To avoid catalyst removal operations and soap formation, much effort has been focused on the development of solid acid or base catalysts that could be used in a heterogeneous catalyzed process for biodiesel production [13]. The use of base heterogeneous catalysts have drawbacks when low-grade feedstocks that contain a high level of free fatty acids because low conversions are obtained and soap formation occurs leading to additional downstream processing steps. Use of low-grade oleaginous feedstocks, such as waste cooking oil, wastes from animal fats and non-edible oils, may enhance profitability of the biodiesel production process, since they constitute a waste source. However, low-grade oleaginous feedstocks contain water and impurities, which can affect the transesterification reaction for systems that use base heterogeneous catalysts. Heterogeneous catalysts with acidic characteristics can overcome these drawbacks, and simultaneously promote both FFA (free fatty acids) esterification and triglycerides transesterification reactions, although harsh reaction conditions are generally required [14, 15]. Bifunctional catalysts, on the other hand, can provide acidic and basic active sites on the catalytic surface and have the possibility to catalyze several reaction pathways in a single process, reducing the number of unit operations and allowing intensification of the biodiesel production process [16, 17]. The use of bifunctional heterogeneous catalysts with acidic sites for promoting esterification and basic sites of appropriate strength for promoting transesterification is a key requirement for biodiesel production from low-grade feedstocks [18–22].

The replacement of homogenous catalysts by solid catalysts (heterogeneous catalysis) shows several advantages, with the most important one being simplified operating procedures. The solid catalyst can be separated from the reaction mixture and it can reduce the environmental pollution [23]. The possibility of using a fixed-bed reactor packed with the solid catalyst is interesting since this approach can minimize catalyst mechanical damage associated with the use of the catalyst particles in a slurry reactor under batch operating conditions [24].

In general, processes under batch operating conditions are characterized by low productivity and high operating costs. The productivity of the biodiesel production process can be increased by operation in a continuous mode. The use of a solid capable of acting as a heterogeneous catalyst for both esterification and transesterification reactions may allow the implementation of a continuous biodiesel production process that operates at high efficiency [25].

Several continuous commercial processes to produce biodiesel from vegetable oils using a solid catalyst have been developed [26] although the conditions require high reaction temperatures and feedstock oils of good quality (low water content and free fatty acids), or require supercritical alcohol at high pressures (170 bar) and high temperatures (300–450 °C) [25] or require conditions of moderate temperature and pressure, but employ stirred reactors with the heterogeneous catalyst solid in powder form. Some researchers have studied heterogeneous catalyst performance in continuous operation with waste oils as feedstock, but good yields (higher than 96%) have only been obtained at high reaction temperatures [27, 28]. Feng et al. [29] and Cheng et al. [30] reported on the use of a catalyst based in an acidic cation-exchange resin evidencing good catalytic activity in continuous esterification reactions, but its performance in transesterification reactions was not assessed.

A biodiesel production process that operates in a continuous mode under moderate temperatures, using a catalytic packed bed reactor would be an ideal system. Such a system is described in this chapter.

Traditionally, a conventional route to improve the performance and durability of the state-of-the-art energy conversion and generation systems consists in the development of novel catalytic materials with outstanding properties. Nevertheless, the integration of micro and nanostructured materials may be considered an adequate strategy to achieve such a goal. The microstructural control, through micro and/or nano-engineering, has proven to be important in several relevant areas such as heterogeneous catalysis for biofuels production.

The production of green fuels and the efficiency of production processes are highly dependent of the microstructure of the catalytic materials used and therefore, the main goal, involves the development of a new concept in the fabrication of microstructured materials, which is the use of 3D printing methods to fabricate any type of complex structures. Controlling precisely the macrostructure and optimizing the microstructure will enhance the available chemical area where the reactions take place. The 3D printing methods have not been widely applied to the design of catalytic microstructures and may certainly find potential applications in any research field involving materials engineering. Such control should allow increased efficiencies of devices assembled with 3D–materials and potentially overcoming drawbacks related to material degradation and high costs of starting materials.

This chapter provides a description of a biodiesel production process using a heterogeneous catalyst in a continuous production reactor system. The objective of this chapter is to provide an example on the development of a 3D microstructured catalyst for the production of biodiesel from sustainable feedstocks (vegetable, non-edible and waste oils) in a packed-bed reactor.

Biodiesel production from waste vegetable oils and non-edible oils are considered a sustainable resource compared with the use of high-priced, food-based vegetable oils as raw materials. The major drawback of using these low-grade substrates as reactants for biodiesel synthesis is their high free fatty acid (FFA) content, which is difficult to convert into biodiesel. In this chapter, the catalytic activity of a bifunctional heterogeneous catalyst is described that converts FFA and triglycerides (TGs) simultaneously into biodiesel when the catalytic system is constituted by a packed-bed in continuous flow operation. In the overall context of green energy, biodiesel can be derived from biomass and from wastes and it can be obtained in an efficient approach using a bifunctional catalyst in particles or 3D microstructured in a packed-bed reactor configuration.

11.2 Bifunctional Catalyst Operating in a Packed Bed Catalytic Reactor

The catalytic material tested in the example continuous biodiesel production process is an amorphous aluminum silicate naturally occurring as volcanic material (Pumice stone granules technical grade, PANREAC) modified by potassium impregnation [31, 32]. Potassium content loaded in the catalytic material can be quantified by inductively coupled plasma optical emission spectrometry (ICP-OES) using an Optima 3300DV Perkin Elmer instrument. The basic and acid sites present on the catalytic surface can be quantified by temperature-programmed desorption (TPD) of carbon dioxide and ammonia measurements, respectively, using a Micromeritics TPR/TPD 2900 apparatus. Chemical state of elements and surface composition can be evaluated by X-ray photoelectron spectroscopy (XPS) using a VG Escalab 200R spectrometer equipped with a hemispherical electron analyzer and Mg Kα (1253.6 eV) X-ray source.

In this example, the transesterification reaction is performed in a packed-bed catalytic reactor with continuous operation. Figure 11.1 shows the particles studied as catalyst and the 3D microstructured catalytic material. The 3D microstructured material was synthetized from an aluminum-silicate material (pumice particles) used as support for potassium loading. Pumice particles were ball-milled and dried

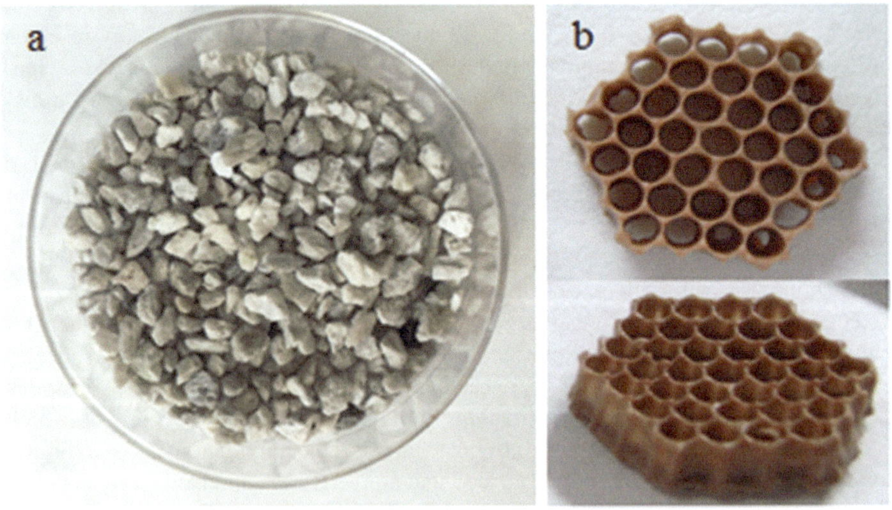

Fig. 11.1 Heterogeneous catalyst: (**a**) particles, (**b**) 3D microstructured

in an oven at 70 °C during 24 h. A suspension of the catalytic source material was prepared by mixing, 10 g. of the previously ball-milled pumice, with 10 g. of a solvent mixture of methyl-ethyl-ketone and ethanol (3:2, w/w), 0,5 g. of the dispersant Triton-Q, 1 g. of polyethylene glycol (PEG400) and 1 g. of dibutyl phthalate as plasticizers, and 1 g. of polyvinyl butyral (Butvar) as binder. The mixture was ball-milled for 2 h, at 200 rpm. An organic template (3D printed) was used as former to be coated with the suspension of the catalytic material. Successive layers of the slurry were added. Once the mold was completely coated, it was calcined into the muffle furnace during 4 h with the following ramp of temperature: 5°/min up to 1100 °C – 5°/min to 100 °C. During this thermal treatment, the organic template was burned, remaining only the microstructured catalytic material obtaining a 3D microstructured material. Finally, the catalytic material was heated at 1400 °C during 4 h to achieve the mechanical resistance required. Before using the catalytic particles or the 3D catalytic pieces as catalysts for biodiesel production reaction, the potassium loading was developed by KOH impregnation as explained in a previous paper [32]. This material constitutes the bifunctional catalytic material studied in this paper as catalyst for transesterification reaction.

Figure 11.2 shows the reaction system used. The reactor (2 cm diameter and 15 cm height) was packed with catalyst particles or 3D microstructured catalyst.

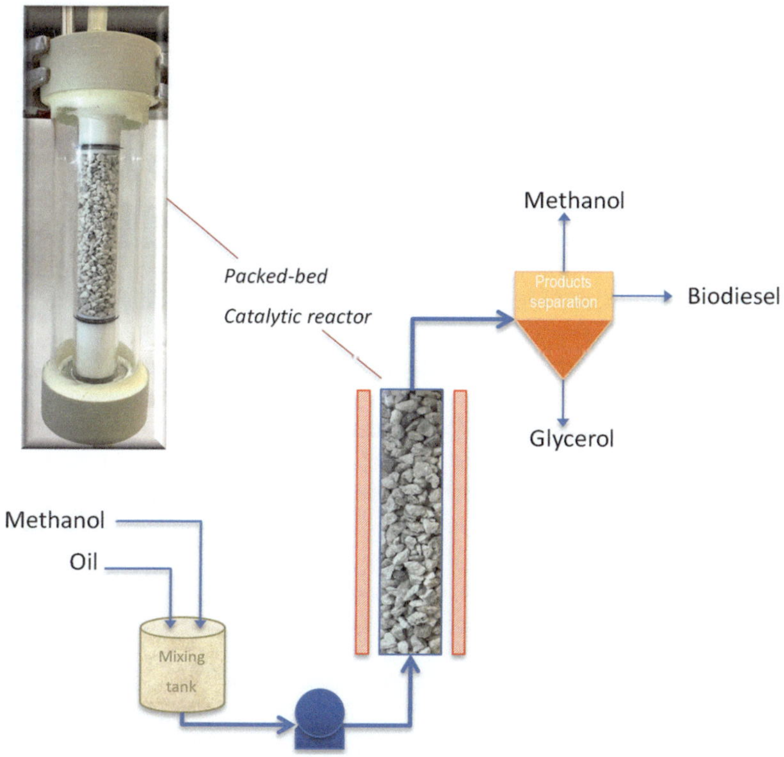

Fig. 11.2 Scheme of the tubular fixed-bed reactor packed with the catalytic particles and the system for continuous biodiesel production

Methanol and feedstock oil (sunflower oil, non-edible oil or waste oil) were mixed and used as reactive for the transesterification reaction for biodiesel production (18:1 methanol/oil molar ratio). The feedstock was preheated and then fed to the reactor using a peristaltic pump with a flow rate of 2 mL/min. The reaction temperature, measured with a thermocouple placed into the catalytic bed, was thermostatically maintained at 55 °C. FAME (Fatty Acid Methyl Esters) content in biodiesel product was analyzed by ^{1}H Nuclear Magnetic Resonance spectrometry (BrukerAvance 400) [33, 34].

To study the reuse of the catalyst, the deactivated catalyst was washed with ethanol to remove all remaining organic compounds (triglycerides and esters tracks) after reaction, and then the catalyst particles were submitted to potassium loading.

11.3 Biodiesel Production from Vegetable, Waste and Non-edible Oil

In this section, a continuous process for biodiesel production from vegetable, non-edible and waste oils using a bifunctional catalyst is described. The transesterification reaction is promoted by the bifunctional solid catalyst packed in a packed-bed reactor capable of operating at low temperatures. The catalyst consists of a natural amorphous aluminum silicate, in particles or 3D microstructured, modified by potassium loading constituting a bifunctional catalyst and characterized as previously reported [32]. The main characteristics of the bifunctional catalyst are shown in the Table 11.1 where the acidic and basic sites of the catalytic surface are quantified. When metal loading is increased, the support is covered by potassium increasing the active sites and modifying the catalytic activity.

The K/Si atomic ratios obtained by XPS analysis are summarized in Table 11.1. From the XPS analysis it is found that the binding energies corresponding to the main Si2p and Al2p peaks remain constant after potassium loading on the support. This means that the impregnation process does not cause substantial changes in the aluminum-silicate support. The aluminum atoms are in the characteristic oxidation state Al (III) in an Al-O-Si network (Al_2O_3) [35] and the Si atoms appears as those typically found in silicates (Si-O-Si). The counter anion of potassium ion is found as an oxide anion (O^{2-}) [36], which indicates that potassium ions strongly interact with the negatively charged [Si-O-Al^-] lattice oxygen [37]. Moreover, as one would expect, the K/Si atomic ratio increases when potassium loading increases

Table 11.1 Characteristics of the bifunctional catalysts: potassium content, total acidity/basicity, potassium/Silica surface atomic ratios, turn over frequency (unpublished data)

	K (% wt)	Acid sites (μmol/g)	Base sites (μmol/g)	K/Si at	TOF_{total} (s^{-1})
Catalyst 1	6.207	3.71	36.29	0.662	0.0078
Catalyst 2	4.410	2.42	20.69	0.388	0.0135
Support	3.174	2.33	18.58	0.058	0.0149

(Table 11.1). A decrease in the Al/Si atomic ratio in samples subjected to potassium loading is found indicating some dealumination of the substrate, which would lead to an increase of surface acid sites strength.

From the NH_3-TPD and CO_2-TPD analyses that can be derived for the aluminum-silicate support and corresponding potassium-loaded catalysts, the presence of acid and basic surface sites are found to be widely distributed on the surface of the catalytic material. It is, however, emphasized that the support material displays only a low catalytic activity in the transesterification reaction of sunflower oil. The impregnation of the aluminum-silicate substrate with potassium hydroxide increases the population of its basic and acid sites, as well as the acid/basic strength, resulting in an increase in activity [29]. Figure 11.3 shows the acidic and basic sites as a function of the potassium loaded in the support material. It may be claimed that not only quantity of acid/basic sites but also the strength of the sites must be considered to explain the catalytic activity based on biodiesel yield. Nevertheless, the increased

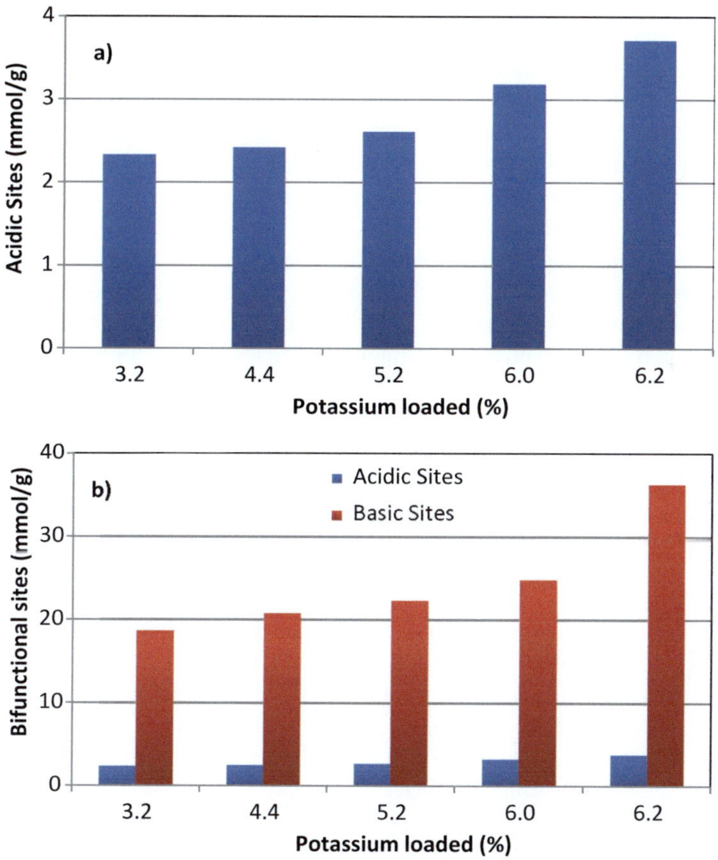

Fig. 11.3 Variation of the (**a**) acidic sites and (**b**) basic sites as a function of the potassium loaded in the support material (unpublished data)

acid sites enhance the esterification of free fatty acids, while the basic sites are central to the reaction of transesterification of lipids constituting a bi-functional catalytic material. Using this catalytic material, the reaction can be performed at low temperatures, similar to those temperatures used in the homogeneous catalyzed process for commercial biodiesel production.

The potential benefits of a granular or 3D configured solid catalyst for the biodiesel industry are apparent. This particulate material as catalyst, contrary to the use of a solid catalyst in powder form, can be placed in a packed bed reactor avoiding an extremely high pressure drop in the reactor for biodiesel production.

11.3.1 Testing the Efficiency of the Bifunctional Catalytic Material for Biodiesel Production

Few 3D–structured or particulate solids with high catalytic activity for biodiesel production reaction have been studied in the literature [27, 32, 38, 39]. In this chapter, the scaling up of a bifunctional catalytic material is shown to demonstrate the feasibility of a catalytic packed-bed reactor with continuous flow operation for biodiesel production. The catalytic reactor constitutes a multiphase reactor because three phases are present in the reactor system: the catalyst in a solid phase and the immiscible reactants (oil and methanol) in two liquid phases. Microstructured catalytic particles [40] based on catalytically-active materials have been made into the shape of an organic template used as the former as an effective way to improve catalytic efficiency in a chemical reactor. The catalytic activity of the 3D microstructured material has been compared to the activity of the catalytic material when it was used in particles, which is its original form. In the next sections, the transesterification of waste oil are shown as studied in the multiphase reactor. Several vegetable oils were employed as reactive or feedstock in the transesterification reaction in the continuous reaction system studied: sunflower oil, non-edible oil and waste oil. The main characteristics of these oils are shown in Table 11.2.

Transesterification reaction of sunflower oil, non-edible oil and waste oil with methanol to fatty acid esters was carried out over the catalyst in the packed-bed

Table 11.2 Properties of raw materials used for biodiesel production and main characteristics of the biodiesel obtained (unpublished data)

		Acid value (mg KOH/g)	$Density_{[15°C]}$ (g/cm^3)	$Viscosity_{[40°C]}$ (cSt)
Feedstock	Sunflower oil	0.40	0.924	27.3
	Non-edible oil	4.60	0.920	27.5
	Waste oil	3.96	0.926	28.9
Biodiesel	Sunflower oil	0.04	0.887	4.4
	Non-edible oil	0.41	0.883	4.2
	Waste oil	0.17	0.844	4.7

reactor (Fig. 11.2). During the continuous runs the temperature and the fed MeOH/oil molar ratio were fixed at those values corresponding to the optimization in batch mode (55 °C, 18:1 MeOH/oil) [32], whereas liquid hourly space velocity (LSHV) was 2.3 h^{-1}. The catalyst bed height was optimized to provide enough reaction time and active sites to promote the reaction between the oil and methanol at a given flow rate. The catalytic reactor constitutes a multiphase reactor because three phases are present in the reactor system: the catalyst in a solid phase and the non-miscible reactants (oil and methanol) in two liquid phases. In this sense, the mixture of methanol/oil must be vigorously agitated to improve the contact between the reactants and the catalyst and favoring the mass transfer in the catalytic reactor.

Figure 11.4 shows sunflower oil conversions versus time on stream for two different catalysts in particles: Catalyst 1 has a potassium loading of 6.2 wt% and Catalyst 2 has a potassium loading of 4.4 wt%. The studied catalytic materials are bifunctional showing the presence of acid and basic surface sites: 3.71 μmol acid sites/g and 36.29 μmol base sites/g for Catalyst 1 and 2.42 μmol acid sites/g and 20.69 μmol base sites/g for Catalyst 2. Less than 20 minutes are necessary to achieve steady state conditions in the packed bed. The catalytic material with high potassium loadings showed higher catalytic activity than the material with low potassium loadings. The high activity of Catalyst 1 compared with Catalyst 2 can be due to the presence of more acidic and basic sites on the surface of that catalytic material as discussed in previous papers [32].

Figure 11.4 shows that the studied Catalyst 1 in particles configuring a catalytic packed-bed can be used for promoting transesterification reactions under a continuous flow mode and demonstrates that high biodiesel yields (>98%) are obtained at low temperatures (ca. 55 °C). In this example, the catalytic activity was maintained up to 60 min on stream, however, at longer times, a gradual decay in activity was observed.

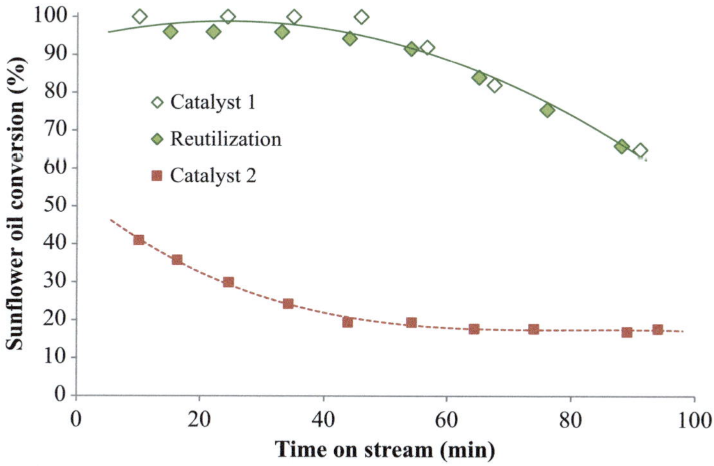

Fig. 11.4 Sunflower oil conversion versus time on stream obtained in the packed bed reactor with continuous flow operation using Catalyst 1 and Catalyst 2 in particles (unpublished results)

Some materials have been tested in the literature as heterogeneous catalysts in the transesterification of vegetable oils for continuous biodiesel production, achieving good yields only when the reaction takes place at high temperatures or pressures [27, 41–43]. However, the studied bifunctional catalyst in particles in the packed-bed reactor configuration shown has been demonstrated as feasible for biodiesel production at low reaction temperatures (55 °C).

The FAME values achieved in this continuous catalytic run was unexpectedly high as compared with previous studies under batch-mode using the optimized values of operational variables such as temperature and MeOH/oil ratio. This enhancement of the catalytic activity can be attributed to good contact of the reactants with the catalyst. The global activity of the catalyst can be improved by varying and optimizing the catalyst amount and the mean residence time or LHSV in the packed-bed reactor.

A more detailed study of the operation of the reactor was performed to evaluate the possibility of industrial application with the catalyst type and its recycle. Once catalytic activity decay was observed in the packed-bed, catalyst particles were reactivated using the procedure explained in a previous section of this chapter. The activity of the reactivated catalytic material was studied carrying out a new reaction cycle in the catalytic packed-bed reactor. The catalytic behavior of the regenerated catalytic material constituting the catalytic packed-bed reactor is presented in Fig. 11.4. Comparatively, it shows that there are no considerable changes in the catalytic behavior of the material with respect to its activity in the first reaction run, demonstrating that the material is stable.

Despite of the catalytic activity decay after some time on stream and due to the fact that high FAME yields were obtained with the catalytic particles in a packed-bed reactor configuration operating with low reaction temperature, the development of an in situ reactivation of the catalyst could be implemented. Two packed-bed reactors operated in parallel allow continuous operation with one reactor being used for the transesterification reaction while the catalyst in the other reactor is being regenerated [44]. When the catalyst contained in the first reactor becomes exhausted, the feed stream is switched to the second reactor so that the process of catalyst regeneration can begin in the first reactor.

The cost of biodiesel production is roughly proportional to the cost of the feedstock and is in the range from 70 to 95% of the total operating costs [45]. An alternative to reduce biodiesel production costs is to use low-grade oils as feedstock (waste oils, non edible oils). However, these oils have undesirable characteristics compared with vegetable oils, such as high acidity, high impurity content and considerable water content as mentioned in the introduction. The activity of the heterogeneous catalyst used for biodiesel production for transesterification can be expected to decrease. Nevertheless, it is important to assess the performance of the catalysts and reactor system described in this chapter for use with low-grade oils as feedstock for transesterification.

Waste oil from cooking and a non-edible oil was used as raw material for biodiesel production to study the behavior of the catalytic packed-bed when low-grade oils, which have high free fatty acids (FFA) content, were used as feedstock inlet

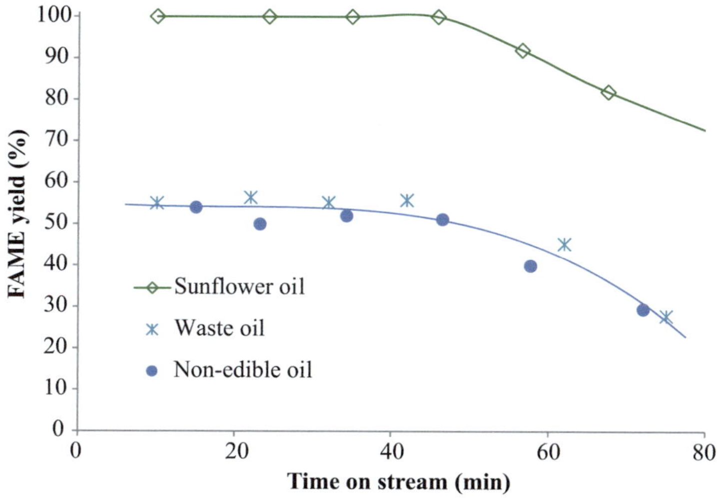

Fig. 11.5 Sunflower oil conversion versus time on stream obtained in the packed bed reactor with continuous flow operation using Catalyst 1 in particles and several oils as feedstock (unpublished results)

for the continuous operation process for biodiesel production. Therefore, oils with different acid values were evaluated: non-edible oil and waste oils, taking sunflower oil transesterification results as a base for comparing the catalytic activity. In Fig. 11.5, the catalytic activity of Catalyst 1 (6.2 wt% potassium content) is compared when the transesterification reaction is carried out in continuous operation using several oils. The FAME yields of the biodiesel product obtained from the low-grade oils were lower than those obtained from sunflower oil and can be attributed to the high FFA content in the low-grade oils (in the range 3–5 KOH/g for non-edible and waste oil). Water and other cooking residues exist in the waste oil and also contribute to low conversions [46]. Similar results have been found when the same catalyst was evaluated in a slurry reactor in batch operation [32, 47].

One line of thought for improving the performance of biodiesel production in a sustainable way depends on the development of novel catalytic materials having outstanding properties that are manufactured with ecologically-friendly processes. However, another and simpler line of thought for improving the performance of biodiesel production is through the integration of micro and nanostructured materials as an strategy to achieve this goal. Hence, the microstructural control, through micro and/or nano-engineering, has proven to be important in several relevant areas such as heterogeneous catalysis for biofuels production.

Few structured or particulate solids with high catalytic activity for biodiesel production reaction have been studied in the literature [27, 32, 38, 47, 48]. In this chapter, the micro-structuration of the studied catalytic material has been presented to evaluate possible improvements in the catalytic packed-bed reactor system with continuous flow operation. Micro-structured catalytic particles [38, 40] based on the

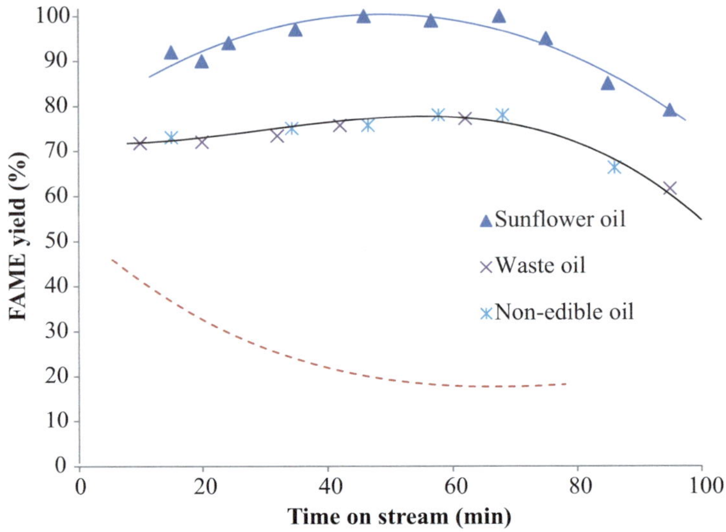

Fig. 11.6 Activity of the Catalyst 2 microstructured. Dashed lines: activity of the Catalyst 2 in particles (unpublished results)

catalytically active material, tested previously in particles, was used in the packed-bed reactor configuration for the continuous production of biodiesel in a multiphase reactor configuration.

Figures 11.6 and 11.7 show the results obtained when the structured material was used as catalyst in the packed-bed for biodiesel production from vegetable, non-edible and waste oils. The same operation variables were used to establish conclusions in a comparative way. As can be seen, the 3D–structured material showed better performance in the catalytic packed bed. The Catalyst 2 (low potassium content), in the 3D structured configuration, showed improvements in activity until 80% with respect to the same catalytic material configured with the packed bed in particles or granules.

Experiments developed in the reactor packed with Catalyst 2 in a 3D microstructured configuration are shown in Fig. 11.7. As shown in Fig. 11.7, yields close to 99% were obtained in continuous flow mode. Experiments on transesterification reaction with high acid-containing vegetable oils have also been conducted with no special acid removal treatment of the raw material. Waste oil and non-edible oil lead to an ester product composition compatible with biodiesel requirements. In that case, esterification and transesterifications reactions occur simultaneously driven by the bifunctional catalytic material used. In that case, esterification and transesterifications reactions occur simultaneously driven by the bifunctional catalytic material used. In this heterogeneous process, the stability of the catalyst was proved with no metal leaching findings in the biodiesel product or in the methanol excess. There is no formation of salts or soaps with the following advantages with respect to the homogeneous catalyzed process: no neutralization step is required, there is no formation of

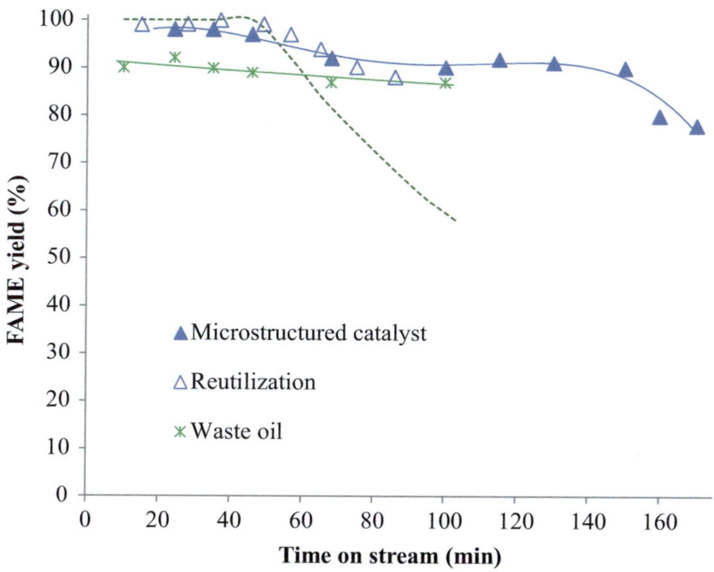

Fig. 11.7 Activity of the Catalyst 1 microstructured. Dashed lines: activity of the Catalyst 1 in particles (unpublished results)

wastewater residues from the cleaning of the biodiesel. With all its features, this heterogeneous catalyzed process can be considered a sustainable and green process.

The microstructured catalyst showed higher stability than the same catalytic material in particles, maintaining the catalytic activity during longer times on stream. The catalyst reuse reactions were performed under the same operation conditions showing that the microstructured material can be regenerated and reused without significant decrease in the reaction yield. Based in these results, the activity decay in the packed-bed over high operation time can be overcome by using several catalytic beds units with alternating reaction/regeneration runs.

11.4 Conclusions and Future Outlook

The catalytic material presented in this chapter is a very interesting heterogeneous material that is active for the production of biodiesel via transesterification and it is active at low temperatures. The catalytic material was found to have favorable characteristics when used in a multiphase packed-bed reactor under continuous flow operation. The bifunctional catalyst presented in this chapter is efficient for the transesterification of triglycerides and free fatty acids from sunflower oil, waste oil and non-edible oils, due to the presence of high catalytic active sites density.

The catalytic material presented in this chapter showed high catalytic activity when feedstocks with low free fatty acid content were employed, being possible to adjust the reactor operation conditions to achieve good conversions when waste and non-edible oils were used. Continuous biodiesel production is feasible in the reactor configurations studied and the studied catalyst was reusable and could be reactivated in situ.

Biodiesel production by means of heterogeneous solid catalysts has been industrialized with processes operating at high temperatures (220–240) °C and high pressures, which consequently concur with high costs. Therefore, more research activities on new heterogeneous catalyst applications that have industrial potential are necessary. Biodiesel production requires optimized processes that are compatible with high production capacities and that have simplified operations, provide high yields and do not require special pretreatments of the feedstock or generate waste streams. The 3D–microstructured catalyst and multiphase reactor system introduced in this chapter may be a good candidate for future industrial processes that use bifunctional catalysts, because it opens the possibility for improving catalytic performance with material and system design that is applicable to using residues and low-grade feedstocks.

Research needs for bifunctional catalyst used in biodiesel production [49] are: design of efficient bifunctional catalysts with adequate physical and chemical properties to carry simultaneous esterification and transesterification reactions of low-grade feedstocks, definition of the catalyst structure-activity correlation of the bifunctional heterogeneous catalysts to obtain an optimum number of active sites and improvement of catalytic activity and stability during biodiesel reactor operation and production.

Waste materials not only provide alternative renewable feedstocks, but also a resource from which to create catalysts. Waste materials such as animal bones and naturally available clays can be used as raw materials for catalyst preparation [50, 51]. The use of raw materials, modified to develop an effective catalyst or as feedstocks, in the biodiesel technology will contribute to low-cost biodiesel production. Currently, the introduction of different metals and groups into several raw materials by the impregnation method [52, 53] is a good way to develop bifunctional catalysts with enhanced catalyst stability and activity for sustainable biodiesel technology. When combined with 3D microstructured techniques described in this chapter, the possibilities for developing new practical catalytic systems are unlimited.

Acknowledgements This work was supported by the Research Projects of the Spanish Ministry of Economy and Competitiveness: ENE2013-47826-C4-1-R and ENE2016-74889-C4-2-R.

References

1. Atadashi I, Aroua M, Abdul Aziz A, Sulaiman N. The effects of water on biodiesel production and refining technologies: a review. Renew Sust Energ Rev. 2012;16:3456–70.
2. Basha SA, Raja Gopal K, Jebaraj S. A review on biodiesel production, combustion, emissions and performance. Renew Sust Energy Rev. 2009;13:1628–34.
3. Demirbas A. Biofuels securing the planet's future energy needs. Energy Convers Manag. 2009;50:2239–49.

4. Demirbas A. Biorefineries: current activities and future developments. Energy Convers Manag. 2009;50:2782–801.
5. Demirbas A. Competitive liquid biofuels from biomass. Appl Energy. 2011;88:17–28.
6. Luque R, Melero JA. Advances in biodiesel production: process and technologies. Cambridge: Woodhead Publishing; 2012.
7. Varanda M, Pinto G, Martins F. Life cycle analysis of biodiesel production. Fuel Process Technol. 2011;92(5):1087–94.
8. Sieira P, Galante E, Boareto Mendes A, Haddad A. Life cycle assessment of a biodiesel production unit. Am J Chem Eng. 2015;3(2):25–9.
9. Meher LC, Sager DV, Naik SN. Technical aspects of biodiesel production by transesterification—a review. Renew Sust Energ Rev. 2006;10:248–68.
10. Fukuda H, Kondo A, Noda HJ. Biodiesel fuel production by transesterification of oils. J Biosci Bioeng. 2001;92:405–16.
11. Tesser R, Di Serio M, Casale L, Sannino L, Ledda M, Santacesaria E. Acid exchange resins deactivation in the esterification of free fatty acids. Chem Eng J. 2010;161:212–22.
12. Yaakob Z, Mohammad M, Alherbawi M, Alam Z, Sopian K. Overview of the production of biodiesel from waste cooking oil. Renew Sust Energ Rev. 2013;18:184–93.
13. Ren Y, He B, Yan F, Wong H, Cheng Y, Lin L, Feng Y, Li J. Continuous biodiesel production in a fixed bed reactor packed with anion-exchange resin as heterogeneous catalyst. Bioresour Technol. 2012;113:19.
14. West AH, Posarac D, Ellis N. Assessment of four biodiesel production processes using HYSYS.Plant. Bioresour Technol. 2008;99:6587.
15. Marchetti JM, Miguel VU, Errazu AF. Techno-economic study of different alternatives for biodiesel production. Fuel Process Technol. 2008;89:740.
16. Qiu Z, Zhao L, Weatherley L. Process intensification technologies in continuous biodiesel production. Chem Eng Process Process Intensif. 2010;49(4):323–30.
17. Chouhan AP, Sarma AK. Biodiesel production from Jatropha curcas L. oil using Lemna perpusilla Torrey ash as heterogeneous catalyst. Biomass Bioenergy. 2013;08(55):386–9.
18. Santacesaria E, Martinez Vicente G, Di Serio M, Tesser R. Main technologies in biodiesel production: state of the art and future challenges. Catal Today. 2012;195:2–13.
19. Zhao Q, Wang H, Zheng H, Sun z, Shi W, Wang S, Wang X, Jiang Z. Acid–base bifunctional HPA nanocatalysts promoting heterogeneous transesterification and esterification. Catal Sci Technol. 2013;3:2204–9.
20. Farooq M, Ramli A. Optimization of process parameters for the production of biodiesel from waste cooking oil in the presence of bifunctional γ- Al2O3-CeO2 supported catalysts. Malays J Anal Sci. 2015;19(1):8–19.
21. Farooq M, Ramli A, Subbarao D. Biodiesel production from waste cooking oil using bifunctional heterogeneous solid catalyst. J Clean Prod. 2013;59:131–40.
22. Veillette M, Giroir-Fendler A, Faucheux N, Heitz M. Esterification of free fatty acids with methanol to biodiesel using heterogeneous catalysts: from model acid oil to microalgae lipids. Chem Eng J. 2017;308:101–9.
23. Baroutian S, Aroua M, Rama A, Sulaiman N. A packed bed membrane reactor for production of biodiesel using activated carbon supported catalyst. Bioresour Technol. 2011;102:1095–102.
24. Ni J, Meunier FC. Esterification of free fatty acids in sunflower oil over solid acid catalysts using batch and fixed bed-reactors. Appl Catal A Gen. 2007;333:122–30.
25. Clayton V, McNeff C, McNeff L, Yan B, Nowlan D, Rasmussen M, Gyberg A, Krohn B, Fedie R, Hoye T. A continuous catalytic system for biodiesel production. Appl Catal A Gen. 2008;343:39–48.
26. Warshaw, B., Heterogeneous catalysts – a step change in biodiesel processing, International, March 2009,55–59. (http://www.evercatfuels.com).
27. Melero J, Fernando Bautista L, Iglesias J, Morales G, Sánchez- Vázquez R. Production of biodiesel from waste cooking oil in a continuous packed bed reactor with an agglomerated Zr-SBA-15/bentonite catalyst. Appl Catal B Environ. 2014;145:197–204.

28. Lin H, Tan C. Continuous transesterification of coconut oil with pressurized methanol in the presence of a heterogeneous catalyst. J Taiwan Inst Chem Eng. 2014;45:495–503.
29. Feng Y, Zhang Z, Li J, He B. A continuous process for biodiesel production in a fixed bed reactor packed with cation- exchange resin as heterogeneous catalyst. Bioresour Technol. 2011;102:3607.
30. Cheng Y, Feng Y, Ren X, Liu A, Gao G, He F, Yang L. Comprehensive kinetic studies of acidic oil continuous esterification by cation-exchange resin in fixed bed reactors. Bioresour Technol. 2012;113:65.
31. Álvarez-Galván MC, Brito A, García-Álvarez FJ, De la Peña O'Shea VA, Borges ME, Pawelec B. Catalytic behaviour of bifunctional pumice-supported and zeolite/pumice hybrid catalysts for *n*-pentane hydroisomerization. Appl Catal A Gen. 2008;350:38–45.
32. Borges ME, Díaz L, Alvarez-Galván MC, Brito A. High performance heterogeneous catalyst for biodiesel production from vegetal and waste oil at low temperature. Appl Catal B. 2011;102:310–5.
33. Gelbard G, Brès O, Vargas R, Vielfaure F, Schuchardt U. ^{1}H nuclear magnetic resonance determination of the yield of the transesterification of rapeseed oil with metanol. J Am Oil Chem Soc. 1995;72:1239–41.
34. Borges ME, Díaz L, Gavín J, Brito A. Estimation of the content of fatty acid methyl esters (FAME) in biodiesel samples from dynamic viscosity measurements. Fuel Process Technol. 2011;92:597–9.
35. Koshizaki N, Umehara H, Oyama T. XPS characterization and optical properties of Si/SiO_2, Si/Al_2O_3 and Si/MgO co-sputtered films. Thin Solid Films. 1998;325(1–2):130–6.
36. Huhler M, Schlögl R, Ertl G. The nature of the iron oxide-based catalyst for dehydrogenation of ethylbenzene to styrene 2. Surface chemistry of the active phase. J Catal. 1992;138:413.
37. Asano K, Ohnishi C, Iwamoto S, Shioya Y, Inoue M. Potassium-doped Co_3O_4 catalyst for direct decomposition of N_2O. Appl Catal B Environ. 2008;78:242–9.
38. Borges ME, Ruiz-Morales JC, Díaz L. Improvement of biodiesel production through microstructural engineering of a heterogeneous catalyst. J Ind Eng Chem. 2013;19(3):791–6.
39. Carvalho F, Ebambi T, Ribeiro-Santos E, Carvalho A, Montero R. K_2MgSiO_4: a novel K^+-trapped biodiesel heterogeneous catalyst produced from serpentine $Mg_3Si_2O_5(OH)_4$. J Mol Catal A Chem. 2016;422:258–65.
40. Ruiz-Morales JC, Peña-Martínez J, Canales-Vázquez J, Marrero-López D, Savaniu C, Núñez P. Cost-effective microstructural engineering of solid oxide fuel cell components for planar and tubular designs. J Am Ceram Soc. 2009;92:276–9.
41. Cheng Z, Cunwen W, Weiguo W, Yuanxin W, Faquan Y, Ruan C, Junfeng Z. Continuous production of biodiesel from Soybean Oil using supercritical methanol in a vertical tubular reactor: I. Phase holdup and distribution of intermediate product along the axial direction. Chin J Chem Eng. 2010;18:626–9.
42. Doná G, Cardozo-Filho L, Silva C, Castilhos F. Biodiesel production using supercritical methyl acetate in a tubular packed bed reactor. Fuel Process Technol. 2013;106:605–10.
43. Di Serio M, Mallardo S, Carotenuto G, Tesser R, Santacesaria E. Mg/Al hydrotalcite catalyst for biodiesel production in continuous packed bed reactors. Catal Today. 2012;195:54–8.
44. Hillion G, Delfort B, Pennec D, Bournay L, Chodorge J. Biodiesel production by continuous process using a heterogeneous catalyst. Prepr Pap Am Chem Soc Div Fuel Chem. 2003;48(2):636.
45. Hasan B, Diya'uddeen B, Abdul Aziz A, Daud W, Chakrabarti MH. Performance evaluation of biodiesel from used domestic waste oils: a review. Process Saf Environ Prot. 2012;90:164–79.
46. Díaz L, Borges ME. Low-quality vegetable oils as feedstock for biodiesel production using K-Pumice as solid catalyst. Tolerance of water and free fatty acids contents. J Agric Food Chem. 2012;60(32):7928–33.
47. Borges ME, Díaz L. Catalytic packed-bed reactor configuration for biodiesel production using waste oil as feedstock. Bioenergy Res. 2013;6(1):222–8.

48. Buasri A, Chaiyut N, Loryuenyong V, Rodkum C, Chaikwan T, Kumphan N. Continuous process for biodiesel production in packed bed reactor from waste frying oil using potassium hydroxide supported on Jatropha curcas fruit shell as solid catalyst. Appl Sci. 2012;2:641–53.
49. Ramli A, Farooq M, Naeem A, Khan S, Hummayun M, Iqbal A, Ahmed S, Shah L. Bifunctional heterogeneous catalysts for biodiesel production using low cost feedstocks: a future perspective, frontiers in bioenergy and biofuels chapter 14, Intechopen; 2017.
50. Bennett J, Wilson K, Lee A. Catalytic applications of waste derived materials. J Mater Chem A. 2016;4:3617–37.
51. Sanjay B. Heterogeneous catalyst derived from natural resources for biodiesel production: a review. Res J Chem Sci. 2013;3(6):95–101.
52. Widayat W, Wicaksono A, Firdaus L, Okvitarini N. Synthesis H-Zeolite catalyst by impregnation KI/KIO3 and performance test catalyst for biodiesel production. IOP Conference Series: Materials Science and Engineering 107; 2016.
53. Wu X, Zhu F, Qi J, Zhao L. Biodiesel production from sewage sludge by using alkali catalyst. Procedia Environ Sci. 2016;31:26–30.

Part IV
Chemoenzymatic Catalysis

Chapter 12
Conversion of Biomass Using Simultaneous Chemo- and Bio-catalysis

Patrick J. Morgan, Fabio Lorenzini, and Andrew C. Marr

Abstract As many countries move towards fossil fuel divestment, greater emphasis is being placed on the use of sustainable sources for the production of fuels and chemicals. Due to the current scale of production of petroleum, a mixed portfolio of alternative sources is needed to supply these demands. Biomass is a potential renewable and sustainable source for the production of biofuels and valuable chemicals. The conversion of biomass into biofuels and bio-derived platform chemicals to be directly placed into chemical production streams has been intensely explored over the last decade. One interesting avenue highlighted in this chapter is the development of systems that combine bio- and chemo-catalysis to convert biomass into value-added chemicals: in this endeavour, the targeted transformations are enabled by the concerted action of both chemical and biological catalysts. The combination of racemization catalysts and bio-catalytic enzymes for the production of optically pure products such as alcohols, amines and acyloins, in dynamic kinetic resolution, is hereby discussed along with several recent examples of biomass valorisation using actual, or potential, protocols involving the simultaneous use of both bio- and chemo-catalytic steps.

12.1 Introduction

The scientific community and the wider world have come to the realisation that our current consumption of the Earth's natural resources is unsustainable, and that new technologies, methodologies and renewable philosophies must be adopted to preserve these resources for future generations [1]. Petroleum derivatives are the primary feedstock for chemical production worldwide. Due to the recent volatility in oil prices, the current geopolitical turmoil in major oil producing regions, the fears of future uncertainty in oil availability, and environmental concerns, the need to move away from this feedstock and towards one that is sustainable and environmentally benign has become acute. One interesting alternative to petroleum for the

P.J. Morgan • F. Lorenzini • A.C. Marr (✉)
School of Chemistry and Chemical Engineering, Queen's University Belfast, Belfast, UK
e-mail: a.marr@qub.ac.uk

Z. Fang et al. (eds.), *Production of Biofuels and Chemicals with Bifunctional Catalysts*, Biofuels and Biorefineries 8, https://doi.org/10.1007/978-981-10-5137-1_12

production of chemicals and fuels is biomass. Biomass includes plant or animal matter, energy crops, food and animal waste, as well as by-products of some industrial and chemical processes. This organic matter is a rich carbon source, and the production of basic platform chemicals and biofuels by biomass processing can be viable. The sheer quantity of biomass produced annually worldwide could potentially satisfy the present and future demands of fuels, materials and chemicals, currently produced from fossil fuels. In the United States alone, nearly 1.4 billion tonnes of biomass, to be converted into commodity chemicals or used in the energy sector, can be obtained from agricultural and forestry sources each year [2].

Research on the synthesis of basic chemicals from accessible biomass sources is growing within the chemical industry. Chemical building blocks including alcohols, diols and acids are produced from biomass *via* aerobic or anaerobic fermentations and bio-enzymatic transformations. Ligno-cellulosic biomass can be processed in bio-refineries to yield bio-based platform chemicals. Other biomass sources are sugar, starch, fats, and oils derived from edible crops, and waste from agricultural production. However, the use of ligno-cellulose is preferred as competition with food production is minimized. Furthermore, ligno-cellulose can also undergo gasification to produce syn-gas that can then be processed in the same way as fossil fuel-derived syn-gas, using well established petrochemical routes. The bio-refinery transformations have been classified according to feedstock and summarised by Vennestrøm et al. [3].

To achieve widespread chemicals and fuels manufacture *via* bio-refineries, initial deoxygenation of the oxygen rich compounds resulting from biomass is required. Complete industrial scale transition from oil to biomass feedstocks remains problematic as the initial bio-derived reagents resulting from biomass are generally impure dilute aqueous solutions containing oxygen rich species. Biomass treatment is therefore incompatible with current processing methods based on crude oil, as the latter requires oxidation during primary treatment. It will only be possible to integrate the platform chemicals synthesised from biomass into traditional chemical processes once issues with initial biomass processing are overcome [4, 5].

To move from a petroleum based industry to a biomass-based one, new technologies and methodologies must be developed, utilizing both biological and chemical means, for the conversion of biomass into desirable commodity and platform chemicals. A mature bio-industry will require the development of an entirely adaptable bio-refinery: in which a large range of products can be synthesised from bio-derived platform chemicals, the source of the latter being flexible and dependent upon the feedstock supply, as well as on other commercial, agricultural and industrial demands.

Attention has been drawn to the production of valuable chemicals from biomass by combining bio- and chemo-catalysis [6–8]. Biomass is first converted into bio-derived platform chemicals *via* biocatalytic means; then, value-added commodity chemicals are delivered *via* chemocatalytic processing of the platform chemicals. One area of specific interest includes the direct transformation of biomass into valuable chemicals *via* a bifunctional catalytic system, using a one-pot, one-step bio- and chemo-catalytic route. The conversion of biomass into foods and drinks enriched with targeted chemicals *via* bio-catalytic routes is well-documented [9, 10]. Brewers have made use of the bio-chemical activity of yeast for thousands of years. Coupling this with the selectivity and efficiency offered by chemo-catalytic routes could lead to the development of procedures that will facilitate the change from a petroleum derived economy to a bio-based one. The dedicated conversion of biomass into valuable chemicals, while relatively new, is a developing sector that holds much promise to revolutionise the way energy and chemicals are currently produced and to dictate the way they will be produced in the future.

12.2 Dynamic Kinetic Resolution (DKR)

Chemo-enzymatic dynamic kinetic resolution (DKR) is operated on the industrial scale by fine chemicals and pharmaceutical companies. For example, DSM in the Netherlands used DKR to resolve racemic mixtures of alcohols [11]. Optically pure isomers are of high value to the pharmaceutical industry, particularly with respect to selective stereochemical control. The selective separation of one isomer within a racemic mixture can be achieved by employing bio-catalysts that will react with one of the two stereoisomers at a much higher rate than the other, leading to the isolation of an optically pure product, this is kinetic resolution (KR). A number of techniques have been developed for the synthesis of important organic substrates involving the use of a bio-catalytic enzyme, such as lipase [12–15]. Enzymes are proteins produced by living organisms that act as catalysts for specific bio-chemical transformations. They require specific conditions and substrates to function, and their use outside their operating conditions can cause the protein to denature. Enzymes have been utilized in KR reactions thanks to their high selectivity towards certain functional groups [16]. Efficient KR reactions allow for $> 99\%$ *ee* of the desired product. However, the maximum theoretical yield from a racemic mixture is 50%, as the enzyme used reacts with only one stereoisomer. This major issue can make KR undesirable on an industrial scale. DKR can provide the solution to this problem. DKR involves the use of a chemical catalyst (e.g. an organometallic complex) that catalyses the racemization of the unwanted stereoisomer: more of the reacting isomer is then provided, thus allowing the enzymatic transformation to proceed further. The maximum theoretical yield of the chemical synthesis, now catalyzed by both the racemization catalyst and the bio-catalytic enzyme, is now 100% (Scheme 12.1). Although the synthesis may be complicated by incompatibilities in the reaction conditions when employing the racemization and biological catalysts simultaneously, the ultimate goal is the successful delivery of the optically pure product by combining bio- and chemo-catalysis in a one-pot, multi-step process. The commercial success of this process makes it an obvious exemplar for combined bio- and chemo-catalysis in one-pot.

It is noteworthy to mention that when bio- and chemo-catalysis are combined in the absence of a racemisation catalyst, the synthesis of target compounds, although successful, has to deal with selectivity limitations. For example, combining

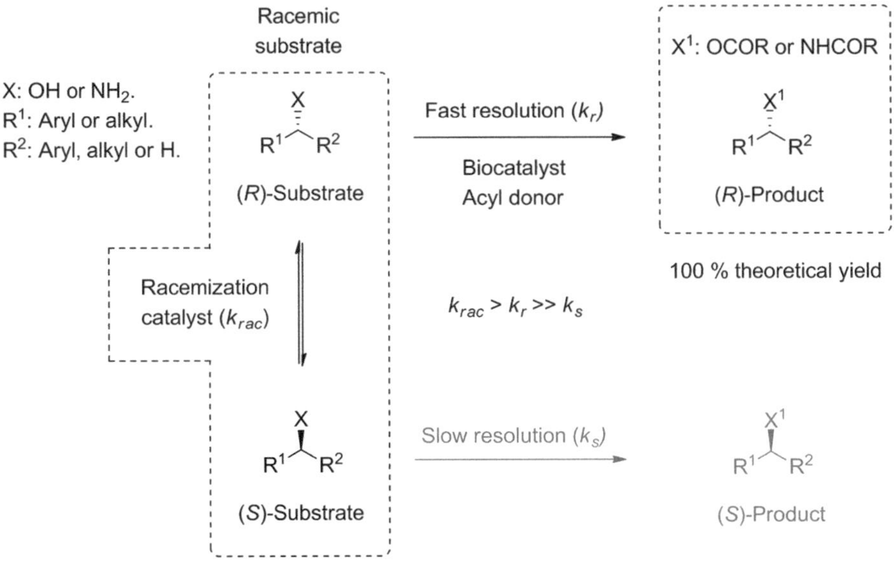

Scheme 12.1 Schematic representation of alcohol and amine acylation *via* dynamic kinetic resolution: increasing reaction yield by combining bio- and chemo-catalysis

hydrogenation of acetophenone in the presence of heterogeneous transition metal catalysts with acylation of the so-formed (*R*)-1-phenylethanol catalyzed by an immobilized enzyme, to produce (*R*)-1-phenylethyl acetate, was recently demonstrated by Murzin [17, 18]. In the presence of Pd–Al_2O_3 and immobilised lipase Novozym 435, using ethyl acetate both as the solvent and the acyl donor, under atmospheric hydrogen, and at 70 °C, (*R*)-1-phenylethyl acetate was synthesised, at 47% conversion, in 45% selectivity (maximum selectivity 50%).

12.2.1 Alcohols

Biomass based feedstocks can deliver highly oxygenated bio-based intermediates that could be directly converted into numerous value-added chemicals, including alcohols. Comprehensive flow-charts of the products of biomass derivatization were produced by the US Department of Energy [19], highlighting routes from biomass feedstock to various valuable products. Of particular note was the prevalence of acids including succinic acid, malic acid, 2,5-furandicarboxylic acid, and 3-hydroxypropionic acid, which can be readily converted into related alcohols. Since then, a number of reviews have highlighted the formation of C4 and C5 alcohols from bio-derived diacids [20], bio-ethanol from biomass derived syngas [21], and phenol derivatives from bio-petroleum [22]. It is evident that biomass feedstocks offer the opportunity to produce alcohols, which can then undergo further derivatization into valuable commodities. The conversion of alcohols into

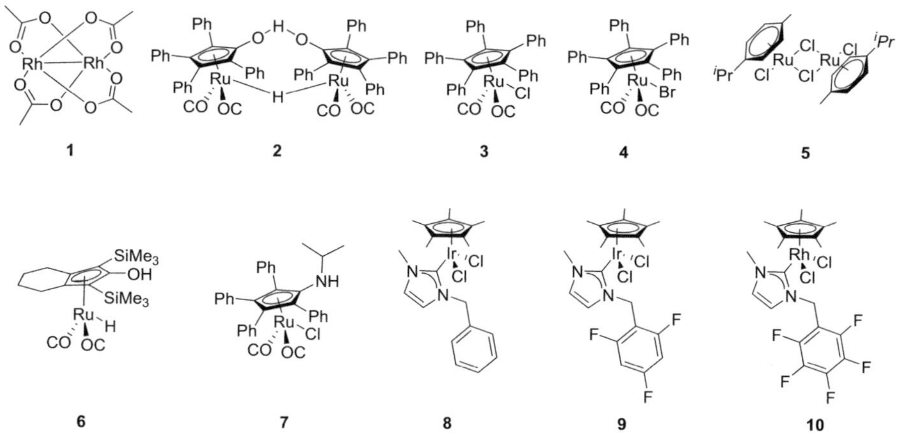

Scheme 12.2 Transition metal catalysts used in bio- chemo-catalytic dynamic kinetic resolution and valorisation of biomass

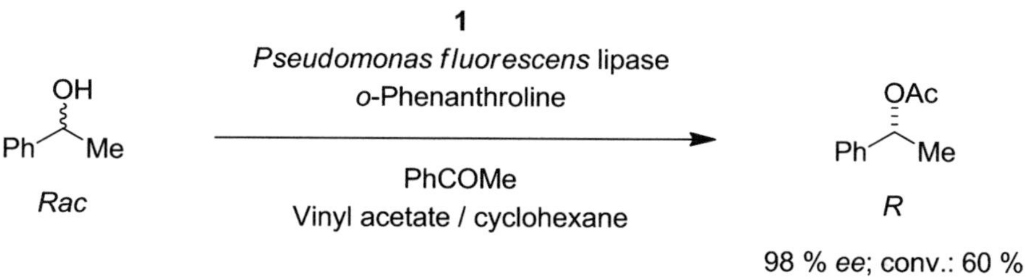

Scheme 12.3 Chemo-enzymatic enantioselective dynamic kinetic resolution of phenylethyl alcohol

value-added chemicals *via* simultaneous bio- and chemo-catalysis is therefore an important field of investigation.

Chemo-enzymatic DKR was reported by Williams and co-workers in 1996 [23]: a transition metal complex was used to catalyse racemisation of (*S*)-(−)-phenylethyl alcohol, followed by enzymatic acylation, yielding (*R*)-1-phenylethyl acetate. Stoichiometric amounts of acetophenone were used as a hydrogen transfer mediator to allow alcohol racemisation to proceed. Among the transition and main group metal racemization catalysts tested, the rhodium complex $Rh_2(OAc)_4$ (**1**) (Scheme 12.2) was found to promote, in the presence of *Pseudomonas* species lipase, the most efficient DKR for the transformation of phenylethyl alcohol to (*R*)-1-phenylethyl acetate. (*R*)-1-Phenylethyl acetate was produced with 98% *ee* and 60% conversion, when reacting phenylethyl alcohol in the presence of 2 mol % of **1**, 6 mol % *o*-phenanthroline, one equivalent of acetophenone, at 20 °C, for 72 h (Scheme 12.3). Williams and co-workers highlighted that the DKR reaction conditions have to be compatible with both the racemisation and enzyme catalysts, while also limiting metal catalyzed esterification side reactions. The authors also proposed that the performance of the combined chemo-enzymatic system may be affected by the interactions between the organometallic catalyst and the enzyme. This may involve the soft metal interacting with electron rich groups on the polypeptide surface, which will detrimentally affect racemization, and may also perturb the enzyme. Two approaches to limit this interaction are the use of a bulkier metal complex that cannot penetrate the protein due to steric hindrance, and the protection of the enzyme or catalyst by entrapping it in a material support (compartmentalization) [24].

In 1997, Bäckvall and co-workers reported that the transformation of methylbenzyl alcohol into (*R*)-*α*-methylbenzyl acetate occurs in high yields and high optical purity using ruthenium-catalyzed alcohol racemization coupled with enzymatic resolution. Combining Shvo's catalyst (**2**) and Novozym 435 (*Candida antarctica* lipase B immobilised on acrylic resin) allowed production of (*R*)-*α*-methylbenzyl acetate in 92% yield, and with optical purity > 99.5% *ee* [25]. The reaction was performed in ^{t}BuOH, at 70 °C, with 2 mol % of **2**, in the presence of one equivalent of acetophenone as the co-catalyst to promote the ruthenium-catalyzed hydrogen transfer allowing efficient isomerization. Immobilised CALB lipase was found to be thermally stable under the reaction conditions required to provide good activity for **2**. The substrate scope of the reaction was extended to diols in 1999 [26]. Still using **2** and Novozym 435 as the chemo- and bio-catalysts, respectively, acylation of aliphatic, aromatic and heteroaromatic based diols, some bearing additional functionalities, resulted in diacetates with enantioselectivity > 96– > 99% *ee* and moderate to high yields (43–90%).

In 2001, Bäckvall and co-workers discovered that the combination of **2** and Novozym 435 also allowed the chemo-enzymatic DKR of azido alcohols, direct precursors of biological, catalytic and synthetic key molecules such as aziridines and vicinal amino alcohols, to the corresponding enantiomerically pure acetates (with *ee* up to 99%, and conversion up to 98%) [27]. This strategy allowed the practical syntheses of (*S*)-propanolol (**13**), a widely used anti-hypertensive drug (Scheme 12.4). DKR of the racemic azido alcohol **11** yielded the azido acetate **12** in 86% *ee* and 71% isolated yield. Then, a one-pot two-step procedure, consisting in hydrolysis with LiOH in methanol followed by azide reduction and *in situ* reductive

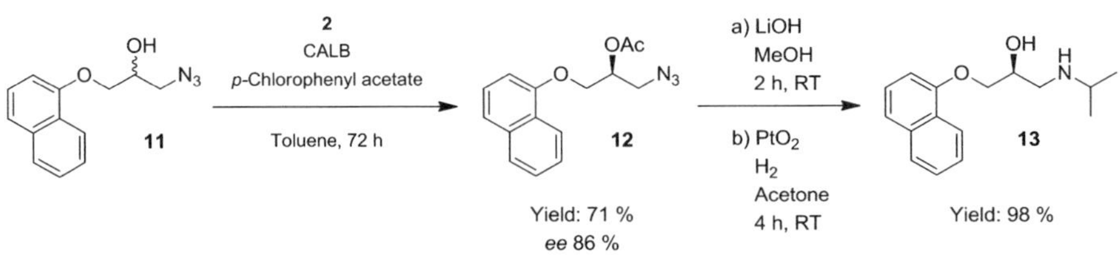

Scheme 12.4 Synthesis of anti-hypertensive drug *(S)*-propanolol via dynamic kinetic resolution of racemic azido alcohols to azido acetates, followed by one-pot, two-step hydrolysis/reduction/reductive alkylation

alkylation using platinum dioxide in the presence of acetone, led to **13**. Recrystallization from cyclohexane afforded **13** of high enantiomeric purity.

Lipase B from *Candida antartica* (CALB) is the most commonly applied enzyme in DKR reactions, but many other enzymes have also been used. Bäckvall and co-workers combined the KR of α-hydroxy acid derivatives such as esters using *Pseudomonas cepacia* lipase and the acyl donor *p*-chlorophenyl acetate with a ruthenium-catalyzed racemization process *via* hydrogen transfer in the presence of **2**, to produce acetates with enantiomeric excess up to 94%, using molecular hydrogen to inhibit formation of ketone by-products [28]. The overall reduction in product yield caused by the formation of small quantities of the associated ketone, was tackled by using the mild and efficient hydrogen source 2,4-dimethyl-3-pentanol. Following the same general trend, running the DKR of α-hydroxy acid derivatives *N,N*-diisopropyl-4-hydroxypentanamide, the acetate was obtained in yield up to 93%, and enantiomeric excess up to 99% when 2,4-dimethyl-3-pentanol was used as the hydrogen donor.

When combining bio- and chemo-catalytic processes, differences in solvent compatibility can be an issue, as experienced by Backvall and co-workers in 2001 when attempting to couple C-C bond formation *via* aldol condensation with metal and enzyme catalyzed racemisation/acylation [29]. The use of a two solvent system allowed production of β-hydroxy ester derivatives with high enantiomeric purity (up to 99% *ee*) and yields up to 82% by combining DKR with an aldol reaction, in a one-pot procedure (Scheme 12.5): the aldol condensation was first carried out in tetrahydrofuran; then, after removal of the latter, addition of *tert*-butyl methyl ether allowed, in the presence of **2** and *Pseudomonas cepacia* lipase, DKR synthesis of β-hydroxy esters, still within a one-pot synthesis.

Although the rapid racemization of alcohols is known to be catalyzed by several metal complexes, the combination of this activity with an enzyme catalyzed KR is not necessarily effective. Racemization by the metal catalyst may be slowed or inhibited by the enzyme, or the metal catalyst may interfere with the enzyme to give poor resolution. Shvo's catalyst (**2**) has an extremely bulky facial ligand that sterically impedes the active site from interactions with the enzyme polypeptide, but this comes at a cost, as the complex requires raised temperatures to activate it. Bäckvall

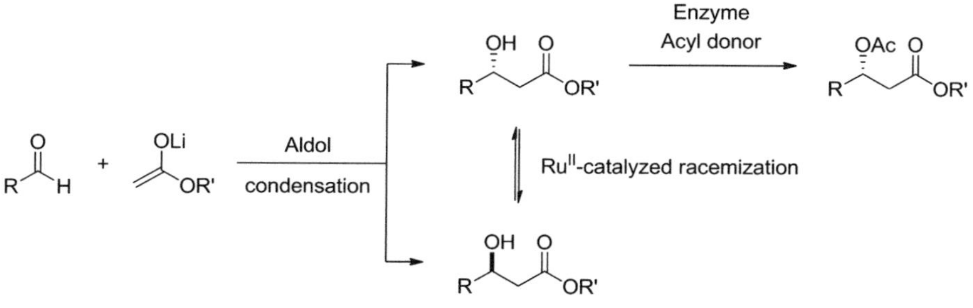

Scheme 12.5 One-pot aldol condensation – chemo-enzymatic synthesis of enantiopure β-hydroxy esters

and co-workers attempted to overcome this process limitation, which can damage the enzyme. A redesigned protocol for DKR of secondary alcohols at room temperature provided enantiopure products in high yields and short reaction times [30, 31]. The solution was simply to design catalyst precursors based on mononuclear complexes (**3** and **4**), but retaining the same class of bulky capping ligand. Pre-activation of catalysts **3** and **4** with KO^{t}Bu, followed by addition of the substrate 1-phenylethanol, and then CALB, since for most of the substrates addition of CALB at the pre-activation stage was found not to affect the effective and selective acetate production, and using isopropenyl acetate as the acyl donor, afforded, (in some cases in 3 h), at room temperature, formation of (*R*)-1-phenylethanol acetate in up to 99% yield with up to >99% *ee*. No supplementary hydrogen source was required under these DKR conditions. A large range of secondary alcohols substrates, with the phenyl ring of 1-phenylethanol variously substituted, were transformed *via* DKR to the corresponding acetate with high yields (83–99%) and good enantioselectivity (91– > 99% *ee*) (Table 12.1). While the addition of electron-donating groups such as the methoxy group resulted in high yield (96%) and selectivity (> 99% *ee*) in 6 h, almost quantitative yields of enantiopure acetates were given by substrates with electron-withdrawing groups on the phenyl ring, but at longer reaction times, for instance in 20 h for nitro- and nitrile-substitution. A similar effect was observed when increasing the secondary alcohol aliphatic chain length. Very high selectivity was also obtained for a variety of aliphatic substrates: for example, 1-cyclohexylethanol afforded the corresponding acetate, after 7 h, in > 99% *ee* and 98% yield. DKR carried out with this procedure also afforded high selectivity for heteroaromatic substrates: benzofuran, furan, piperidine, 4-pyridine, and thiophene substituted alcohols gave excellent yields (93– > 99%) and *ee* 96– > 99%. Useful functionalised alcohols, building blocks for the synthesis of chiral alcohols derivatives, were also resolved to enantiopure acetates at room temperature. For example, the DKR of 3-hydroxy-3-phenylpropanenitrile afforded the acetate, in 6 h, in 85% yield and 97% *ee*. DKR of the diol 1,1′-(1,4-phenylene)bis(ethan-1-ol) was also successful, although at higher temperatures (50 °C): 90% yield of the corresponding diester were isolated after 10 h, with > 99% *ee*.

The scale-up of the optimised system for the ruthenium and enzyme-catalyzed DKR of (*rac*)-1-phenylethanol was investigated by Bäckvall and co-workers [32]. 159 g (97% yield) of (*R*)-1-phenylethanol acetate could be produced on a 1 mol-scale with enantioselectivity > 99.8% *ee*, at 70 °C, in 20 h, under controlled atmosphere, and employing only 0.05 mol % of **3**, 0.5 mg of CALB per mmol of substrate, and 150 mL of toluene as the solvent. On a 10 mol-scale, the DKR of (*rac*)-1-phenylethanol led to 1.43 kg (87% yield) of enantiomerically enriched acetate (97% *ee*), in 21 days, using only 0.01 mol % of **3** and 0.1 mg of CALB per mmol of substrate.

The resolution of cycloalkanediols, precursors to various building blocks in asymmetric and pharmaceutical synthesis, *via* enzyme-catalyzed desymmetrization to monoacetates coupled with dynamic kinetic asymmetric transformation (DYKAT), was found to allow the synthesis of diacetates in high *trans*/*cis* ratio and excellent enantioselectivity [33]. An efficient desymmetrization-DYKAT sequence

Table 12.1 Selected dynamic kinetic resolution transesterification reactions of alcohols, as described in Sect. 12.2.1

Entry	Substrate	Chemo-catalyst	Bio-catalyst (mg/mmol)	Acyl donor	Time (h)	Temperature (°C)	Yield (%)[a]	Product *ee* (%)	Refs.
1[b]	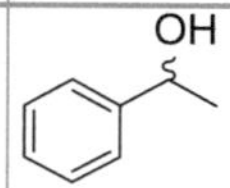	2 mol % **2**	Nov-435 (25)	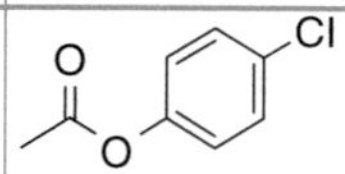	87	70	100 (92)	99.5	[25]
2	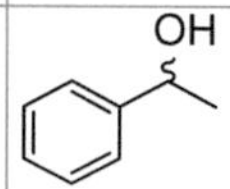	2 mol % **2**	Nov-435 (30)	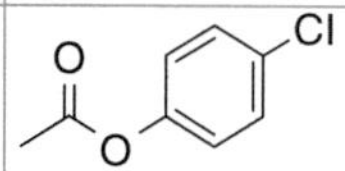	46	70	80	>99	[51]
3[c]	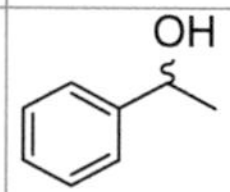	0.05 mol % **3**	CALB (0.5)	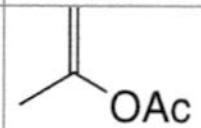	20	70	97	>99	[32]
4[d]	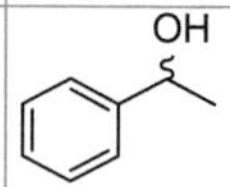	5 mol % **4**	CALB (6)	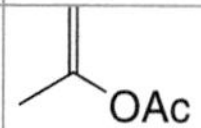	3	RT	98	>99	[31]
5[d]	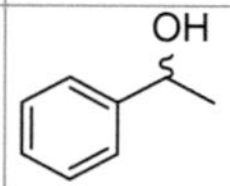	5 mol % **3**	CALB (6)	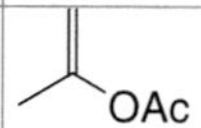	3	RT	98	>99	[30]
6[d]	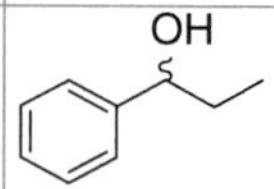	5 mol % **2**	PS-D (0.5)	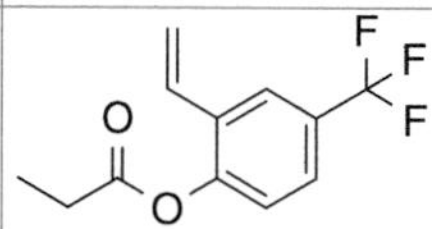	24	80	79	92	[34]
7	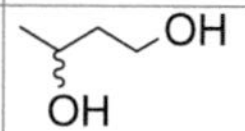	4 mol % **2**	Nov-435 (60)	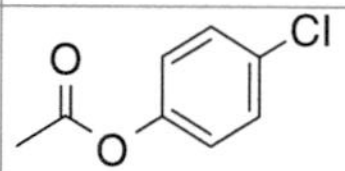	24	70	90	>99	[26]

(continued)

Table 12.1 (continued)

Entry	Substrate	Chemo-catalyst	Bio-catalyst (mg/mmol)	Acyl donor	Time (h)	Temperature (°C)	Yield (%)[a]	Product *ee* (%)	Refs.
8[d]	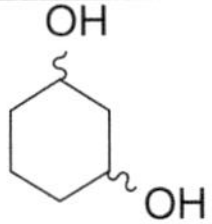	6 mol % **2**	CALB (1.3)	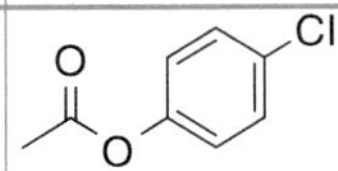	72	70	>95	94[c]	[33]
9[d]	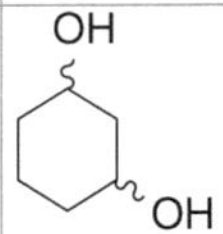	6 mol % **3**	CALB (1.3)	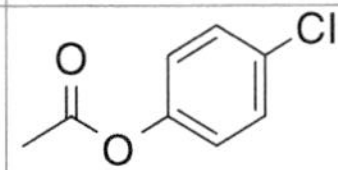	24	70	41	96[e]	[33]
10[d]	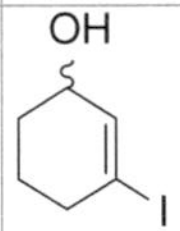	5 mol % **3**	PS-IM (25)	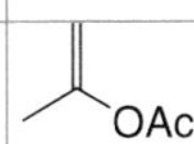	24	30	>99 (90)	99	[38]
11	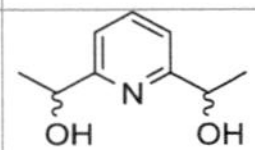	4 mol % **2**	Nov-435 (60)	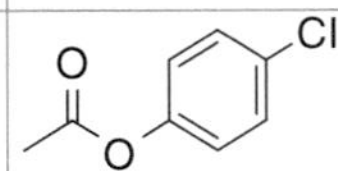	24	70	78	>99	[26]
12[d]	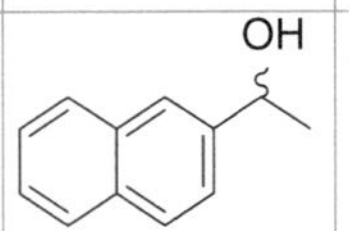	5 mol % **3**	CALB (6)	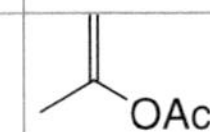	6	RT	93	>99	[30]
13	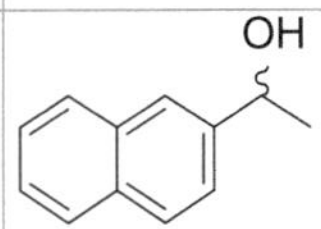	2 mol % **2**	Nov-435 (30)	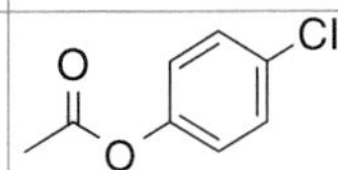	48	70	79	>99	[51]

14[d]	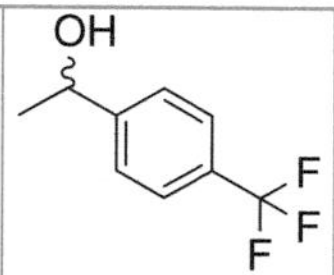	5 mol % **3**	CALB (6)	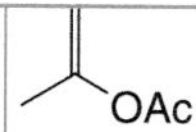	24	RT	>98 (98)	>99	[31, 37]
15[d]	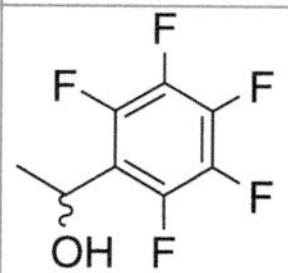	5 mol % **3**	CALB (6)	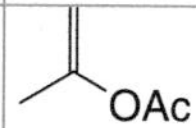	24	RT	98 (97)	>99	[31, 37]
16	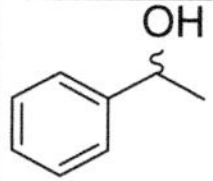	0.1 mol % **10**	Nov-435 (40.5)	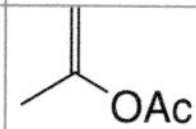	18	70	>99	97	[39]
17	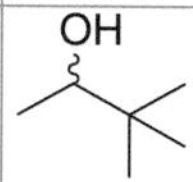	0.1 mol % **8**	Nov-435 (40.5)	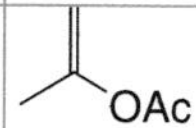	18	70	89	99	[39]
18	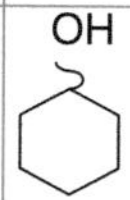	0.1 mol % **10**	Nov-435 (40.5)	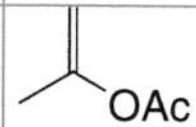	8	70	>99	99	[39]
19[f]	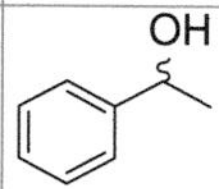	CBV400 (500 mg)	Nov-435 (1 g)	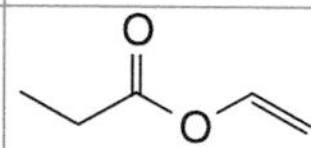	6	50	98	97	[41]
20	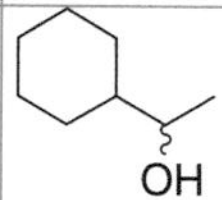	5 mol % **4**	CALB (6)	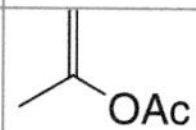	17	RT	98	>99	[31]
21	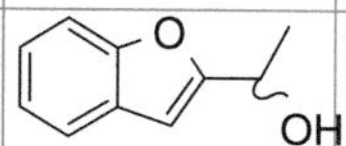	5 mol % **4**	CALB (6)	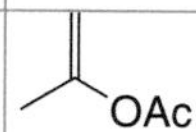	6	RT	98 (92)	96	[31]

(continued)

Table 12.1 (continued)

Entry	Substrate	Chemo-catalyst	Bio-catalyst (mg/mmol)	Acyl donor	Time (h)	Temperature (°C)	Yield (%)[a]	Product *ee* (%)	Refs.
22	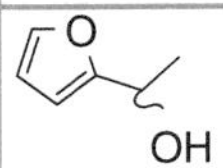	5 mol % **4**	CALB (6)	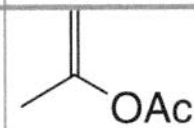	6	RT	93	96	[31]
23	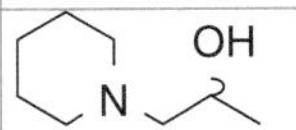	5 mol % **4**	CALB (6)	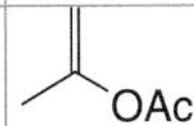	5	RT	>99	99	[31]
24	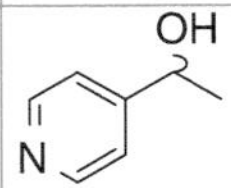	5 mol % **4**	CALB (6)	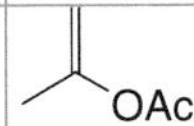	20	RT	97 (96)	99	[31]
25	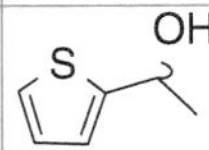	5 mol % **4**	CALB (6)	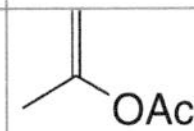	6	RT	>98 (98)	>99	[31]
26	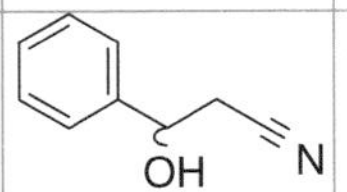	5 mol % **4**	CALB (6)	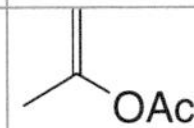	6	RT	85	97	[31]
27	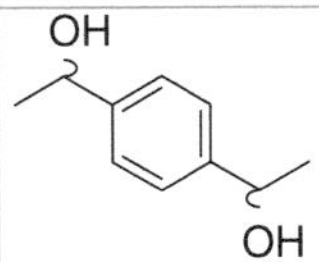	5 mol % **4**	CALB (6)	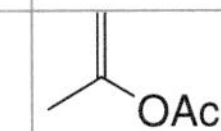	10	RT	94 (90)	97	[31]

All reactions were carried out in toluene, with the exception of entry 1, which was carried out in ^{t}BuOH and 19, which was carried out in an ionic liquid/scCO_2 biphasic system

[a]Isolated yield in parenthesis

[b]Acetophenone (1 equiv.) additive

[c]Na_2CO_3 and KO^{t}Bu additives present

[d]1000 mmol scale

[e]Diastereomeric ratio 56:43, *cis*:*trans*

[f]Bio- and chemo-catalysts coated in ionic liquid [Btma][NTf_2]

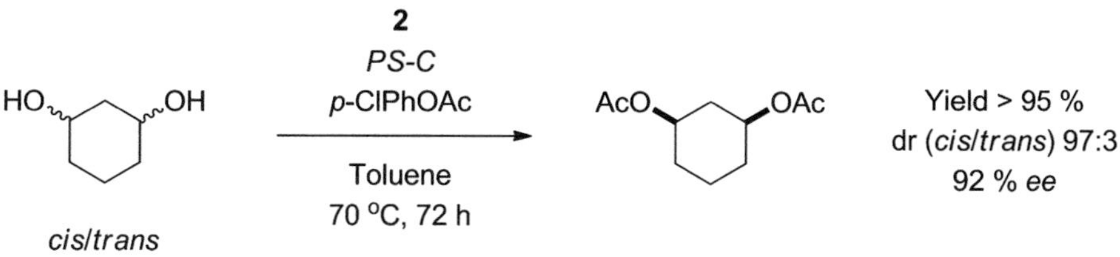

Scheme 12.6 Metal- and enzyme-catalyzed diastereoselective dynamic kinetic asymmetric transformation (DYKAT) of *cis/trans*-1,3-cycloalkanediol to *cis*-diacetate

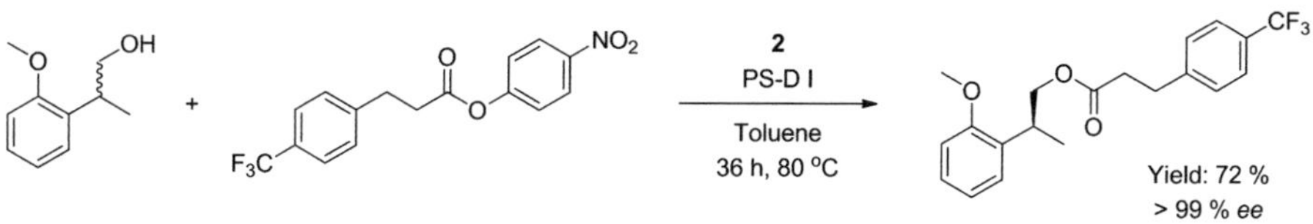

Scheme 12.7 Enzyme- and metal-catalyzed dynamic kinetic resolution of primary alcohols bearing an unfunctionalized stereogenic center in the *β*-position

was accomplished by reacting the *cis/trans*-diol 1,3-cyclohexanediol, the enzyme *Pseudomonas cepacia* lipase PS-C, **2** and *p*-chlorophenyl acetate, in the presence of 2,4-dimethyl-3-pentanol, at 70 °C: the corresponding (*R*,*S*)-diacetate was produced in high yields (> 95%), with 92% *ee*, and dr *cis/trans* 97:3 (Scheme 12.6).

The DKR of an unfunctionalized *β*-stereogenic centre was reported in 2007 [34]. The extension of the DKR protocol outlined to a chiral, non-functionalized carbon centre adjacent to an alcohol (or amine) functionality is important with regards to its potential application to the related structure of a number of biologically active compounds, and to that of *β*-aryl-substituted alcohols that are, either by themselves (Naproxol), or after oxidation to the corresponding acid (Naproxen and Ibuprofen), part of important non-steroidal anti-inflammatory drugs. The racemization was found to proceed exploiting the dehydrogenase/hydrogenase activity of **2**, generating an equilibrium between the alcohol and an aldehyde, where the latter can then undergo enolization. The DKR process is completed by the enzyme Amano Lipase PS-D I from *Pseudomonas cepacia*, following previous KR work by Kawasaki [35, 36]. Numerous primary *β*-racemic alcohols were transformed into optically pure alcohol esters in good to high yields (70–87%), with up to > 99% *ee* (Scheme 12.7).

DKR was applied to fluoro-substituted aryl and cyclic allylic alcohols, both structural elements in pharmaceutical and agrochemical products, to yield the respective acetates in high yields and enantioselectivity. Various perfluoro-substituted benzylic secondary alcohols have been transformed, at room temperature, to their respective (*R*)-acetates in excellent yields (97–99%) and enantioselectivity (98– > 99%) using **3** as the isomerisation catalyst and CALB as the enzyme, adding the substrate alcohol, and then the isopropenyl acetate, after catalyst activation [37]. Using CALB or *Pseudomonas cepacia* lipase, and **2** or **3** as the ruthenium isomerisation catalysts, the cyclic allylic alcohol 3-phenylcyclohex-2-enol has been converted by DKR, at 60–70 °C, to the corresponding (*R*)-acetate with high enantioselectivity, varying in the range 80–97% *ee* [38].

Within the literature on DKR of alcohols, there is a predominance of Ru catalysts for racemization. This does not mean that other complexes and materials cannot be effectively applied to this transformation, and in some cases, provide advantages compared to Ru based protocols. Marr and Saunders showed that Rh and Ir piano stool *N*-heterocyclic carbene complexes support efficient DKR reactions to yield acetate esters with excellent enantioselectivity [39]. Cp*Ir complex **8** and Cp*Rh complex **10** were found to be active towards a range of substrates working in tandem with Novozyme 435. **8** was active for the DKR of (*rac*)-3,3-dimethyl-2-butanol affording 89% yield of the (*R*)-acetate with 99% ee. Some DKR reactions could be affected without the addition of base (K_2CO_3), this is advantageous as the presence of base is detrimental to the enzyme and complicates separations. **8** and **10** enabled the DKR of (*rac*)-1-cyclohexylethanol under base free conditions to yield >99% with 99% ee. **10** could affect the DKR of (*rac*)-1-phenylethanol without base to afford >99% yield with 97% ee.

The continuous operation of chemo-enzymatic DKR was demonstrated by Lozano, Iborra and co-workers [40]. The organometallic chemo-catalyst was replaced by a solid acid (benzenesulfonic acid functionalised silica) and used in conjunction with Novozym 435 in ionic liquid media. Supercritical CO_2 was used as the carrier solvent. After optimisation, and using an acidic zeolite for racemization, (*rac*)-1-phenylethanol could be resolved into (*R*)-phenylethyl propionate at 50 °C and 100 bar, with yields of up to 98.0% and enantioselectivity of up to 97% ee. No loss of activity was measured during 14 days of operation [41].

A recyclable, metal-free DKR protocol for the synthesis of enantiopure, synthetically valuable secondary carbocyclic esters, in high yields, from tertiary carbinols, combining resin-bound acidic catalysts and lipases was developed in 2014 by Bäckvall, Deska and co-workers [42]. At room temperature, in a biphasic isooctane/water system, where water is used to prevent formation of the dehydrated condensation etheral side-product, the combination of the racemization/isomerization catalyst, the acid ion-exchange resin Dowex 50Wx4, with a nylon basket filled with immobilized CALB, or the lipase from *Pseudomonas cepacia*, immersed into the organic top layer, using a lipophilic acyl donor such as vinyl caprate, synthetically valuable carbocyclic building blocks were obtained in moderate to high yields (67–90%) and enantioselectivity 84–99% *ee* (Scheme 12.8). This procedure was found to be tolerant of multiple aryl and allyl functional groups adjacent to the tertiary alcohol centre of the substrate, while the presence of an alkyl group lowered the yields.

Another chemo-enzymatic protocol moving away from expensive and less abundant metal systems was developed by Rueping in 2016 [43]. Combining the iron dehydrogenation–hydrogenation catalyst **6** with CALB as the bio-catalyst, and

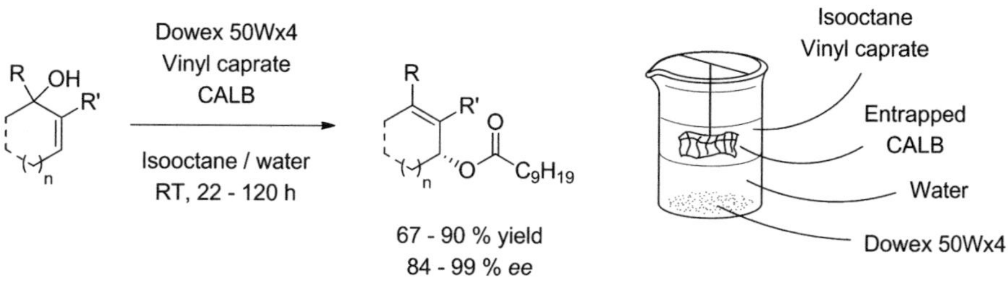

Scheme 12.8 Migratory dynamic kinetic resolution of carbocyclic allylic alcohols *via* a teabag approach

p-chlorophenyl acetate as the acyl donor, afforded, in 24 h, at 60 °C, DKR of benzylic and aliphatic secondary alcohols to the corresponding enantioenriched (*R*)-acetates. Using this protocol, a large array of secondary benzylic, heteroaromatic and aliphatic alcohols bearing electron-withdrawing and electron-donating groups underwent successful DKR transesterification, with good to high yields (68–95%), and high enantioselectivities (92–99% *ee*), showing the potential of this iron/lipase catalyzed resolution.

Only a few reports on (*S*)-selective DKR of alcohols can be found in the literature. The bio- chemo-catalytic protocols allowing resolution of alcohol substrates were limited to the (*R*)-configuration until Kim, Park and co-workers reported in 2003 the (*S*)-selective DKR of alcohols delivering esters (92–99% *ee*) in good to high yields (77–95% yield) by combining, at room temperature, surfactant-treated subtilisin and the aminocyclopentadienylruthenium complex **7** [44]. A few years later, Backvall and co-workers reported improved (*S*)-esters yields and enantioselectivities by using a similar chemo-catalytic protocol involving complex **3** [45].

In 2011, the combination of the ruthenium catalyst **3** with a variant of CALB, (CALB W104A), with a mutation in one of the residues, a tryptophan amino acid being replaced with alanine, afforded, using isopropenyl acetate as the acyl donor, (*S*)-selective resolution of secondary alcohols. 1-phenylpropanol was converted to its respective (*S*)-acetate, at 60 °C, after 3 days, in toluene, with 99% conversion and 96.5% *ee* [46]. (*S*)-selective DKR was also successfully performed on longer alkylic chain secondary alcohols. Efficient acylation of 1-phenylheptanol could be achieved at just over room temperature (30 °C), but at longer reaction times (6 days), to give the product with 99% conversion and 97% *ee*.

The examples described above, which are representative of the advances in chemo-enzymatic DKR protocols since 1996, show how aliphatic, aromatic, benzylic, azido, perfluorinated, allylic, and homoallylic primary, secondary and tertiary alcohols, including diols, can undergo resolution, *via* the appropriate selection of bio-catalyst, chemo-catalyst, and reaction conditions, with excellent enantioselectivity and yields. DKR of alcohol substrates has also been shown to be a viable technique for the induction of enantioselectivity in natural product synthesis [47–50].

12.2.2 Amines

Optically pure amines are highly valued chiral building blocks and intermediates, which are used extensively in the pharmaceutical industry [52]. They are often prepared through asymmetric kinetic resolution, amino acid reduction, carboxylic acid crystallization [53] and the reductive amination of ketones [54].

The DKR of amines is challenging [55]. Chemo-enzymatic resolution of amines was reported in 1996 by Reetz and co-workers [56]. Using Novozym SP 435 and palladium on carbon as the racemization catalyst, a racemic mixture of phenylethylamine underwent resolution *via* *N*-acylation: (*R*)-*N*-(1-phenylethyl)acetamide was isolated, after 8 days, in 64% yield with excellent enantioselectivity (99% *ee*) at 50–55 °C. This represented a potentially powerful procedure for the production of optically pure acetamides directly from biomass [57].

The coupling of metal-catalyzed racemization with enzymatic KR for the DKR of racemic amines and amino acids providing single enantiomeric products in high yields has been reviewed by Kim and Park [58].

Following work by Kim [59], Jacobs [60], and Bäckvall [61], and their co-workers, the DKR of several benzylic and aliphatic primary amines, and an amino acid amide, combining palladium nanoparticles entrapped in aluminium oxyhydroxide (Pd/AlO(OH)) with the lipase CALB, was reported in 2007 by Kim and Park [62]. Using ethyl acetate or methoxyacetate as the acyl donor, benzylic amines were resolved into the corresponding amides at 70 °C, using 1 mol % of the Pd catalyst, in good to high yields (84–99%) and high enantioselectivity (97–99% *ee*). *p*-Methoxy and *p*-trifluoromethyl substitution of the benzylic group, as well as different open and cyclic aliphatic substituent at the benzylic carbon, were tolerated. DKR of aliphatic amines proved to be more difficult: 92–95% amides yield, with 98–99% *ee*, were obtained at higher catalyst loadings (12 mol % Pd/AlO(OH) and temperatures (100 °C). The combination of Pd/AlO(OH) with Novozym 435 also displayed robust recyclability, with high thermostability: minimal reduction in yield and enantioselectivity were observed in the DKR of **14** (Scheme 12.9) over 10 reaction

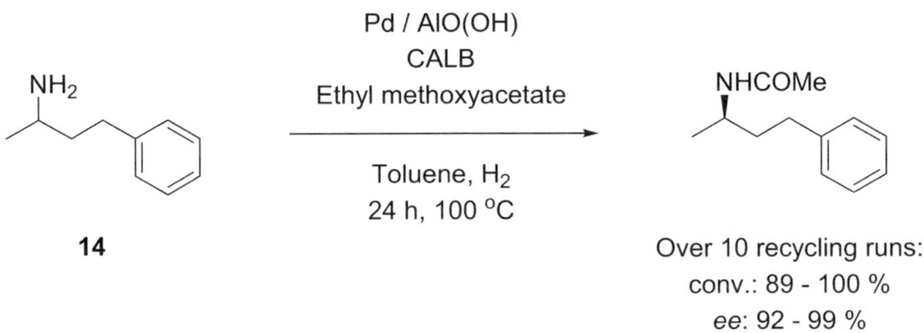

Scheme 12.9 Recyclable chemo-enzymatic dynamic kinetic resolution of a primary amine combining a palladium racemisation nanocatalyst and a lipase resolution catalyst

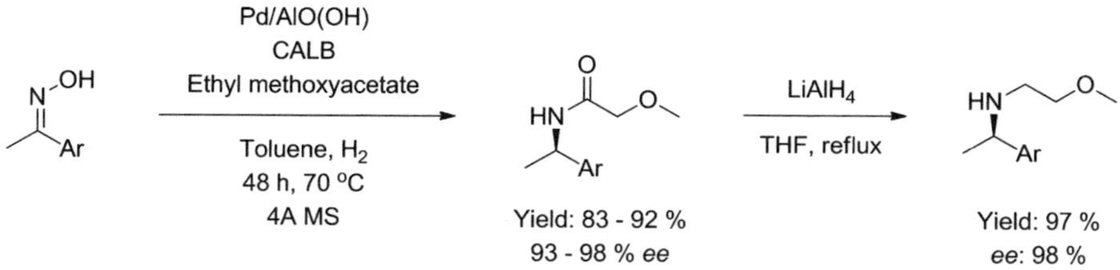

Scheme 12.10 Bio- chemo-catalytic asymmetric reductive ketoxime acylation, followed by amide reduction to chiral secondary amines

cycles. The amino acid amide phenylalanine amide could also be resolved by the chemo-enzymatic DKR protocol developed by Park: the corresponding amide was produced in 96% yield and 98% *ee*.

The chemo-enzymatic protocol described for the resolution of primary amines was exploited in 2008 by Kim, Park and co-workers for the synthesis of α-chiral amides, and ultimately chiral amines, from achiral ketoximes [63]. In this one-pot procedure, the Pd nanocatalyst and the lipase are employed for three consecutive transformations, namely hydrogenation, racemization, and acylation. Pd/AlO(OH) acts as both the racemization and hydrogenation catalyst. Using ethyl methoxyacetate as the acyl donor, a low pressure of hydrogen, and molecular sieves, acyclic and cyclic ketoximes can be converted, at 70 °C, to the corresponding amides in good to high yields (83–92%) and enantioselectivities 93–98% *ee* (Scheme 12.10). While electron-donating groups on the ketoxime benzylic ring lower the amide yield (i.e. *p*-OMe: 83%), side reactions are prevented by electron-withdrawing groups (i.e. *p*-CF_3: 90% yield). Kim, Park and co-workers also proved that secondary chiral amines, to be used as important chiral building blocks or ligands, could be readily produced, quantitatively and without loss in enantiopurity, by the easy reduction of the isolated amides.

The introduction of fluorine into a molecule can impart significant biological, chemical and physical changes within a compound. Due to its electronegativity, a fluorine atom or a trifluoromethylic group can alter and enhance the activity of biologically important molecules [64]. Fluorinated and trifluoromethylated amines are particularly important in the synthesis of pharmaceutically active molecules. For example, Liang, Lin and co-workers proved that (*S*)-3-trifluoromethyl-tetrahydroisoquinoline, an important (*S*)-inhibitor of phenylethanolamine *N*-methyltransferase, can be synthesised in a one-pot cascade process also involving the lipase, nanometal co-catalyzed enantioselective transformation of an α-trifluoromethylated ketoxime to an optically active amide (78% yield, 98% *ee*) [65]. Under a hydrogen atmosphere, in the presence of the palladium or nickel nanocatalysts Pd/Al_2O_3 or Ni/Al_2O_3, prepared *via* atomic layer deposition, as the hydrogenation and racemization chemo-catalysts, and CALB as the acylation biocatalyst, using isopropyl acetate and triethylamine, a range of α-trifluoromethylated ketoximes were transformed, in 6 days, at 70 °C, into the corresponding chiral amides in yields 62–95% and enantioselectivity 90–99% *ee*.

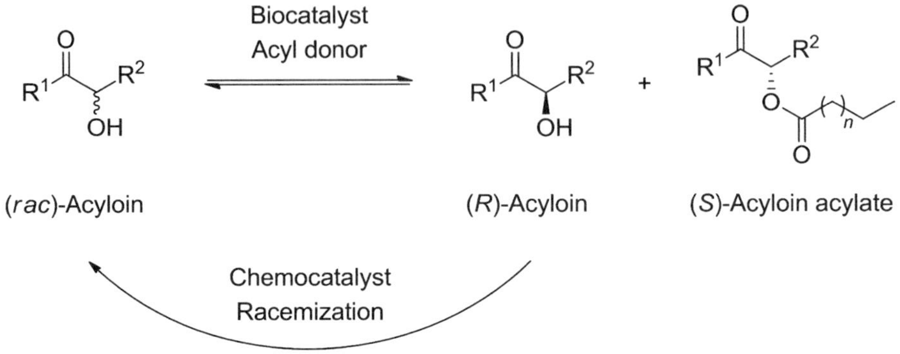

Scheme 12.11 General scheme for the bio- chemo-catalytic resolution of racemic acyloins to (*S*)-esters (See Table 12.2 for substrate scope)

12.2.3 Acyloins

Acyloins are important natural molecular precursors for many pharmaceutical drugs in the treatment of cancer, HIV and Alzheimer's disease among others [66, 67] and they can be used as building blocks for the synthesis of large specialized heterocyclic rings [68–70]. General preparative methods for these involve the reduction of α-diketones [71], the deacylation of acyloin esters and benzoin condensations [72, 73]. Combination of bio- and chemo-catalysis has allowed the preparation of enantiopure acyloins, particularly of the benzoin motif (Scheme 12.11).

12.2.3.1 Benzoins

The combination of bio- and chemo-catalysis for the enantioselective acylation of benzoinic substrates was reported by Alcantara and co-workers in 2006. Employing a KR of the racemic substrates catalyzed by the lipase TL from *Pseudomonas stutzeri*, combined with an *in situ* substrate racemization catalyzed by the ruthenium complex **2**, in the presence of trifluoroethyl butyrate as the acyl donor, the (*S*)-acylated products were produced, at 40–50 °C, in 48–72 h, in conversions up to 90%, with enantiomeric excess values > 99% (Scheme 12.12) [74]. Two years later, Alcantara and co-workers found that the activation of the bio-catalyst by immobilisation in a hydrophobic material by silicone elastomer entrapment allows use of the lipase at higher temperature, affording higher yields in shorter times. The recyclable one-pot benzoin DKR run in the presence of the immobilised lipase TL and **2** allowed conversions up to 87% and 99% *ee*, at 60 °C, in 20 h [75].

In 2014, Ansorge-Schumacher and co-workers showed how the combined bio-chemo-catalytic resolution of racemic benzoin could be improved by heterogenising both the enzyme and the racemization catalysts [76]. The lipase TL from *Pseudomonas stutzeri* was immobilised by adsorption on the porous polypropylene resin Accurel

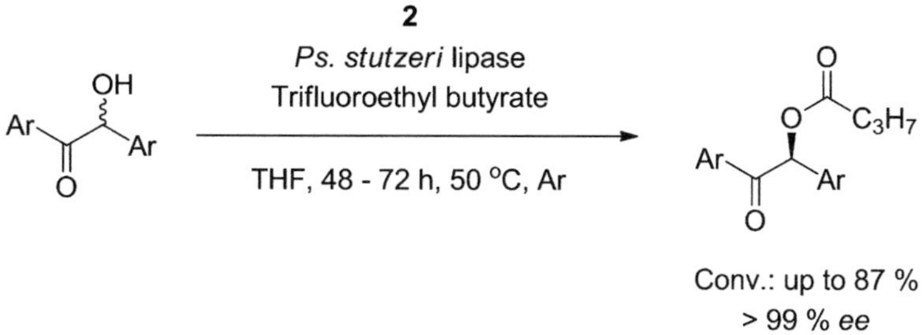

Scheme 12.12 Combining lipase TL from *Pseudomonas stutzeri* with Shvo's catalyst **2**: one-pot enantioselective acylation of benzoins

MP1001 (Acc-LipTL), while the metal catalysts were entrapped in the mesoporous materials metal-TUD-1, metal-associated acidic meso-porous silicates. In 2-MeTHF or toluene, the one-pot DKR of (*rac*)-benzoin, using a combination of Acc-LipTL and Al- or Zr-TUD-1 as the bio- and chemo-catalysts, respectively, afforded, in the presence of vinyl butyrate as the acyl donor, at 50 °C, in 5 h, enantioselective production of (*S*)-benzoin with 85.9–98.4% conversions and 83.3–98.5% *ee*. In the presence of Acc-LipTL and Zr-TUD-1, in 2-MeTHF, the resolution of (*rac*)-benzoin was proved to be recyclable. Stabilization of Acc-LipTL by prior incubation in a polyethylene imine buffer solution enabled full conversion (> 99%) over five cycles, with product *ee* remaining >98%.

In 2014 Martin-Matute and co-workers developed a one-pot, bio- chemo-catalytic system that allowed the enantioselective synthesis of α-hydroxy ketones from benzoins at room temperature [77]. Combining lipase TL with a ruthenium catalyst formed from [Ru(*p*-cymene)Cl_2]$_2$ (**5**) and 1,4-bis(diphenylphosphino) butane at room temperature, in the presence of vinyl butyrate, enabled good to high yields (73–93%) of (*S*)-esters of aromatic α-hydroxy ketones, with enantioselectivity 94–99% *ee*. The substrate scope and reaction conditions for this reaction are described in Table 12.2.

Between 2005 and 2007 Faber and co-workers discovered that the bio-catalytic racemization of a range of variously functionalized acyloins [78] and α-hydroxycarboxylic acids [79] can be accomplished under mild reaction conditions using a range of microbes, including *Lactobacillus* spp. such as *Lactobacillus paracasei* DSM 20207, with limited occurrence of side reactions.

The examples discussed in this section highlight how, by the simultaneous use of a racemization chemo-catalyst and a biological catalyst, DKR enables the effective, enantioselective transformation of various functionalities, including a number of biologically important moieties, to target compounds. Several techniques allowing the conversion of racemic or enantiopure substrates to the desired products have been discussed, showing that high conversion and selectivity can be achieved under mild conditions. It has been shown that the bio- chemo-catalytic protocol DKR can be a powerful technique for the synthesis of optically pure compounds. The production of alcohols from biomass and renewable resources has been reviewed [5, 80–82]. The majority of nitrogen containing compounds, in particular aromatic amines,

Table 12.2 Transesterification and dynamic kinetic resolution producing benzoin acylates, as described in Sect. 2.3.1

	Substrate				Product *ee*	
Entry	R[1]	R[2]	Chemo-catalyst	Conv. (%)[a]	(%)	Refs.
1[a]	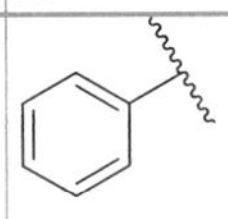	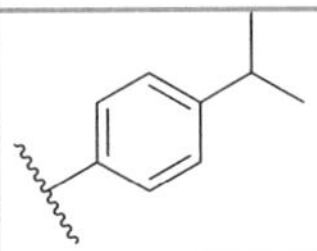	2.5 mol % **2**	94	> 99	[87]
2[a]	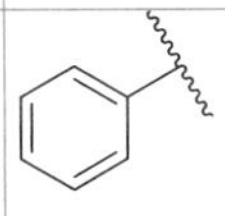	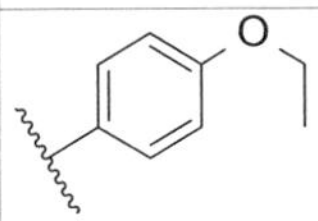	2.5 mol % **2**	95	> 99	[87]
3[a]	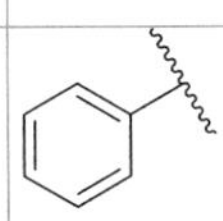	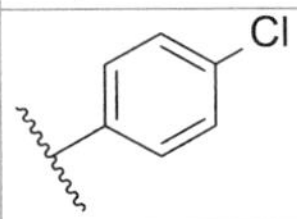	2.5 mol % **2**	79	> 99	[87]
4[a]	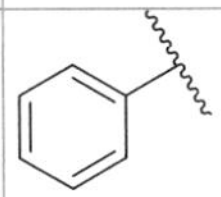	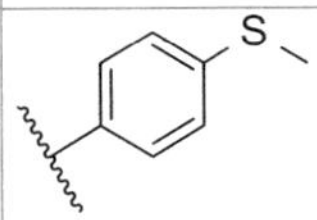	2.5 mol % **2**	63	> 99	[87]
5[a]	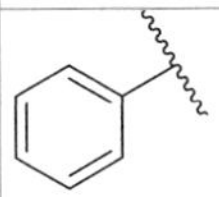	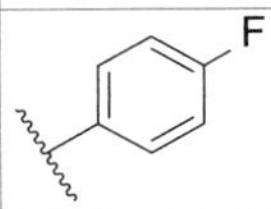	2.5 mol % **2**	93	> 99	[87]
6[a]	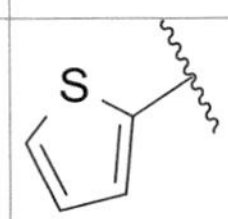	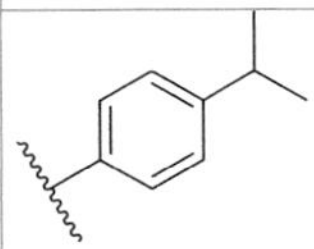	2.5 mol % **2**	80	> 99	[87]
7[a]	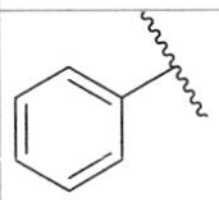	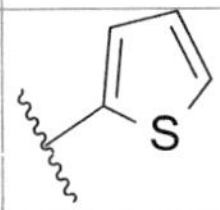	2.5 mol % **2**	89	> 99	[87]
8[b]	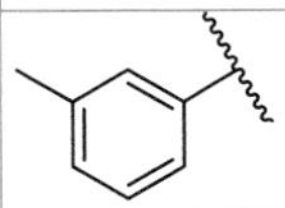	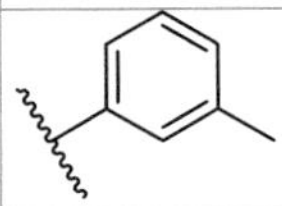	1.25 mol % **5**	91	98	[77]
9[b]	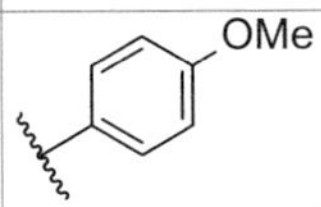	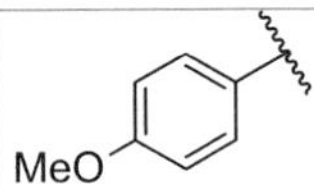	1.25 mol % **5**	93	> 99	[77]
10[b]	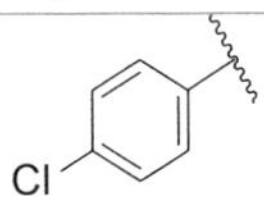	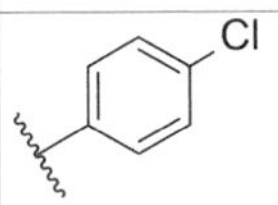	1.25 mol % **5**	87	95	[77]
11[b]	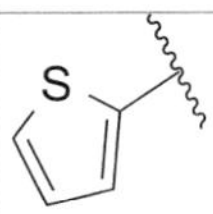	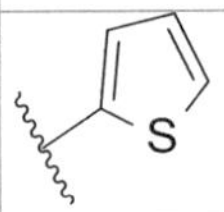	1.25 mol % **5**	90	96	[77]

[a]Reaction conditions: KR step: 0.2 mmol of substrate, lipase TL (10 mg/ml), 1.2 mmol of vinyl butyrate, 2 mL THF, 50 °C; DKR step: product and substrate from KR step, lipase TL (20 mg/mL), 0.005 mmol catalyst **2**, 1.2 mmol trifluoroethyl butyrate, THF, 50 °C, Ar atmosphere

[b]Reaction conditions: 0.2 mmol of substrate, lipase TL (40 mg), 0.0025 mmol catalyst **5**, 0.005 mmol ligand dppb, 0.0025 mmol ^{t}BuOK, 0.6 mmol of vinyl butyrate, 2 mL THF, RT, Ar atmosphere

are currently reliant on the petrochemical industry, but bio-based synthetic routes have started to emerge [83]. Amines, and related nitrogen containing compounds, can now be synthesised through derivatization of bio-based platform chemicals [19], and their formation *via* bio-catalytic routes is of growing interest [84–86]. The combination of these biomass transformations with the chemo-enzymatic DKR protocols can result in the production of optically pure, value-added compounds in high yield *via* simultaneous bio- and chemo-catalysis.

12.3 Valorisation of Biomass

The conversion of biomass into value-added chemicals has been an area of major academic and industrial interest in recent years [85–87]. Significant attention has been drawn to the alternative production of fossil based chemicals from renewable and sustainable sources, in particular as direct fuel replacements, [88]. Enzymatic and whole cell bio-catalysis allows conversion of biomass feedstock into bio-based platform chemicals, which can be then converted to the desired products by a chemo-catalytic process. Combining bio-catalytic fermentation of biomass with chemo-catalysis is expensive: the design of an economically viable chemo-enzymatic protocol, in particular when aimed at direct fuel replacement, represents a challenge. Therefore, biomass valorization should focus on the combination of bio- and chemo-catalysis to target high value commodity chemicals [89].

This section focuses on the transformation of biomass into value-added chemicals and highlights the commercial viability of such routes. Around the world, biomass is often used as a source of energy for the generation of electricity [90]. Biomass is used as a cheap and available source of combustible material, and it has been shown that its conversion into feedstock chemicals can be economically viable, with a value ten times higher than that resulting from its use for energy production [91]. This coupling of energy and chemicals production can be vital to the economics of fuel production, and is a unique benefit of biomass compared to other renewable energies.

A major issue with the valorisation of biomass is the impure, frequently poorly defined, nature of the biomass feed source, often containing a mixture of compounds that are challenging to separate. Valorisation techniques must be able to deal with these impurities without loss of activity or performance nor requiring extensive processing. From a sustainability perspective it would be ideal if the transformation of biomass to the value-added chemical could be carried out in a one-pot synthesis, combining catalytic cascade reactions. This approach would limit and reduce the use of solvents, reagents and energy, as compared to conventional multi-step synthesis. One such class of one-pot reactions includes the simultaneous use of both bio- and chemo-catalyzed steps working in tandem [92].

A downside associated with the use of biomass for the production of value-added chemicals is the impure, dilute, aqueous, and highly oxygenated nature of the compounds that tend to be produced. One method to exploit these solutions is by using bacteria, algae or fungi in whole cell bio-catalysis. These organisms can convert the bio-derived, impure, aqueous solutions into target compounds *via* the cell metabolism. In addition, the parallel use of chemo-catalysts can further diversify the chemicals produced. A number of alcohols and carboxylic acids are common products of cell metabolism. Among them, 1,3-propanediol (**15**) has been prepared from the bacterial fermentation of both pure and crude glycerol derived from biomass waste generated in biofuel production. **15** is already a bio-based chemical as the production of Bio-PDO, commercialised by DuPont and Tate & Lyle, and operated on the 60,000 tonnes/anum scale, is more efficient than the petrochemical route. The selective amination of **15** so-formed *via* a hydrogen transfer (HT) catalyst was reported in 2009 by Stephens, Marr and co-workers [93]. This represented the combination of whole cell bio-catalysis and chemo-catalytic homogeneous catalysis for the valorisation of biomass waste. In the presence of the HT catalyst Cp*Ir(NHC) complex **8** (Scheme 12.2), **15** first undergoes oxidation to the corresponding aldehyde; the reaction with an amine and subsequent hydrogenation leads to a potential mixture of mono- and di-substituted amines. If the reaction is performed in ionic liquid dehydration activity is observed, despite the presence of water. Combining bio-catalytic treatment of crude glycerol with *Clostridium butyricum* with Cp*Ir(NHC)-catalyzed HT amination, in $N_{1,8,8,8}NTf_2$, a biologically non-toxic hydrophobic ionic liquid, at 42 °C, one only amination product, *N*-propyl aniline, was formed. This procedure outlines a method to utilise waste glycerol from biodiesel, a waste biomass source, to deliver a value-added chemical from a one-pot process combining bio- and chemo-catalytic protocols.

One of the challenges to overcome with a bio-derived platform chemical such as **15** is the difficulty in extracting it, prior to chemo-catalysis, from the fermentation medium resulting from bio-catalysis, due to the high water solubility of the alcohol. That can be dealt with by using hydrophobic ionic liquids, which will form a biphasic system with the fermentation broth, while extracting the desired chemical platform. Once the latter has been isolated in the extraction medium, further chemo-catalytic reactions can occur, ultimately leading to the target molecules.

In 2016, Marr and co-workers demonstrated the chemo-catalytic hydrogen transfer initiated dehydrogenation (HTID) of **15** catalyzed by **8** and its analogues [4]. This represented the dehydration of a bio-alcohol in the presence of water. The bio-alcohol can be derived from whole cell bio-catalytic routes (Scheme 12.13). The selective transformation of **15** to propanal was reported, highlighting the potential

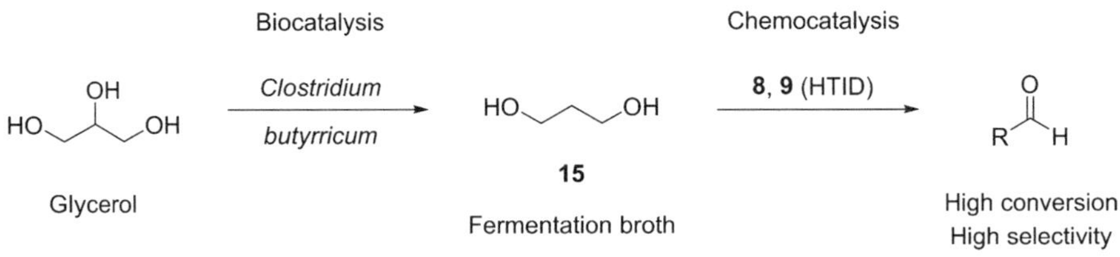

Scheme 12.13 Valorisation of glycerol to aldehydes *via* bio-catalysis coupled with hydrogen transfer initiated dehydrogenation (HTID) of **15** catalyzed by Cp*Ir(NHC) complexes

of this bio-derived platform chemical as a precursor for further derivatization towards functionalities including carboxylic acids, esters, amines, amides and higher olefins. Propanal was targeted due to its large global production (> 100,000 tonnes per annum) from petroleum and its use in the manufacturing of lubricants, resins, plasticisers and coatings. The recyclable synthetic distillation of propanal, with high selectivity (up to 87.5%) and in high crude yields (up to 99%), was performed by reacting **15** in ionic liquids, in the presence of **8** or **9** [94] as the chemo-catalysts and different HT-assisting bases, at 100–150 °C, for only 6 h. The selectivity towards the target propanal was found to be effected by the ionic liquid used, the reaction temperature, the catalyst and base loading, as well as the pressure. The moderate dynamic vacuum, along with elevated temperatures, allows the selective isolation of the volatile target value-added chemical, preventing aldol condensation and hydrogenation side-reactions, leading to C3 and C6 by-products. The HTID of **15** can also be successfully driven towards the selective production of 2-methylpent-2-enal, an important intermediate in the synthesis of plasticisers and pharmaceuticals as well as a commercially important compound widely used in the cosmetic, fragrance, and flavour industries, by reacting **15** in the presence of the chemo-catalysts **8** and **9** coupled with a base, in ionic liquids, at temperature varying in the range 100–180 °C, but in a sealed reaction tube [95]. The formation of C6 products is initiated by the self-aldol condensation of propanal: by running the reaction in the closed reaction vessel, propanal is forced to remain in the reacting mixture after its formation, and then dimerises *via* self-aldol condensation. The successful synthesis of value-added chemicals such as propanal and 2-methylpent-2-enal out of the ionic liquid solutions of the potentially biomass-derived platform chemical **15**, mimicking the product of extraction of aqueous glycerol fermentation broth, proves that the combination of Cp*Ir(NHC) catalyzed HTID of **15** in ionic liquids with bio-catalysis using *Clostridium butyricum* has the potential to allow the direct bio- chemo-catalytic transformation of waste glycerol into valuable chemicals, with the capability for further chemical derivatization.

Adipic acid is a commercially important chemical and a precursor for the production of 6,6-nylon; its global production is estimated to increase to 3.6 million tonnes per annum by 2022 [96]. Commercial adipic acid is derived from benzene based compounds; therefore, a more sustainable, environmentally friendly, and economically viable approach has been sought. In 2016, a route to the formation of adipic acid from waste sugar beet residue combining bio- and chemo-catalysis was discovered by Zhang, Zhao and co-workers [97]. Sugar beet residue is an inexpensive, large scale waste product from agricultural processing: 25 million tonnes of sugar beet residue are produced every year. Following enzymatic breakdown of sugar beet residue leading to *D*-galacturonic acid, the combination of bio-catalytic production of mucic acid using an engineered whole cell *Escherichia coli* bacterium with chemo-catalytic deoxydehydration using the rhenium catalyst NH_4ReO_4 followed by hydrogenation *via* platinum on carbon, afforded adipic acid (Scheme 12.14) [98]. The combination of bio- and chemo-catalysis has enabled the production of bio-based adipic acid, from waste sugar beet residue, without competing with food production, and adding significant value to this waste biomass.

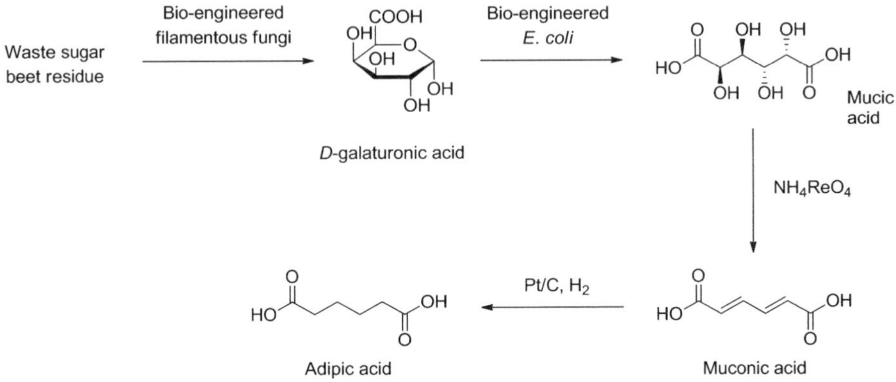

Scheme 12.14 Bio- chemo-catalytic route for the production of adipic acid from waste sugar beet residue

Cellulose, the most abundant biopolymer on the planet, has vast potential as a bio-based feedstock for the production of a diverse range of commodity chemicals. A number of companies have developed commercially viable pilot plants for the production of succinic acid from cellulose [99–101]. 5-hydroxymethylfurfural (5-HMF) is an industrially attractive chemical derived from cellulose. A number of commodity and fine chemicals, including 2,5-dimethylfuran and 2,5-furandicarboxylic acid, can be obtained from 5-HMF [102]. Large scale production of 5-HMF is complicated by its thermal and chemical instability. The conversion of cellulose to 5-HMF is challenging and requires isomerization of glucose to fructose to proceed.

A protocol combining bio- and chemo-catalysis affording the synthesis of 5-HMF from cellulose-derived glucose was described by Gimbernat et al in 2017 [7]. A triphasic system keeps the incompatible bio- and chemo-catalytic systems separate. Glucose is bio-catalytically isomerised to fructose in the presence of glucose isomerise, in an aqueous feed phase. Fructose then accesses an organic phase, in which it is transformed into a fructose-boronate carrier species to facilitate its transport through the organic layer to reach a second, chemo-catalytic, receiving aqueous phase. In it, after hydrolysation of the fructose-boronate species back to fructose at the organic/aqueous interlayer, the acid catalyzed dehydration of fructose to 5-HMF product occurs (Scheme 12.15). Despite low yields due to limitations caused by poor fructose transport between the two aqueous phases, this process is a powerful proof of concept offering a methodology towards cellulose valorisation *via* combination of bio- and chemo-catalysis.

An integrated chemo-enzymatic process for the production of 5-HMF from glucose was reported by Alfonso and co-workers in 2013. The biphasic ionic liquid/water system, in which sweetzyme and nitric acid catalyzed glucose isomerization and fructose dehydration, respectively, afforded up to 91% yield of 5-HMF, in up to 99% purity [103]. The chemo-catalytic step was carried out after removal of the enzyme from the reaction mixture.

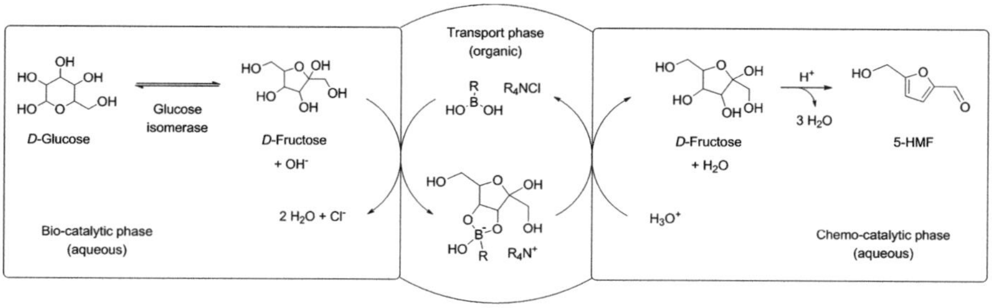

Scheme 12.15 Bio- chemo-catalytic route for the production of 5-hydroxymethylfurfural (5-HMF) from bio-derived glucose

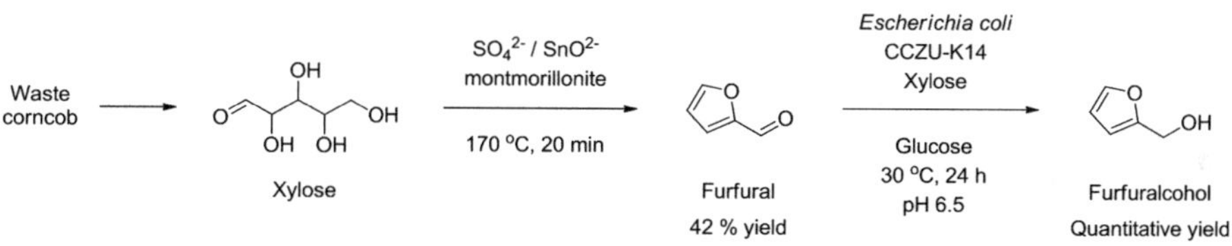

Scheme 12.16 Integrated chemo-enzymatic synthesis of furfuralcohol from bio-derived xylose

He and co-workers invented a bio- and chemo-catalytic protocol to transform bio-derived xylose into furfuralcohol incorporating a montmorillonite solid acid catalyst. This comprised a chemo-catalytic system for the conversion of pre-treated corncob and corncob hydrosylate biomass, rich in xylose, into the important bio-based platform chemical furfural [104]: Zhang and co-workers reported that, using a biphasic water/toluene system, in the presence of the SO_4^{2-}/ SnO_2^- montmorillonite solid acid catalyst, furfural was produced in 82% yield, with 94% xylose conversion [6]. He and co-workers then developed an integrated one-pot protocol. Zhang and co-workers' chemo-catalytic process was combined with the whole cell bio-catalytic reductive transformation of furfural into furfuralcohol, using recombinant *Escherichia coli* CCZU-K14 [105]. Quantitative yields of the alcohol were obtained (Scheme 12.16). This system represents an effective bio- chemo-catalytic method for the valorisation of hemicellulose rich agricultural waste towards bio-based platform chemicals.

The successful combination of bio- and chemo-catalysis for the transformation of biomass into chemicals compatible with the refinery infrastructure and suitable for blending with petrol, diesel and jet fuels could have an enormous economic impact. The replacement of transportation fuels with renewable products is desirable; however, biomass fermentation often leads to compounds that are only suitable as fuel additives. Therefore, the incorporation of a chemical process is required to convert these bio-based products into compounds suitable as transportation fuels for the land, nautical and aviation transport industries. This can be achieved, for example, by coupling a chemo-catalytic route to the bio-catalytic production of

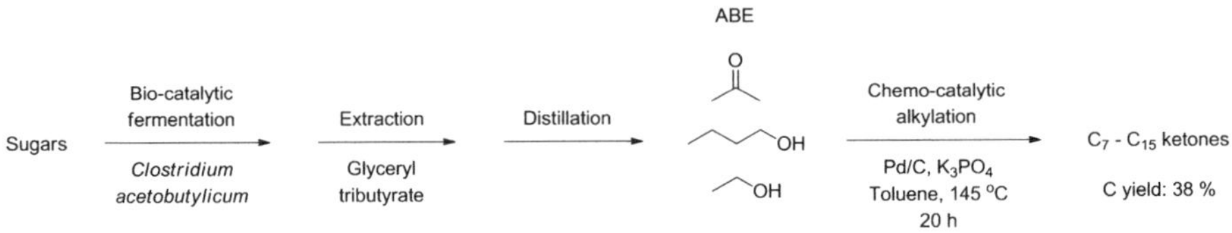

Scheme 12.17 Production of long chain ketones from sugars combining extractive, *Clostridium acetobutylicum* catalyzed fermentation with palladium catalyzed alkylation of acetone/*n*-butanol/ethanol (ABE)

acetone/*n*-butanol/ethanol (ABE). ABE can be produced by fermentation of a number of biomass derived sources, including sugars, in the presence of *Clostridium acetobutylicum* [106].Toste and co-workers coupled the bio-chemical production of ABE from glucose with Pd catalyzed functionalisation of acetone for the production of longer chain molecules. After extraction of ABE with glyceryl tributyrate and distillation, alkylation, performed in the presence of Pd/C, for 20 h, in toluene, at 145 °C, with 1.28 molar equivalents of K_3PO_4, afforded C_7-C_{15} products representing 38% of the carbon contained in the glucose feed (Scheme 12.17) [107]. The overall carbon yield including C_4-C_6 products was 58%. Further reduction of the long chain chemicals would yield the desired hydrocarbon fuel blend stocks suitable as liquid transportation fuels.

Many renewable compounds offer numerous possibilities for further valorisation, including vegetable oils or animal unsaturated fatty acids and terpenes [108]. Given their relatively high carbon content and olefin functional groups, these compounds can be further valorised, for instance by selective oxidation of the alkene functionality towards the formation of aldehydes and carboxylic acids. The formation of aldehydes from alkenes, for instance, has been explored in the presence of ruthenium [109, 110], osmium [111] and tungsten [112] chemocatalysts, and hetero- and homo-geneous catalytic systems for the oxidative double bond cleavage with interest towards biomass valorisation have been reviewed [113].

12.4 Conclusions and Future Outlook

In this chapter a series of combined bio- chemo-catalytic routes that have the potential to facilitate the transition away from petroleum feedstock and towards sustainable and environmentally benign feeds have been highlighted. The production of valuable products *via* one-pot chemo-enzymatic protocols of different types, contemplating chemo-catalytic treatment of chemical platforms including alcohols, diols, acids, amines, and acyloins, potentially resulting from aerobic/anaerobic fermentation and bio-enzymatic transformation of biomass, has been presented. The use of chemo-enzymatic dynamic kinetic resolution (DKR) for numerous substrates has been detailed, showing the wide range of functionalities currently accessible and how the substrate scope has expanded over the last two decades, allowing high yield acylation, good substrate tolerance, and enantioselectivity *via* combined bio- and chemo-catalysis. It has also been noted that DKR can facilitate the formation and acylation of several biologically important molecules. Specific examples showing how the incorporation of bio- and chemo-catalysis can simplify the delivery of value-added chemicals directly from renewable, biomass derived feedstock have also been presented.

The future of combined bio- and chemo-catalytic routes will be largely dictated by rapid developments in whole cell and isolated enzyme biocatalysis. Innovations such as synthetic biology and directed evolution are vastly increasing the scope of bio-catalysis and this will open up exciting new possibilities. Developments in protein synthesis and purification have led to the emergence of the field of artificial metalloenzymes, within which the biocatalyst and chemocatalyst are no longer separate, but rather bound as one [114].

References

1. Sheldon RA. Engineering a more sustainable world through catalysis and green chemistry. R Soc Interface. 2016;1:1–7.
2. Perlack RD, Wright LL, Turhollow AF, Graham RL, Stokes BJ, Erbach DC. Biomass as feedstock for a bioenergy and bioproducts industry: the technical feasibility of a billion-ton annual supply. Oak Ridge: Oak Ridge National Laboratory; 2005.
3. Vennestrøm PNR, Christensen CH, Pedersen S, Grunwaldt JD, Woodley JM. Next-generation catalysis for renewables: combining enzymatic with inorganic heterogeneous catalysis for bulk chemical production. ChemCatChem. 2010;2:249–58.
4. Wang Y-M, Lorenzini F, Rebros M, Saunders GC, Marr AC. Combining bio- and chemo-catalysis for the conversion of bio-renewable alcohols: homogeneous iridium catalysed hydrogen transfer initiated dehydration of 1,3-propanediol to aldehydes. Green Chem. 2016;18:1751–61.
5. Zakzeski J, Bruijnincx PCA, Jongerius AL, Weckhuysen BM. The catalytic valorization of lignin for the production of renewable chemicals. Chem Rev. 2010;110:3552–99.
6. Qing Q, Guo Q, Zhou L, Wan Y, Xu Y, Ji H, Gao X, Zhang Y. Catalytic conversion of corncob and corncob pretreatment hydrolysate to furfural in a biphasic system with addition of sodium chloride. Bioresour Technol. 2017;226:247–54.
7. Gimbernat A, Guehl M, Capron M, Ferreira NL, Froidevaux R, Girardon J-S, Dhulster P, Delcroix D, Dumeignil F. Hybrid catalysis: a suitable concept for the valorization of biosourced saccharides to value-added chemicals. ChemCatChem. 2017;9:2080–4.
8. Liu B, Zhang Z. One-pot conversion of carbohydrates into furan derivatives *via* furfural and 5-hydroxylmethylfurfural as intermediates. ChemSusChem. 2016;9:1–23.
9. Pérez-Torrado R, Gamero E, Gómez-Pastor R, Garre E, Aranda A, Matallana E. Yeast biomass, an optimised product with myriad applications in the food industry. Trends Food Sci Technol. 2015;46:167–75.
10. Ungaro VA, Liria CW, Romagna CD, Costa NJS, Philippot K, Rossi LM, Machini MT. A green route for the synthesis of a bitter-taste dipeptide combining biocatalysis, heterogeneous metal catalysis and magnetic nanoparticles. RSC Adv. 2015;5:36449–55.
11. De Vries JG. Twenty-five years of homogeneous catalysis for the production of bulk and fine chemicals: a personal account. Top Catal. 2014;57:1306–17.
12. Carvalho ACLDM, Fonseca TDS, De Mattos MC, De Oliveira MDCF, De Lemos TMLG, Molinari F, Romano D, Serra I. Recent advances in lipase-mediated preparation of pharmaceuticals and their intermediates. Int J Mol Sci. 2015;16:29682–716.

13. Langvik O, Saloranta T, Murzin DY, Leino R. Heterogeneous chemoenzymatic catalyst combinations for one-pot dynamic kinetic resolution applications. ChemCatChem. 2015;7:4004–15.
14. Odjadjare EC, Mutanda T, Olaniran AO. Potential biotechnological application of microalgae: a critical review. Crit Rev Biotechnol. 2017;37:37–52.
15. Clouthier CM, Pelletier JN. Expanding the organic toolbox: a guide to integrating biocatalysis in synthesis. Chem Soc Rev. 2012;41:1585.
16. Ghanem A, Aboul-Enein HY. Application of lipases in kinetic resolution of racemates. Chirality. 2005;17:1–15.
17. Mäki-Arvela P, Sahin S, Kumar N, Heikkilä T, Lehto V-P, Salmi T, Murzin DY. Cascade approach for synthesis of *R*-1-phenyl ethyl acetate from acetophenone: effect of support. J Mol Catal A Chem. 2008;285:132–41.
18. Mäki-Arvela P, Sahin S, Kumar N, Mikkola JP, Eränen K, Salmi T, Murzin DY. One-pot utilization of heterogeneous and enzymatic catalysis: synthesis of *R*-1-phenylethyl acetate from acetophenone. Catal Today. 2009;140:70–3.
19. Werpy T, Petersen G. Top value added chemicals from biomass volume I – results of screening for potential candidates from sugars and synthesis gas. Washington, DC: US Department of Energy; 2004.
20. Sun D, Sato S, Ueda W, Primo A, Garcia H, Corma A. Production of C4 and C5 alcohols from biomass-derived materials. Green Chem. 2016;18:2579–97.
21. Subramani V, Gangwal SK. A review of recent literature to search for an efficient catalytic process for the conversion of syngas to ethanol. Energy Fuels. 2008;22:814–39.
22. Li J, Wang C, Yang Z. Production and separation of phenols from biomass-derived biopetroleum. J Anal Appl Pyrolysis. 2010;89:218–24.
23. Dinh PM, Howarth JA, Hudnott AR, Williams JMJ, Harris W. Catalytic racemisation of alcohols: applications of enzymatic resolution reactions. Tetrahedron Lett. 1996;37:7623–6.
24. Pollock CL, Fox KJ, Lacroix SD, McDonagh J, Marr PC, Nethercott AM, Pennycook A, Qian S, Robinson L, Saunders GC, Marr AC. Minimizing side reactions in chemoenzymatic dynamic kinetic resolution: organometallic and material strategies. Dalt Trans. 2012;41:13423–8.
25. Larsson ALE, Person BA, Backvall J-E, To D, Kagan HB. Enzymatic resolution of alcohols coupled with ruthenium-catalyzed racemization of the substrate alcohol. Angew Chem Int Ed. 1997;36:1211–2.
26. Persson BA, Huerta FF, Bäckvall JE. Dynamic kinetic resolution of secondary diols *via* coupled ruthenium and enzyme catalysis. J Org Chem. 1999;64:5237–40.
27. Pàmies O, Bäckvall JE. Dynamic kinetic resolution of β-azido alcohols. An efficient route to chiral aziridines and β -amino alcohols. J Org Chem. 2001;66:4022–5.
28. Runmo A, Pàmies O, Faber K, Bäckvall J. Dynamic kinetic resolution of α -hydroxy acid derivatives. Tetrahedron Lett. 2002;43:2983–6.
29. Huerta FF, Bäckvall JE. Enantioselective synthesis of β-hydroxy acid derivatives *via* a one-pot aldol reaction-dynamic kinetic resolution. Org Lett. 2001;3:1209–12.
30. Martín-Matute B, Edin M, Bogár K, Bäckvall JE. Highly compatible metal and enzyme catalysts for efficient dynamic kinetic resolution of alcohols at ambient temperature. Angew Chemie - Int Ed. 2004;43:6535–9.
31. Martín-Matute B, Edin M, Bogár K, Kaynak FB, Bäckvall JE. Combined ruthenium(II) and lipase catalysis for efficient dynamic kinetic resolution of secondary alcohols. Insight into the racemization mechanism. J Am Chem Soc. 2005;127:8817–25.
32. Bogár K, Martín-Matute B, Bäckvall JE. Large-scale ruthenium- and enzyme-catalyzed dynamic kinetic resolution of (*rac*)-1-phenylethanol. Beilstein J Org Chem. 2007;3:50.
33. Fransson ABL, Xu Y, Leijondahl K, Bäckvall JE. Enzymatic resolution, desymmetrization, and dynamic kinetic asymmetric transformation of 1,3-cycloalkanediols. J Org Chem. 2006;71:6309–16.

34. Strübing D, Krumlinde P, Piera J, Bäckvall JE. Dynamic kinetic resolution of primary alcohols with an unfunctionalized stereogenic center in the β-position. Adv Synth Catal. 2007;349:1577–81.
35. Kawasaki M, Goto M, Kawabata S, Kodama T, Kometani T. Lipase-catalyzed transesterification of 2-phenyl-1-propanol with vinyl esters having aromatic ring in acyl moiety. Tetrahedron Lett. 1999;40:5223–6.
36. Kawasaki M, Goto M, Kawabata S, Kometani T. The effect of vinyl esters on the enantioselectivity of the lipase-catalysed transesterification of alcohols. Tetrahedron Asymmetry. 2001;12:585–96.
37. Bogár K, Bäckvall JE. High-yielding metalloenzymatic dynamic kinetic resolution of fluorinated aryl alcohols. Tetrahedron Lett. 2007;48:5471–4.
38. Lihammar R, Millet R, Bäckvall JE. Enzyme- and ruthenium-catalyzed dynamic kinetic resolution of functionalized cyclic allylic alcohols. J Org Chem. 2013;78:12114–20.
39. Marr AC, Pollock CL, Saunders GC. Base-free dynamic kinetic resolution of secondary alcohols using "piano-stool" complexes of N-heterocyclic carbenes. Organometallics. 2007;26:3283–5.
40. Lozano P, De Diego T, Larnicol M, Vaultier M, Iborra JL. Chemoenzymatic dynamic kinetic resolution of *rac*-1-phenylethanol in ionic liquids and ionic liquids/supercritical carbon dioxide systems. Biotechnol Lett. 2006;28:1559–65.
41. Lozano P, De Diego T, Mira C, Montague K, Vaultier M, Iborra JL. Long term continuous chemoenzymatic dynamic kinetic resolution of *rac*-1-phenylethanol using ionic liquids and supercritical carbon dioxide. Green Chem. 2009;11:538.
42. Manzuna Sapu C, Görbe T, Lihammar R, Bäckvall JE, Deska J. Migratory dynamic kinetic resolution of carbocyclic allylic alcohols. Org Lett. 2014;16:5952–5.
43. El-Sepelgy O, Alandini N, Rueping M. Merging iron catalysis and biocatalysis-iron carbonyl complexes as efficient hydrogen autotransfer catalysts in dynamic kinetic resolutions. Angew Chemie - Int Ed. 2016;55:13602–5.
44. Kim MJ, Chung YI, Choi YK, Lee HK, Kim D, Park J. (S)-selective dynamic kinetic resolution of secondary alcohols by the combination of subtilisin and an aminocyclopentadienylruthenium complex as the catalysts. J Am Chem Soc. 2003;125:11494–5.
45. Borén L, Martín-Matute B, Xu Y, Córdova A, Bäckvall JE. (S)-Selective kinetic resolution and chemoenzymatic dynamic kinetic resolution of secondary alcohols. Chem A Eur J. 2005;12:225–32.
46. Engström K, Vallin M, Syrén P-O, Hult K, Bäckvall J-E. Organic & biomolecular chemistry mutated variant of Candida antarctica lipase B in (*S*)-selective dynamic kinetic resolution of secondary alcohols. Org Biomol Chem. 2011;9:81–2.
47. Deska J, Bäckvall J-E. Enzymatic kinetic resolution of primary allenic alcohols. Application to the total synthesis and stereochemical assignment of striatisporolide A. Org Biomol Chem. 2009;7:3379–81.
48. Johnston EV, Bogár K, Bäckvall J-E. Enantioselective synthesis of (*R*)-bufuralol *via* dynamic kinetic resolution in the key step. J Org Chem. 2010;75:4596–9.
49. Träff A, Lihammar R, Bäckvall JE. A chemoenzymatic dynamic kinetic resolution approach to enantiomerically pure (*R*)- and (*S*)-duloxetine. J Org Chem. 2011;76:3917–21.
50. Warner MC, Shevchenko GA, Jouda S, Bogár K, Bäckvall JE. Dynamic kinetic resolution of homoallylic alcohols: application to the synthesis of enantiomerically pure 5,6-dihydropyran-2-ones and δ-lactones. Chem A Eur J. 2013;19:13859–64.
51. Anders Persson B, E Larsson AL, Ray L, Backvall J-E. Ruthenium-and enzyme-catalyzed dynamic kinetic resolution of secondary alcohols. J Am Chem Soc. 1999;121:1645–50.
52. Constable DJC, Dunn PJ, Hayler JD, Humphrey GR, Leazer JL Jr, Linderman RJ, Lorenz K, Manley J, Pearlman BA, Wells A, Zaks A, Zhang TY. Key green chemistry research areas: a perspective from pharmaceutical manufacturers. Green Chem. 2007;9:411–20.
53. Breuer M, Ditrich K, Habicher T, Hauer B, Keßeler M, Stürmer R, Zelinski T. Industrial methods for the production of optically active intermediates. Angew Chemie Int Ed. 2004;43:788–824.

54. Nugent TC, El-Shazly M, Wakchaure VN. Ytterbium acetate promoted asymmetric reductive amination: Significantly enhanced stereoselectivity. J Org Chem. 2008;73:1297–305.
55. Verho O, Bäckvall JE. Chemoenzymatic dynamic kinetic resolution: A powerful tool for the preparation of enantiomerically pure alcohols and amines. J Am Chem Soc. 2015;137:3996–4009.
56. Reetz MT, Schimossek K. Lipase-catalyzed dynamic kinetic resolution of chiral amines: use of palladium as the racemization catalyst. Chimia. 1996;50:668–9.
57. Masuo S, Zhou S, Kaneko T, Takaya N, Sousa AC, Martins LO, Robalo MP, Krishnaswamy K, Nair KM, He J, Magarvey N, Piraee M, Vining LC, Blanc V, Liaw DJ, García JM, García FC, Serna F, de la Peña JL, Sapurina I, Pacheco JJ, Davis ME, Lin Z, Nikolakis V, Ierapetritou M, Suvannasara P, Mehl RA, Tolia NH, Joshua-Tor L, Patnaik R, Liao JC, Sprenger GA, Rodriguez A, Kikuchi Y, Tsujimoto K, Kurahashi O, Grossman TH, Kawasaki ES, Punreddy SR, Osburne MS, Gerigk M, Förberg C, Eliaeson T, Häggström L, Kim B, Cho BR, Hahn JS, Ferrandez A, Prieto MA, Garcia JL, Diaz E, Küberl A, Masuo S, Osada L, Zhou S, Fujita T, Takaya N, Tieman D, Li W, Xie D, Frost JW, Kogure T, Wakisaka N, Takaku H, Takagi M, Lee LTC, Ueda M, Morishima M, Kakuta M, Wang HH, Lin MF, Thompson B, Machas M, Nielsen DR, Hutner SH. Bacterial fermentation platform for producing artificial aromatic amines. Sci Rep. 2016;6:25764.
58. Kim Y, Park J, Kim MJ. Dynamic kinetic resolution of amines and amino acids by enzyme-metal cocatalysis. ChemCatChem. 2011;3:271–7.
59. Choi YK, Kim MJ, Ahn Y, Kim MJ. Lipase/palladium-catalyzed asymmetric transformations of ketoximes to optically active amines. Org Lett. 2001;3:4099–101.
60. Parvulescu A, De Vos D, Jacobs P. Efficient dynamic kinetic resolution of secondary amines with Pd on alkaline earth salts and a lipase. Chem Commun. 2005;42:5307–9.
61. Fransson A-BL, Borén L, Pàmies O, Bäckvall J-E. Kinetic resolution and chemoenzymatic dynamic kinetic resolution of functionalized gamma-hydroxy amides. J Org Chem. 2005;70:2582–7.
62. Kim MJ, Kim WH, Han K, Yoon KC, Park J. Dynamic kinetic resolution of primary amines with a recyclable Pd nanocatalyst for racemization. Org Lett. 2007;9:1157–9.
63. Han K, Park J, Kim MJ. Asymmetric reductive acylation of aromatic ketoximes by enzyme-metal cocatalysis. J Org Chem. 2008;73:4302–4.
64. Böhm H-J, Banner D, Bendels S, Kansy M, Kuhn B, Müller K, Obst-Sander U, Stahl M. Fluorine in medicinal chemistry. Chembiochem. 2004;5:637–43.
65. Cheng G, Wu Q, Shang Z, Liang X, Lin X. Stereoselective transformations of α-trifluoromethylated ketoximes to optically active amines by enzyme-nanometal cocatalysis: Synthesis of (*S*)-inhibitor of phenylethanolamine N-methyltransferase. ChemCatChem. 2014;6:2129–33.
66. Burgey CS, Robinson KA, Lyle TA, Nantermet PG, Selnick HG, Isaacs RCA, Lewis SD, Lucas BJ, Krueger JA, Singh R, Miller-Stein C, White RB, Wong B, Lyle EA, Stranieri MT, Cook JJ, McMasters DR, Pellicore JM, Pal S, Wallace AA, Clayton FC, Bohn D, Welsh DC, Lynch JJ, Yan Y, Chen Z, Kuo L, Gardell SJ, Shafer JA, Vacca JP. Pharmacokinetic optimization of 3-amino-6-chloropyrazinone acetamide thrombin inhibitors. implementation of P3 pyridine N-Oxides to deliver an orally bioavailable series containing P1 N-Benzylamides. Bioorg Med Chem Lett. 2003;13(7):1353–7.
67. Fang QK, Han Z, Grover P, Kessler D, Senanayake CH, Wald SA. Rapid access to enantiopure bupropion and its major metabolite by stereospecific nucleophilic substitution on an α-ketotriflate. Tetrahedron Asymmetry. 2000;11:3659–63.
68. Davis F, Chen B. Asymmetric hydroxylation of enolates with n-sulfonyloxaziridines. Chem Rev. 1992;92:919–34.
69. Kidwai M, Mothsra P. A one-pot synthesis of 1,2,4,5-tetraarylimidazoles using molecular iodine as an efficient catalyst. Tetrahedron Lett. 2006;47:5029–31.
70. Wildemann H, Dunkelmann P, Muller M, Schmidt B. A short olefin metathesis-based route to enantiomerically pure arylated dihydropyrans and α,β-unsaturated γ-valero lactones. J Org Chem. 2003;68:799–804.

71. Hoyos P, Sansottera G, Fernández M, Molinari F, Sinisterra JV, Alcántara AR. Enantioselective monoreduction of different 1,2-diaryl-1,2-diketones catalysed by lyophilised whole cells from *Pichia glucozyma*. Tetrahedron. 2008;64:7929–36.
72. Correia J. Isolation of the intermediates in a benzoin-type condensation. J Org Chem. 1983;48:3343–4.
73. Hanson RW. The preparation of furoin-A biomimetic reaction. J Chem Educ. 1993;70:257.
74. Hoyos P, Fernández M, Sinisterra JV, Alcántara AR. Dynamic kinetic resolution of benzoins by lipase-metal combo catalysis. J Org Chem. 2006;71:7632–7.
75. Hoyos P, Buthe A, Ansorge-Schumacher MB, Sinisterra JV, Alcántara AR. Highly efficient one pot dynamic kinetic resolution of benzoins with entrapped pseudomonas stutzeri lipase. J Mol Catal B Enzym. 2008;52:133–9.
76. Nieguth R, Ten Dam J, Petrenz A, Ramanathan A, Hanefeld U, Ansorge-Schumacher MB. Combined heterogeneous bio-and chemo- catalysis for dynamic kinetic resolution of (*rac*)- benzoin. RSC Adv. 2014;4:45495–503.
77. Agrawal S, Martínez-Castro E, Marcos R, Martín-Matute B. Readily available ruthenium complex for efficient dynamic kinetic resolution of aromatic α-hydroxy ketones. Org Lett. 2014;16:2256–9.
78. Nestl BM, Bodlenner A, Stuermer R, Hauer B, Kroutil W, Faber K. Biocatalytic racemization of synthetically important functionalized α-hydroxyketones using microbial cells. Tetrahedron Asymmetry. 2007;18:1465–74.
79. Glueck SM, Pirker M, Nestl BM, Ueberbacher BT, Larissegger-schnell B, Csar K, Hauer B, Stuermer R, Kroutil W, Faber K. Biocatalytic racemization of aliphatic, arylaliphatic, and aromatic α-hydroxycarboxylic acids. J Org Chem. 2005;70:4028–32.
80. Bozell JJ, Petersen GR. Technology development for the production of biobased products from biorefinery carbohydrates—the US Department of Energy's "Top 10" revisited. Green Chem. 2010;12:539–4.
81. Sheldon RA. Green and sustainable manufacture of chemicals from biomass: state of the art. Green Chem. 2014;16:950–63.
82. Tuck CO, Perz E, Horvath I, Sheldon RA, Poliakoff M. Valorization of biomass: deriving more value from waste. Science. 2012;337:695–9.
83. Masuo S, Zhou S, Kaneko T, Takaya N. Bacterial fermentation platform for producing artificial aromatic amines. Sci Rep. 2016;6:25764.
84. Kohls H, Steffen-Munsberg F, Hö M. Recent achievements in developing the biocatalytic toolbox for chiral amine synthesis. Curr Opin Chem Biol. 2014;19:180–92.
85. Narancic T, Davis R, Nikodinovic-Runic J, O' Connor KE. Recent developments in biocatalysis beyond the laboratory. Biotechnol Lett. 2015;37:943–54.
86. Panke S, Held M, Wubbolts M. Trends and innovations in industrial biocatalysis for the production of fine chemicals. Curr Opin Biotechnol. 2004;15:272–9.
87. Hoyos P, Pace V, Sinisterra JV, Alcántara AR. Chemoenzymatic synthesis of chiral unsymmetrical benzoin esters. Tetrahedron. 2011;67:7321–9.
88. Naik SN, Goud VV, Rout PK, Dalai AK. Production of first and second generation biofuels: a comprehensive review. Renew Sust Energ Rev. 2010;14:578–97.
89. Schwartz TJ, O'Neill BJ, Shanks BH, Dumesic JA. Bridging the chemical and biological catalysis gap: challenges and outlooks for producing sustainable chemicals. ACS Catal. 2014;4:2060–9.
90. Environmental Data Centre of Natural Resources – Eurostat. http://ec.europa.eu/eurostat/web/environmental-data-centre-on-natural-resources/natural-resources/energy-resources/energy-from-biomass. Accessed 2 Feb 2017.
91. McCoy M. Solvay commercializes new route to epichlorohydrin. Chem Eng News. 2012;90:15.
92. Marr AC, Liu S. Combining bio- and chemo-catalysis: from enzymes to cells, from petroleum to biomass. Trends Biotechnol. 2011;29:199–204.
93. Liu S, Rebros M, Stephens G, Marr AC. Adding value to renewables: a one pot process combining microbial cells and hydrogen transfer catalysis to utilise waste glycerol from biodiesel production. Chem Commun. 2009;17:2308–10.

94. Ma Y, Wang Y, Morgan PJ, Jackson RE, Liu X, Saunders GC, Lorenzini F, Marr AC. Designing effective homogeneous catalysis for glycerol valorisation: selective synthesis of a value-added aldehyde from 1,3-propanediol via hydrogen transfer catalysed by a highly recyclable, fluorinated Cp*Ir(NHC) catalyst. Catal Today. 2017. Accepted Manuscript: https://doi.org/10.1016/j.cattod.2017.09.036.
95. Ma Y, Wang Y-M, Lorenzini F, Marr AC. Unpublished results.
96. Global Industry Analysts I. Stable Demand from End-Use Sectors to Sustain Growth in the Global Adipic Acid Market. 2016.
97. Zhang H, Li X, Su X, Ang EL, Zhang Y, Zhao H. Production of adipic acid from sugar beet residue by combined biological and chemical catalysis. ChemCatChem. 2016;8:1500–6.
98. Li X, Wu D, Lu T, Yi G, Su H, Zhang Y. Highly efficient chemical process to convert mucic acid into adipic acid and DFT studies of the mechanism of the rhenium-catalyzed deoxydehydration. Angew Chemie Int Ed. 2014;53:4200–4.
99. van Heerden CD, Nicol W. Continuous succinic acid fermentation by actinobacillus succinogenes. Biochem Eng J. 2013;73:5–11.
100. Bomgardner M. Myriant to build succinic acid plant in Louisiana. Chem Eng News. 2011;89:7.
101. Yuzbashev TV, Yuzbasheva EY, Laptev IA, Sobolevskaya TI, Vybornaya TV, Larina AS, Gvilava IT, Antonova SV, Sineoky SP. Is it possible to produce succinic acid at a low pH? Bioeng Bugs. 2011;2:115–9.
102. Van Putten RJ, Van Der Waal JC, De Jong E, Rasrendra CB, Heeres HJ, De Vries JG. Hydroxymethylfurfural, a versatile platform chemical made from renewable resources. Chem Rev. 2013;113:1499–597.
103. Simeonov SP, Coelho JAS, Afonso CAM. Integrated chemo-enzymatic production of 5-hydroxymethylfurfural from glucose. ChemSusChem. 2013;6:997–1000.
104. Peleteiro S, Rivas S, Alonso JL, Santos V, Parajó JC. Furfural production using ionic liquids: a review. Bioresour Technol. 2016;202:181–91.
105. He Y-C, Jiang C-X, Jiang J-W, Di J-H, Liu F, Ding Y, Qing Q, Ma C-L. One-pot chemo-enzymatic synthesis of furfuralcohol from xylose. Bioresour Technol. 2017;238:698–705.
106. Ren C, Gu Y, Hu S, Wu Y, Wang P, Yang Y, Yang C, Yang S, Jiang W. Identification and inactivation of pleiotropic regulator CcpA to eliminate glucose repression of xylose utilization in *Clostridium acetobutylicum*. Metab Eng. 2010;12:446–54.
107. Anbarasan P, Baer ZC, Sreekumar S, Gross E, Binder JB, Blanch HW, Clark DS, Toste FD. Integration of chemical catalysis with extractive fermentation to produce fuels. Nature. 2012;491:235–9.
108. Severino A, Esculcas A, Rocha J, Vital J, Lobo LS. Effect of extra-lattice aluminium species on the activity, selectivity and stability of acid zeolites in the liquid phase isomerisation of α-pinene. Appl Catal A Gen. 1996;142:255–78.
109. Dean FM, Knight JC. The oxidation of 3-alkylidenegrisens to grisen-3-ones by ruthenium tetroxide. J Chem Soc. 1962:4745–52.
110. Berkowitz LM, Rylander PN. Use of ruthenium tetroxide as a multi-purpose oxidant. J Am Chem Soc. 1958;80:6682–4.
111. Pappo R, Allen DS Jr, Lemieux RU, Johnson WS. Notes – osmium tetroxide-catalyzed periodate oxidation of olefinic bonds. J Org Chem. 1956;21:478–9.
112. Oguchi T, Ura T, Ishii Y, Ogawa M. Oxidative cleavage of olefins into carboxylic acid with hydrogen peroxide by tungstic acid. Chem Lett. 1989;18:857–60.
113. Spannring P, Bruijnincx PCA, Weckhuysen BM, Klein Gebbink RJM. Transition metal-catalyzed oxidative double bond cleavage of simple and bio-derived alkenes and unsaturated fatty acids. Cat Sci Technol. 2014;4:2182.
114. Köhler V, Wilson YM, Dürrenberger M, Ghislieri D, Churakova E, Quinto T, Knörr L, Häussinger D, Hollmann F, Turner NJ, Ward TR. Synthetic cascades are enabled by combining biocatalysts with artificial metalloenzymes. Nat Chem. 2012;5:93–9.

Index

Z. Fang et al. (eds.), *Production of Biofuels and Chemicals with Bifunctional Catalysts*, Biofuels and Biorefineries 8, https://doi.org/10.1007/978-981-10-5137-1

E

F

I

J

K

L

M

R

S

T

MIX
Papier aus verantwortungsvollen Quellen
Paper from responsible sources
FSC® C105338

If you have any concerns about our products,
you can contact us on
ProductSafety@springernature.com

In case Publisher is established outside the EU,
the EU authorized representative is:
Springer Nature Customer Service Center GmbH
Europaplatz 3, 69115 Heidelberg, Germany

Printed by Libri Plureos GmbH
in Hamburg, Germany